THE THREE-BODY PROBLEM

STUDIES IN ASTRONAUTICS

Volume 1	Optimal Space Trajectories (J. P. Marec)
Volume 2	Optimal Trajectories in Atmospheric Flight (N. X. Vinh)
Volume 3	Optimal Spacecraft Rotational Maneuvers (J. L. Junkins and J. D. Turner)
Volume 4	The Three-Body Problem (C. Marchal)

Picture on front cover:
The moon in the earth's shadow. Total eclipse of August 17, 1989, 3 hours 45 (universal time), 7 minutes before the end of totality.

STUDIES IN ASTRONAUTICS 4

THE THREE-BODY PROBLEM

CHRISTIAN MARCHAL

Office National d'Études et de Recherches Aérospatiales,
Châtillon, France

ELSEVIER
Amsterdam – Oxford – New York – Tokyo 1990

ELSEVIER SCIENCE PUBLISHERS B.V.
Sara Burgerhartstraat 25
P.O. Box 211, 1000 AE Amsterdam, The Netherlands

Distributors for the United States and Canada:

ELSEVIER SCIENCE PUBLISHING COMPANY INC.
655 Avenue of the Americas
New York, N.Y. 10010, U.S.A.

Library of Congress Cataloging-in-Publication Data

Marchal, Christian.
The three-body problem / Christian Marchal.
p. cm. -- (Studies in astronautics ; v. 4)
Includes index.
ISBN 0-444-87440-2
1. Three-body problem. 2. Mechanics, Celestial. I. Title.
II. Series.
QB362.T5M37 1990
521--dc20 90-39071
CIP

ISBN: 0-444-87440-2 (Vol. 4)
ISBN: 0-444-41813-X (Series)

Printed in The Netherlands.

FOREWORD

It is a pleasure to present this monograph written about the problem of three bodies. The king of sciences is known as astronomy and the queen as celestial mechanics ; princes and princesses are the fields of astronomy, astrophysics, cosmogony, stellar dynamics, observational astronomy, planetary sciences, astrodynamics, lunar theories, etc... And we can ask where is the problem of three bodies in this beautiful family tree ? Many aspects of our field are reversible so we might go in two directions to find the answer and investigate the descendents of the princes and of the princesses or alternately, we might consider the ancestry of the kings and queens. We shall find the problem of three bodies in the group of ancestors since no universe can originate without three body dynamics. It is of course generally accepted that the first 10^{-n} seconds (where the value of n is between 100 and ∞) is governed by particle physics, which science is just now maturing to the level of the problem of three bodies. Nevertheless, this growing process is natural and unavoidable and once the basic problems of particle physics is solved, our colleagues will apply the stability and singularity criteria of the problem of three bodies to explain the origin of the universe.

Newton complained of headaches when he worked on his three body problem (the lunar theory) and found to his surprise that the Sun had greater effect on the Moon's orbit than the Earth. Poincaré showed non-integrability, Lagrange and Euler found equilibrium solutions, Jacobi and Hill restricted and simplified the basic model and the list goes on and on with the most famous names who contributed to dynamics and celestial mechanics in the history of human culture. New discoveries are still made today and the problem of three bodies leads dynamics into new fields. The most significant step for man-kind, the step from two to three bodies separates predictability from uncertainties, Kepler from Newton, Laplace's demon from nondeterminism and in general, artificiality from reality. The trivial problems without possible applications, which we invent and assign as problems to our students in our dynamic classes can be easily solved but offer little pleasure. The subject of this book is alive today and was alive since the origin of the universe.Supercolliders and supercomputers do not "solve" this problem but using the model of the problem of three bodies makes discoveries in the related area of science.

In my 45 years dedicated to the problem of three bodies I never had a dull moment, excepting when I investigated other, easily solvable problems as regrettable deviations. Indeed, my professor announced the subject of my dissertation saying : "Young man, go and solve the problem of three bodies". I soon found out about the problems of non-integrability and in my first despararation I considered leaving my academic career. What change my mind and made me to fall in love with the subject was my youthfull audacity to argue with Poincaré's theorem. (After some time my professor compromised and instructed me to find a force-law which allowed the change of this problem from non-integrability to integrability using a special transformation).

Satisfaction in science is not associated with solving a simple problem but comes from understanding complex situations and enjoying the small steps made to penetrate the mysteries of nature. The subject of this book is one of the best in science since the problem can be simply stated and instead of obtaining its simple solution we are guaranteed by Poincaré's dictum that the complete and general solution will remain hidden. Indeed, is there a better challenge for the human intellect than to engage in an "unsolvable" and at the same time very important problem? And consequently, is there a better and more satisfying book to read and to enjoy than Dr. Marchal's ?

Dr Marchal is probably better qualified to write this book than anyone else in our field since he combines high level of mathematical background with intense familiarity of the litterature of dynamics and celestial mechanics. He is the author of many papers of fundamental importance in our field and is unquestionably one of the scientific descendants of Poincaré. His rich cultural and scientific background is clearly visible and most enjoyable in this book.

Victor SZEBEHELY

DEDICATION

A mon épouse Françoise, la tête et le coeur de mon foyer, et au Liban, mon pays natal abandonné par les autres démocraties à l'invasion, la violence, la barbarie.

To my wife and love Françoise, the head and the heart of our home, and to Lebanon my native country, forsaked by the other democracies to invasion, violence, barbary.

الى رفيقة العمر زوجتي الغالية فرانسواز ،قلب بيتي وعماده .
الى بلدي الاصل ومسقط راسي لبنان ،الذي اضحى تشيكوسلوفاكيا جديده بعد ان اسلمتها الديمقراطيات الاخرى وبقية الدول العربية الى العنف والبربرية .

ACKNOWLEDGMENTS

Je suis heureux de remercier mes amis de partout qui m'ont aidé à écrire ce livre, et en particulier mon épouse Françoise, le Professeur Nguyen Xuan Vinh de l'Université du Michigan, le Professeur Victor Szebehely de l'Université du Texas, Mesdames R.M. Burke et Rina van Diemen, et Messieurs A. van der Avoird et A.J. Oxley d'Elsevier Science Publishers B.V., ainsi que Mesdames Raban et Josse qui ont tapé ces pages sans toujours les comprendre...

Ils ont tous été très patients !

I am happy to thank my friends from so many countries and so many centuries who helped me to write this book, and especially my wife Françoise, Professor Nguyen Xuan Vinh of Michigan University, Professor Victor Szebehely of Texas University, Mrs R.M. Burke, Mrs R. van Diemen, Mr A. van der Avoird, Mr A.J. Oxley of Elsevier Science Publishers B.V., as well as Mrs Raban and Mrs Josse who typed these pages without always understanding them...

They have all been very patient !

Gaarne wil ik mijn dank uitspreken tegenover al die vrienden uit vele landen, die mij in de diverse perioden van mijn werk hebben geholpen dit boek te schrijven.
Speciaal bedank ik mijn echtgenote Françoise, professor Nguyen Xuan Vinh van de universiteit van Michigan, professor Victor Szebehely van de universiteit van Texas, mevrouw R.M. Burke en mevrouw R. van Diemen, de heren A.W. van der Avoird en A.J. Oxley, alsmede de dames Raban en Josse, die het vele tikwerk voor dit boek hebben verricht zonder het overigens allemaal te begrijpen..

Zij allen hebben veel geduld betoond !

SHORT TABLE OF CONTENTS

CONTENTS

Chapter 1

SUMMARIES

THE THREE-BODY PROBLEM

After a short historical presentation the first chapters recall the usual formulations of the three-body problem, the main classical results and the corresponding questions and conjectures.

The theory of perturbations, the analytical approach and the quantitative analysis of the three-body problem have recently reached a high degree of perfection and have been described in several outstanding books such as references 1 to 3. The fantastic progress of computers has also led to many improvements in the quantitative analysis and have disclosed the extreme complexity of the set of solutions. These studies, as presented in the central chapters, provided both impetus and new orientations to the qualitative analysis that is so complementary to the quantitative analysis.

The final chapters describe the remarkable recent progress of the qualitative analysis and deals with questions such as stability, instability, escapes, singularities, regularizations, final evolutions, periodic motions, Arnold tori, asymptotic motions, oscillatory motions, quasi-collision motions etc...

The recent criteria (or tests) of escape are very efficient and approach very close to the true limits of escape. As a result the domain of bounded motions is much smaller than was previously expected. The stability of a three or n-body system seems brittle and the small masses seem under a great risk of being expelled.

The Arnold diffusion conjecture and the near resonance theorem lead to a revision of the notion of stability. The essential property is no longer an illusory indefinite stability, but rather the certainty of only small modifications over billions of years. This seems to be the case of the solar system -except for comets- provided that outer stars remain very far.

Quasi-collision motions, also called oscillatory motions of the second kind, are not rare and give a large probability to the formation of novae by the collision of two stars belonging to a multiple system or, more likely, to the formation of very close binary stars.

The classical Hill stability appearing in the circular restricted three-body problem can be partly extended to the general three-body problem.

The remaining open conjectures and the possible further investigations are discussed in the last chapter.

An entirely new image of the three-body problem is emerging from the latest progress.

LE PROBLEME DES TROIS CORPS

Résumé

Après une courte présentation historique, les premiers chapitres rappellent les formulations usuelles du problème des trois corps, les principaux résultats classiques, les grandes questions et les conjectures correspondantes.

La théorie des perturbations, l'approche analytique et l'analyse quantitative ont récemment atteint un haut degré de perfection et ont été exposées dans plusieurs livres remarquables comme ceux des références 1 à 3. Les fantastiques progrès des ordinateurs ont eux aussi conduit à beaucoup d'améliorations de l'analyse quantitative et ont dévoilé la complexité extrême de l'ensemble des solutions. Tout ceci est exposé dans les chapitres centraux et a donné une nouvelle impulsion et de nouvelles orientations à l'analyse qualitative qui est si complémentaire de l'analyse quantitative.

La dernière partie est orientée vers les progrès récents et remarquables de l'analyse qualitative, vers les questions d'allures finales du mouvement, évasions, singularités, régularisations, mouvements périodiques, tores d'Arnold, mouvements asymptotiques, bornés, oscillatoires, mouvements de quasi-collisions, stabilité et instabilité etc...

Les plus récents tests d'évasion sont très efficaces, ils approchent très près de la limite d'évasion véritable et le domaine des mouvements bornés est beaucoup plus petit qu'on ne le croyait généralement. La stabilité d'un système de trois corps ou davantage semble fragile et les masses les plus petites ont souvent un grand risque d'être rejetées à l'infini.

L'hypothèse "de diffusion d'Arnold" et le théorème de quasi-résonance conduisent à une révision de la notion de stabilité. L'essentiel n'est plus une illusoire stabilité indéfinie mais la certitude de modifications restreintes pour des durées de plusieurs milliards d'années. Ceci semble être le cas du système solaire - hormis les comètes - pourvu qu'aucune étoile ne vienne le perturber.

- Les mouvements de "quasi-collision" ou mouvements oscillatoires du second type ne sont pas rares et donnent une grande probabilité soit à la formation de "nova" par la collision de deux étoiles appartenant à un système multiple, soit, plus vraissemblablement, à la formation de binaires très serrées.

La stabilité classique de Hill qui apparaît dans le problème restreint circulaire des trois corps peut être étendue au problème général des trois corps.

Les conjectures restées ouvertes et les directions futures de recherche sont présentées dans le dernier chapitre.

Les progrès de ces dernières années ont fait émerger une image entièrement nouvelle du problème des trois corps.

ЗАДАЧА ТРЕХ ТЕЛ

Резюме

После краткого классического исторического введения книга начинается с изложения обычных различных постановок задачи трех тел, главных классических результатов, основных вопросов и гипотез.

В последнее время высокой степени совершенства достигли теория возмущений, аналитические методы и численный анализ - они представлены в нескольких недавно опубликованных монографиях. Фантастический прогресс вычислительной техники привел к значительным успехам численного анализа и раскрыл крайнюю сложность строения множества решений. Все это придало новые ориентации и новый импульс старому качественному анализу, который так хорошо дополняет численный анализ.

Основная часть монографии посвящена недавнему замечательному прогрессу качественного анализа, в ней затронуты вопросы распада, сингулярности, регуляризации, финальных движений, периодических, асимптотических, ограниченных и осциллирующих движений, квазистолкновений, хаотических движений и т.д.

Новые критерии распада весьма эффективны и очень близко приближаются к истинной границе распада; область ограниченных движений много меньше, чем ожидалось из общих соображений. Редки случаи устойчивости систем трех и n тел, и тела малой массы диссипируют с большой вероятностью.

Нередки квазистолкновения, которые обеспечивают большую вероятность вспышки "новой" при столкновении двух звезд в кратной системе.

Классическая устойчивость по Хиллу, взникающая в круговой ограниченной задаче трех тел, в последнее время была распространена на случай общей задачи трех тел.

В заключительной главе обсуждаются вопросы, остающиеся открытыми, и возможные дальнейшие исследования.

Здесь формулируется совершенно новый подход к проблеме трёх тел, следующий из результатов самых последних исследований.

Translated into Russian by Professors K.V. Kholshevnikov, J.P. Anosova and V.V. Orlov of Leningrad State University and by Professor Y.S. Ryazantsev of the USSR Academy of Science

DAS DREIKÖRPERPROBLEM

Das Buch beginnt nach einer kurzen historischen Einführung mit den verschiedenen klassischen Formulierungen des Dreikörperproblems und geht auf die großen Fragestellungen und Hypothesen ein und erläutert die hauptsächlichen Resultate.

Die Störungstheorie, der analytische Ansatz und die quantitative Analyse wurden bereits sehr weit entwickelt und in einigen neueren Publikationen abgehandelt. Die großartige Weiterentwicklung der Computertechnik hat auch zu vielen Verbesserungen in der quantitativen Analyse der Probleme geführt und die extreme Komplexheit der Lösungen des Dreikörperproblems enthüllt.

Im Hauptteil des Buches wird der bemerkenswerte Fortschritt der qualitativen Analyse eingehend beschrieben: die Fluchtbahnen, Singularitäten, Regularisierungen, Endstadien der Entwicklung, periodische Bahnen, Oszillationsbahnen, Quasikollisionsbahnen, chaotische Bahnen etc.

Die allerneuesten Tests der Bahnen bezüglich deren Stabilität erweisen sich als sehr effizient und scheinen sehr nahe den tatsächlichen Grenzen zu sein: es zeigt sich, daß die Menge der gebundenen Bahnen viel kleiner als allgemein erwartet ist. Tatsächlich scheint ein System von 3 oder auch n Körpern sehr zerbrechlich zu sein, wobei die kleineren Massen mit hoher Wahrscheinlichkeit aus dem System hinausgeworfen werden.

Die Quasi-Kollisionsbahnen sind nicht selten, wodurch sich eine hohe Wahrscheinlichkeit der Entstehung von Novae durch Zusammenstöße von zwei Sternen in einem Mehrfachsystem ergibt.

Die klassische Hill Stabilität wurde in letzter Zeit auch auf das allgemeine Dreikörperproblem ausgedehnt.

Die verbleibenden offenen Fragen bzw. die möglichen künftigen Forschungen werden im abschließenden Kapitel erörtert.

Das wissenschaftliche Bild des Dreikörperproblems ist also ein ganz neues geworden.

Translated into German by Doktor Rudolf Dvorak of the Institut für Astronomie, Vienna University

El Problema de tres cuerpos

Resumen

Tras una breve presentación clásica e histórica el libro comienza con las distintas formulaciones ordinarias del problema de tres cuerpos, los principales resultados clásicos y las grandes cuestiones y conjeturas.

La teoría de perturbaciones, el enfoque analítico, y el análisis cuantitativo han alcanzado recientemente un alto grado de perfección y han sido expuestos recientemente en algunos libros destacados. El extraordinario avance de los ordenadores ha permitido grandes mejoras en el análisis cuantitivo y ha desvelado la extraordinaria complejidad del conjunto de soluciones. Todo esto ha proporcionado nuevas orientaciones y un nuevo impulso al antiguo análisis cualitativo que completa al análisis cuantitativo.

La mayor parte del libro describe los progresos recientes más notables del análisis cualitativo y trata de problemas tales como escapes, singularidades, regularización, evoluciones finales, movimientos periódicos, movimientos asintóticos, movimientos acotados, movimientos oscilatorios, movimientos de cuasi-colisión, movimientos caóticos, etc.

Las pruebas más recientes de escape son muy eficaces y dan una aproximación muy cercana al verdadero límite de escape siendo el dominio de los movimientos acotados mucho más reducido de lo esperado.

La estabilidad de un sistema de tres o n-cuerpos parece frágil con lo que las pequeñas masas corren un gran riesgo de ser expulsadas.

Los movimientos de cuasi-colisión no son raros lo que provoca con una probabilidad muy elevada la formación de una "nova" por colisión de dos estrellas pertenecientes a un sistema múltiple.

La estabilidad clásica de Hill que aparece en el problema restringido circular de tres cuerpos ha sido recientemente extendida al problema general de tres cuerpos.

En el capítulo final se tratan las restantes conjeturas abiertas y las futuras posibles investigaciones.

Translated into Spanish by Professor Vicente Camarena Badia of Zaragoza University

三 体 問 題

梗 概

簡単な歴史と基礎的な準備の後，本書は三体問題の二三の標準的な基礎方程式，主要な古典的結果，及びこれまでの大きな問題点や予想などで始まる。

摂動論や解析的取り扱いなど定量的議論は近年大いにその完成の度を高め，最近の注目すべきいくつかの著書にも取り込まれている。コンピューターの驚異的進歩がまた定量的議論を大幅に改め，解の集りの極めて複雑な様相を明らかにした。こうして定量的議論を補う古くからの定性的議論に新しい方向付けと新しい活気とが与えられた。

本書の主要部分は定性的議論の最近の目覚しい発展について当てられており，特異点，正則化，三体系の時間的発展，系からの離脱，周期運動，漸近運動，有界運動，振動，近衝突運動，カオス的運動等の問題を取り扱う。

離脱に関する最新の判定条件は非常に有効で離脱の真の限界に迫っている。有界運動の範囲は一般に期待されていたよりはずっと小さい。三体，あるいはｎ体系の安定性の問題は未だ不完全な様で，小質点は常に系から追い出される危険性を持っている。

近衝突運動は決して希ではなく，多重星系で所属する二星が衝突して『新星』となる大きな可能性が存在する。

円型の制限三体問題における古典的なヒルの安定性は最近一般三体問題に拡張された。

その他の未解決の予想や更に追求可能な問題が最後の章で論じられる。

Translated into Japanese by Professor
J. Yoshida, Kyoto Sangyo University

三 体 问 题

摘 要

简短地回顾了历史的和经典的概况后，本书以三体问题通常被采用的各种形式，主要的经典结论和一些重要的问题和猜想着手。

摄动理论，分析方法和定量分析最近已达到相当完善的程度，并且在新近的一些杰出专著中得到了阐述。计算机科学的飞速发展，改进了许多定量分析中的结果，并揭示了介集合的极端复杂性。所有这些已经给古老的定性分析指出了新的方向并给以新的推动力，定性分析也是对定量分析的补充。

本书的主要部分介绍定性分析的最近一些卓越的进展和论述如逃逸、奇点、正规化、最终演化、周期运动、渐近运动、有界运动、振动运动、拟碰撞运动、混沌运动等问题。

有关逃逸的最新检验标准是非常有效的，它非常接近于真正的逃逸极限，有界运动的区域比通常所期望的要小得多。三体或多体系统的稳定性似乎是脆弱的，小质量天体永远有被排斥的极大危险性。

拟碰撞运动并不稀少，它给出由于属于一个多重系统的两颗恒星的碰撞形成一颗“新星”的一个很大的可能性。

出现在圆型限制性三体问题中的希耳稳定性，最近已经被推广到一般三体问题。

在最后一章中，讨论了余下尚未解决的猜想和可能进一步进行的研究。

Translated into Chinese by Professor Yi-Sui Sun of Nanking University

مسائلة الاجسام الثلاثة

مقدمة

يتعرض هذا الكتاب لاحدى المسائل الهامة في الميكانيك السماوي والمعروفة باسم " مسائلة الاجسام الثلاثة "
يبداء المؤلف بعرض تاريخي موجز ينتقل بعده الى صياغة التعابير الشائعة ،وسرد النتائج الاساسية المعروفة والمتعلقة بهذه المسائلة ،واخيرا يتعرض الى امهات الاسئلة والمواضيع التي لاتزال مفتوحة .
لقد بلغت بعض الدراسات درجة رفيعة من الكمال كنظرية الاضرابات ،والتقريب التحليلي ، ونظرية التحليل الكمي ،واصبحت معظم الكتب البارزه في العصر الحاضر تتعرض لهذه المواضيع وقد ادى التقدم الرائع الذي طراء على الحاسب الالي الى تحقيق تحسينات ملحوظة ايضا على نظرية التحليل الكمي ،كما انه ساهم في كشف النقاب عن مدى تعقيد مجموعة الحلول . ولقد كان من نتائج ذلك كله ظهور اتجاهات جديده تولي اهمية كبيرة لطرق التحليل الكيفي الذي يكمل الى حد كبير التحليل الكمي .
يهتم الكتاب بشكل رئيسي بالتقدم الحديث والملحوظ الذي طراء على التحليل الكيفي ، ويتوجه من اجل ذلك نحو دراسة : الاشكال النهائية للحركة ، الانفلات ، النقاط او الحلول الشاذه وازالتها ،وبالاضافة الى ذلك يهتم بدراسة بعض الحركات الهامة :

الدورية ،المقاربة ، المحدوده ، الاهتزازية ، شبه التصادمية ...... الخ .

تتميز احدث اختبارات ظاهرة " الانفلات او الهروب " بفعالية كبيرة جدا ، اذ انها تعطي نتائج قريبة جدا من تلك التي يوءول اليها الانفلات الفعلي . هذا وان المنطقة التي تشغلها الحركات المحدوده هي في الحالة العامة اصغر بكثير مما كان يعتقد .
وعلى مايبدو فان استقرار الجملة الموءلفة من ثلاثة اجسام او اكثر هو استقرار ضعيف ، وغالبا ماتكون الاجسام ذات الكتل الصغيرة اكثر تعرضا لاءن يقذف بها الى اللانهاية .
ان الحركات " شبه التصادمية " ليست نادرة ، وهي تعطي احتمالا كبيرا لكيفية تكوين نجم وقتي جديد " نوڤا " وذلك نتيجة لحدوث اصطدام بين نجمين في جملة مضاعفة .
فيما يتعلق بالاستقرار الكلاسيكي " هيل " والذي كان يظن عند دراسة " المسائلة القسرية " الدورانية لثلاثة اجسام فقد تم حديثا توسيع دراسة الاستقرار بحيث اصبح يشمل مسائلة الاجسام الثلاثة بشكلها العام .
يهتم الفصل الاخير من هذا الكتاب بذكر المواضيع والاسئلة التي لاتزال مفتوحة ، بالاضافة الى التنويه عن الاتجاهات المستقبلية للبحث العلمي في هذا المجال .
واخيرا يمكن القول باءن التقدم الذي حدث في السنوات الاخيرة كان له اثرا فعالا في ابراز " مسائلة الاجسام الثلاثة " واعطائها صورة حية وشكلا جديدا .

Translated into Arabic by Professor Ahmed Karabelli of Kuwait University.

Chapter 2

HISTORY

The problem of the motion of the planets is one of the oldest of mankind, and the observers of Antiquity made Astronomy the Mother of Science. They understood the essential features of our terrestrial condition : the Earth is isolated in space, it has a daily rotation and its shape is spherical. They opened the way for Columbus and Magellan to try to reach China by sailing to the west.

However, mankind had to wait for Tycho-Brahé, Kepler and Newton for accurate observations, accurate laws of planetary motion and an accurate explanation of these motions : the law of universal attraction, which was accurate enough to lead to the discovery of unknown planets such as Neptune and Pluto, even if today we accept General Relativity as a much better approximation.

Newton demonstrated that Kepler's laws of planetary motions correspond to the two-body problem, i.e. the free motions of two spherical bodies moving under the influence of their mutual attraction, and led to the modern form of the three-body problem : "What are the free motions of three given spherical bodies moving under the influence of their mutual gravitational attraction ?"

Throughout the last three centuries the three-body problem has played a major role in the development of science. It has triggered many mathematical studies, methods and theories, as illustrated by Euler, Lagrange, Laplace, Jacobi, Leverrier, Adams, Newcomb, Hamilton, Delaunay, Poincaré, Sundman and many others. It was also the essential part of the famous "problem of the motion of the Moon" that competed during most of the eighteenth century with the progress of clock-making for measuring longitudes.

The difficulties of the three-body problem were the reason for the introduction of new qualitative methods by Poincaré, Birkhoff, Sundman and Chazy ; methods that have since been extended to almost all other branches of science. Finally, we point out that while the meteorologist E.N. Lorenz has recently discovered the "strange attractors" by the use of modern computers for analysis of an atmospheric system (Ref. 4), the astronomer Michel Hénon discovered the "chaotic motions" by the application of modern computers to the three-body problem (Section 9.4 and Ref. 5).

These recent discoveries are related to the theoretical analysis of Poincaré and Birkhoff on the ergodic theorem and that of Kolmogorov, Arnold and Moser on the behaviour of orbits close to periodic motions. But, for the first time, we understand how general are these new types of motion that appear in all kind of problems and questions.

The three-body problem continues to lead the way. Its mathematical simplicity, its singularities and its sensitivity to inital conditions make it and ideal example for each new numerical method of integration. This was shown by the surprises of the "Pythagorean problem" (three planets of masses 3, 4 and 5 initially without velocity at the vertices of a 3, 4, 5 right-angled triangle (Section 10.9.2. and Ref. 6).

We have now reached the point where most theoricians believe that when all three masses are non-zero all solutions are unstable, in the sense that escape solutions are everywhere dense. However, this assertion is contested by almost all numerical analysts. If these theoreticians are correct, a new type of stability will appear : a kind of very long-term stability which is finally destroyed by very small and long-lasting resonances. Perhaps this would imply that the ultimate future of the Solar system is dispersion ; however, this is perhaps as far into the future as phenomena going against the Second Law of thermodynamics !

Chapter 3

THE LAW OF UNIVERSAL ATTRACTION

"Everything happens as if matter attracts matter in direct proportion to the products of masses and in inverse proportion to the square of the distances".

This famous proposition of Newton's "Principia" (1687) was a decisive step in our understanding of the Universe, but notice that the usual differential equations of planetary motion also need the principle of equality of action and reaction and the Newtonian law of inertia that relates forces and accelerations.

The audacity of Newton must be admired : he was launching through empty space between the Sun and the Earth a force strong enough to break a cylindrical steel cable as wide as the Earth ! But his prudence must also be admired ("Everything happens as if ...") ; moreover he waited sixteen years before publishing his ideas because inaccurate measurements of the radius of the Earth had led to a 20% discrepancy between the Earth's gravity and his own calculations based on lunar motion.

Fortunately, in 1682 Newton learned of the new accurate geodesic measurements of the French astronomer Picard and, feeling close to success, he was so overcome by emotion that he was unable to complete the verification. He had to call for the help of a friend...

Chapter 4

EXACT FORMULATIONS OF THE THREE-BODY PROBLEM

Two point masses m_1 and m_2 are separated by the distance r and are attracted by each other (Fig. 1).

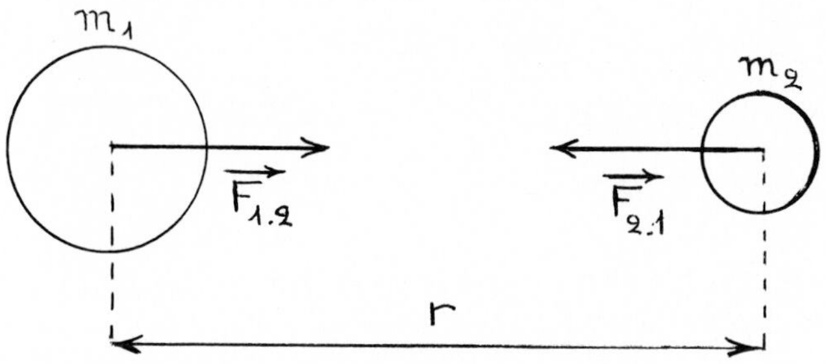

Fig. 1. The law of universal attraction.

The mass m_1 experiences the force $\vec{F}_{1.2}$ and the mass m_2 the opposite force $\vec{F}_{2.1}$:

$$||\vec{F}_{1.2}|| = ||\vec{F}_{2.1}|| = Gm_1m_2r^{-2} \qquad (1)$$

G is a constant of the law of universal attraction, the Newtonian constant which is equal to 6.672×10^{-11} $m^3/s^2.kg$.

Newton demonstrated that the forces $\vec{F}_{1.2}$ and $\vec{F}_{2.1}$ remain the same if the masses m_1 and m_2 have spherical symmetry instead of being point masses (r being the distance between centers).

The equations of the three-body problem have many forms ; we will consider the most usual.

4.1 THE CLASSICAL FORMULATION

Let us use a "Galilean" set of axes (Fig. 2), that is a set of axes in which the law of inertia is the classical Newtonian law :

$$\text{acceleration} = \text{force/mass} \qquad (2)$$

If $\vec{r}_1$ is the radius-vector of the mass m_1 the corresponding three-body equation of motion is :

$$d^2\vec{r}_1/dt^2 = (\vec{F}_{1.2} + \vec{F}_{1.3})/m_1 \qquad (3)$$

In this equation the parameter of description t is the "absolute time" of Newton, we will simply call it the time, and $\vec{F}_{1.2}$ and $\vec{F}_{1.3}$ are the gravitational forces of attraction of m_1 towards the masses m_2 and m_3 respectively.

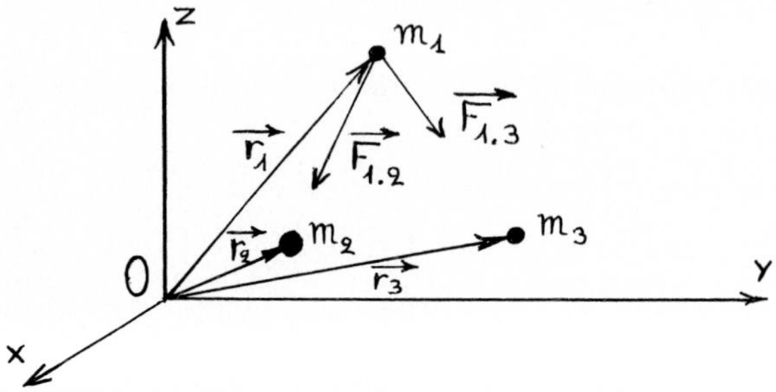

Fig. 2. The classical formulation.

It is traditional to write $\overrightarrow{r_{ij}}$ for the vector from the mass m_i to the mass m_j and to write r_{ij} for its length :

$$\overrightarrow{r_{ij}} = (\overrightarrow{m_i, m_j}) = \vec{r_j} - \vec{r_i} \ ; \ r_{ij} = ||\overrightarrow{r_{ij}}|| \tag{4}$$

hence :

$$||\overrightarrow{F_{1.2}}|| = Gm_1m_2/r^2_{12} \tag{5}$$

and, $\overrightarrow{F_{1.2}}$ being in the direction of the unit vector $\overrightarrow{r_{12}}/r_{12}$:

$$\overrightarrow{F_{1.2}} = Gm_1m_2\overrightarrow{r_{12}}/r^3_{12} \tag{6}$$

We thus arrive at the classical equations of motion for the three masses :

$$d^2\vec{r_i}/dt^2 = G(m_j\overrightarrow{r_{ij}}r^{-3}_{ij} + m_k\overrightarrow{r_{ik}}r^{-3}_{ik}) \ ; \ \{i,j,k\} = \{1,2,3\} \tag{7}$$

When all masses are non-infinitesimal these three equations are sometimes written :

$$m_i d^2\vec{r}_i/dt^2 = \partial U/\partial\vec{r}_i \quad ; \quad i = \{1,2,3\} \tag{8}$$

where :

$$U = \text{"potential"} = U(\vec{r_1},\vec{r_2},\vec{r_3}) = G(\frac{m_1m_2}{r_{12}} + \frac{m_1m_3}{r_{13}} + \frac{m_2m_3}{r_{23}}) \tag{9}$$

4.2 THE LAGRANGIAN FORMULATION

The Lagrangian variables are the vectors $\vec{r}_{12}$, $\vec{r}_{23}$, and $\vec{r}_{31}$. They are labelled according to the index of the opposite mass in the triangle of the three bodies.

In order to avoid any confusion, we will call m_A, m_B, m_C the three masses and $\vec{r}_A$, $\vec{r}_B$, $\vec{r}_C$ the three corresponding vectors (Fig. 3), with $\vec{r}_A = \vec{r}_{BC}$, $\vec{r}_B = \vec{r}_{CA}$ and $\vec{r}_C = \vec{r}_{AB}$.

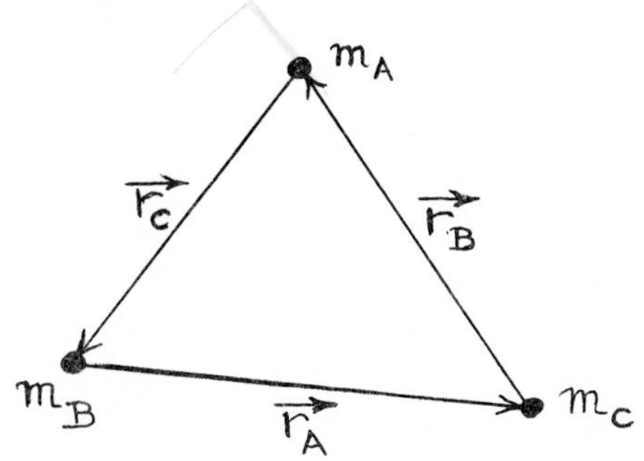

Fig. 3. The Lagrangian variables.

Since $\vec{r}_{12} = \vec{r}_2 - \vec{r}_1$ we have $d^2\vec{r}_{12}/dt^2 = d^2\vec{r}_2/dt^2 - d^2\vec{r}_1/dt^2$.
It is easy to deduce the Lagrangian equations of motion from the equations (7).

$$d^2\vec{r}_A/dt^2 = G\left(m_A\left(\frac{\vec{r}_B}{r_B^3} + \frac{\vec{r}_C}{r_C^3}\right) - (m_B + m_C)\,\frac{\vec{r}_A}{r_A^3}\right) \tag{10}$$

$$d^2\vec{r}_B/dt^2 = G\left(m_B\left(\frac{\vec{r}_C}{r_C^3} + \frac{\vec{r}_A}{r_A^3}\right) - (m_B + m_A)\,\frac{\vec{r}_B}{r_B^3}\right) \tag{11}$$

$$d^2\vec{r}_C/dt^2 = G\left(m_C\left(\frac{\vec{r}_A}{r_A^3} + \frac{\vec{r}_B}{r_B^3}\right) - (m_A + m_B)\,\frac{\vec{r}_C}{r_C^3}\right) \tag{12}$$

with of course :

$$\vec{r}_A + \vec{r}_B + \vec{r}_C = 0 \tag{13}$$

$$r_A = ||\vec{r}_A|| \ ; \ r_C = ||\vec{r}_B|| \ ; \ r_C = ||\vec{r}_C|| \tag{14}$$

The Lagrangian equations of motion (10), (11), (12) are sometimes written :

$$d^2\vec{r}_J/dt^2 = G(m_J\vec{W} - M\vec{r}_J r_J^{-3}) \quad ; \ J = \{A,B,C\} \tag{15}$$

where :

$$\vec{W} = \vec{r}_A r_A^{-3} + \vec{r}_B r_B^{-3} + \vec{r}_C r_C^{-3} \tag{16}$$

$$M = m_A + m_B + m_C = \text{total mass.} \tag{17}$$

4.3 THE JACOBI FORMULATION

The Jacobi decomposition of the three-body problem is not symmetrical (Fig. 4) but it is also the most useful.

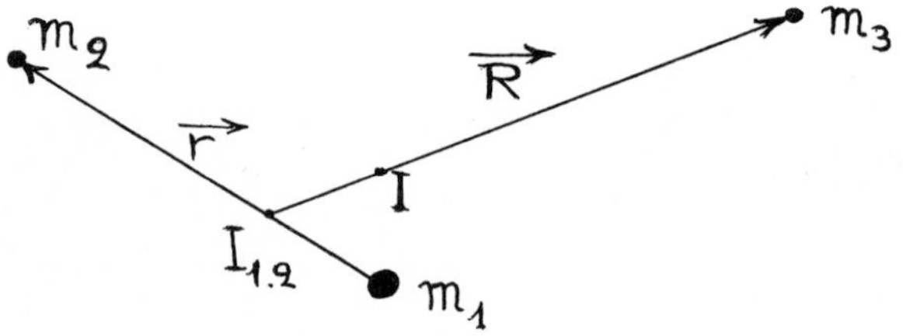

Fig. 4. The Jacobi decomposition of the three-body problem.

Jacobi used two main vectors : the vector $\overrightarrow{r_{1.2}}$ called $\vec{r}$ and the vector $\vec{R}$ from $I_{1.2}$ (center of mass of m_1 and m_2) to m_3 ; hence $\vec{R}$ passes through I, the general center of mass.

Let :

$$m_1/(m_1 + m_2) = \alpha \; ; \; m_2/(m_1 + m_2) = \beta \tag{18}$$

then :

$$\alpha + \beta = 1 \; ; \; \overrightarrow{r_{13}} = \vec{R} + \beta\vec{r} \; ; \; \overrightarrow{r_{23}} = \vec{R} - \alpha\vec{r} \tag{19}$$

and also :

$$\vec{R} = \alpha\overrightarrow{r_{13}} + \beta\overrightarrow{r_{23}} \tag{20}$$

Hence, from (10), (11), (12), the Jacobi equations for the three-body motions are :

$$d^2\vec{r}/dt^2 = -G(m_1+m_2)r^{-3}\vec{r} + Gm_3\{r_{23}^{-3}\vec{r}_{23} - r_{13}^{-3}\vec{r}_{13}\} \tag{21}$$

$$d^2\vec{R}/dt^2 = -\,GM\{\alpha r_{13}^{-3}\vec{r}_{13} + \beta r_{23}^{-3}\vec{r}_{23}\} \tag{22}$$

where again :

$$M = m_1 + m_2 + m_3 = \text{total mass.} \tag{23}$$

It is when r/R is small ("lunar case") or when $(m_2 + m_3)/M$ is small ("planetary case") that the Jacobi decomposition is the most interesting. This is because then :

$$d^2\vec{r}/dt^2 \simeq -G(m_1 + m_2)\vec{r}.r^{-3} \quad (24)$$

$$d^2\vec{R}/dt^2 \simeq -GM\vec{R}.R^{-3} \quad (25)$$

If these equations were exact the motions of $\vec{r}$ and $\vec{R}$ would be ordinary Keplerian motions. Thus, in the favourable cases, the Jacobi decomposition leads to two slowly perturbed Keplerian motions.

As usual we will call the corresponding osculating orbital elements n_i, a_i, e_i, i_i, Ω_i, ω_i, M_i for the "inner orbit" (orbit of $\vec{r}$) and n_e, a_e, e_e, i_e, Ω_e, ω_e, M_e for the outer or "exterior" orbit (orbit of $\vec{R}$ and of m_3 with respect to $I_{1.2}$).

These osculating orbital elements are defined by the following usual Keplerian expressions.

$\vec{v}$ and $\vec{V}$ are the velocity vectors :

$$\vec{v} = d\vec{r}/dt \ ; \ \vec{V} = d\vec{R}/dt \quad (26)$$

The semi-major axes a_i and a_e are given by :

$$a_i = \left\{\frac{2}{r} - \frac{v^2}{G(m_1 + m_2)}\right\}^{-1} ; \ a_e = \left\{\frac{2}{R} - \frac{V^2}{GM}\right\}^{-1} \quad (27)$$

The mean angular motions n_i and n_e are given by :

$$n_i = \{G(m_1 + m_2)/a_i^3\}^{1/2} \ ; \ n_2 = \{GM/a_e^3\}^{1/2} \quad (28)$$

The eccentricities e_i and e_e are given by :

$$e_i = \left\{1 + \frac{(\vec{r}.\vec{v})^2 - r^2v^2}{G(m_1 + m_2)a_i}\right\}^{1/2} ; \ e_e = \left\{1 + \frac{(\vec{R}.\vec{V})^2 - R^2V^2}{GMa_e}\right\}^{1/2} \quad (29)$$

The osculating orbital planes are the $(\vec{r},\vec{v})$ and $(\vec{R},\vec{V})$ planes, they are related to the inclinations i_i and i_e and to the longitudes of ascending nodes Ω_i and Ω_e.

The arguments of pericenters ω_i and ω_e and the mean anomalies M_i and M_e are given by classical and somewhat more complex Keplerian expressions.

4.4 THE HAMILTON AND DELAUNAY FORMULATION

Delaunay found a set of parameters such that the three-body equations of motion have the Hamiltonian form.

Let us define the two "reduced masses" m and $\mathcal{M}$ by :

$$m = m_1m_2/(m_1 + m_2) \ ; \ \mathcal{M} = m_3(m_1 + m_2)/M \quad (30)$$

With the above orbital elements of the inner and outer orbits of the Jacobi decomposition, the Delaunay elements are the following :

$$\left.\begin{aligned}
\mathcal{L}_i &= m n_i a_i^2 && ; \ \ell_i = M_i \\
\mathcal{G}_i &= m n_i a_i^2 \sqrt{(1-e_i^2)} && ; \ g_i = \omega_i \\
\mathcal{H}_i &= m n_i a_i^2 \sqrt{(1-e_i^2)} \cos i_i && ; \ h_i = \Omega_i \\
\mathcal{L}_e &= \mathcal{M} n_e a_e^2 && ; \ \ell_e = M_e \\
\mathcal{G}_e &= \mathcal{M} n_e a_e^2 \sqrt{(1-e_e^2)} && ; \ g_e = \omega_e \\
\mathcal{H}_e &= \mathcal{M} n_e a_e^2 \sqrt{(1-e_e^2)} \cos i_e && ; \ h_e = \Omega_e
\end{aligned}\right\} \quad (31)$$

The Hamiltonian H of the problem is identical to the energy integral that will be presented in the Section 5.1.3.

$$H = - \frac{G^2 m^3 (m_1 + m_2)^2}{2\mathcal{L}_i^2} - \frac{G^2 \mathcal{M}^3 M^2}{2\mathcal{L}_e^2} + G m_3 \left(\frac{m_1 + m_2}{R} - \frac{m_1}{r_{1.3}} - \frac{m_2}{r_{2.3}}\right) \quad (32)$$

As usual in Hamiltonian problems, the Hamiltonian H must be expressed in terms of the twelve conjugate parameters $\mathcal{L}_i$, $\mathcal{G}_i$ ... h_e. The weakest part of this method is that expression of lengths R, r_{13}, r_{23} in terms of the Delaunay elements is not simple.

The differential equations of motion are the following usual Hamiltonian expressions.

$$\left.\begin{aligned}
d\mathcal{L}_i/dt &= -\partial H/\partial \ell_i \ ; & d\mathcal{G}_i/dt &= -\partial H/\partial g_i \ ; & d\mathcal{H}_i/dt &= -\partial H/\partial h_i \\
d\ell_i/dt &= \partial H/\partial \mathcal{L}_i \ ; & dg_i/dt &= \partial H/\partial \mathcal{G}_i \ ; & dh_i/dt &= \partial H/\partial \mathcal{H}_i \\
d\mathcal{L}_e/dt &= -\partial H/\partial \ell_e \ ; & d\mathcal{G}_e/dt &= -\partial H/\partial g_e \ ; & d\mathcal{H}_e/dt &= -\partial H/\partial h_e \\
d\ell_e/dt &= \partial H/\partial \mathcal{L}_e \ ; & dg_e/dt &= \partial H/\partial \mathcal{G}_e \ ; & dh_e/dt &= \partial H/\partial \mathcal{H}_e
\end{aligned}\right\} \quad (33)$$

This Hamiltonian problem with six degrees of freedom will, by a very simple method, be reduced to a problem with only three degrees of freedom (see Chapter 7, the elimination of the nodes and the time).

Many other formulations of the three-body problem are possible. The "heliocentric coordinates" compete with the Jacobi variables in the planetary case, while the "Lagrange planetary equations" are often considered simpler than the Delaunay equations.

Chapter 5

THE INVARIANTS IN THE THREE-BODY PROBLEM

5.1 THE TEN CLASSICAL INTEGRALS AND THE LAGRANGE-JACOBI IDENTITY

The classical integrals are the integral of the center of mass (six parameters), the integral of the angular momentum (three parameters) and the integral of the energy (one parameter).

5.1.1 The integral of the center of mass

Since $\vec{r}_{ij} = -\vec{r}_{ij}$ the three classical equations of motion (7) give the following identity.

$$d^2(m_1\vec{r}_1 + m_2\vec{r}_2 + m_3\vec{r}_3)/dt^2 = 0 \tag{34}$$

The vector $\vec{r}_I = (m_1\vec{r}_1 + m_2\vec{r}_2 + m_3\vec{r}_3)/M$ is the radius-vector for the center of mass and integration of (34) gives :

$$\vec{r}_I = \vec{A}t + \vec{B} \tag{35}$$

The center of mass has a uniform rectilinear motion.

In most cases $\vec{A}$ and $\vec{B}$ are chosen equal to zero and the resulting set of axes is "the center of mass coordinate system".

Note that the Lagrangian, Jacobi and Delaunay formulations only concern the relative motions of the three bodies ; they have already used the simplification given by the integral of the center of mass.

5.1.2 The integral of the angular momentum

The equations of motion (7) give a second identity :

$$\sum_{j=1}^{3} m_j\vec{r}_j \times (d^2\vec{r}_j/dt^2) = 0 \tag{36}$$

Let us call $\vec{v}_1$, $\vec{v}_2$, $\vec{v}_3$ and also $\vec{v}_A$, $\vec{v}_B$, $\vec{v}_C$ the velocity vectors of the classical and of the Lagrangian formulations :

$$\vec{v}_j = d\vec{r}_j/dt \ ; \ j = \{1,2,3,A,B,C\} \tag{37}$$

The equation (36) can be integrated to obtain :

$$\vec{c} = \text{angular momentum} = \sum_{j=1}^{3} m_j\vec{r}_j \times \vec{v}_j \tag{38}$$

The constant angular momentum $\vec{c}$ is especially interesting when it is expressed in the center of mass system, as we will do henceforth. It is then equal to the following expressions given in the Lagrangian, Jacobi and Delaunay formulations :

Lagrangian formulation :

$$\vec{c} = \frac{m_A m_B m_C}{M} \left(\frac{\vec{r}_A \times \vec{v}_A}{m_A} + \frac{\vec{r}_B \times \vec{v}_B}{m_B} + \frac{\vec{r}_C \times \vec{v}_C}{m_C} \right) \tag{39}$$

Jacobi formulation (with the reduced masses given in (30)) :

$$\vec{c} = m\vec{r} \times \vec{v} + \mathcal{M}\vec{R} \times \vec{V} \tag{40}$$

Delaunay formulation :

Let :

$$\sqrt{(\mathcal{G}_i^2 - \mathcal{H}_i^2)} = \mathcal{K}_i \quad ; \quad \sqrt{(\mathcal{G}_e^2 - \mathcal{H}_e^2)} = \mathcal{K}_e \tag{41}$$

then :

$$\vec{c} = \{\mathcal{K}_i \sin h_i + \mathcal{K}_e \sin h_e \; ; \; - \mathcal{K}_i \cos h_i - \mathcal{K}_e \cos h_e \; ; \; \mathcal{H}_i + \mathcal{H}_e\} \tag{42}$$

The plane containing the center of mass and normal to $\vec{c}$ is the "invariable plane", all collisions and all collinear alignments occur in this plane (Ref. 7) (note that if $\vec{v}$ and $\vec{V}$ are bounded and if $\vec{r}$ is parallel to $\vec{R}$, then (40) implies that $\vec{c}.\vec{r} = \vec{c}.\vec{R} = 0$).

If $\vec{c} = 0$ and $m\mathcal{M} \neq 0$ the equation (40) shows that all four vectors $\vec{r}$, $\vec{v}$, $\vec{R}$, $\vec{V}$ belong to the same plane and the three-body motion remains forever in this plane. (Ref. 8).

5.1.3 The integral of the energy

This famous integral has a well known expression :

$$h = \frac{1}{2} (m_1 v_1^2 + m_2 v_2^2 + m_3 v_3^2) - G\left(\frac{m_1 m_2}{r_{1.2}} + \frac{m_1 m_3}{r_{1.3}} + \frac{m_2 m_3}{r_{2.3}}\right) \tag{43}$$

and it is easy to verify with the equations of motion (7), (8), (9) that its derivative dh/dt is indeed zero.

If h is expressed in the center of mass coordinate system, as we will do henceforth, it is equal to the following expressions.

Lagrangian formulation :

$$h = \frac{m_A m_B m_C}{M} \left\{ \frac{1}{2} \left(\frac{v_A^2}{m_A} + \frac{v_B^2}{m_B} + \frac{v_C^2}{m_C} \right) - GM\left(\frac{1}{m_A r_A} + \frac{1}{m_B r_B} + \frac{1}{m_C r_C} \right) \right\} \tag{44}$$

Jacobi formulation :

$$h = \frac{1}{2} (mv^2 + \mathcal{M}V^2) - G\left(\frac{m_1 m_2}{r} + \frac{m_1 m_3}{r_{13}} + \frac{m_2 m_3}{r_{23}} \right) \tag{45}$$

Delaunay formulation :

The energy h is identical to the Hamiltonian H written in (32), however its expression in terms of the twelve Delaunay elements is complex.

5.1.4 The Lagrange-Jacobi identity

It is customary to present the energy integral in the following way.

$$h = T - U \tag{46}$$

where :

$$T = \text{kinetic energy} = \frac{1}{2}\sum_{j=1}^{3} m_j v_j^2 = \frac{1}{2}(mv^2 + \mathcal{M}V^2) \tag{47}$$

and :

$$U = \text{potential} = G\left(\frac{m_1 m_2}{r_{12}} + \frac{m_1 m_3}{r_{13}} + \frac{m_2 m_3}{r_{23}}\right) \tag{48}$$

On the other hand let us call I the semi-moment of inertia (in the center of mass axes) :

$$I = \frac{1}{2}\left(m_1 r_1^2 + m_2 r_2^2 + m_3 r_3^2\right) = \frac{m_1 m_2 m_3}{2M}\left(\frac{r_{12}^2}{m_3} + \frac{r_{13}^2}{m_2} + \frac{r_{23}^2}{m_1}\right) = \frac{1}{2}(mr^2 + \mathcal{M}R^2) \tag{49}$$

Lagrange noted that the second derivative of I is of simple form (and Jacobi extended this remark to the n-body problem) :

$$dI/dt = m_1 \vec{r}_1 \vec{v}_1 + m_2 \vec{r}_2 \vec{v}_2 + m_3 \vec{r}_3 \vec{v}_3 \tag{50}$$

$$d^2I/dt^2 = \sum_{j=1}^{3} m_j (v_j^2 + \vec{r}_j \cdot d^2\vec{r}_j/dt^2) = 2T - U \tag{51}$$

which with (46) leads to the usual form of the Lagrange-Jacobi identity :

$$d^2I/dt^2 = U + 2h \tag{52}$$

These equations (46) to (52) have led to the first qualitative results of the three-body problem.

A) Case $h < 0$

(46) and (47) imply :

$$U = T - h \geqq -h > 0 \tag{53}$$

U is given in (48) and its positive lower bound (-h) implies :

$$\inf(r_{12}, r_{13}, r_{23}) \leqq -\frac{G}{h}(m_1 m_2 + m_1 m_3 + m_2 m_3) \tag{54}$$

Thus if the energy integral h is negative the smallest mutual distance remains bounded and it is impossible that all mutual distances r_{ij} be larger than the length written in the right-hand side of (54). Let this length be 2a, we will call it "generalized major-axis" in the Chapter 11 of the qualitative analysis.

B) Case $h \geqq 0$

Since $U > 0$ equation (52) leads to :

$$d^2I/dt^2 = U + 2h > 0 \tag{55}$$

I(t) is concave up and has only one minimum, furthermore if at some initial time t_1 we have $I = I_1$ and $dI/dt = I'_1$ the condition $d^2I/dt^2 > 2h$ implies at all time at which the motion is defined :

$$I \geq I_1 + I_1'(t - t_1) + h(t - t_1)^2 \tag{56}$$

If $h > 0$ and/or $I_1' > 0$ the semi-moment of inertia I cannot remain forever bounded. This is also the case if $h = 0$ and $I_1' \leq 0$. Indeed then $d^2I/dt^2 = U$ and when I is bounded U is positive and bounded away from zero.

Thus, if the energy integral h is positive or zero, the largest mutual distance cannot remain forever bounded, but goes to infinity with t if the motion remains forever defined ; the corresponding orbit is said to be "open".

5.2 THE UNSUCCESSFUL RESEARCHS OF NEW INTEGRALS

Many people have tried to obtain new integrals for the three-body problem especially Bruns, Poincaré, Siegel, Painlevé (Ref.5,9 - 19) but with only negative results. For instance : "No new integral of motion can be algebraic with respect to the velocities of the three bodies (and arbitrary with respect to the positions)" (Ref. 11).

The extreme complexity of modern numerical results (Section 9.4 and Ref. 5) seems to confirm this absence of supplementary integral of motion, however the meaning of this proposition must be stated precisely.

There exists of course integrals of motion that are functions of both the present state $\vec{X}$ and the time t, for instance those expressing the value x_{j1} of some component x_j at some given time t_1 :

$$x_{j1} = f_1\big(\vec{X},(t - t_1)\big) \tag{57}$$

The Sundman series (Ref. 20) can be considered as belonging to this class of integrals, but unless $t - t_1$ is very small their convergence is so slow that they are almost useless.

There are also many integrals independant of the time but only valid for a part of the orbits.

Let us consider for instance the Lagrange-Jacobi identity :

$$d^2I/dt^2 = U + 2h \tag{58}$$

For orbits with a positive or zero energy integral h the second derivative d^2I/dt^2 is always positive and I(t) is concave up. The semi-moment of inertia I has one and only one minimum I_m that can theoretically be expressed in terms of the present state $\vec{X}$ of the three-body system :

$$I_m = f(\vec{X}) \tag{59}$$

I_m is an excellent integral of motion that is continuous and even analytic in terms of $\vec{X}$ when $h > 0$.

What happens if we try to extend the integral (59) to orbits of negative energy integral h ?

When h is negative the semi-moment of inertia I can have several minima, and even an infinite number of minima if the orbit is bounded or oscillatory, hence either :

A) I_m is the absolute minimum of I for the orbit of interest, it is then not a continuous function of the present state $\vec{X}$ and its expression is extremely complex, or

B) I_m is the nearest minimum of I, it is then a "transitory integral" whose value changes at each maximum of I, or

C) I_m is a particular minimum of I, the corresponding manifold of phase space has then an infinite number of sheets and is dense in some parts of phase space, where it has of course no interest. The integral is said to be "non isolating".

Thus the conjectured absence of new non-classical integrals means absence of integrals that would be

A) Independant of time,

B) Continuous in terms of the present state,

C) Non-transitory,

D) Isolating,

E) Useful even for bounded and oscillatory orbits.

5.3 THE SCALE TRANSFORMATION, THE VARIATIONAL THREE-BODY PROBLEM AND THE ELEVENTH "LOCAL INTEGRAL"

From a given solution $\vec{r}_j(t)$, $j = \{1,2,3\}$ of the equations of motion (7), (8), (9) it is easy to deduce other solutions by a rotation, a symmetry or a translation (in space and/or in time). Khilmi (Ref. 21) has demonstrated the value of considering a dimensional analysis. This is obtained by the following scale transformation :

If $\vec{r}_j(t)$; $j = \{1,2,3\}$ is a solution of the equations (7), (8), (9) of the three-body problem, then $\vec{\rho}_j(t) = k^2\vec{r}_j(t/k^3)$; $j = \{1,2,3\}$ is another solution.
This new solution has :
Velocities : $d\vec{\rho}_j/dt = k^{-1}\ \vec{v}_j(t/k^3)$
Energy integral $= k^{-2}h$
Angular momentum $= k\vec{c}$ (60)

Note that this scale transformation can be extended immediately to the n-body problem, also that the product c^2h is constant in this transformation. This will be an essential parameter in qualitative analysis of the motion. In a two-body case c^2h is directly related to the eccentricity e of the motion by $2(m_1 + m_2)c^2h = G^2m_1^3m_2^3(e^2 - 1)$.

This scale transformation is related to the eleventh "local integral" via the variational three-body problem in the following way.

The variational three-body problem appears in the study of the vicinity of a given solution (for instance in stability studies).

Let us call $\vec{r}_j(t)$; $j = \{1,2,3\}$ the radius-vectors of the solution of interest and $\vec{r}_j(t) + \vec{\delta}_j(t)$ those of a neighbouring solution.

The first-order differential equations of the three $\vec{\delta}_j$ can be deduced from the general equations (7), (8), (9) :

$$d^2\vec{\delta}_1/dt^2 = Gm_2\{r_{12}^{-3}\vec{\delta}_{12} - 3r_{12}^{-5}(\vec{r}_{12}.\vec{\delta}_{12})\vec{r}_{12}\} + Gm_3\{r_{13}^{-3}\vec{\delta}_{12} - 3r_{13}^{-5}(\vec{r}_{13}.\vec{\delta}_{13})\vec{r}_{13}\} \quad (61)$$

and two other similar equations, where of course $\overrightarrow{\delta_{ij}} = \vec{\delta}_j - \vec{\delta}_i$.

These variational equations can also be presented in the Lagrangian or Jacobi forms.

The usual integrals of motion give :

$$\left.\begin{array}{l} d^2(m_1\vec{\delta}_1 + m_2\vec{\delta}_2 + m_3\vec{\delta}_3)/dt^2 = 0 \\ m_1\vec{\delta}_1 + m_2\vec{\delta}_2 + m_3\vec{\delta}_3 = \vec{D}t + \vec{E} \end{array}\right\}\text{integral of the center of mass} \quad (62)$$

$$\left.\begin{array}{l} \sum_{j=1}^{3} m_j(\vec{r}_j \times \vec{\delta}'_j + \vec{\delta}_j \times \vec{v}_j) = \delta\vec{c} \\ \text{with } \vec{\delta}'_j = d\vec{\delta}_j/dt \end{array}\right\}\text{integral of the angular momentum} \quad (63)$$

$$\sum_{j=1}^{3} m_j\vec{v}_j.\vec{\delta}'_j + G \sum_{1\leq i<j\leq 3} m_i m_j \frac{\vec{r}_{ij}\vec{\delta}_{ij}}{r_{ij}^3} = \delta h \quad \Big\}\text{integral of the energy} \quad (64)$$

These integrals can be related to the Hamiltonian character of the three-body problem and to the corresponding symplecticity.

Let us consider an arbitrary Hamiltonian $H(\vec{p},\vec{q},t)$ and the corresponding equations of motion :

$$d\vec{q}/dt = \partial H/\partial\vec{p} \; ; \; d\vec{p}/dt = - \partial H/\partial\vec{q} \quad (65)$$

The symplecticity is equivalent to the following property : if we consider equations (65) and the corresponding first-order variational equations, if furthermore we consider a solution $\vec{p}(t)$, $\vec{q}(t)$ and two arbitrary neighbouring solutions $\vec{p} + \delta\vec{p}_A$, $\vec{q} + \delta\vec{q}_A$ and $\vec{p} + \delta\vec{p}_B$, $\vec{q} + \delta\vec{q}_B$; if finally the $\delta\vec{p}_A$, $\delta\vec{q}_A$ and $\delta\vec{p}_B$, $\delta\vec{q}_B$ are exact solutions of the first-order variational equations they satisfy :

$$\delta\vec{p}_A(t) \bullet \delta\vec{q}_B(t) - \delta\vec{p}_B(t) \bullet \delta\vec{q}_A(t) = \text{constant} \quad (66)$$

The three-body problem has many possible Hamiltonian formulations, of which the Delaunay formulation of section 4.4. is one. We here choose the following :

$$\left.\begin{array}{l} \vec{p} = (\vec{p}_1,\vec{p}_2,\vec{p}_3) = (m_1\vec{v}_1, m_2\vec{v}_2, m_3\vec{v}_3) \\ \vec{q} = (\vec{q}_1,\vec{q}_2,\vec{q}_3) = (\vec{r}_1,\vec{r}_2,\vec{r}_3) \\ H = \frac{1}{2}\sum_{j=1}^{3}(p_j^2/m_j) - G(\frac{m_1m_2}{r_{12}} + \frac{m_1m_3}{r_{13}} + \frac{m_2m_3}{r_{23}}) \end{array}\right\} \quad (67)$$

hence :

$$\left.\begin{array}{l}\delta\vec{p} = (m_1\vec{\delta}'_1\ ,\ m_2\vec{\delta}'_2\ ,\ m_3\vec{\delta}'_3)\ ;\ \delta\vec{q} = (\vec{\delta}_1\ ,\ \vec{\delta}_2\ ,\ \vec{\delta}_3) \\ \text{with, as in (61) :} \\ \vec{\delta}_j = \delta\vec{r}_j\ ;\ \vec{\delta}'_j = \delta\vec{v}_j\ ;\ j = \{1,2,3\}\end{array}\right\}\quad(68)$$

and the Hamiltonian H is identical to the energy integral h.

The integrals (62), (63), (64) are easily related to the symplecticity relation (66), it is sufficient to choose suitable $\delta\vec{p}_A$, $\delta\vec{q}_A$. For instance if $\delta\vec{p}_A$, $\delta\vec{q}_A$ correspond to a small translation of the set of axes :

$$\overrightarrow{\delta_{1A}} = \overrightarrow{\delta_{2A}} = \overrightarrow{\delta_{3A}} = \vec{\varepsilon}t + \vec{\phi}\ ;\ \vec{\delta}'_{1A} = \vec{\delta}'_{2A} = \vec{\delta}'_{3A} = \vec{\varepsilon} \qquad (69)$$

then, for any constant vectors $\vec{\varepsilon}$ and $\vec{\phi}$ and any neighbouring solution, the relation (66) leads to the following :

$$(m_1\vec{\delta}_1 + m_2\vec{\delta}_2 + m_3\vec{\delta}_3)\vec{\varepsilon} - (m_1\vec{\delta}'_1 + m_2\vec{\delta}'_2 + m_3\vec{\delta}'_3)(\vec{\varepsilon}t + \vec{\phi}) = \text{constant} \qquad (70)$$

This condition (70) is identical to the condition (62) given by the integral of the center of mass.

Similarly, if $\vec{\delta}p_A$, $\vec{\delta}q_A$ correspond to a small difference of orientation :

$$\vec{\delta}_{jA} = \vec{\varepsilon}\times\vec{r}_j\ ;\ \vec{\delta}'_{jA} = \vec{\varepsilon}\times\vec{v}_j\ ;\ j = \{1,2,3\} \qquad (71)$$

and we obtain for any constant vector $\vec{\varepsilon}$ and any neighbouring solution :

$$\vec{\varepsilon}\cdot\{\sum_{j=1}^{3} m_j(\vec{r}_j\times\vec{\delta}'_j + \vec{\delta}_j\times\vec{v}_j)\} = \text{constant} \qquad (72)$$

that is precisely the condition (63).

The condition (64) corresponds to the following differences :

$$\vec{\delta}_{jA} = \varepsilon\vec{v}_j\ ;\ \vec{\delta}'_{jA} = \varepsilon d^2\vec{r}_j/dt^2\ ;\ j = \{1,2,3\} \qquad (73)$$

that is to a constant small advance ε (also called time-translation).

Thus the variational forms (62), (63), (64) of the usual integrals of motion are related to the neighbouring solutions given by a small rotation or translation (in space and/or in time). However, other simple neighbouring solutions are given by the scale transformation (60), it is indeed sufficient to choose the scale factor k in the vicinity of unity and with $k = 1 + \varepsilon$ we obtain the following :

$$\vec{\delta}_{jA} = \varepsilon(2\vec{r}_j - 3t\vec{v}_j)\ ;\ \vec{\delta}'_{jA} = -\ \varepsilon(\vec{v}_j + 3td^2\vec{r}_j/dt^2)\ ;\ j = \{1,2,3\} \qquad (74)$$

and thus, with (66), the "eleventh local integral" :

$$\sum_{j=1}^{3} m_j(2\vec{r}_j\vec{\delta}'_j + \vec{v}_j\vec{\delta}_j) - 3t\delta h = \text{constant} \qquad (75)$$

Unfortunately this integral is only an integral of the first-order variational equation (61) (as are those of (62), (63), (64) themselves) and, if we consider the exact equations of motion, the left-hand member of (75) has a second-order derivative.

The situation can be improved in the following way.

A) Instead of the left member of (75), let us use an equivalent expression that is anti-symmetrical with respect to the two solutions $\vec{p}(t)$, $\vec{q}(t)$ and $\vec{p} + \delta\vec{p}$, $\vec{q} + \delta\vec{q}$. This procedure always leads to an odd-order derivative (here to a derivative of the third-order).

Our "eleventh local integral" then becomes :

$$Q = \sum_{j=1}^{3} m_j(2\vec{r}_j\vec{\delta}'_j + \vec{v}_j\vec{\delta}_j) - 3t\delta h + \sum_{j=1}^{3} m_j\{2(\vec{r}_j + \vec{\delta}_j)\vec{\delta}'_j + (\vec{v}_j + \vec{\delta}'_j)\vec{\delta}_j\} - 3t\delta h \tag{76}$$

Of course, δh is here the exact difference between the energy integrals of the two neighbouring solutions (and not the approximate first-order expression written in (64)).

If we write :

$$I' = \sum_{j=1}^{3} m_j\vec{r}_j\vec{v}_j = \text{derivative of the semi-moment of inertia } I \tag{77}$$

we can simplify the "eleventh local integral" Q into :

$$Q = \sum_{j=1}^{3} m_j(\vec{r}_j\vec{\delta}'_j - \vec{v}_j\vec{\delta}_j) + 3\delta I' - 6t\delta h \tag{78}$$

With (58) it is easy to verify that dQ/dt, that is $3\delta U + \sum_{j=1}^{3} m_j(\vec{r}_j\vec{\delta}''_j - \vec{r}''_j\vec{\delta}_j)$, is of the third-order and this expression Q can easily be extended to the general n-body problem.

B) It is possible to go further, to add to Q a suitable small correcting third-order term and to build, step by step and order after order, a series that will represent a new integral of motion. This series will be an exact integral, but it will remain local and will converge only in a small zone around the orbit of interest.

C) For the Delaunay variables of Section 4.4 a simple Q_D equal to (− Q + third order) is given by :

$$\left.\begin{aligned} Q_D = {} & (2\mathcal{L}_i + \delta\mathcal{L}_i)\delta\ell_i + (2\mathcal{G}_i + \delta\mathcal{G}_i)\delta g_i + (2\mathcal{H}_i + \delta\mathcal{H}_i)\delta h_i + (2\mathcal{L}_e + \delta\mathcal{L}_e)\delta\ell_e + \\ & + (2\mathcal{G}_e + \delta\mathcal{G}_e)\delta g_e + (2\mathcal{H}_e + \delta\mathcal{H}_e)\delta h_e + 6t.\delta h \end{aligned}\right\} \tag{79}$$

the main terms of its derivative are usually $-2G\left[\dfrac{m^3(m_1+m_2)^2(\delta\mathcal{L}_i)^3}{\mathcal{L}_i^5} + \dfrac{M^3M^2(\delta\mathcal{L}_e)^3}{\mathcal{L}_e^5}\right]$

Expressions taking account of a possible difference δt are sometimes considered :

$$
\left.\begin{aligned}
Q &= \sum_{j=1}^{3} m_j(\vec{r}_j\vec{\delta}_j' - \vec{v}_j\vec{\delta}_j) + 3\delta I' - 6t\delta h - 4h\delta t - 5\delta h\delta t \\
Q_D &= (2\mathcal{L}_i + \delta\mathcal{L}_i)\delta\ell_i + (2\mathcal{G}_i + \delta\mathcal{G}_i)\delta g_i + (2\mathcal{H}_i + \delta\mathcal{H}_i)\delta h_i + (2\mathcal{L}_e + \delta\mathcal{L}_e)\delta\ell_e + \\
&\quad + (2\mathcal{G}_e + \delta\mathcal{G}_e)\delta g_e + (2\mathcal{H}_e + \delta\mathcal{H}_e)\delta h_e + 6t\delta h + 4h\delta t + 5\delta h\delta t
\end{aligned}\right\} \quad (80)
$$

5.4 THE INTEGRAL INVARIANTS

The symplecticity relation (66) and the corresponding integrals along a curve of a surface of phase space are sometimes called "integral invariants" (Ref. 22). However usually the integral invariant is the "volume in phase space" (another consequence of (66), that will lead to the ergodic properties of the three-body problem (see Section 11.7.10.2 and Ref.23-29).

The integral invariant corresponds to the following property that is very common among Hamiltonian systems. Let us consider a set S of solutions of a three-body problem with three given masses. The set S corresponds to a domain D_1 in phase space at the time t_1 and to a domain D_2 at the time t_2 and if the domain D_1 is measurable the domain D_2 is also measurable and has the same measure.

$$\int_{D_1} d\Omega = \int_{D_2} d\Omega \quad (81)$$

This property is true for all the presentations of the three-body problem with the following volume element $d\Omega$.

A) Classical presentation (equations (4)-(9)).

With $\vec{r}_j = (x_j, y_j, z_j)$ and $\vec{v}_j = d\vec{r}_j/dt = (x_j', y_j', z_j')$ we obtain a volume element $d\Omega$ with 18 dimensions.

$$d\Omega = dx_1dy_1dz_1dx_2dy_2dz_2dx_3dy_3dz_3\,dx_1'dy_1'dz_1'dx_2'dy_2'dz_2'dx_3'dy_3'dz_3' \quad 82)$$

B) Lagrangian presentation (equations (10)-(17)).

Since $\vec{r}_A + \vec{r}_B + \vec{r}_C = 0$ the number of dimensions is reduced to twelve and we can choose, for instance, the following :

$$d\Omega = dx_Ady_Adz_Adx_Bdy_Bdz_B\,dx_A'dy_A'dz_A'dx_B'dy_B'dz_B' \quad (83)$$

(The choice of A and C or B and C instead of A and B would have led to the same volume element and the same measure).

C) Jacobi presentation (equations (18)-(23)).

$\vec{r} = (x,y,z)$ and $\vec{R} = (X,Y,Z)$, hence we obtain the following :

$$d\Omega = dxdydzdXdYdZdx'dy'dz'dX'dY'dZ' \quad (84)$$

This volume element gives measures identical to those given by (83).

D) Delaunay presentation (equations (31)-(33)).

$$d\Omega = d\mathcal{L}_i d\mathcal{G}_i d\mathcal{H}_i d\ell_i dg_i dh_i d\mathcal{L}_e d\mathcal{G}_e d\mathcal{H}_e d\ell_e dg_e dh_e \quad (85)$$

In all cases $d\Omega$ is the usual volume element of the phase-space of interest.

Chapter 6

EXISTENCE AND UNIQUENESS OF SOLUTIONS. BINARY AND TRIPLE COLLISIONS. REGULARIZATION OF SINGULARITIES.

The usual three-body equations of motion (4)-(9) or (10)-(17) or (18)-(23) are regular except for "collisions", i.e. when one or several mutual distances r_{ij} are zero ; hence the usual theorems on differential equations lead to the following.

"The solution of a given three-body problem remains uniquely defined at least as long as all three mutual distances remain bounded away from zero".

We will, of course, always choose initial conditions where all r_{ij} are non-zero and we must consider the singularities when one or several r_{ij} approach zero.

The three-body problem is much simpler than the general n-body problem in that it ignores the "infinite expansion in a bounded interval of time". It has only two singularities(Ref.30-36) the binary collision and the triple collision.

If there is a collision at time t_c all three bodies go to a definite position when $t \to t_c$. If the collision is a binary collision the motion of the isolated body is continuous and twice continuously differentiable at t_c while the two colliding bodies m_i and m_j are such that when $t \to t_c$:

$$\left.\begin{array}{ll} \vec{r}_{ij}(t_c - t)^{-2/3} \to \vec{A} & \\ \vec{v}_{ij}(t_c - t)^{1/3} \to -2\vec{A}/3 & \text{The limit vector } \vec{A} \text{ satisfies :} \\ (d^2\vec{r}_{ij}/dt^2).(t_c - t)^{4/3} \to -2\vec{A}/9 & ||\vec{A}|| = \{4.5G(m_i + m_j)\}^{1/3} \end{array}\right\} \quad (86)$$

If the collision is a triple collision all three bodies go to the center of mass when $t \to t_c$ and they then satisfy (except for the very singular case presented at the end of this Chapter 6) :

$$\left.\begin{array}{l} \vec{r}_j(t_c-t)^{-2/3} \to \vec{A}_j \\ \vec{v}_j(t_c-t)^{1/3} \to -2\vec{A}_j/3 \\ (d^2\vec{r}_j/dt^2(t_c-t)^{4/3} \to -2\vec{A}_j/9 \end{array}\right\} j = \{1,2,3\} \qquad (87)$$

The three limit vectors $\vec{A}_j$ satisfy the relation of the center of mass $(m_1\vec{A}_1 + m_2\vec{A}_2 + m_3\vec{A}_3 = 0)$ and the relations of universal attraction :

$$-\frac{2\vec{A}_1}{9} = G\left(\frac{m_2\vec{A}_{12}}{A_{12}^3} + \frac{m_3\vec{A}_{13}}{A_{13}^3}\right) \qquad (88)$$

and two other similar equations with $\vec{A}_{ij} = \vec{A}_j - \vec{A}_i$

The solutions of (88) give "central configurations" (see Section 8.1). Thus, "when three bodies are going to a triple collision they approach a central configuration".

The central configurations are of two types :

A) Collinear (or Euler) central configurations.

The three $\vec{A}_j$ are collinear, and these triple collisions occur only if the three-body motion is rectilinear : the three bodies move along a fixed straight line (in the center of mass axes).

B) Triangular (or Lagrange) central configurations.

The three $\vec{A}_j$ are at the tips of an equilateral triangle (all $\vec{A}_{ij}$ satisfy $||\vec{A}_{ij}|| = \{4.5GM\}^{1/3}$) and this second kind of triple collision requires planar and non rectilinear three-body motion.

The equations (38) and (87) show that the angular momentum $\vec{c}$ must be zero for triple collisions which is obvious in the case of rectilinear motions and which is a necessary condition in the other case (we have already noticed that $\vec{c} = 0$ and $m_1m_2m_3 \neq 0$ imply planar three-body motion).

Binary and triple collisions have been the subject of many studies (for instance, Ref. 30-52) and the motions going down to a triple collision have been developed into "Siegel series" (Ref. 37, 48) the first term of which is given by (87) and (88).

Is it possible to give a natural extension to a solution after a collision ? That question is called "regularization",it has a theoretical interest and is also very useful for the numerical integrations of solutions with close approaches.

Two types of regularizations have been defined :

A) Analytical or Siegel-regularization (Ref. 37).

The function $x = t^{1/3}$ is extended into the past (i.e. for negative t) by $x = -(-t)^{1/3}$ while $y = t^{2/3}$ is extended into $y = (-t)^{2/3}$ because they both correspond to the general equations $x^3 = t$ and $y^3 = t^2$.

By various methods Siegel extends his definition of analytical regularization, but he cannot extend $x = t^{p/q}$ to the past when p is odd and q is even.

B) Topological or Easton-regularization (Ref. 53).

If the differential system $d\vec{x}/dt = f(\vec{x},t)$ has a singularity at $\vec{x}_s, t_s$ the solution leading to that point is called singular. However generally most of the neighbouring solutions are regular and if they remain together after t_s the extension of the singular solution can be defined by continuity.

If the regular neighbouring solutions diverge after t_s the singularity is not Easton-regularizable.

The examples given by Mc-Gehee (Ref. 39) show that the two regularizations are independant.

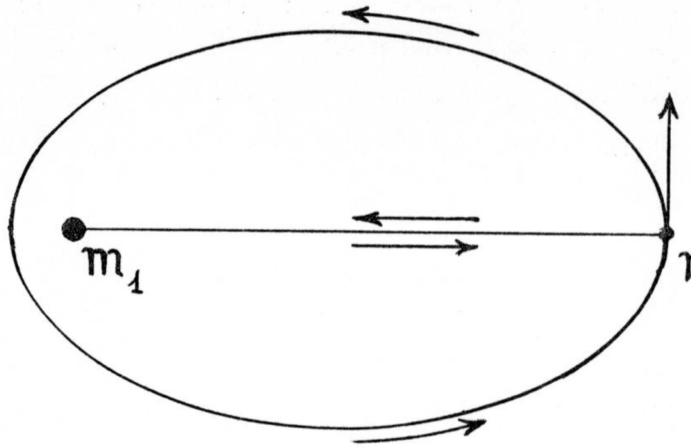

Fig. 5. The continuity of the two-body problem at collision.

It is well-known that the two-body problem is continuous at collision (Fig. 5) and Sundman (Ref. 20) has similarly regularized the binary collisions of the three-body problem*.

This regularization is both a Siegel and an Easton-regularization, so it is natural that we consider that the three-body motions without triple collision are uniquely defined from $t = -\infty$ to $t = +\infty$.

Sundman has also demonstrated (Ref. 20) that if the angular momentum $\vec{c}$ is not zero the time t and the coordinates of the three bodies can be developed into always converging power series of a suitable parameter τ, even if the number of binary collisions is very large.

This Sundman result can be extended to all three-body motions without triple collisions, even if $\vec{c} = 0$, provided that the perimeter $(r_{12} + r_{13} + r_{23})$ remains bounded away from zero. However, unfortunately, the Sundman series converge so slowly that they are almost useless.

The triple collisions are never Easton- and almost never Siegel-regularizable (Ref. 47, 54). The neighbouring solutions undergo a triple close approach followed by an expansion and a wide dispersion. Most of these neighbouring solutions lead to the formation of a very small binary escaping at a very large velocity in a direction opposite to that of the third body, and that isolated escaping body is generally either the smallest mass or the second smallest (Ref. 38, 49).

The following conjecture can give an idea of the importance of the dispersion effect in the vicinity of triple collisions.

---

*More sophisticated regularizations of binary collisions have recently been proposed (Ref. 55-58).

Let us consider, in the center of mass axes, a three-body motion leading from some initial conditions to a triple collision at the time t_c (Fig. 6). Let us also consider an arbitrary "triple explosion" at the time t_c with the same three masses m_1, m_2, m_3 and the same energy integral h.

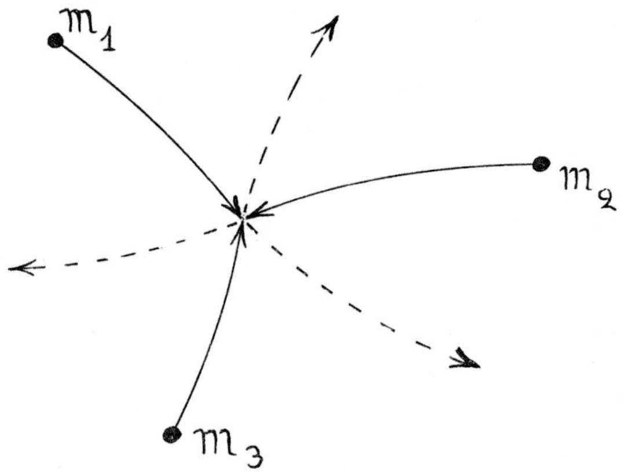

Fig. 6. The total dispersion conjecture.
Full line : triple collision, dotted line : triple explosion.

The "total dispersion conjecture" is then the following : In any neighbourhood of the initial conditions of the Fig. 6 there always exist initial conditions leading as nearly as desired to the final conditions of the triple explosion.

This conjecture is false if we restrict the analysis to rectilinear motions and/or if the smallest mass is infinitely small with respect to the total mass, but it seems to be true in the other cases either for two or for three-dimensional motions.

An exhaustive review of the singularities of the three-body problem leads finally to the following.

Let us consider two large masses falling towards each other along a straight line and colliding at time t_c. If a third infinitesimal mass moves back and forth between the two large masses along their straight line of motion it will undergo an infinite number of binary collisions with each of the two large masses before the triple collision at t_c.

This last type of triple collision is exceptional and requires an infinitesimal mass, however notice that there are also similar plane solutions asymptotic when $t \to t_c$ to the above rectilinear motion. These plane asymptotic triple collisions have no binary collision in a sufficiently small neighbourhood of t_c.

This last type of triple collision is extremely sensitive to initial conditions, and is neither Siegel- nor Easton-regularizable. Hence it is natural to consider that a triple collision (of any type) is the end of a three-body motion, while this motion can naturally be extended after a binary collision.

Thus most three-body motions will be uniquely defined from $t = -\infty$ to $t = +\infty$ and especialy all those with a non-zero angular momentum.

A simple regularization of the binary collisions of the masses m_1 and m_2 has been proposed by Burdet and Heggie (Ref.57,58) it is the following with the Jacobi vectors $\vec{r}$ and $\vec{R}$ and their derivatives $\vec{v}$ and $\vec{V}$ (Fig. 6.1).

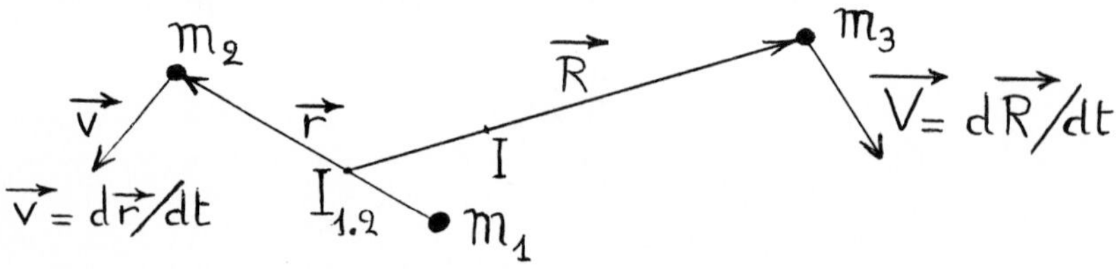

Fig. 6.1. The Jacobi vectors $\vec{r}$, $\vec{R}$, $\vec{v}$, $\vec{V}$.

A) Initial equations of motion (Jacobi equations of Section 4.3) :

$$\left.\begin{aligned}
&m_1/(m_1+m_2) = \alpha\ ;\ m_2/(m_1+m_2) = \beta = 1-\alpha\ ;\ m_1+m_2+m_3 = M\\
&\vec{r}_{13} = \vec{R} + \beta\vec{r}\ ;\ \vec{r}_{23} = \vec{R} - \alpha\vec{r}\\
&d^2\vec{r}/dt^2 = -G(m_1+m_2)\frac{\vec{r}}{r^3} + Gm_3\left(\frac{\vec{r}_{23}}{r_{23}^3} - \frac{\vec{r}_{13}}{r_{13}^3}\right)\\
&d^2\vec{R}/dt^2 = -GM(\alpha\vec{r}_{13}\, r_{13}^{-3} + \beta\vec{r}_{23}\, r_{23}^{-3})
\end{aligned}\right\}\quad(88.1)$$

B) New variables s, k, $\vec{e}$

$$\left.\begin{aligned}
&s \text{ is defined by } dt/ds = r\\
&k = (v^2/2) - \left(G(m_1+m_2)/r\right)\ ;\quad k \text{ is proportional to the mechanical energy of the binary}\\
&\vec{e} = \vec{v}\times(\vec{r}\times\vec{v}) - G(m_1+m_2)\frac{\vec{r}}{r} = \text{"Laplacean vector" of the binary}
\end{aligned}\right\}\quad(88.2)$$

C) Regularized equations of motion

$$\left.\begin{aligned}
&\left.\begin{aligned}
&d^2\vec{r}/ds^2 = 2k\vec{r} - \vec{e} + r^2\vec{\varepsilon}\\
&dk/ds = \vec{\varepsilon}\cdot\frac{d\vec{r}}{ds}\\
&d\vec{e}/ds = \vec{\varepsilon}\times\left(\vec{r}\times\frac{d\vec{r}}{ds}\right) + \frac{d\vec{r}}{ds}\times(\vec{r}\times\vec{\varepsilon})
\end{aligned}\right\}\text{ with } \vec{\varepsilon} = Gm_3\left(\frac{\vec{r}_{23}}{r_{23}^3} - \frac{\vec{r}_{13}}{r_{13}^3}\right)\\
&d\vec{R}/ds = r\vec{V}\\
&d\vec{V}/ds = -GMr(\alpha\vec{r}_{13}r_{13}^{-3} + \beta\vec{r}_{23}r_{23}^{-3})
\end{aligned}\right\}\quad(88.3)$$

These equations are smooth at $\vec{r} = 0$, they can easily be extended to the n-body problem, they have the two following integrals of motion related to the definitions of k and $\vec{e}$:

$$2G(m_1 + m_2)r + 2kr^2 - (d\vec{r}/ds)^2 = 0 \qquad (88.4)$$

$$G(m_1 + m_2)r\vec{r} + r^2\vec{e} + (\vec{r} \times \frac{d\vec{r}}{ds}) \times \frac{d\vec{r}}{ds} = 0 \qquad (88.5)$$

Chapter 7

FINAL SIMPLIFICATIONS, THE ELIMINATION OF NODES, THE ELIMINATION OF TIME.

The Lagrangian and the Jacobi formulations of the three-body problem (Chapter 4) are twelfth order systems of differential equations. These systems can be reduced to eighth order with the integrals of motion. Lagrange has shown the possibility of a supplementary reduction through "elimination of the nodes" (Ref. 59). However, his system of equations is very complex.

The nodes can be eliminated in a very simple way with the Hamiltonian Delaunay formulation, provided that the reference plane be the invariable plane normal to $\vec{c}$.

The components of $\vec{c}$ are then (0,0,c) and, with (41), (42) we have :

$$\left.\begin{aligned} &\mathcal{K}_i = \mathcal{K}_e \\ &\mathcal{G}_i^2 - \mathcal{H}_i^2 = \mathcal{G}_e^2 - \mathcal{H}_e^2 \\ &h_i + \pi = h_e \\ &\mathcal{H}_i + \mathcal{H}_e = c \end{aligned}\right\} \quad (89)$$

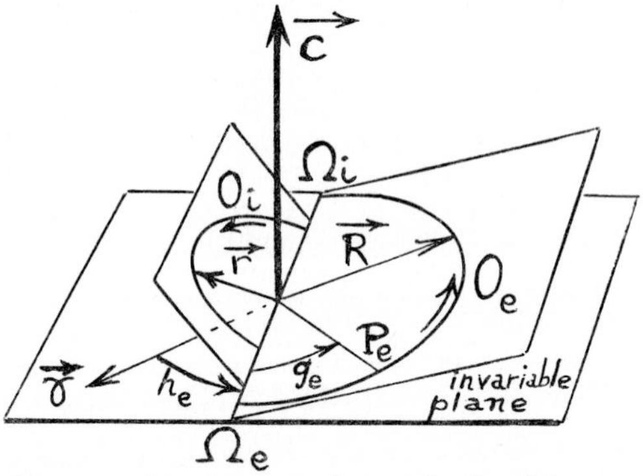

Fig. 7. The inner orbit O_i, the outer or "exterior" orbit O_e and the invariable plane.

Let us consider the inner and outer orbits of Fig. 7, i.e. the two-body orbits of the exact equations (24) and (25) (also called osculating orbits).

The line of nodes $\Omega_e\Omega_i$ of the two orbital planes is always in the invariable plane and h_e is its longitude, while g_i and g_e are the arguments of the pericenters P_i and P_e with respect to this line of nodes.

For three given masses m_1, m_2, m_3 the Hamiltonian H of the problem is known as soon as the eight Delaunay parameters $\mathcal{L}_i$, $\mathcal{G}_i$, $\mathcal{L}_e$, $\mathcal{G}_e$, ℓ_i, g_i, ℓ_e, g_e (see (31)) and the modulus c of the angular momentum are given, indeed :

A) $\mathcal{L}_i$, $\mathcal{G}_i$, $\mathcal{L}_e$, $\mathcal{G}_e$ give the size and the shape of the inner and outer orbits.

B) ℓ_i and ℓ_e give the positions along the two orbits.

C) The relative orientation of the two orbits is given by the two angles g_i, g_e and by the mutual inclination $i_i + i_e$ which is a function of $\mathcal{G}_i$, $\mathcal{G}_e$, c (Fig. 8).

D) The Hamiltonian H is given in (32), it is identical to the energy integral h and is independent of the absolute orientation of the system of the two orbits (angles h_i, h_e, directions of $\vec{c}$ and $\vec{\gamma}$).

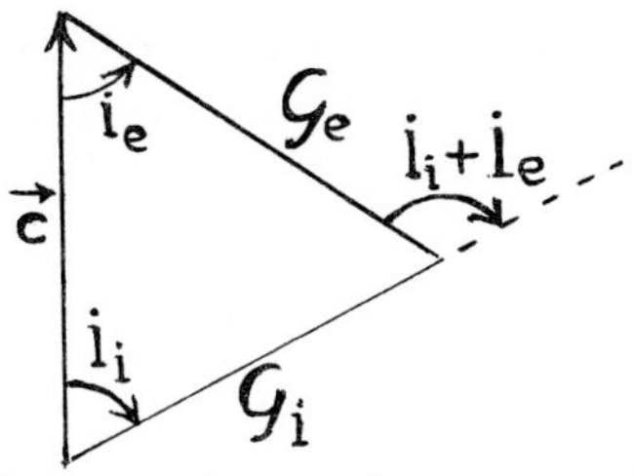

Fig. 8. Relations between the angular momenta $\mathcal{G}_i$, $\mathcal{G}_e$, $\vec{c}$ and the inclinations i_i, i_e.

$$c^2 = \mathcal{G}_i^2 + \mathcal{G}_e^2 + 2\mathcal{G}_i\mathcal{G}_e \cos(i_i + i_e).$$

Hence for given values of the Newtonian constant G and the three masses m_1, m_2, m_3 we can write :

$$H = H(\mathcal{L}_i, \mathcal{G}_i, \mathcal{L}_e, \mathcal{G}_e, \ell_i, g_i, \ell_e, g_e, c) \tag{90}$$

This leads to the following Hamiltonian system with only four degrees of freedom.

$$\left.\begin{aligned} \frac{d\mathcal{L}_i}{dt} &= -\frac{\partial H}{\partial \ell_i} ; \quad \frac{d\mathcal{G}_i}{dt} = -\frac{\partial H}{\partial g_i} ; \quad \frac{d\mathcal{L}_e}{dt} = -\frac{\partial H}{\partial \ell_e} ; \quad \frac{d\mathcal{G}_e}{dt} = -\frac{\partial H}{\partial g_e} \\ \frac{d\ell_i}{dt} &= \frac{\partial H}{\partial \mathcal{L}_i} ; \quad \frac{dg_i}{dt} = \frac{\partial H}{\partial \mathcal{G}_i} ; \quad \frac{d\ell_e}{dt} = \frac{\partial H}{\partial \mathcal{L}_e} ; \quad \frac{dg_e}{dt} = \frac{\partial H}{\partial \mathcal{G}_e} \end{aligned}\right\} \tag{91}$$

Six of these eight equations have exactly the meaning they had in (33), the partial derivatives $\partial H/\partial \mathcal{G}_i$ and $\partial H/\partial \mathcal{G}_e$ have different meanings but the same value.

The ignorable parameters $\mathcal{H}_i$, $\mathcal{H}_e$, h_i, h_e are given by the following :

$$\left.\begin{aligned} &\left.\begin{aligned} \mathcal{H}_i &= (c^2 + \mathcal{G}_i^2 - \mathcal{G}_e^2)/2c \\ \mathcal{H}_e &= (c^2 - \mathcal{G}_i^2 + \mathcal{G}_e^2)/2c \end{aligned}\right\} \text{from equations (89)} \\ &h_i + \pi = h_e ; \quad dh_i/dt = dh_e/dt = \partial H/\partial c \end{aligned}\right\} \tag{92}$$

The next simplification is the elimination of time t. Let us choose, for instance, a fast-moving parameter such as ℓ_i as the description parameter. Then the new Hamiltonian will be L_i and we must reverse equation (90) and express L_i in terms of the other parameters :

$$L_i = L_i(H, G_i, L_e, G_e, \ell_i, g_i, \ell_e, g_e, c) \tag{93}$$

This leads classically to a new Hamiltonian system with one less degree of freedom, but also with a non-autonomous Hamiltonian :

$$\left.\begin{aligned} dG_i/d\ell_i &= \partial L_i/\partial g_i \; ; \; dL_e/d\ell_i = \partial L_i/\partial \ell_e \; ; \; dG_e/d\ell_i = \partial L_i/\partial g_e \\ dg_i/d\ell_i &= -\partial L_i/\partial G_i \; ; \; d\ell_e/d\ell_i = -\partial L_i/\partial L_e \; ; \; dg_e/d\ell_i = -\partial L_i/\partial G_e \end{aligned}\right\} \tag{94}$$

The integrals of motion H and c are constant, L_i is given by (93) and the ignorable parameters H_i, H_e, h_i, h_e, t are given by the following :

$$\left.\begin{aligned} &H_i = (c^2 + G_i^2 - G_e^2)/2c \; ; \; H_e = c - H_i \\ &h_i + \pi = h_e \; ; \; dh_i/d\ell_i = dh_e/d\ell_i = -\partial L_i/\partial c \\ &dt/d\ell_i = \partial L_i/\partial H \end{aligned}\right\} \tag{95}$$

It seems that in the formulation (93), (94), (95) the general three-body problem reaches its simplest system of differential equations, being then an Hamiltonian system with three degrees of freedom.

Remarks :

A) Elimination of the nodes is very interesting and the system (91) is much simpler than the system (33) ; on the contrary the elimination of time is of much less interest and an autonomous Hamiltonian system with four degrees of freedom is almost as simple as the corresponding non-autonomous system with three degrees of freedom.

B) The Delaunay parameters have many singularities, for instance L_i, ℓ_i, L_e, ℓ_e become complex for hyperbolic osculating orbits.

Many methods have been discovered to avoid these singularities. For instance if L_i and ℓ_i are complex, or if they become complex in the motion of interest, then it is customary to substitute for these conjugate parameters two other equivalent conjugate parameters : the energy integral of the inner orbit $(h_i = mv^2/2 - Gm_1m_2/r)$ and the difference $t - t_p$ equal to $n_i\ell_i$ (the time t_p is the time of pericenter passage on the inner osculating orbit of interest).

C) Another main singular case is that of planar motions.

In this case :

$$\left.\begin{aligned} &i_i = 0 \text{ or } \pi \; ; \; i_e = 0 \text{ or } \pi \\ &H_i = G_i \cos i_i = \pm G_i \; ; \; H_e = G_e \cos i_e = \pm G_e \\ &g_i, h_i, g_e, h_e \text{ are not defined.} \end{aligned}\right\} \tag{96}$$

Let us call h_i and h_e the longitudes of the pericenters of inner and outer orbits respectively.

We can then write the following :

$$H = \text{Hamiltonian of the problem} = \text{energy integral} = \left.\begin{aligned} &= -\frac{G^2m^3(m_1+m_2)^2}{2\mathcal{L}_i^2} - \frac{G^2\mathcal{M}^3M^2}{2\mathcal{L}_e^2} + Gm_3\left(\frac{m_1+m_2}{R} - \frac{m_1}{r_{13}} - \frac{m_2}{r_{23}}\right) \\ &= H(\mathcal{L}_i,\mathcal{H}_i,\mathcal{L}_e,\mathcal{H}_e,\ell_i,h_i,\ell_e,h_e) \end{aligned}\right\} \quad (97)$$

and we obtain the following Hamiltonian system with four degrees of freedom :

$$\left.\begin{aligned} \frac{d\mathcal{L}_i}{dt} &= -\frac{\partial H}{\partial \ell_i} ; \frac{d\mathcal{H}_i}{dt} = -\frac{\partial H}{\partial h_i} ; \frac{d\mathcal{L}_e}{dt} = -\frac{\partial H}{\partial \ell_e} ; \frac{d\mathcal{H}_e}{dt} = -\frac{\partial H}{\partial h_e} \\ \frac{d\ell_i}{dt} &= \frac{\partial H}{\partial \mathcal{L}_i} ; \frac{dh_i}{dt} = \frac{\partial H}{\partial \mathcal{H}_i} ; \frac{d\ell_e}{dt} = \frac{\partial H}{\partial \mathcal{L}_e} ; \frac{dh_e}{dt} = \frac{\partial H}{\partial \mathcal{H}_e} \end{aligned}\right\} \quad (98)$$

The sum $\mathcal{H}_i + \mathcal{H}_e$ is equal to the angular momentum c and on the other hand h_i and h_e appear in H only through the difference $h_e - h_i$.

Hence we can write :

$$h_e - h_i = \Delta \quad (99)$$

and :

$$H = H(\mathcal{L}_i,c,\mathcal{L}_e,\mathcal{H}_e,\ell_i,\ell_e,\Delta) \quad (100)$$

With this new expression of the Hamiltonian we obtain a system with three degrees of freedom :

$$\left.\begin{aligned} d\mathcal{L}_i/dt &= -\partial H/\partial\ell_i ; d\mathcal{L}_e/dt = -\partial H/\partial\ell_e ; d\mathcal{H}_e/dt = -\partial H/\partial\Delta \\ d\ell_i/dt &= \partial H/\partial\mathcal{L}_i ; d\ell_e/dt = \partial H/\partial\mathcal{L}_e ; d\Delta/dt = \partial H/\partial\mathcal{H}_e \end{aligned}\right\} \quad (101)$$

and the ignorable parameters are given by :

$$\mathcal{H}_i = c - \mathcal{H}_e ; dh_i/dt = \partial H/\partial c ; h_e = h_i + \Delta \quad (102)$$

Finally, the system (101) can be reduced to an Hamiltonian system with only two degrees of freedom by eliminating the time and choosing another parameters of description such as ℓ_i.

We must first reverse (100) :

$$\mathcal{L}_i = \mathcal{L}_i(H,c,\mathcal{L}_e,\mathcal{H}_e,\ell_i,\ell_e,\Delta) \quad (103)$$

and we obtain, with $\mathcal{L}_i$ as Hamiltonian :

$$\left.\begin{aligned} d\mathcal{L}_e/d\ell_i &= \partial\mathcal{L}_i/\partial\ell_e ; d\mathcal{H}_e/d\ell_i = \partial\mathcal{L}_e/\partial\Delta ; \\ d\ell_e/d\ell_i &= -\partial\mathcal{L}_i/\partial\mathcal{L}_e ; d\Delta/d\ell_i = -\partial\mathcal{L}_i/\partial\mathcal{H}_e \end{aligned}\right\} \quad (104)$$

H and c remain constant (enery integral and angular momentum) and the ignorable parameters are given by the following :

$$\left.\begin{aligned} &\mathcal{H}_i = c - \mathcal{H}_e \\ &dh_i/d\ell_i = -\partial\mathcal{L}_i/\partial c ; h_e = h_i + \Delta \\ &dt/d\ell_i = \partial\mathcal{L}_i/\partial H \end{aligned}\right\} \quad (105)$$

D) Finally a well-known singularity happens for circular or near circular orbits. In this case the parameter ℓ_i is not a good description parameter and it is preferable to use $\ell_i + g_i$ in the three dimensional case (equations

(93) and (94)) and either $\ell_i - \Delta\cos i_i$ or $\ell_i - \Delta\cos i_i - \ell_e \cos i_i \cos i_e$ in the plane case with $i_i = 0$ or π and $i_e = 0$ or π (equations (103) and (104).

Chapter 8

SIMPLE SOLUTIONS OF THE THREE-BODY PROBLEM

8.1 THE LAGRANGIAN AND EULERIAN SOLUTIONS. THE CENTRAL CONFIGURATIONS

Lagrange found three-body motions in which the mutual distances are constant, and Euler extended them and found solutions in which the ratios of mutual distances are constant. The corresponding configurations are called "central configurations" because the accelerations of the three bodies are directed toward the center of mass and are proportional to the corresponding radius-vectors.

Let us consider the Lagrangian equations of motion (13)-(17).

A first simple case is the "triangular case". Assume that the three bodies are at the corners of an equilateral triangle (Fig. 9). Thus $r_A = r_B = r_C$, hence (16) and (13) imply $\vec{W} = 0$ and the equations of motion (15) become :

$$d^2\vec{r}_J/dt^2 = - GM\vec{r}_J r_J^{-3} \quad ; \; J = \{A,B,C\} \tag{106}$$

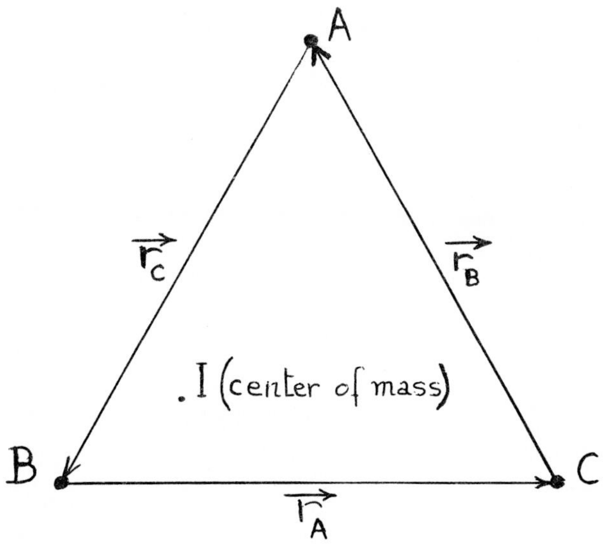

Fig. 9. A triangular (or Lagrangian) central configuration. The points A,B,C, are at the corners of an equilateral triangle.

These equations are those of the vector $\vec{r}_{12}$ of relative motion in a two-body system in which the total mass is M, and we can keep the equilateral triangle configuration of our three-body system (and the equations (106)) if the three mutual distances r_A, r_B, r_C have the same two-body evolution.

In the axes of the center of mass the three bodies then describe, in the same direction, three coplanar Keplerian orbits with the same eccentricity, the same period, the same time of pericenter passage and the same "attracting focus" at the center of mass (Fig. 10). These motions are called Lagrangian three-body motions.

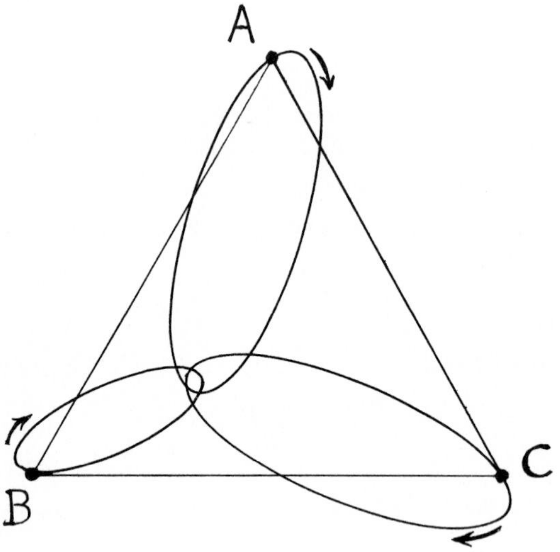

Fig. 10. Elliptic Lagrangian three-body motion.

The remaining solutions with constant ratios of mutual distances use the "collinear central configurations" (Fig. 11) and in the corresponding "Eulerian motions" the three masses describe three homothetic Keplerian orbits with the same "attracting focus" at the center of mass. These motions require of course three initial parallel and homothetic velocities, but they also require <u>homothetic initial accelerations</u> and we thus obtain the following condition

$$\frac{d^2\vec{r}_1/dt^2}{\vec{r}_1} = \frac{d^2\vec{r}_2/dt_1^2}{\vec{r}_2} = \frac{d^2\vec{r}_3/dt_1}{\vec{r}_3} \qquad (107)$$

which is called the "condition of relative equilibrium".

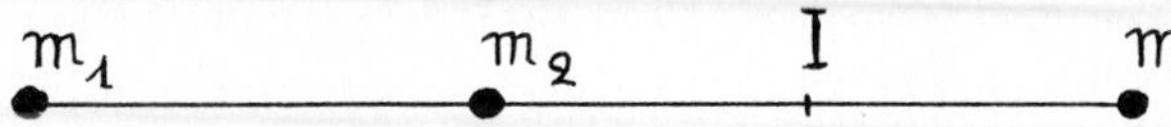

$$r_{23}/r_{12} = x$$

$$(m_1+m_2)x^5+(3m_1+2m_2)x^4+(3m_1+m_2)x^3 =$$
$$= (m_2+3m_3)x^2+(2m_2+3m_3)x+(m_2+m_3)$$

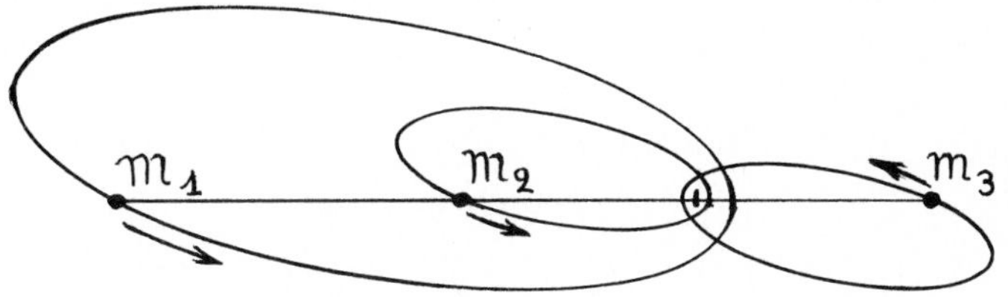

Fig. 11. A collinear (or Eulerian) central configuration and a corresponding elliptic Eulerian three-body motion.

Note the following :

A) The condition is always satisfied for the equilateral triangles and never for the other triangles.

B) Since $m_1\vec{r}_1 + m_2\vec{r}_2 + m_3\vec{r}_3 = 0$ and $\frac{d^2}{dt^2}(m_1\vec{r}_1 + m_2\vec{r}_2 + m_3\vec{r}_3) = 0$ the condition (107) express only one condition and not two conditions.

C) By differences we can express (107) in terms of the relative positions only :

$$\frac{d^2\vec{r}_{12}/dt^2}{\vec{r}_{12}} = \frac{d^2\vec{r}_{23}/dt^2}{\vec{r}_{23}} \tag{108}$$

that is, for the collinear configuration of Fig. 11 with m_2 between m_1 and m_3 :

$$\frac{1}{r_{12}}\left[m_3\left(\frac{1}{r_{23}^2}-\frac{1}{r_{13}^2}\right)-\frac{m_1+m_2}{r_{12}^2}\right]=\frac{1}{r_{23}}\left[m_1\left(\frac{1}{r_{12}^2}-\frac{1}{r_{13}^2}\right)-\frac{m_2+m_3}{r_{23}^2}\right] \tag{109}$$

Finally, since $r_{13} = r_{12} + r_{23}$, we obtain the following slightly asymmetrical expression for the condition of relative equilibrium :

$$m_3 r_{12}^2 (r_{23}^3 - r_{31}^3) + m_1 r_{23}^2 (r_{31}^3 - r_{12}^3) = m_2 r_{31}^2 (r_{12}^3 - r_{23}^3) \quad (110)$$

It is customary to present this relation in the following way :

$$\left.\begin{array}{l} r_{23}/r_{12} = x \; ; \; r_{13}/r_{12} = 1 + x \\ x^5(m_1+m_2) + x^4(3m_1+2m_2) + x^3(3m_1+m_2) = x^2(m_2+3m_3) + x(2m_2+3m_3) + (m_2+m_3) \end{array}\right\} (111)$$

This fifth-degree equation has always one and only one positive root and hence, for three given masses, there are always three and only three collinear central configurations, according to which mass is between the other two masses.

Note that the condition (110) or (111) is equivalent to the following :

"In a collinear central configuration <u>the position</u> of the center of mass is a function of <u>the positions</u> of the three masses only (and is independent of the masses themselves), indeed, with (111), the ratio $r_1/r_{1.2}$ (that is $\left[m_2 + m_3(1 + x)\right]/M$) is given by :

$$r_1/r_{12} = (x^5 + 3x^4 + 3x^3)/(x^4 + 2x^3 + x^2 + 2x + 1) \quad (112)"$$

A simpler equivalent expression is given by the following :

"Let us assign the abscissae (-1) and (+1) to the extreme masses m_1 and m_3 of the Fig. 11, the abscissae x_2 of the mass m_2 and x_I of the center of mass are related by :

$$x_I = (x_2^5 - 2x_2^3 + 17x_2)/(x_2^4 - 10x_2^2 - 7) \quad (113)"$$

x_I decreases from + 1 to - 1 when x_2 increases from - 1 to + 1 (Fig. 12).

8.2 STABILITY OF THE EULERIAN AND LAGRANGIAN MOTIONS

Let us consider a three-body system almost exactly in the conditions of a given Eulerian or Lagrangian motion, will it forever remain in the vicinity of this motion ?

For parabolic Eulerian or Lagrangian motions the answer is almost always no, indeed when the time t goes to infinity the parabolic Eulerian and Lagrangian motions have mutual distances r_{ij} increasing as $t^{2/3}$ while for the neighbouring motions we have the following :

A) If the energy integral h is negative the smallest mutual distance remains bounded.

B) If $h > 0$ the largest mutual distance increases as t and not as $t^{2/3}$.

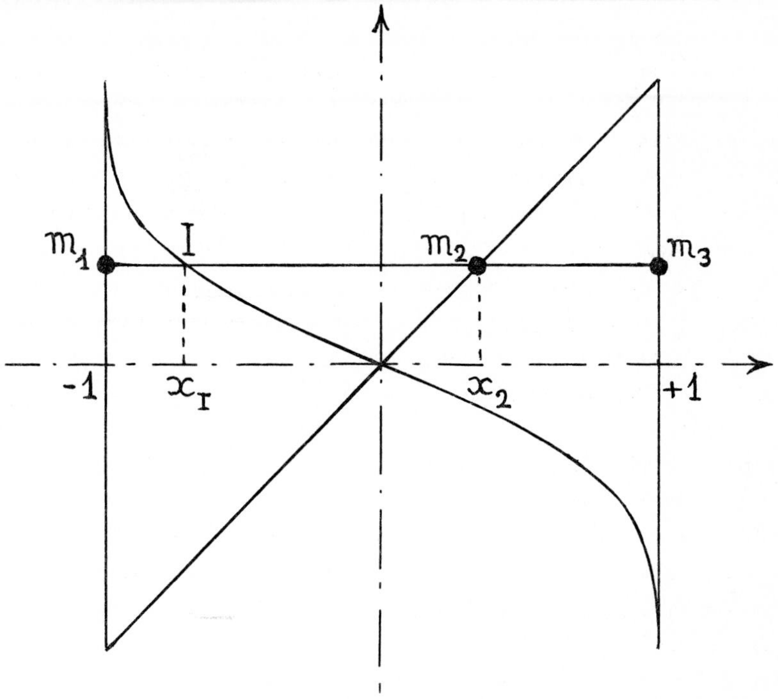

Fig. 12. All collinear central configurations in a single figure. The abscissas of m_1, m_2, m_3 are -1, x_2, $+1$ (with $|x_2|<1$) and the abscissa x_I of the center of mass I is given by :

$$x_I = \frac{x_2^5 - 2x_2^3 + 17x_2}{x_2^4 - 10x_2^2 - 7}$$

This necessary and sufficient condition for a collinear central configuration with m_2 between m_1 and m_3 is equivalent to the following simple and beautiful condition :

$$\frac{\vec{r_1}}{r_{23}^2} + \frac{\vec{r_3}}{r_{12}^2} = \frac{\vec{r_2}}{r_{13}^2}$$

C) If h = 0 almost all neighbouring motions have an hyperbolic-elliptic final evolution : their smallest mutual distance remain bounded and their largest mutual distance increases as t.

If the Eulerian or Lagrangian motion of interest is hyperbolic the answer depends on the definition of the word "vicinity" : the configuration, the ratios of mutual distances, the velocities will remain forever very close to that of the Eulerian or Lagrangian motion but the deviations of the positions will generally increase as the time t.

The most interesting cases are of course these of circular or elliptic Eulerian or Lagrangian motions and we will consider them in the two next Sections.

8.2.1 Fisrt-order analysis

The elliptic or circular Eulerian or Lagrangian motions are periodic and the first-order analysis of their vicinity can be done using Floquet theory. Let us consider the small differences between the Eulerian or Lagrangian motion and the neighbouring motion. After each period these differences have undergone a constant linear transformation and if one of the eigenvalues of this linear transformation has a modulus larger than unity the corresponding difference increases exponentially in terms of the number periods and the Eulerian or Lagrangian motion of interest is not stable.

Let us call $\vec{r}_1(t)$, $\vec{r}_2(t)$, $\vec{r}_3(t)$ the three radius-vectors of the Eulerian-or Lagrangian motion of interest and $\vec{r}_j + \vec{\delta}_j$, $\{j = 1,2,3\}$ those of the neighbouring motion.

Take x_j, y_j, z_j as the components of $\vec{\delta}_j$ in the radial, circumferential and out-of-plane axes of the vector $\vec{r}_{23}$ (i.e. $x_j = \vec{\delta}_j (\vec{r}_{23}/r_{23})$, etc...).

In the Eulerian case with m_2 between m_1 and m_3 there are neighbouring motions satisfying the following to first order :

$$\vec{\delta}_1/m_2m_3r_{23} \equiv - \vec{\delta}_2/m_1m_3r_{13} \equiv \vec{\delta}_3/m_1m_2r_{12} \tag{114}$$

and the other neighbouring motions can always be put in this form after, if necessary, the choice of a suitable slightly different Eulerian reference motion.

We will call x, y, z the components of the vector given in (114) (hence $x = x_1/m_2m_3r_{23}$, etc...) and the first-order differential equations of x, y, z are the following (Ref. 47, 60)

$$(1 + e \cos v)(d^2x/dv^2 - 2dy/dv) = (2K + 3)x \tag{115}$$

$$(1 + e \cos v)(d^2y/dv^2 + 2dx/dv) = - Ky \tag{116}$$

$$(1 + e \cos v)(d^2z/dv^2 + z) = - Kz \tag{117}$$

where :

$$\left.\begin{array}{l} e = \text{eccentricity} \\ v = \text{true anomaly} \end{array}\right\} \text{of the Eulerian motion of interest}$$

$$K = \frac{m_3 r_{12}(r_{13}^3 - r_{23}^3) + m_1 r_{23}(r_{13}^3 - r_{12}^3)}{(m_1 + m_3) r_{12}^2 r_{23}^2 + m_2 r_{13}^2 (r_{12}^2 + r_{23}^3)} \tag{118}$$

The constant K is always in the range of $0 < K \leqq 7$ ($K = 0$ for $m_1 = m_3 = 0 < m_2$ and $K = 7$ for $m_1 = m_3 > 0 = m_2$).

Bennett (Ref. 61) has studied the stability of the Eulerian motions numerically for plane three-body systems with an infinitesimal mass (plane restricted three-body problem) and he found that these motions are always exponentially unstable for all values of the eccentricity e and the ratio of the two major masses. This conclusion can be extended to all Eulerian motions since the three masses only appear through the above constant K whose range of variation, from 0 to 7, is the same for the restricted three-body problem and for the general three-body problem.

In the Lagrangian case the analysis is similar (Ref.47, 60).

A) The first-order analysis of the out-of-plane components z_1, z_2, z_3 is simple and leads to the following :

$$z_k = (a_k \sin v + b_k \cos v)/(1 + e \cos v) \; ; \; k = \{1,2,3\} \tag{119}$$

with of course :

$$\left.\begin{array}{l} e = \text{eccentricity} \\ v = \text{true anomaly} \end{array}\right\} \text{of the Lagrangian motion of interest} \tag{120}$$

The a_k and b_k are six constants of integration and the out-of-plane motion is then first-order stable.

B) For the in-plane components x_k and y_k we can always (to the first-order) reach the following relations, if necessary after the choice of a suitable slightly different Lagrangian reference motion.

$$m_1(x_1 + iy_1) \equiv m_2 j(x_2 + iy_2) \equiv m_3 j^2(x_3 + iy_3) \tag{121}$$

where :

$$\left\{\begin{array}{l} i = \sqrt{-1} \; ; \; j = (i\sqrt{3} - 1)/2 \; ; \; j^3 = 1 \\ \text{The succession } m_1, m_2, m_3 \text{ is in the direction of motion} \end{array}\right. \tag{122}$$

(hence, in Fig. 10, A is m_1, C is m_2 and B is m_3)

C) The differential equations for the x_k, y_k are :

$$\left.\begin{array}{l} 2(1 + e \cos v)(d^2x/dv^2 - 2dy/dv) = 3x(1 + N) \\ 2(1 + e \cos v)(d^2y/dv^2 + 2dx/dv) = 3y(1 - N) \end{array}\right\} \tag{123}$$

where x and y are real and given by :

$$\left\{\begin{array}{l} x + iy = \left[R_1(x_1+iy_1)+R_2(x_2+iy_2)+R_3 j^2(x_3+iy_3)\right](1+e\cos v)(m_1+jm_2+j^2m_3)^{1/2} \\ \text{with } R_1,\ R_2,\ R_3 \text{ three arbitrary real constants} \end{array}\right\} \quad (124)$$

and :

$$N = \left[m_1^2 + m_2^2 + m_3^2 - m_1m_2 - m_1m_3 - m_2m_3\right]^{1/2} \Big/ (m_1 + m_2 + m_3) \quad (125)$$

which implies :

$0 \leq N \lesssim 1$

Note that when $N = 1$ the system (123) is equivalent to the system (115)-(116) for $K = 0$ (These systems are then easily integrable, they are also integrable for $K = -1$ and/or $e = 0$).

A surprising property of the fourth-order linear system (123) is that it may be decomposed.

Let us consider the following second-order linear system :

$$\begin{aligned} 4N(1+e\cos v).dx/dv &= \left[e^2\sin 2v - 2Ne\sin v\right]x + \left[e^2\cos 2v + 4N(1+e\cos v) - 3N^2 - Q\right]y \\ 4N(1+e\cos v).dy/dv &= \left[e^2\cos 2v - 4N(1+e\cos v) - 3N^2 + Q\right]x - \left[e^2\sin 2v + 2Ne\sin v\right]y \end{aligned} \quad (126)$$

The solutions of (126) are particular solutions of (123) when the constant Q satisfies :

$$Q = \pm\left[e^4 + 2N^2e^2 + 9N^4 - 8N^2\right]^{1/2} \quad (127)$$

With the two possible values of Q the two linear systems (126) give four independent solutions of (123) and the reduction is thus complete.

The stability of elliptic Lagrangian motions has been studied numerically in the restricted planar case by Danby and Bennett (Ref.61, 62). They give the hatched zones of stable motion presented in Fig. 13 in terms of the mass ratio R and the eccentricity e.

These results can be extended to the general three-body problem since the out-of-plane motion is stable and the masses appear only through the parameter N given in (125), hence the relation between N and the ratio R of Fig. 13 is :

$$N = (1 - 3R + 3R^2)^{1/2}, \text{ that is } R = \left[3 - \sqrt{(12N^2 - 3)}\right] \Big/ 6 \quad (128)$$

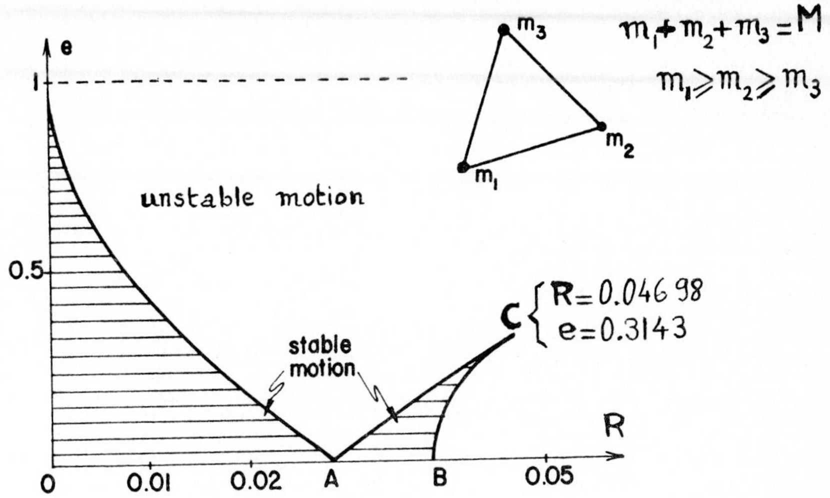

Fig. 13. Zones of first order stability for elliptic Lagrangian motions in terms of the parameter R and the eccentricity e.

In the restricted case, when $m_3 = 0$, the parameter R is the mass ratio m_2/M.

In the general case R is a function of N of (125):

$$R = \left(3-\sqrt{12N^2-3}\right)/6 = (m_2+m_3)/M + m_2 m_3/m_1^2 + O\{m_2^3 m_3/m_1^4\}$$

$$R_A = 0.02860\ldots = (3-\sqrt{8})/6 \ ; \ N_A^2 = 11/12$$
$$R_B = 0.03852\ldots = (9-\sqrt{69})/18 \ ; \ N_B^2 = 8/9$$

Notes :

A) The limit curve BC of Fig. 13 corresponds to Q = 0 that is to : $e^2 = \sqrt{8(N^2-N^4)} - N^2$ or else to : $e^2 = \left[24R(1-R)(1-3R+3R^2)\right]^{1/2} - 1+3R-3R^2$; (Ref. 63,54. I thank H.Yoshida,ref 64,who indicated me the work or J.Tschauner).

B) The restricted case corresponds to $0 < R \leq 1/2$ and then to $1 > N \geq 1/2$; however in the general case N can have any value in the range $0 \leq N \leq 1$. Fortunately it is easy to show that when $N \leq 1/2$ the system (123) is always exponentially unstable.

Hence the first-order study leads to the following.

A) All elliptic or circular Eulerian motions are exponentially unstable.

B) Elliptic or circular Lagrangian motions are generally exponentially unstable and stability requires a very large asymmetry : at least 95.3% of the total mass must be in the largest mass.

8.2.2 Complete analysis of the stability

Few studies go further than the first-order analysis (Ref.65-67) and they all deal with the simplest case, the case of circular Lagrangian motions in the plane circular restricted three-body problem.

Let us consider the system (123) in the circular case, that is when $e = 0$, and let us try to obtain its periodic solutions $x = A.\exp(i\omega v)$; $y = B.\exp(i\omega v)$.

We obtain :

$$\left.\begin{aligned} -2A\omega^2 - 4Bi\omega &= 3A(1+N) \\ -2B\omega^2 + 4Ai\omega &= 3B(1-N) \end{aligned}\right\} \qquad (129)$$

that is :

$$(3 + 3N + 2\omega^2)i/4\omega = B/A = 4i\omega/(3 - 3N + 2\omega^2) \qquad (130)$$

hence ω is a root of the following equation :

$$4\omega^4 - 4\omega^2 + 9 - 9N^2 = 0 \qquad (131)$$

that is, in terms of the mass ratio R equal to $m_2/(m_1 + m_2)$:

$$4\omega^4 - 4\omega^2 + 27R - 27R^2 = 0 \qquad (132)$$

Here ω is real when R is in the range $0 \leqq R \leqq 0.03852...$ and it can have the following four values :

$$\text{either } \omega = \pm\,\omega_1 \text{ with } \omega_1 = \left\{\frac{1}{2}\left[1 - \sqrt{(1 - 27R + 27R^2)}\right]\right\}^{1/2} \qquad (133)$$

$$\text{or } \omega = \pm\,\omega_2 \text{ with } \omega_2 = \left\{\frac{1}{2}\left[1 + \sqrt{(1 - 27R + 27R^2)}\right]\right\}^{1/2} \qquad (134)$$

Second- and third-order analysis of the vicinity of Lagrangian motions in the plane circular restricted three-body problem leads to the following.

A) The motion is unstable at the two following resonances.

A_1 : $\omega_2/\omega_1 = 2$; $R = 0.02429... = (45 - \sqrt{1833})/90$ (135)

A_2 : $\omega_2/\omega_1 = 3$; $R = 0.01352... = (15 - \sqrt{213})/30$ (136)

B) The third-order analysis is insufficient when the "first Birkhoff invariant" is zero, that is when :

$$644\,\omega_1^4\omega_2^4 - 541\,\omega_1^2\omega_2^2 + 36 = 0$$

$$R = 0.01091... = \frac{1}{2} - \sqrt{\frac{3265 + \sqrt{799780}}{17388}} \qquad (137)$$

C) For the other values of the mass ratio R in the open range {0 ; 0.03852...} the Lagrangian motions of the plane circular restricted three-body problem are stable.

Note that for the limit values R = 0 and R = 0.03852... the Lagrangian motions are unstable* but these instabilities are not exponential as they are for larger R.

The fifth-order analysis shows that when the "first Birkhoff invariant" is zero (i.e. when R = 0.01091...) the "second Birkhoff invariant" is not zero and the Lagrangian motions are stable. Hence the stability analysis is complete for Lagrangian motions of the plane circular restricted three-body problem.

In the other cases, i.e. non-circular and/or non-restricted and/or non-planar, the stability analysis is very difficult (see Section 10.7.10). However if the "Arnold diffusion conjecture" is true all these motions would be unstable, but it would be an extremely slow instability that would not contradict the motions presented in the next Section.

8.3 THE EULERIAN AND LAGRANGIAN MOTIONS IN NATURE AND IN ASTRONAUTICS

Being exponentially unstable the Eulerian motions are never met in nature, but since 1906 fifteen asteroids have been discovered near the triangular Lagrangian points of the Sun-Jupiter system. These asteroids remain in the zone of stability about the Lagrangian points and are called the "Trojans", their names, either greek or trojan, are chosen from the Iliad (Fig 14).

As in Mythology the Greeks are ahead of Jupiter and the Trojans are behind him, with the exceptions of Hector and Patroclus : these two unfortunate warriors are thus surrounded by their enemies and more than one billion kilometers away from their friends !

The space probes Voyager I and II have discovered two small asteroids in the saturnian system at the Lagrangian points of the large satellite Thetys and another asteroid at a Lagrangian point of Dione.

All these asteroids remain in the vicinity of the orbit of the corresponding minor primary (i.e. Jupiter, Thetys or Dione) and "librate" very slowly around the corresponding triangular Lagrangian point. For the asteroid Diomedes the amplitude of the libration exceeds 50° !

---

*No, error! For R = 0.03852.. the Lagrangian motion is stable (see the thesis of my student Mr El Bakkali Larbi).

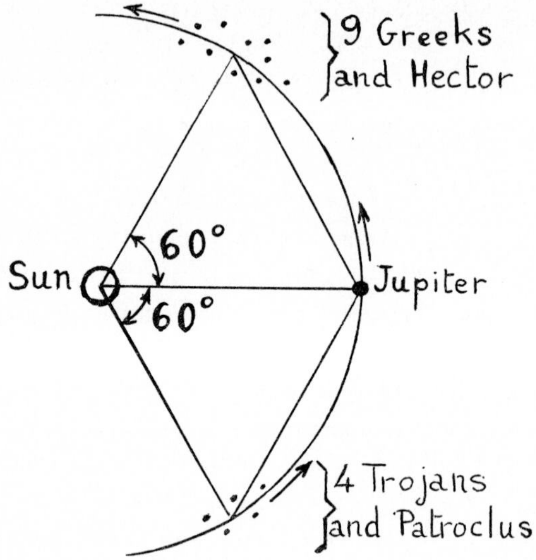

Number	French name
588	Achille
624	Hector
659	Nestor
911	Agamemnon
1143	Ulysse
1404	Ajax
1437	Diomède
1583	Antilochus
1647	Ménélas
1749	Télamon
617	Patrocle
884	Priam
1172	Enée
1173	Anchise
1208	Troïle

Fig. 14. The "Trojans" asteroids.

In astronautics the triangular Lagrangian points are of little interest, being very far from the major bodies. But the collinear Lagrangian points (also called Eulerian points) can be used for particular missions.

Let us consider for instance the three Eulerian points of the Earth-Moon system (Fig. 15). Two of them are in the vicinity of the Moon, L_1 is at an average distance of 58 000 km and L_2 at 64 500 km.

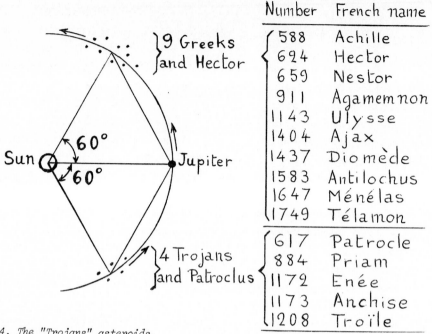

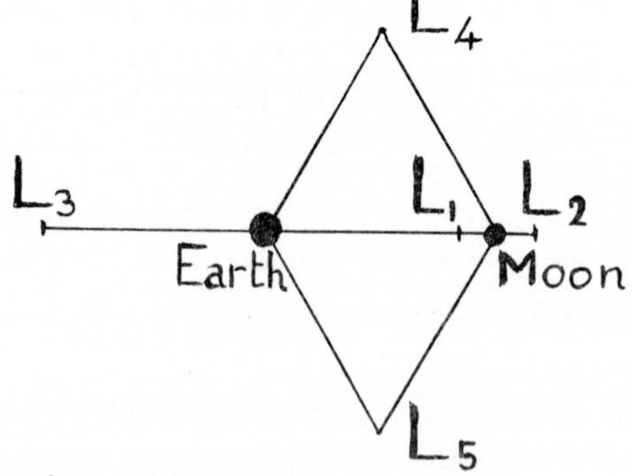

Fig. 15. The Lagrangian points of the Earth-Moon system.
Mass of the Moon/mass of the Earth = 0.012 300 02
Earth-Moon distance = 1 gives: L_1-Moon = 0.150 934 2
L_2-Moon = 0.167 832 7 ; L_3-Earth = 0.992 912 1

L_1 is always above the near side of the Moon and L_2 above the far side, they are "seleno-stationary" and can be used for seleno-stationary satellites as useful for the Moon as are the geostationary satellites for the Earth (L_2 will be especially useful for radio-comunications of the far side).

Being unstable the motions at L_1 and L_2 will require a station-keeping procedure very similar to that already used for geostationary satellites. The corresponding loss of propellant will remain very small (Ref. 62).

The two nearest collinear Lagrangian points of the Sun-Earth system are at 1.5 million kilometers. They will be used for several missions, for instance for the continuous observation of the Sun from beyond the Earth-magnetosphere (and hence in better measurement conditions).

8.4 OTHER EXACT SOLUTIONS OF THE THREE-BODY PROBLEM

Apart from the Eulerian and Lagrangian motions only two very particular exact solutions of the three-body problem have been found (see also the Appendix 1 for negative masses).

8.4.1 The isoceles solutions

Two equal masses A and B rotate on the same circular orbit in the Oxy plane around their center of mass 0 (Fig. 16). The third mass C is infinitesimal and moves along the z-axis.

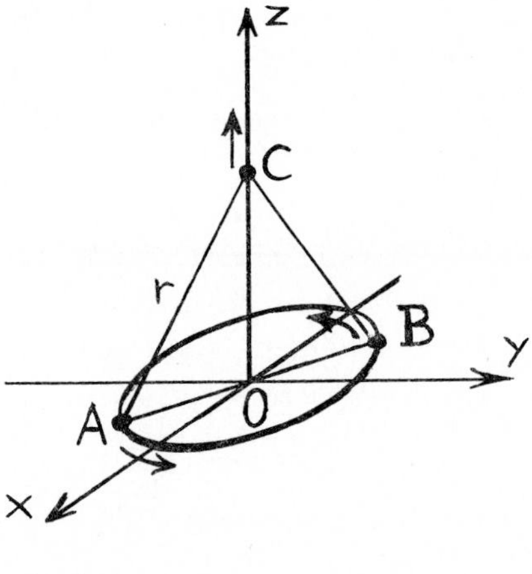

Fig. 16. The isosceles solution.

The motion of C has a simple integral of motion (with r = distance AC) :

$$(dz/dt)^2 - 2GM/r = \text{constant} \qquad (138)$$

and since $r^2 - z^2 = (AO)^2$ = constant, the time t is given in terms of z by a simple quadrature.

8.4.2. The z-axis Hill solutions

The Hill problem is a very particular case of the three-body problem (see Section 9.2), whose equations of motion can be written :

$$\left.\begin{aligned} x'' &= 2y' + 3x - x/r^3 \\ y'' &= -2x' - y/r^3 \\ z'' &= -z - z/r^3 \end{aligned}\right\} \quad \text{with } x' = dx/dt,\quad x'' = d^2x/dt^2, \text{ etc ...},\quad r = (x^2+y^2+z^2)^{1/2} \tag{139}$$

The Hill problem has the Jacobi integral of motion :

$$\Gamma = \frac{2}{r} + 3x^2 - z^2 - x'^2 - y'^2 - z'^2 \tag{140}$$

For the z-axis Hill solutions x and y are identically zero while z and z' are related by :

$$z'^2 = \frac{2}{|z|} - \Gamma - z^2 \tag{141}$$

hence finally time t is given in terms of z by a simple quadrature.

8.5 OTHER SIMPLE SOLUTIONS OF THE THREE-BODY PROBLEM

Several simple families of solutions of the three-body problem are usually presented in terms of suitable series.

A) The Siegel series provide the solutions for an impending triple collision (Ref.37) of for a triple close approach (Ref. 48), their first terms appear in (86)-(88).

B) The Brown series are used in the lunar problem and give the simplest periodic solutions of the Hill equations (139) (see Section 9.2.1.).

This technique can be used in many other cases, for instance for the direct or retrograde "pseudo-circular orbits" (Fig. 17 and Section 10.8.1) or for the "Halo orbits" about the Lagrangian points (Section 10.8.3 and Ref. 68–74)

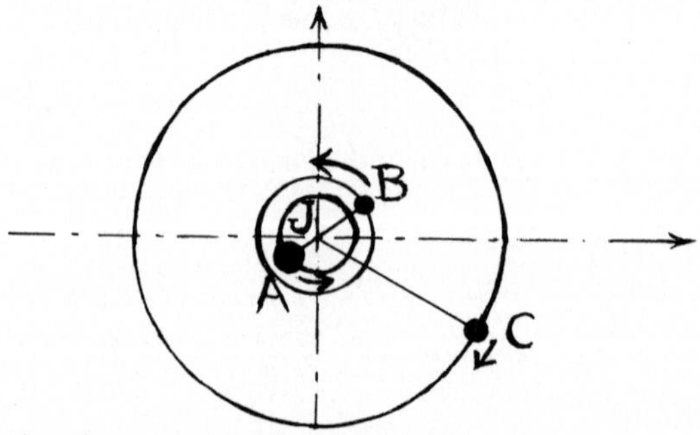

Fig. 17. A "pseudo-circular retrograde orbit" in the non-rotating axes of the center of mass J of the small binary.

However the power of numerical methods has led to accurate numerical integrations and to the numerical presentation of a great number of families of periodic three-body orbits (e.g.Ref 68-137 with successive surveys 131, 132).

For three given masses the families of periodic orbits are one-non-trivial-parameter families that can be constructed step by step by continuity and that meet each other at the "bifurcations" where exchanges of stability take place.

In Reference 104 are given many numerical and graphical informations about the simplest periodic orbits of the plane Hill problem (equations (139), (140) with $z \equiv 0$) and especially the three below Figs. 18, 19 and 20. Notice at orbit g_1 the exchange of stability between the families g and g'.

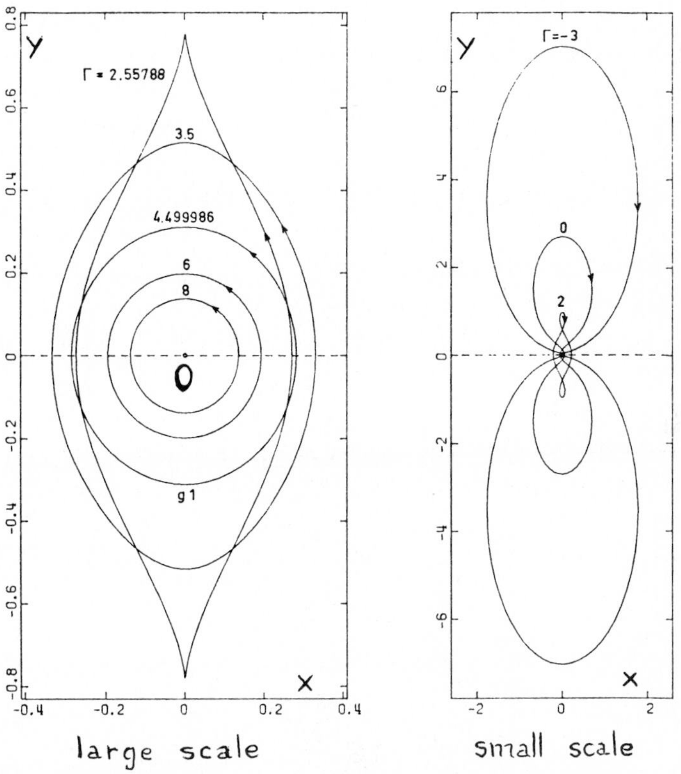

Fig. 18. Strömgren family g of direct "pseudo circular" periodic orbits of the plane Hill problem. Source : Hénon, M. [104].
These orbits are given by the series of Brown. They have the "in-plane" and "out-of-plane" first order stability when $\Gamma \geq 4.499\,986$ *but they loose the in-plane stability for all smaller* Γ .

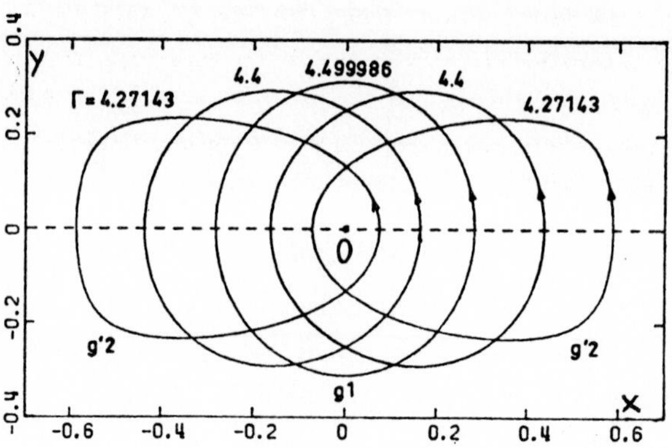

Fig. 19. Stromgren family g' of direct and simple-periodic orbits of the plane Hill problem. Source : Hénon, M. [104].
The orbits of the figure have the "in-plane" stability, but this stability is lost when $\Gamma < 4.271\ 43$ *except when* $-4.692\ 19 \geq \Gamma \geq -4.704\ 79$.

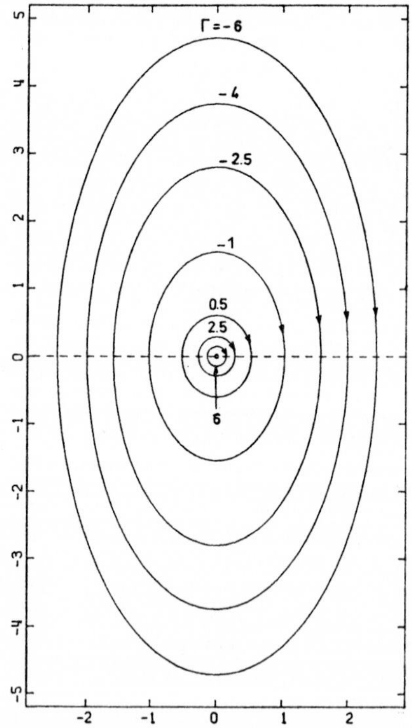

Fig. 20. Stromgren family of retrograde "pseudo-circular" periodic orbits of the plane Hill problem. Source : Hénon, M. [104].
These orbits are given by the Brown series ; they have both the "in-plane" and the "out-of-plane" first order stability for any value of Γ.

The periodic "pseudo-circular" orbits of the plane three-body problem have been computed in the references 133 and 134 for the case of three equal masses. These orbits remain almost circular even for large perturbations but the extreme end of the retrograde family is the Schubart rectilinear orbit (Section 10.9.1 and Ref. 133,135 .

Chapter 9

THE RESTRICTED THREE-BODY PROBLEM

Let us consider the following problem (Fig. 21). A space probe P is launched towards the Moon in the cislunar space with some given initial conditions, it undergoes the attraction of the Earth and the Moon and we will neglect other effects. What will be its motion ?

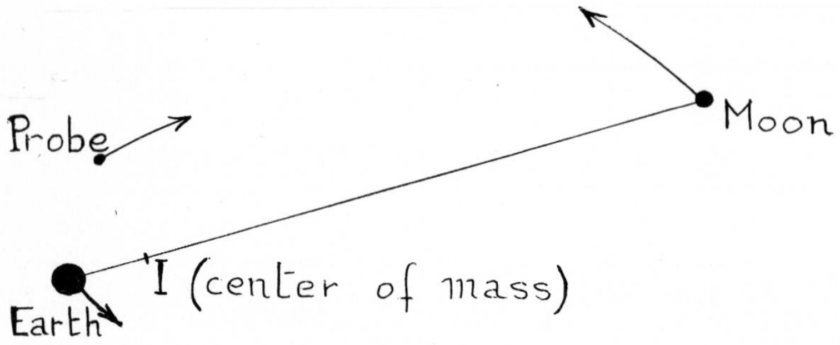

Fig. 21. The restricted three-body problem.

The Earth and the Moon have of course themselves their own motions and we must take account of them but the probe does not perturb the Earth or the Moon and that three-body problem is thus restricted.

The restricted three-body problem is then the following :

A) The two "primaries" have masses m_1 and m_2, they move under their mutual attraction and have an ordinary two-body motion.

B) The third body has an infinitesimal mass m_3, its equation of motion was given in (7) as :

$$d^2\vec{r}_3/dt^2 = G(m_1\vec{r}_{31}r_{31}^{-3} + m_2\vec{r}_{32}r_{32}^{-3}) \tag{142}$$

and we want to know the motion of m_3 for given initial conditions and given motions of the primaries.

The motion of the primaries can be circular, elliptic, parabolic, hyperbolic, rectilinear ; the motion of the third body can be rectilinear, planar or three-dimensional so many possible restricted three-body problems are thus defined. The most commonly considered is the plane circular restricted three-body problem.

The circular restricted three-body problem has a well-known integral of motion the "Jacobi integral" already presented in (140) for the Hill case. No other integral of motion have ever been found in the circular, elliptic, parabolic or hyperbolic cases, but notice that if the primaries have a rectilinear motion the third body has a constant angular momentum with respect to the axis along which the two primaries move.

9.1 THE CIRCULAR RESTRICTED THREE-BODY PROBLEM

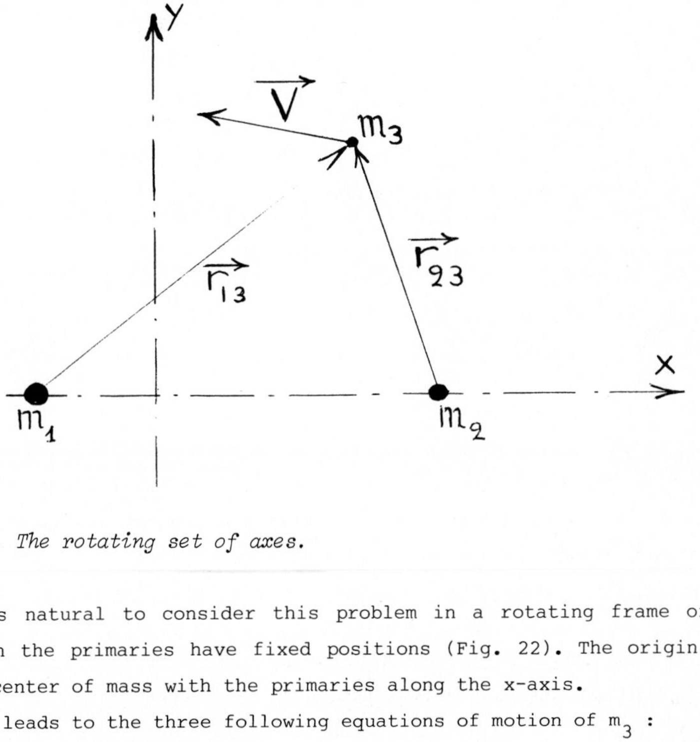

Fig. 22. The rotating set of axes.

It is natural to consider this problem in a rotating frame of reference in which the primaries have fixed positions (Fig. 22). The origin is usually at the center of mass with the primaries along the x-axis.

This leads to the three following equations of motion of m_3 :

$$\left.\begin{aligned} x''_3 &= G\left(m_1 \frac{x_1 - x_3}{r_{13}^3} + m_2 \frac{x_2 - x_3}{r_{23}^3}\right) + 2\omega y'_3 + \omega^2 x_3 \\ y''_3 &= -Gy_3\left((m_1/r_{13}^{\,3}) + (m_2/r_{23}^{\,3})\right) - 2\omega x'_3 + \omega^2 y_3 \\ z''_3 &= -Gz_3\left((m_1/r_{13}^{\,3}) + (m_2/r_{23}^{\,3})\right) \end{aligned}\right\} \quad (143)$$

In these equations the constant ω is the rate of rotation of the frame of reference. It is related to the other constants by :

$$\omega^2 r_{12}^{\ 3} = G(m_1 + m_2) \tag{144}$$

while of course :

$$r_{12} = x_2 - x_1 \ ; \ x_1 = -\ m_2 r_{12}/(m_1 + m_2) \ ; \ x_2 = m_1 r_{12}/(m_1 + m_2) \tag{145}$$

The units of length, mass and time are usually chosen such that :

$$r_{12} = 1 \ ; \ m_1 + m_2 = 1 \ ; \ G = 1 \tag{146}$$

This implies $\omega = 1$ and the system (143) can be written in the following vectorial form :

$$\vec{r}_3'' = m_1 \vec{r}_{13}\left(1 - \frac{1}{r_{13}^{\ 3}}\right) + m_2 \vec{r}_{23}\left(1 - \frac{1}{r_{23}^{\ 3}}\right) + \begin{pmatrix} 2y_3' \\ -\ 2x_3' \\ -\ z_3 \end{pmatrix} \tag{147}$$

If $\vec{V}$ is the velocity of m_3 with respect to the rotating frame of reference (Fig. 22), the Jacobi integral of motion is then :

$$\Gamma = m_1\left(\frac{2}{r_{13}} + r_{13}^{\ 2}\right) + m_2\left(\frac{2}{r_{23}} + r_{23}^{\ 2}\right) - z_3^{\ 2} - V^2 \tag{148}$$

Let us put :

$$J = \text{Jacobi function} = m_1\left(\frac{2}{r_{13}} + r_{13}^{\ 2}\right) + m_2\left(\frac{2}{r_{23}} + r_{23}^{\ 2}\right) - z_3^{\ 2} \tag{149}$$

J is a function of the position of m_3 and is equal to $\Gamma + V^2$, hence $J \geq \Gamma$ and for large Γ a part of the x,y,z space is forbidden to the third body.

The surfaces of constant J are the Hill surfaces and their intersections with the Oxy plane are the Hill curves (Fig. 23).

The properties of the Hill curves and surfaces and of the corresponding motions are classical and have been described in detail in many books (see for instance "Theory of orbits", Ref. 1 , pages 141-207).

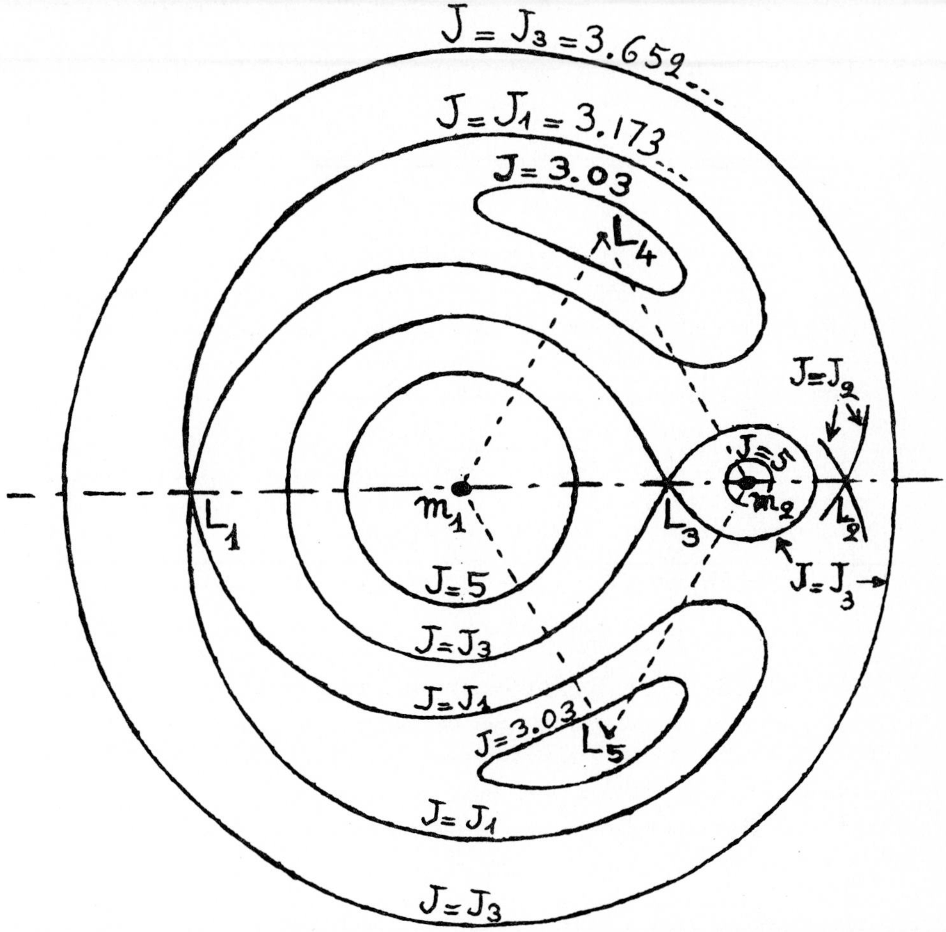

Fig. 23. Hill curves in the case $m_1 = 10m_2$ in terms of the position of m_3 with respect to m_1 and m_2:

$$r_{12} = 1 \; ; \; J(m_1 + m_2) = m_1\left(r_{13}^2 + \frac{2}{r_{13}}\right) + m_2\left(r_{23}^2 + \frac{2}{r_{23}}\right)$$

The function J is minimum and equal to 3 at the triangular Lagrangian points L_4 and L_5, it is infinite at m_1, at m_2 and at infinity and finally it has three saddle points at the collinear Lagrangian points L_1, L_2 and L_3 (the curve $J = J_2$ remains very close to the curve $J = J_3$ except near L_2).

$$m_1 = 10m_2 \text{ implies } \begin{cases} J_1 = 3.173222 \; ; \; (m_1,L_1) = 0.946926 \\ J_2 = 3.534182 \; ; \; (m_2,L_2) = 0.346992 \\ J_3 = 3.652916 \; ; \; (m_2,L_3) = 0.282487 \end{cases}$$

Note that the subscripts of L_1, L_2, L_3 are the subscript of the mass that is between the two other masses (this figure is different from Fig. 15) and hence $m_1 \geqq m_2 \geqq m_3$ implies $J_1 \leqq J_2 \leqq J_3$. This property will be extended to the general three-body problem in Section 11.6.3.

Let us only recall that J is stationary at the five Lagrangian points or "equilibrium points" L_1 to L_5 (Fig. 23). If, for a given motion, the Jacobi constant Γ is larger than J_3, that is $J(L_3)$, the condition $J \geqq \Gamma$ divides the zone of possible motion of m_3 into three disconnected parts : either near m_1, or near m_2, or very far away. The third body remains forever in one of these three parts and in the two first cases it cannot escape, it is the Hill stability.

This partition of space can be extended to the case $\Gamma = J_3$, since then the point L_3 only corresponds to an Eulerian motion with m_3 forever at L_3.

If (for $m_1 \geqq m_2$) Γ satisfies $J_2 \leqq \Gamma < J_3$ there is a reduced Hill stability : the zone of possible motion of m_3 is divided into two disconnected parts. The central part surrounds m_1 and m_2 and is bounded.

The circular restricted three-body problem has been the subject of innumerable analytical or numerical studies, especially in the plane case, and numerous families of its periodic orbits have been computed (e.g.ref 1-3,5,68-72,74,77-81,84-92,103-110,116,118-126,138-140).

A special mention must be given to the periodic orbits about the Lagrangian points.

A) When m_2/M is smaller than the ratio R_B of the figure 13 (i.e. 0.03852...) the Lagrangian motion at L_4 is stable. There are then two families of periodic orbits around L_4 or L_5 : the short-period family that is not very remarkable, and the long-period family that exhibits clockwise orbits very similar to the Hill curves of Fig. 23 (however, these orbits are thicker than the Hill curves and almost correspond to the Hill curves of a ratio m_2/M four times greater).

This similarity goes very far and the largest orbits have the horseshoe shape of the Hill curves surrounding L_4, L_1 and L_5*. Brown has even conjectured (Ref. 143) that the orbit asymptotic to L_1 in the past belongs to this family and, as the Hill curve of L_1, comes back to L_1 and is asymptotic to L_1 in the future (homoclinic orbit).

However Deprit,Henrard and Garfinkel(Ref 138,139,141,142)have shown that this Brown conjecture is not true and the homoclinic orbits of the family have a past and also future limit which is a small, almost elliptic, orbit around L_1, with a short period (see Section 11.5).

Surprisingly, a conjecture similar to that of Brown is true for the point L_2 for some suitable values of the ration m_2/M (Ref. 141, part 6).

B) Among the motions about the collinear Lagrangian points the "Halo orbits" are the most famous (Section 10.8.3 and Appendix 2 and also ref 68-74).

The "Halo orbits" are simple periodic orbits. They are three-dimensional and symmetrical with respect to the Oxz plane. They will be very useful for

* e.g. see ref. 140.

many practical purposes, for instance for the radio-communications with the far side of the Moon (Fig. 24).

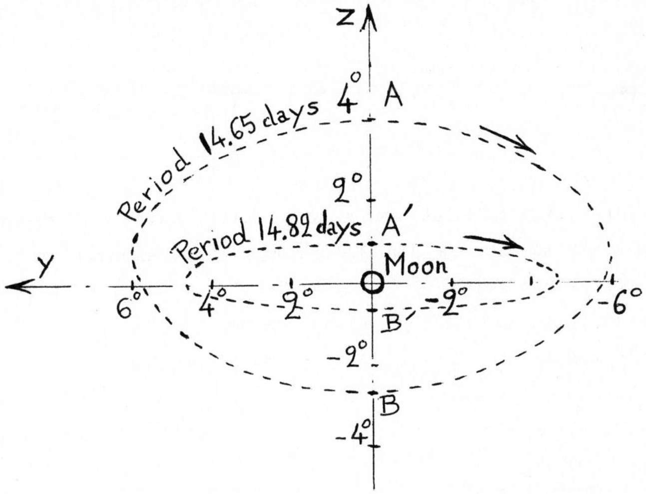

Fig. 24. Two possible clockwise Halo orbits beyond the Moon (as seen from Earth).
The points A, A', B', B are respectively 72 800 km, 74 600 km, 49 500 km, 41 300 km beyond the Moon.
The average lunar angular diameter is only 0.517°

9.2 THE HILL PROBLEM

Let us consider the equations (143)-(145) of the circular restricted three-body problem.

If the ratio m_1/m_2 is very large and if we need to study the motion of m_3 in the vicinity of m_2 (motion of a probe in the vicinity of a planet but with a still large solar influence) we proceed as follows.

A) Choose the units of length, mass and time such that :

$$m_2 = 1 \; ; \; G = 1 \; ; \; \omega = 1 \tag{150}$$

B) Choose at m_2 the origin of the frame of reference :

$$\left. \begin{aligned} x &= x_3 - x_2 \\ y &= y_3 \\ z &= z_3 \end{aligned} \right\} \tag{151}$$

The equations of motion then become :

$$\left.\begin{aligned} x'' &= -x/r^3 + 3x + 2y' + \varepsilon_1 && \text{with : } r = (x^2 + y^2 + z^2)^{1/2} \\ y'' &= -y/r^3 - 2x' + \varepsilon_2 && \text{and :} \\ z'' &= -z/r^3 - z + \varepsilon_3 && \varepsilon_1, \varepsilon_2, \varepsilon_3 = O\left(\frac{r}{m_1} ; \frac{r^2}{\sqrt[3]{m_1}}\right) \end{aligned}\right\} \quad (152)$$

If m_1/m_2 is very large (332946 in the Sun, Earth case) the terms ε_1, ε_2, ε_3 can be considered negligible and there remain the equations of the Hill problem already written in (139) :

$$\left.\begin{aligned} x'' &= -x/r^3 + 3x + 2y' \\ y'' &= -y/r^3 - 2x' \\ z'' &= -z/r^3 - z \end{aligned}\right\} \quad (153)$$

The corresponding Jacobi integral of motion is :

$$\Gamma = \frac{2}{r} + 3x^2 - z^2 - x'^2 - y'^2 - z'^2 \quad (154)$$

and the nearest Lagrangian points have the coordinates : $\{\pm 3^{-1/3}, 0, 0\}$.

Notice that the equations (153) are also the equations of the relative motion of m_2 and m_3 in the rotating set of axes if :

- m_1 is very far and very large
- $G = 1$
- $m_2 + m_3 = 1$
- r_{23} is of order one or smaller.
- The center of mass of m_2 and m_3 describe a circular orbit around m_1 with an angular velocity ω equal to one.

Thus the Hill problem is more than a particular case of the circular restricted three-body problem (Ref. 144), it has the same degree of generality.

Figures 18, 19, 20 of Section 8.5 present some simple periodic orbits of the planar Hill problem.

9.2.1 The Brown series

The plane periodic orbits of Figs. 18 and 20 can be developed into Brown series (Ref. 145).

Let us put :

$$u = x + iy = \text{complex affix of the } x,y \text{ point} \tag{155}$$

In the plane case the Hill equations (153) become :

$$\left.\begin{aligned} r &= |u| \\ u'' &= -u/r^3 - 2iu' + 1.5(u + \bar{u}) \end{aligned}\right\} \tag{156}$$

and the Fourier analysis of a periodic orbit of Figs. 18 or 20 gives :

$$u = \sum_{n=-\infty}^{+\infty} a_n \exp\{(2n+1)i\theta\} \tag{157}$$

In this expression the a_n are real Fourier constants and the angle θ is the mean elongation with respect to the x-axis :

$$\left.\begin{aligned} &\theta = \theta_0 + (q-1)t \\ &q = \text{mean angular velocity of } m_3 \text{ with respect to the absolute axes} \end{aligned}\right\} \tag{158}$$

The constants a_n of (157) can be obtained in terms of q by identification

$$a_n = |q|^{-2/3} . b_n \quad ; \; n = \{\ldots -2, -1, 0, 1, 2, \ldots\} \tag{159}$$

$$\left.\begin{aligned} b_o &= 1 - \frac{1}{6q^2} + \frac{1559}{2304q^4} + \frac{347}{192q^5} + 0(q^{-6}) \\ b_1 &= \frac{3}{16q^2} + \frac{7}{8q^3} + \frac{251}{96q^4} + \frac{899}{144q^5} + 0(q^{-6}) \\ b_2 &= \frac{25}{256q^4} + \frac{1553}{1920q^5} + 0(q^{-6}) \\ b_{-1} &= -\frac{19}{16q^2} - \frac{97}{24q^3} - \frac{2753}{288q^4} - \frac{8369}{432q^5} + 0(q^{-6}) \\ b_{-2} &= \frac{23}{640q^5} + 0(q^{-6}) \\ \text{Other } b_n &= 0(q^{-2|n|}) \end{aligned}\right\} \tag{160}$$

In the lunar case q is 13.368 (the number of sideral revolutions per year) and (160) gives a "first variation orbit" with a relative accuracy better than 10^{-6}. The average distance of New Moon and Full Moon is thus more than 1% smaller than the average distance of Quarters.

9.2.2 The lunar motion within 1000 km

The problem of the motion of the Moon is essentially a Hill problem. It is one of the worst problems of Celestial Mechanics because the solar perturbations are very large and the planetary perturbations are not negligible.

This problem has given rise to many new mathematical methods and has competed during most of the eighteenth century with the progress of clock-making for the measurement of longitudes.

Some astronomers have devoted the major part of their life to this problem (for instance Ref. 146) and to-day the best theory of lunar motion can be found in the Ref. 147 . This theory needs 15000 terms for the expression of the lunar motion with an accuracy of about 10 meters, many of these terms have a planetary origin and some, among the smallest, are given by the relativistic effects.

For an expression of the position of the Moon with an accuracy of only 1000 km (i.e. an angular accuracy of 0°15) we need the five following mean angles in terms of the time T :

A) Mean longitude of the Moon (in degrees) :

$$L' = 270.434164 + 481267.8831T - 0.001133T^2 + 0.0000019T^3 \qquad (161)$$

B) Mean anomaly of the Earth (in its motion about the Sun) :

$$M = 358.475833 + 35999.0498T - 0.000150T^2 - 0.0000033T^3 \qquad (162)$$

C) Mean anomaly of the Moon (in degrees) :

$$M' = 296.104608 + 477198.8491T + 0.009192T^2 + 0.0000144T^3 \qquad (163)$$

D) Mean elongation of the Moon (i.e. the difference between the lunar and solar longitudes) :

$$D = 350.737486 + 445267.1142T - 0.001436T^2 + 0.0000019T^3 \qquad (164)$$

E) Mean angular distance between the Moon and its ascending node :

$$F = 11.250889 + 483202.0251T - 0.003211T^2 - 0.0000003T^3 \qquad (165)$$

These five angles are given in degrees with a time T expressed in "Julian centuries" after the time "1900,0".

A Julian century is 36525 days and the time 1900,0 is 1899 December 31 at 12 hours (ephemeris time) ; hence, because of the absence of February 29 in the year 1900, the time T = 0.84 corresponds to 1984, January the first at 12 hours (ephemeris time).

After more than 25 years of comparison the full parallelism of the ephemeris time and the international atomic time has been verified up to 10^{-10} and since 1984 the ephemeris time is called "Temps dynamique terrestre" (terrestrial dynamic time) and is defined by :

TDT = TAI + 32.184 seconds

TAI : ("Temps Atomique International" (International Atomic Time")).

The usual civil time of Greenwich is now the UTC (universal time coordinated), it always remains at less than one second from the old universal time related to the irregular Earth rotation (and so useful for the navigators) but its difference to the TAI is always an integer number of seconds and is readjusted from time to time according to the rotation of the Earth :

in 1980 : TAI = UTC + 19 seconds

in 2060 : TAI = UTC + about 2 minutes.

The position of the Moon is given by the following :

A) Earth-Moon distance (center to center) in kilometers :

$r_{minimum} = 356\ 000$ km

$r_{maximum} = 407\ 000$ km

At the quarters : $r_{minimum} = 369\ 500$ km, $r_{maximum} = 405\ 000$ km

$$
\left\{
\begin{aligned}
r = {} & 384\ 990 \\
& - 20\ 900 \cos M' \\
& - 3\ 680 \cos(2D - M') \\
& - 2\ 940 \cos 2D \\
& - 570 \cos 2M' \\
& - 170 \cos(2D + M') \\
& \pm 900
\end{aligned}
\right\} \quad (166)
$$

B) Ecliptic latitude of the Moon (in degrees) :

$|\phi|_{maximum} = 5.31°$

At New Moon and at Full Moon : $|\phi|_{maximum} = 5.01°$

$$
\left\{
\begin{aligned}
\phi = {} & 5.128 \sin F \\
& + 0.281 \sin(M' + F) \\
& + 0.278 \sin(M' - F) \\
& + 0.173 \sin(2D - F) \\
& + 0.055 \sin(2D + F - M') \\
& + 0.046 \sin(2D - F - M') \\
& + 0.033 \sin(2D + F) \\
& \pm 0.09
\end{aligned}
\right\} \quad (165)
$$

C) Ecliptic longitude of the Moon (in degrees) :

$$
\left.
\begin{aligned}
\lambda = {} & L' & & + 0.059 \sin(2D - 2M') \\
& + 6.298 \sin M' & & + 0.057 \sin(2D - M - M') \\
& + 1.274 \sin(2D - M') & & + 0.053 \sin(2D + M') \\
& + 0.658 \sin 2D & & + 0.046 \sin(2D - M) \\
& + 0.214 \sin 2M' & & + 0.041 \sin(M' - M) \\
& - 0.186 \sin M & & - 0.035 \sin D \\
& - 0.114 \sin 2F & & - 0.030 \sin(M' + M) \\
& & & \pm 0.12
\end{aligned}
\right\} \quad (168)
$$

The terms whose arguments are M', 2M', F, M' + F, M' - F, 2F represent the "equation of the center", they correspond to the eccentricity and the inclination of the lunar orbit.

The terms whose arguments are 2D, 2D ± M', 2D - 2M', 2D ± F, 2D ± F - M' correspond to the phenomena called "evection" and "variation" and, as terms of the equation of the center, they are given by solutions of the Hill problem. Note for instance that the terms a_1 and a_{-1} of (157) and (159) correspond to - 2940 cos2D in (166) and 0.658° sin2D in (168).

The remaining terms only appear here in the expression of the longitude, those with M are related to the eccentricity of the Earth orbit ; the term - 0.035° sin D is the parallactic term and corresponds to the differences between the Hill problem and the circular restricted three-body problem (differences ε_1, ε_2, ε_3 in (152)).

These expressions of the motion of the Moon will remain accurate for several millenia, but the tidal effects and the planetary perturbations will very slowly modify all coefficients and sooner or later new expressions will become necessary.

9.3 THE ELLIPTIC, PARABOLIC AND HYPERBOLIC RESTRICTED THREE-BODY PROBLEMS

Because of their greater complexity, the non-circular restricted three-body problems have been much less analysed than the circular restricted three-body problem (for instance Ref. 40,61-66,73-74,105,114,128,130).

This complexity has two main causes : the motion of the primaries is much less simple and the Jacobi integral disappears.

Fortunately an elegant simplification has been found : the "rotating-pulsating coordinates", let us proceed as follows.

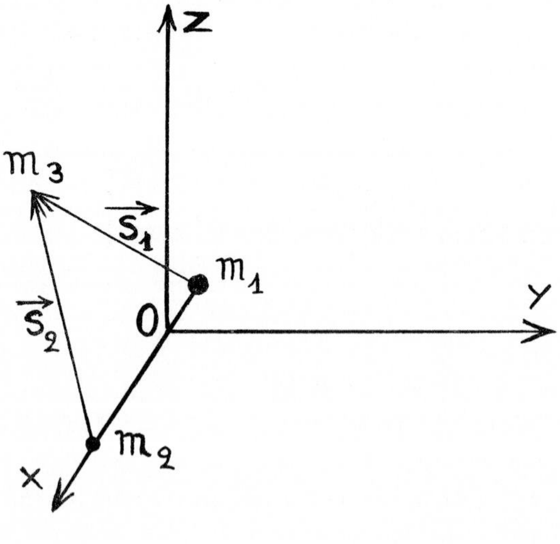

Fig. 25. The "rotating-pulsating" set of axes.

A) The Oxy plane will be the plane of the orbits of the two primaries (with positive direction of revolution).

B) At any time the unit of length will be the mutual distance between the two primaries ; hence we will have a variable unit of length.

C) The x-axis will join the two primaries which will have fixed abscissae and then fixed positions in this very particular set of axes.

D) Usually the origin is either the center of mass (Fig. 25) or one of the two primaries but it can be another point, for instance one of the collinear Lagrangian points (see Section 10.8.3).

E) The parameter of description will not be the time but the true anomaly v of the primaries along their orbits.

F) We will call $\vec{s}_1$ and $\vec{s}_2$ the vectors $(\overrightarrow{m_1, m_3})$ and $(\overrightarrow{m_2, m_3})$ in this set of axes, hence :

$$\left.\begin{array}{l} \vec{s}_1 = \begin{pmatrix} x_3 - x_1 \\ y_3 \\ z_3 \end{pmatrix} ; \quad \vec{s}_2 = \begin{pmatrix} x_3 - x_2 \\ y_3 \\ z_3 \end{pmatrix} \\ x_2 = 1 + x_1 \; ; \; x_1 \text{ and } x_2 \text{ are fixed} \; ; \; s_1 = \dfrac{r_{1.3}}{r_{1.2}} \; ; \; s_2 = \dfrac{r_{2.3}}{r_{1.2}} \end{array}\right\} \quad (169)$$

In these conditions the equation of motion of the infinitesimal mass m_3 will be the following, with dots for the differentiations with respect to the true anomaly v.

$$\left.\begin{array}{l} \dfrac{d^2}{dv^2}\begin{pmatrix} x_3 \\ y_3 \\ z_3 \end{pmatrix} = \begin{pmatrix} \ddot{x}_3 \\ \ddot{y}_3 \\ \ddot{z}_3 \end{pmatrix} = \dfrac{m_1\vec{s}_1(1-s_1^{-3}) + m_2\vec{s}_2(1-s_2^{-3})}{(m_1 + m_2)(1 + e\cos v)} + \begin{pmatrix} 2\dot{y}_3 \\ -2\dot{x}_3 \\ -z_3 \end{pmatrix} \\ \text{with : } e = \text{eccentricity of the orbits of the primaries.} \end{array}\right\} \quad (170)$$

Notice that if $e = 0$ we find again the equations (146), (147) of the circular case with $v = t$.

This equation (170) can be presented in the following Hamiltonian form.

$$\left.\begin{aligned}
&\vec{q} = \begin{pmatrix} x_3 \\ y_3 \\ z_3 \end{pmatrix} \; ; \; \vec{p} = \begin{pmatrix} p_x \\ p_y \\ p_z \end{pmatrix} = \begin{pmatrix} \dot{x}_3 - y_3 \\ \dot{y}_3 + x_3 \\ \dot{z}_3 \end{pmatrix} \\
&H = \frac{1}{2}\left[(x_3 - p_y)^2 + (y_3 + p_x)^2 + z_3^2 + p_z^2\right] - \frac{m_1\left(\frac{s_1^2}{2} + \frac{1}{s_1}\right) + m_2\left(\frac{s_2^2}{2} + \frac{1}{s_2}\right)}{(m_1 + m_2)(1 + e\cos v)}
\end{aligned}\right\} \qquad (171)$$

with of course :

$$\left.\begin{aligned}
&s_1 = \left[(x_3 - x_1)^2 + y_3^2 + z_3^2\right]^{1/2} \; ; \; s_2 = \left[(x_3 - x_2)^2 + y_3^2 + z_3^2\right]^{1/2} \\
&H = H(\vec{p}, \vec{q}, v\} \; ; \; d\vec{q}/dv = \partial H/\partial \vec{p} \; ; \; d\vec{p}/dv = -\partial H/\partial \vec{q}
\end{aligned}\right\} \qquad (172)$$

Notice :

1) In the case $e = 0$ the above Hamiltonian system is autonomous : $\partial H/\partial v = 0$. It implies H = constant along the solutions and indeed then the Jacobi integral of motion Γ given in (148) is equal to $-2H$.

2) This Hamiltonian formulation of the restricted three-body problem has many similarities with the general Hamiltonian formulation of the equations (65), (67) in Section 5.3. However notice that the latter is not adapted to the restricted three-body problem, its parameter $\vec{p}_3$ is identically zero and $\partial H/\partial \vec{p}_3$ is not defined.

The main drawback of the non-circular restricted three-body problem is the disappearance of the Jacobi integral of motion written in (148). Fortunately some of the qualitative conclusions related to this integral can be preserved.

Let us put :

$$\Gamma = \frac{m_1}{m_1 + m_2}\left[\frac{2}{s_1} + s_1^2\right] + \frac{m_2}{m_1 + m_2}\left[\frac{2}{s_2} + s_2^2\right] - (1 + e\cos v)(z_3^2 + \dot{x}_3^2 + \dot{y}_3^2 + \dot{z}_3^2) \qquad (173)$$

This leads to :

$$d\Gamma/dv = e \sin v(z_3^2 + \dot{x}_3^2 + \dot{y}_3^2 + \dot{z}_3^2) \qquad (174)$$

Hence :

A) In the circular case e is zero, Γ is constant and is indeed the Jacobi integral.

B) When e is not zero the direction of variation of Γ is given by the sign of sin v.

Let us consider for instance a parabolic or hyperbolic case. The angle v will increase from - Arc cos(-1/e) to Arc cos(-1/e) and sin v is negative before the passage of the primaries at their periapsis and positive after that passage. The function Γ has a minimum Γ_m at v = 0, it is decreasing for negative v and increasing for positive v.

These circumstances are favourable to an extension of the notion of Hill stability presented with the figure 23 in Section 9.1.

A) If $\Gamma_m > J_3$ the function Γ is forever larger than J_3 and we have the same Hill stability as in the circular case : m_3 remains forever either near m_1 or near m_2 or very far away. In the two first cases the motion m_3 remains bounded (in the "rotating-pulsating" set of axes).

B) The same conclusions arise if $\Gamma_m = J_3$ since then the point L_3 only corresponds to an Eulerian motion with m_3 forever at L_3.

C) A reduced Hill stability exists in the range $J_2 \leqq \Gamma_m < J_3$ (if we assume $m_1 \geqq m_2$) with two disconnected zones of possible motion.

D) Finally, if $\Gamma_m < J_2$ the function Γ is increasing for positive v and if for some v it reaches J_2 or even J_3 we can apply the above conclusions for all subsequent time (a symmetrical property being true for negative v).

For the elliptic restricted three-body problem the results are less simple : the function Γ has a minimum at each $v = 2k\pi$ and a maximum at each $v = (2k + 1)\pi$. the above conlusions are valid for at most one revolution of the primaries and the escapes generally occur when the distance of the primaries is small.

In spite of this loss of stability numerical computations show that in the elliptic restricted three-body problem the stability of satellites is not very different from that of the circular restricted three-body problem.

9.4 THE COPENHAGEN PROBLEM AND THE COMPUTATIONS OF MICHEL HÉNON

The "Copenhagen problem" is a particular case of the plane circular restricted three-body problem : the case in which the two primaries have equal masses.

This problem was one of the main subjects of E. Strömgren and the Copenhagen school during the begining of the twentieth century and they computed a great number of periodic orbits (Ref. 1 pages 455-497).

During the 1960s Michel Hénon,Director of the Nice observatory, came back to this problem with modern computers and found an unexpected phenomenon : the profusion of "chaotic motions" (called "semi-ergodic motions" in his early papers, Ref. 5,88,89.

These chaotic motions were already known since the early works of Poincaré (Ref. 148) but they were considered merely as a curiosity and not at all as an essential phenomenon. This phenomenon now appears in almost all domains of science and seems to be related to phenomena such as turbulence in fluid flows, uncontrolled spinning fall of a plane, difficulty of long-time predictions in meteorology, indeterminism in systems with a large number of parameters (kinetic theory of gases) etc...

Hénon uses three bodies A, B, C (Fig. 26) with masses $m_A = m_B = 0.5$; m_C infinitesimal and his set of axes Oxy is rotating with the primaries ($x_A = -0.5$ $x_B = 0.5$; $y_A = y_B = 0$).

The constant G of the law of Newton is unity and the Jacobi integral Γ is :

$$\Gamma = \frac{1}{r_A} + \frac{1}{r_B} + x^2 + y^2 - \dot{x}^2 - \dot{y}^2 \tag{175}$$

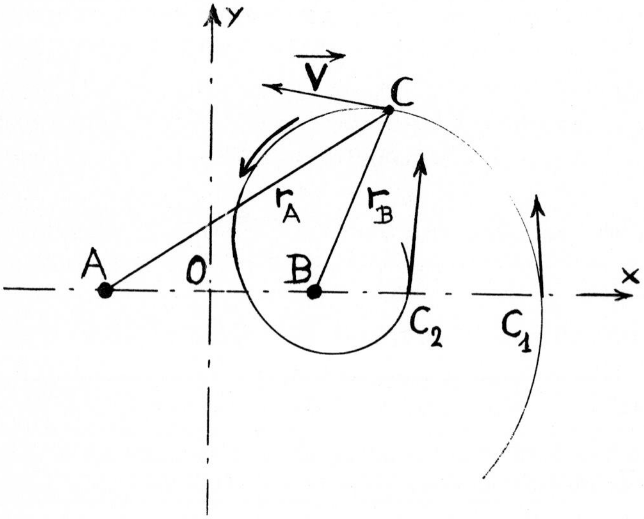

Fig. 26. The Copenhagen problem in the rotating set of axes Oxy.

In order to simplify the analysis Hénon uses the Poincaré method of the "surface of section". The "surface of section" used is the x axis and each times the orbit of C crosses that axis with a positive $\dot{y}$ (points C_1, C_2, ...) the values of x and $\dot{x}$ are picked up (y is then zero and $\dot{y}$ is given by the Jacobi integral).

Thus a solution gives a trajectory in the Oxy plane and a succession of points $x,\dot{x}$ in the $x,\dot{x}$ plane.

For large values of the Jacobi integral Γ there is Hill stability, the perturbations are weak and most orbits are periodic or "quasi-periodic".

Fig. 27 presents two orbits in the vicinity of A for Γ = 4.5 : an almost circular clockwise dotted periodic orbit (its center is almost at A) and a clockwise quasi-periodic orbit (full line with many loops).

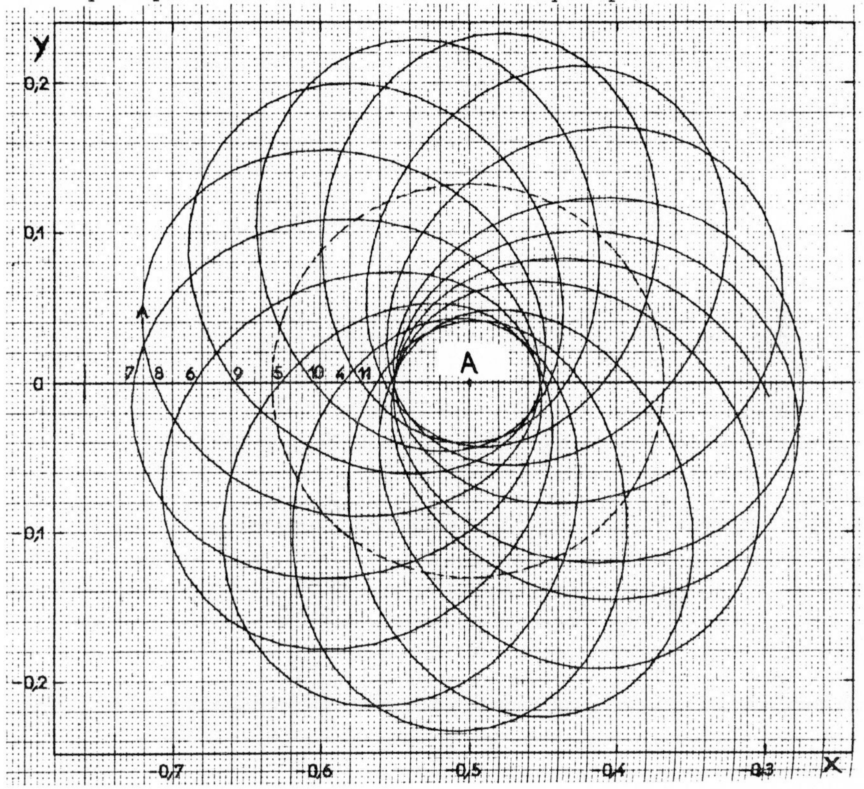

Fig. 27. A dotted periodic orbit and a quasi-periodic orbit in the vicinity of A. Source : Hénon, M. [5, 88, 89].
These orbits are clockwise and correspond to Γ = 4.5. The second orbit is close to an elliptic Keplerian motion seen in a rotating set of axes.

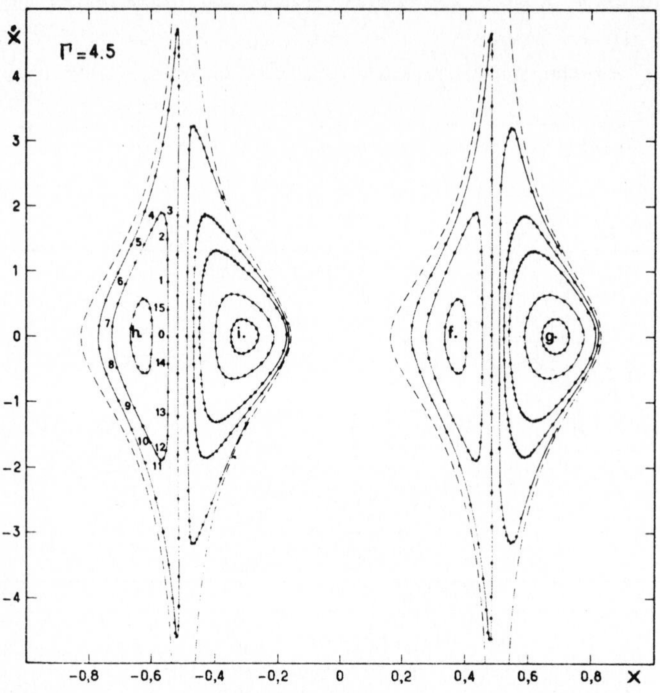

Fig. 28. The solutions in the x,x' plane for Γ *= 4.5*
Source : Hénon, M. [5, 88, 89].

These two orbits are also represented in the $x,\dot{x}$ plane of Fig. 28.

The dotted periodic orbit of Fig. 27 always crosses the x-axis in the same conditions, it corresponds only to the point h while the isolated points i, f, g correspond to three other simple periodic orbits.

The quasi-periodic orbit of Fig. 27 (full line) crosses the x-axis at the points 4, 5, 6, 7 ..., these points correspond to the points 4, 5, 6, 7 ... of the Fig. 28. The successive points are dense along the closed curve drawn in the figure, and the other closed curves corresponds to similar orbits of Γ = 4.5 while the dotted limits correspond to the limit given by the Jacobi integral.

The Fig. 28 seems very regular, but that is an illusion. Look at the Fig. 29 that is an enlargement of the Fig. 28 in the vicinity of the point i : between two closed curves of the family of the Fig. 28 we find a series of nine "islands". These nine islands of the $x,\dot{x}$ plane correspond to a unique solution similar to that of Fig. 30 ; a quasi-periodic solution remaining in the vicinity of a periodic solution with several loops.

The periodic solution corresponds to nine points at the centers of the islands and for the quasi-periodic solution the successive points jump from one island to the next and come back to the initial island after nine jumps. Thus with one point from each set of nine jumps, each island becomes densely filled exactly as the quasi-periodic solution of Fig. 27 fills its closed curve of Fig. 28.

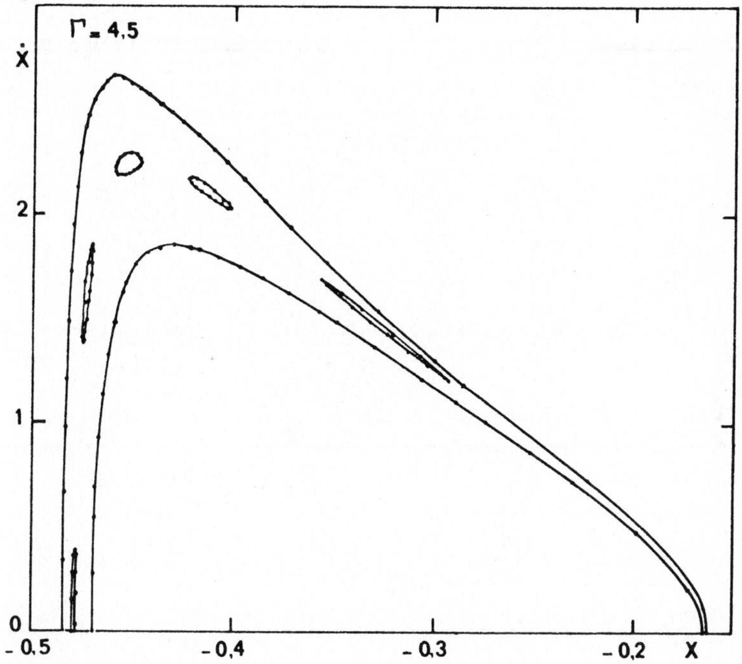

Fig. 29. Enlargement of Fig. 28 in the vicinity of the point i.
Source : Hénon, M. [5, 88, 89].

This chain of islands is not at all an isolated phenomenon. On the contrary each stable periodic orbit is surrounded by many such chains ; there are even many "chains of chains of islands" along the large islands and the subdivision goes to infinity ! The final image is extremely complex.

Another phenomenon is present between the curves and the chain of islands : the continuity is broken in small zones of "chaotic motions".

These zones are dense everywhere but their total measure is very small in Figs. 28, 29, in which the perturbations are small.

For $\Gamma = 3.5$ the perturbations are large and the chaotic motion of Fig. 31 corresponds to all the isolated $x, \dot{x}$ points of Fig. 32. These seem to fill neither a curve nor a chain of islands but a surface, that is here greater than half the attainable domain.

The first points have been numbered, they seem to jump here and there at random, or at least "chaotically", even though as a matter of fact the succession is deterministic.

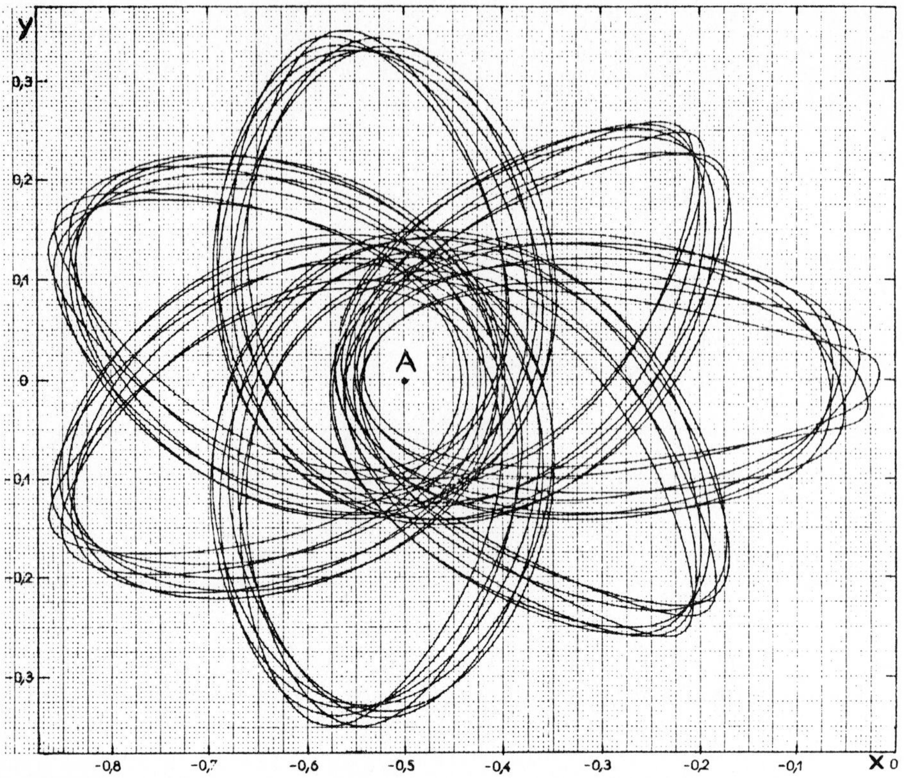

Fig. 30. A quasi-periodic orbit remaining in the vicinity of a periodic orbit with seven loops. Source : Hénon, M. [5, 88, 89].
This counter-clockwise orbit with $\Gamma = 4$ *corresponds, in the* x,x' *plane, to a chain of islands similar to that of Fig. 29.*

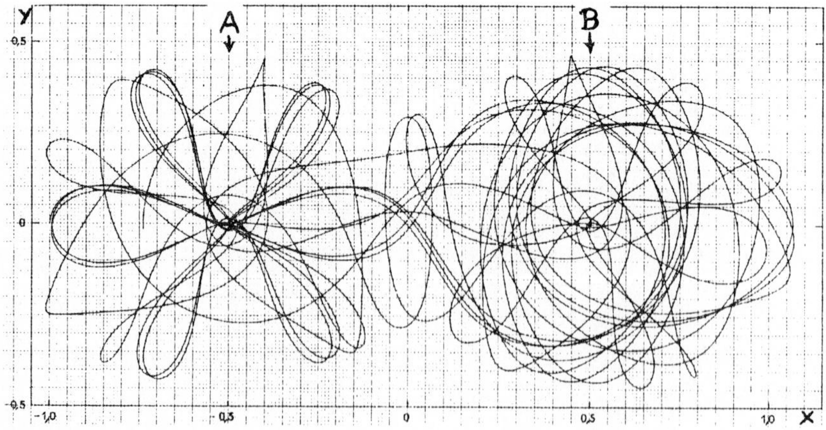

Fig. 31. A chaotic or "semi-ergodic" orbit for $\Gamma = 3.5$.
Source : Hénon, M. [5, 88, 89].

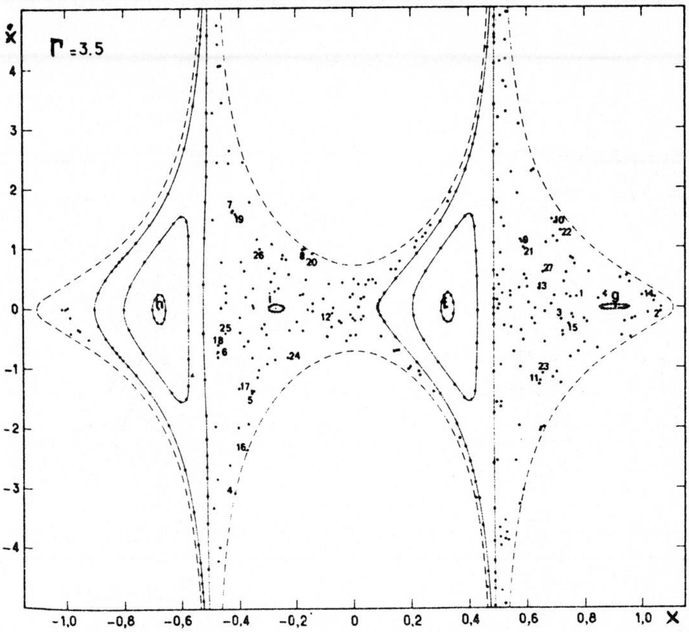

Fig. 32. The chaotic orbit of Fig. 31 corresponds to all the isolated points of this figure in the x,x' plane (the first points are numbered There remain quasi-periodic orbits corresponding to the closed curves. Source : Hénon, M. [5, 88, 89].

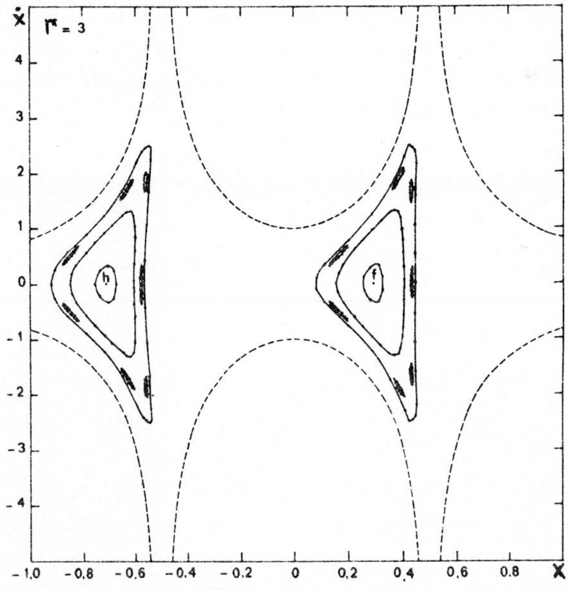

Fig. 33. The x,x' plane when $\Gamma = 3$*. There remain quasi-periodic motions corresponding to the closed curves or to the chains of islands, but the motions are no longer Hill stable and almost all motions outside the largest closed curves lead to an escape of the small mass.*
Source : Hénon, M. [5, 88, 89].

When Γ is below 3.456796... a new phenomenon appears. The motions are no longer Hill stable and almost all motions that are outside the largest closed curves of Fig. 33 escape to infinity.

In the $x, \dot{x}$ plots the course of these escapes is that $\dot{x}$ goes to a bounded and negative or zero limit $\dot{x}_\infty$ while x goes to $-\infty$ with a jump at each step going to $2\pi\dot{x}_\infty$.

The remaining closed curves of islands show the possibility of bounded motions even below the limit of Hill stability.

We will met similar phenomena in the general three-body problem.

A) Let us consider an arbitrary three-body system and an arbitrary value of the angular momentum, there will be "Arnold tori" of quasi-periodic solutions for all negative values of the energy integral.

The n-dimensional tori are the n-dimensional generalization of closed curves and these "Arnold tori" will always have a positive measure in phase space.

B) It seems that between the Arnold tori of periodic or quasi-periodic motions we always find the following.

B.1) A set of measure zero of "abnormal orbits" (unstable periodic orbits, orbits asymptotic to a periodic or quasi-periodic solution, etc...).

B.2) Chaotic motions if, for the given values of the integrals of motion, the attainable domain between the nearest Arnold tori is bounded. These chaotic motions densely fill the attainable domain.

B.3) Escape motions if the attainable domain is unbounded : the two largest mutual distances will go to infinity either as the time t (hyperbolic escape) or at least as $t^{2/3}$ (parabolic escape).

However in the general three-body problem very few possibilities of bounded attainable domains are known. They generally require some special symmetry (see Section 11.7.8) and thus, unlike the restricted case, chaotic motions will remain exceptional in the general three-body problem (see Section 11.11).

Chapter 10

THE GENERAL THREE-BODY PROBLEM. QUANTITATIVE ANALYSIS

Quantitative analysis has its two main bases in the old analytical methods, essentially the theories of perturbations, and the new numerical methods that are improving so rapidly.

10.1 THE ANALYTICAL METHODS

A review of perturbation theories is a review of the greatest of mathematical works and Celestial Mechanics has posed the hardest challenge to mathematicians.

Karl-Friedrich Gauss, "the King of mathematicians" became famous for solving the problem of the motion of the minor planet Ceres that was lost in the radiance of the Sun, and the discoveries of Neptune and Pluto are among the greatest mathematical achievements.

If we consider only French mathematicians of the past centuries we find Lagrange, Laplace, Clairaut, d'Alembert, Poisson, Leverrier, Delaunay, Poincaré... all working on Celestial Mechanics.

The most famous methods of perturbations are those of Lagrange, Encke, Cowell, Hansen, Delaunay, Hill, Von Zeipel, Hori... They are described in many outstanding books and especially these of Tisserand (Ref. 59), Szebehely (Ref. 1), Sarychev (Ref. 3) and the series of five books of Hagihara (Ref. 2). We suggest the reader refer to these books if necessary and we will only develop a short analysis and present a few examples.

In a perturbation method a "main effect" is defined (e.g. the solar attraction) that generally gives either pure Keplerian motions or uniformly perturbed Keplerian motions, and many "small effects" slowly modify the simple main motion.

The slowness of the modification allows approximation of the small effects and consideration of their influence at the neigbouring simple main motion instead of at the complex true motion.

The small effects are always related to some small parameters. In the planetary methods the small parameters are the planetary masses when compared to that of the Sun. In the lunar methods the small parameter is the ratio between the Earth-Moon distance and the Sun-Earth distance, but the solar perturbation is large and the lunar problem is much more difficult than the planetary problem.

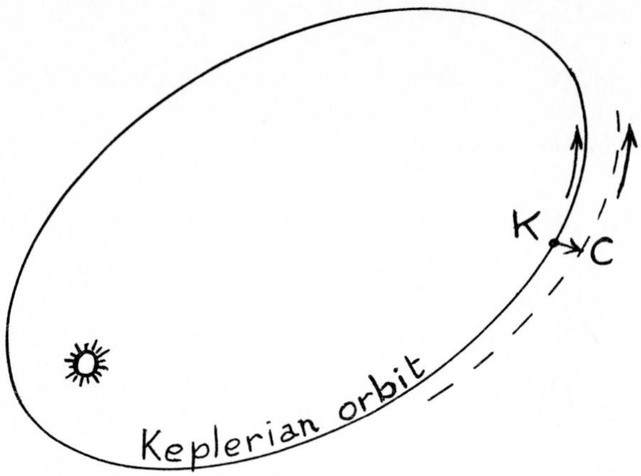

Fig. 34. The Encke method. Study of the variations of the vector $\overrightarrow{KC}$.

The Encke method (Fig. 34) is especially suited for large eccentricities and cometary motions. Encke does not consider the usual orbital elements but studies directly the variations of the vector $\overrightarrow{KC}$ between the point K (with a suitable Keplerian motion) and the comet C of interest.

The acceleration of $\overrightarrow{KC}$ is easy to write and as long as the distance KC remains small a first-order analysis is sufficient and leads to simple quadratures.

Delaunay found the "Delaunay elements" (see Section 4.4) that allow an easy Hamiltonian presentation and, like Hansen some years before, he has applied his own method to the problem of the Moon.

In the Hamiltonian methods of Delaunay (1860), Von Zeipel (1916) and Hori (1966) a succession of canonical transformations eliminate firstly the short-period terms (period of the order of the revolution) and secondly the long-period terms. The remaining Hamiltonian, called the "secular Hamiltonian", has no more angular variables and is easily integrable.

The method of Delaunay is effective but very lengthy and eliminates the short- and long-period terms one by one. The Von Zeipel method (Ref.149) eliminates them in only two steps but was almost forgotten when Dirk Brouwer used it in 1959 for the motion of artificial satellites (Ref. 129). The method of Hori (Ref.150) is even faster and simpler but it uses Lie series instead of Taylor series for expression of the Hamiltonian.

10.2 AN EXAMPLE OF THE VON ZEIPEL METHOD.
INTEGRATION OF THE THREE-BODY PROBLEM TO FIRST ORDER

In most triple stellar systems the Jacobi ratio r/R is small, there is a close binary and an isolated third body (Fig. 35).

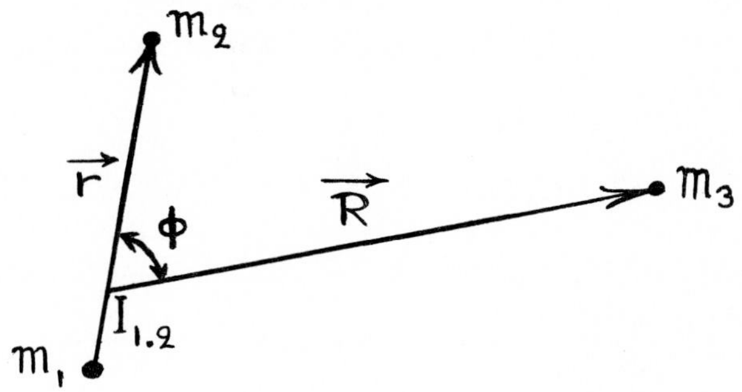

Fig. 35. The Jacobi parameters $\vec{r}$, $\vec{R}$ and Φ.
$\Phi = (\vec{r},\vec{R})$; $I_{1.2}$ is the center of mass of m_1 and m_2.

If $(r/R)^3$ is much smaller than $(m_1 + m_2)/M$, with $M = m_1 + m_2 + m_3$, the three-body problem is integrable to first order and even to second order for any inclinations and any eccentricities.

Let us consider that interesting particular case of the three-body problem. We will use the Jacobi osculating orbital elements n_i, a_i, e_i i_i, Ω_i, ω_i, M_i of the "inner orbit" (orbit of $\vec{r}$) and n_e, a_e, e_e, i_e, Ω_e, ω_e, M_e of the outer or "exterior orbit" (orbit of $\vec{R}$). These elements are defined by usual Keplerian relations as (26)-(29).

We will also use the corresponding Delaunay elements $\mathcal{L}_i$ to h_e given in (30), (31) and the set of axes of Fig. 7, i.e. with Oxy as the invariable plane. These axes allow easy elimination of the nodes as described in Chapter 7.

We thus arrive at the following Hamiltonian system already presented in Sections 4.4 and 7.

$$M = m_1 + m_2 + m_3 \;;\; m = m_1 m_2/(m_1+m_2) \;;\; \mathcal{M} = m_3(m_1+m_2)/M \tag{176}$$

$$H = \text{Hamiltonian} = -\frac{G^2 m^3 (m_1+m_2)^2}{2\mathcal{L}_i^{\,2}} - \frac{G^2 \mathcal{M}^3 M^2}{2\mathcal{L}_e^{\,2}} + G m_3 \left[\frac{m_1 + m_2}{R} - \frac{m_1}{r_{1.3}} - \frac{m_2}{r_{2.3}}\right] \tag{177}$$

For given Newtonian constant G and masses m_1, m_2, m_3 the hamiltonian H must be expressed in terms of the angular momentum c and eigth Delaunay elements :

$$H = H(\mathcal{L}_i, \mathcal{G}_i, \mathcal{L}_e, \mathcal{G}_e, \ell_i, g_i, \ell_e, g_e, c) \tag{178}$$

This expression is complex but also uniquely determined, as shown in Chapter 7, and it leads to the following Hamiltonian system with four degrees of freedom.

$$\left.\begin{aligned} & d\mathcal{L}_i/dt = -\,\partial H/\partial \ell_i \;;\; d\mathcal{G}_i/dt = -\,\partial H/\partial g_i \;;\; d\mathcal{L}_e/dt = -\,\partial H/\partial \ell_e \;;\; d\mathcal{G}_e/dt = -\,\partial H/\partial g_e \\ & d\ell_i/dt = \partial H/\partial \mathcal{L}_i \;;\; dg_i/dt = \partial H/\partial \mathcal{G}_i \;;\; d\ell_e/dt = \partial H/\partial \mathcal{L}_e \;;\; dg_e/dt = \partial H/\partial \mathcal{G}_e \end{aligned}\right\} \tag{179}$$

The four ignorable Delaunay elements $\mathcal{H}_i$, $\mathcal{H}_e$, h_i, h_e are given by the following :

$$\left.\begin{aligned} & \mathcal{H}_i = (c^2 + \mathcal{G}_i^2 - \mathcal{G}_e^2)/2c \\ & \qquad\qquad\qquad\qquad \mathcal{H}_i + \mathcal{H}_e = c \\ & \mathcal{H}_e = (c^2 - \mathcal{G}_i^2 + \mathcal{G}_e^2)/2c \\ & h_e = h_i + \pi \;;\; dh_i/dt = dh_e/dt = \partial H/\partial c \end{aligned}\right\} \tag{180}$$

Let us consider the Hamiltonian H given in (177). If r << R the third term is small and can be developed in terms of r, R and the angle ϕ equal to $(\vec{r},\vec{R})$:

$$\left.\begin{aligned} G m_3 \left[\frac{m_1+m_2}{R} - \frac{m_1}{r_{13}} - \frac{m_2}{r_{23}}\right] = \frac{G m_1 m_2 m_3}{m_1+m_2}\Bigg[& \frac{r^2(1-3\cos^2\phi)}{2R^3} + \\ & + \frac{m_1-m_2}{m_1+m_2}\,\frac{r^3(3\cos\phi-5\cos^3\phi)}{2R^4} + O\left(\frac{r^4}{R^5}\right)\Bigg] \end{aligned}\right\} \tag{181}$$

If that third term of H is considered as negligible the three-body problem is simple and is decomposed into two unperturbed Keplerian motions : ℓ_i and ℓ_e are two linear functions of the time and the other Delaunay elements are constant. Hence the real motions are two slowly perturbed Keplerian motions.

We will not develop the third term of the Hamiltonian H in terms of the Delaunay elements, as must be done, but will use directly a Von Zeipel transformation leading to a "long-period three-body problem" that will be integrable to first order.

10.2.1 Principle of the method of Von Zeipel

Von Zeipel (Ref. 149) uses a "generating function" S and the corresponding canonical tranformation from the initial Hamiltonian elements (i.e. here $\mathcal{L}_i, \mathcal{G}_i, \mathcal{L}_e, \mathcal{G}_e, \ell_i, g_i, \ell_e, g_e$) to the new "long-period" elements (we will call them here $\mathcal{L}_s, \mathcal{G}_s, \mathcal{L}_T, \mathcal{G}_T, \ell_s, g_s, \ell_T, g_T$; the subscript s corresponds to the long-period orbit of the <u>second</u> body (with respect to the first) and the subscript T corresponds to the long-period orbit of the <u>third</u> body).

The purpose of Von Zeipel is to derive a new Hamiltonian system without short period effects (to first order) by a suitable choice of the generating function S.

The unavoidable complexity of the Von Zeipel method is caused by the mixed character of the function S that is given in terms of one half of the initial variables and one half of the new variables :

$$S = S(\mathcal{L}_i, \mathcal{G}_i, \mathcal{L}_e, \mathcal{G}_e, \ell_s, g_s, \ell_T, g_T, c) \tag{182}$$

and the canonical transformation from the "osculating elements" $\mathcal{L}_i, \mathcal{G}_i, \ell_i, g_i, \mathcal{L}_e, \mathcal{G}_e, \ell_e, g_e$ to the "long-period elements" $\mathcal{L}_s, \mathcal{G}_s, \ell_s, g_s, \mathcal{L}_T, \mathcal{G}_T, \ell_T, g_T$ is defined by the following :

$$\left.\begin{aligned} &\mathcal{L}_s = \partial S/\partial \ell_s \;;\; \mathcal{G}_s = \partial S/\partial g_s \;;\; \mathcal{L}_T = \partial S/\partial \ell_T \;;\; \mathcal{G}_T = \partial S/\partial g_T \\ &\ell_i = \partial S/\partial \mathcal{L}_i \;;\; g_i = \partial S/\partial \mathcal{G}_i \;;\; \ell_e = \partial S/\partial \mathcal{L}_e \;;\; g_e = \partial S/\partial \mathcal{G}_e \end{aligned}\right\} \tag{183}$$

Under these conditions the equations of motion (179) become :

$$\left.\begin{aligned} &\frac{d\mathcal{L}_s}{dt} = -\frac{\partial H}{\partial \ell_s} \;;\; \frac{d\mathcal{G}_s}{dt} = -\frac{\partial H}{\partial g_s} \;;\; \frac{d\mathcal{L}_T}{dt} = -\frac{\partial H}{\partial \ell_T} \;;\; \frac{d\mathcal{G}_T}{dt} = -\frac{\partial H}{\partial g_T} \\ &\frac{d\ell_s}{dt} = \frac{\partial H}{\partial \mathcal{L}_s} \;;\; \frac{dg_s}{dt} = \frac{\partial H}{\partial \mathcal{G}_s} \;;\; \frac{d\ell_T}{dt} = \frac{\partial H}{\partial \mathcal{L}_T} \;;\; \frac{dg_T}{dt} = \frac{\partial H}{\partial \mathcal{G}_T} \end{aligned}\right\} \tag{184}$$

The new system again has the Hamiltonian form, its Hamiltonian H is again identical to the energy integral h, and it must be expressed in terms of the angular momentum c and the new elements :

$$H = H(\mathcal{L}_s, \mathcal{G}_s, \mathcal{L}_T, \mathcal{G}_T, \ell_s, g_s, \ell_T, g_T, c) \tag{185}$$

It is customary to write :

$$S = \mathcal{L}_i \ell_s + \mathcal{G}_i g_s + \mathcal{L}_e \ell_T + \mathcal{G}_e g_T + S_1 \tag{186}$$

If S_1 were zero the canonical transformation (183) would be an identity. S_1 will be of first order and so will be the differences between the osculating elements and the long-period elements.

A suitable choice of S_1 will lead to an Hamiltonian H independant of ℓ_s, ℓ_T, g_T (to first order) and thus to an integrable Hamiltonian system.

10.2.2 Application of the Von Zeipel method to the three-body problem

Like S in (182), S_1 must theoretically be expressed in terms of L_i, G_i, L_e, G_e, ℓ_s, g_s, ℓ_T, g_T (and also in terms of such constants as G, m_1, m_2, m_3 and the angular momentum c). However, in order to have a simpler and finite expression of S_1 we will use the ordinary orbital elements of the two intermediate orbits defined by L_i, G_i, ℓ_s, g_s and L_e, G_e, ℓ_T, g_T.

$$S_1 = S_1(G,m_1,m_2,m_3,n_i,a_i,b_i,e_i,E_2,g_s,n_e,b_e,p_e,e_e,\ell_T,v_3,g_T,j) \qquad (187)$$

where :

n_i, a_i, e_i, n_e, e_e are the osculating orbital elements defined in Section 4.3.

b_i, b_e, p_e are the usual auxiliary orbital elements given by :

$b_i = a_i\sqrt{(1 - e_i^2)}$; $b_e = a_e\sqrt{(1 - e_2^2)}$: semi-minor axes

$p_e = a_e(1 - e_2^2)$: semi-latus rectum of the exterior orbit

E_2 is the eccentric anomaly of the inner intermediate orbit :

$$\ell_s = E_2 - e_i \sin E_2$$

v_3 is the true anomaly of the exterior intermediate orbit :

$$\tan(v_3/2) = \left(\frac{1 + e_e}{1 - e_e}\right)^{1/2} \tan(E_3/2) \; ; \; \ell_T = E_3 - e_e \sin E_3$$

j is the mutual inclination of the osculating orbits and also of the intermediate orbits :

$$j = i_i + i_e$$

Note that n_i, a_i, e_i, n_e, e_e, b_i, b_e p_e are also orbital elements of the two intermediate orbits. (188)

With these notations a usual identification procedure leads to the following very long but finite expression for S_1.

$$S_1 = \frac{Gm_1m_2m_3}{m_1 + m_2}\left[\frac{a_i^2}{16n_e b_e^3} S_T + \frac{(1 + e_e\cos v_3)^3}{24n_i p_e^3} S_s\right] \tag{189}$$

where :

$$\left.\begin{aligned} S_T = {} & 2(v_3 + e_e\sin v_3 - \ell_T)\left[3\sin^2 j - 2 + 3e_i^2(5\sin^2 j\sin^2 g_s - 1 - \sin^2 j)\right] + \\ & + 5e_i^2\cos j\sin 2g_s\left[3\cos(2g_T+2v_3) + 3e_e\cos(2g_T+v_3) + e_e\cos(2g_T+3v_3)\right] + \\ & + \left[5e_i^2(\cos^2 j\sin^2 g_s - \cos^2 g_s) + (e_i^2 - 1)\sin^2 j\right]\cdot\left[3\sin(2g_T+2v_3) + 3e_e\sin(2g_T+v_3) + \right. \\ & \left. + e_e\sin(2g_T+3v_3)\right] \end{aligned}\right\} \tag{190}$$

and :

$$\left.\begin{aligned} S_s = {} & a_i^2\left[-24e_i\sin E_2 + 9e_i^2\sin 2E_2 + e_i^3(9\sin E_2 - \sin 3E_2)\right] + \\ & + a_i^2F_1^2\left[-9\sin 2E_2 + e_i(81\sin E_2 + 3\sin 3E_2) - 18e_i^2\sin 2E_2 - 36e_i^3\sin E_2\right] + \\ & + a_ib_iF_1F_2\left[18\cos 2E_2 - e_i(90\cos E_2 + 6\cos 3E_2) + e_i^2(18\cos 2E_2 - 45)\right] + \\ & + b_i^2F_2^2\left[9\sin 2E_2 - e_i(9\sin E_2 + 3\sin 3E_2)\right] \end{aligned}\right\} \tag{191}$$

where :

$$F_1 = \cos g_s\cos(g_T + v_3) + \cos j\sin g_s\sin(g_T + v_3) \tag{192}$$

$$F_2 = \cos j\cos g_s\sin(g_T + v_3) - \sin g_s\cos(g_T + v_3) \tag{193}$$

Since the function S_1 is given in terms of the parameters of (187) and not those of (182) the transformation formulas (183), (186) become :

$$\mathcal{L}_s = \mathcal{L}_i + \frac{1}{1 - e_i\cos E_2}\frac{\partial S_1}{\partial E_2} \tag{194}$$

$$\mathcal{G}_s = \mathcal{G}_i + (\partial S_1/\partial g_s) \tag{195}$$

$$\mathcal{L}_T = \mathcal{L}_e + (\partial S_1/\partial \ell_T) + \frac{b_e^3}{p_e^3}(1 + e_e\cos v_3)^2\frac{\partial S_1}{\partial v_3} \tag{196}$$

$$\mathcal{G}_T = \mathcal{G}_e + (\partial S_1/\partial g_T) \tag{197}$$

$$\ell_i = \ell_s - \frac{3n_i}{\mathcal{L}_i}\frac{\partial S_1}{\partial n_i} + \frac{2a_i}{\mathcal{L}_i}\frac{\partial S_1}{\partial a_i} + \frac{b_i}{\mathcal{L}_i}\frac{\partial S_1}{\partial b_i} + \frac{1-e_i^2}{\mathcal{L}_i e_i}\left[\frac{\partial S_1}{\partial e_i} + \frac{\sin E_2}{1-e\cos E_2}\frac{\partial S_1}{\partial E_2}\right] \tag{198}$$

$$g_i = g_s + \frac{b_i}{\mathcal{G}_i}\frac{\partial S_1}{\partial b_i} - \frac{1-e_i^2}{\mathcal{G}_i e_i}\left[\frac{\partial S_1}{\partial e_i} + \frac{\sin E_2}{1-e_i\cos E_2}\frac{\partial S_1}{\partial E_2}\right] + \frac{1}{\sin j}\left[\frac{1}{\mathcal{G}_e} + \frac{\cos j}{\mathcal{G}_i}\right]\frac{\partial S_1}{\partial j} \tag{199}$$

$$\ell_e = \ell_T - \frac{3n_e}{\mathcal{L}_e}\frac{\partial S_1}{\partial n_e} + \frac{b_e}{\mathcal{L}_e}\frac{\partial S_1}{\partial b_e} + \frac{(1-e_e^2)}{\mathcal{L}_e e_e}\frac{\partial S_1}{\partial e_e} + \frac{\sin v_3(2+e_e\cos v_3)}{\mathcal{L}_e e_e}\frac{\partial S_1}{\partial v_3} \tag{200}$$

$$\left.\begin{aligned} g_e = g_T &+ \frac{b_e}{\mathcal{G}_e}\frac{\partial S_1}{\partial b_e} + \frac{b_e}{\mathcal{G}_e}\frac{\partial S_1}{\partial p_e} - \frac{(1-e_e^2)}{\mathcal{G}_e e_e}\frac{\partial S_1}{\partial e_e} - \frac{\sin v_3(2+e_e\cos v_3)}{\mathcal{G}_e e_e}\frac{\partial S_1}{\partial v_3} + \\ &+ \frac{1}{\sin j}\left[\frac{1}{\mathcal{G}_i} + \frac{\cos j}{\mathcal{G}_e}\right]\frac{\partial S_1}{\partial j} \end{aligned}\right\} \tag{201}$$

These expressions are of course more complex than the relations (183) but allow exact expression, with a finite number of terms, of the canonical transformation of interest.

Note that these differences between osculating and long-period elements are of first order and are given in terms of the elements of the intermediate orbits (the elements of (187) and (188)).

It remains to express the Hamiltonian H in terms of the long-period elements :

$$H = -\frac{G^2 m^3 (m_1 + m_2)^2}{2\mathcal{L}_s^2} - \frac{G^2 \mathcal{M}^3 M^2}{2\mathcal{L}_T^2} + Q + \varepsilon \tag{202}$$

The two first and main terms of the right-hand side are similar to those of the initial Hamiltonian (177).

Q is the first-order term and the canonical transformation has especially been chosen in order that Q be independent of ℓ_s and ℓ_T, it is also independent of g_T, and the first-order long-period approximation is thus integrable.

Q can be expressed in terms of the angular momentum c and the following long-period Delaunay elements : $\mathcal{L}_s$, $\mathcal{G}_s$, $\mathcal{L}_T$, $\mathcal{G}_T$, g_s. We will write it in terms of the following correspondig long-period orbital elements.

$$\left.\begin{aligned} a_s &= (m_1 + m_2)\mathcal{L}_s^2/Gm_1^2 m_2^2 = \text{inner semi-major axis} \\ b_T &= M\mathcal{L}_T\mathcal{G}_T/Gm_3^2(m_1 + m_2^2) = \text{outer semi-minor axis} \\ e_s &= (1 - \mathcal{G}_s^2/\mathcal{L}_s^2)^{1/2} = \text{inner eccentricity} \\ g_s &= \text{inner (long-period) argument of pericenter} \\ j_L &= \text{long-period mutual inclination :} \\ &\quad c^2 = \mathcal{G}_s^2 + \mathcal{G}_T^2 + 2\mathcal{G}_s\mathcal{G}_T\cos j_L \end{aligned}\right\} \tag{203}$$

Q is then given as :

$$Q = \frac{Gm_1m_2m_3a_S^2}{8(m_1+m_2)b_T^3}\left[-2 - 3e_S^2 + 3\sin^2 j_L(1 - e_S^2 + 5e_S^2\sin^2 g_S)\right] \qquad (204)$$

In (202) ε is the error term :

$$\varepsilon = O\left[\frac{Gm_1m_2m_3a_S^3}{(m_1 + m_2)p_T^4} \cdot \sup\left\{\frac{|m_1 - m_2|}{m_1 + m_2} ; \left[\frac{Ma_S}{(m_1 + m_2)p_T}\right]^{1/2}\right\}\right] \qquad (205)$$

where :

$p_T = M\mathcal{G}_T^2/Gm_3^2(m_1 + m_2)^2$ = outer (long-period) semi-latus rectum

In (205) the coefficient $|m_1 - m_2|/(m_1 + m_2)$ comes of course from the right-hand side of (181).

10.2.3 First-order integration of the three-body problem.

If in (202) the error term ε can be considered as negligible the remaining Hamiltonian problem is integrable. Indeed :

A) The remaining Hamiltonian H is a function of $\mathcal{L}_S$, $\mathcal{G}_S$, $\mathcal{L}_T$, $\mathcal{G}_T$, g_S, c only, and is independant of ℓ_S, ℓ_T, g_T.

B) The equations of motion (184) lead to the following.

B.1 $\mathcal{L}_S$, $\mathcal{L}_T$, $\mathcal{G}_T$ are constant and so are the corresponding long-period orbital elements : the semi-major axes a_S and a_T, the mean angular motions n_S and n_T, the outer eccentricity e_T, the outer semi-minor axis b_T and the outer semi-latus rectum p_T (as shown from the usual relations (203) and $n_S^2a_S^3 = G(m_1 + m_2)$; $n_T^2a_T^3 = GM$; $b_T^2/a_T^2 = p_T/a_T = 1 - e_T^2 = \mathcal{G}_T^2/\mathcal{L}_T^2$).

B.2 The parameters ℓ_S, ℓ_T (mean anomalies) and g_T (outer argument of pericenter) are "ignorable". They never appear in the right-hand members of the equations of motion (184) and are then obtained at the end by the three following final quadratures $d\ell_S/dt = \partial H/\partial\mathcal{L}_S$; $d\ell_T/dt = \partial H/\partial\mathcal{L}_T$ and $dg_T/dt = \partial H/\partial\mathcal{G}_T$.

We will see that $d\ell_T/dt$ is constant and very close to n_T, the derivative $d\ell_S/dt$ is generally not constant but remains very close to n_S, finally dg_T/dt is small, of first order, and has generally large relative variations.

C) It remains to integrate the variations of $\mathcal{G}_S$ and g_S, they are easily integrable because Q is an integral of the first-order approximation (indeed, in (202), ε is negligible, $\mathcal{L}_S$ and $\mathcal{L}_T$ are constant and the Hamiltonian H is also constant since it is independent of the time t).

Let us examine that quastion ; the equations of motion (184) and the expression (204) for Q lead to the following.

$$\frac{d\mathcal{L}_S}{dt} = -\frac{\partial H}{\partial \ell_S} = 0 \; ; \; \frac{d\mathcal{L}_T}{dt} = -\frac{\partial H}{\partial \ell_T} = 0 \; ; \; \frac{d\mathcal{G}_T}{dt} = -\frac{\partial H}{\partial g_T} = 0 \qquad (206)$$

$$\frac{d\mathcal{G}_S}{dt} = -\frac{\partial H}{\partial g_S} = -\frac{\partial Q}{\partial g_S} = -\frac{15Gm_1m_2m_3a_S^2}{8(m_1+m_2)b_T^3} e_S^2 \sin^2 j_L \sin 2g_S \qquad (207)$$

$$\left.\begin{aligned}\frac{d\ell_S}{dt} &= \frac{\partial H}{\partial \mathcal{L}_S} = n_S + \frac{2a_S}{\mathcal{L}_S}\frac{\partial Q}{\partial a_S} + \frac{(1-e_S^2)}{\mathcal{L}_S e_S}\cdot\frac{\partial Q}{\partial e_S} = \\ &= n_S + \frac{2Q}{\mathcal{L}_S} + \frac{5Gm_3}{4n_S b_T^3}(3\sin^2 g_S \sin^2 j_L - 1)\end{aligned}\right\} (208)$$

$$\left.\begin{aligned}\frac{dg_S}{dt} &= \frac{\partial H}{\partial \mathcal{G}_S} = \left[\frac{1}{\mathcal{G}_T \sin j_L} + \frac{\cot a n\, j_L}{\mathcal{G}_S}\right]\frac{\partial Q}{\partial j_L} - \frac{(1-e_S^2)}{\mathcal{G}_S e_S}\cdot\frac{\partial Q}{\partial e_S} = \\ &= \frac{3m_1m_2n_Ta_S^2}{4(m_1+m_2)^2p_T^2}\cos j_L(1-e_S^2+5e_S^2\sin^2 g_S) + \\ &+ \frac{3Gm_3a_S}{4n_Sb_Sb_T^3}\left[2-2e_S^2+5\sin^2 g_S(e_S^2-\sin^2 j_L)\right]\end{aligned}\right\} (209)$$

$$\frac{d\ell_T}{dt} = \frac{\partial H}{\partial \mathcal{L}_T} = n_T + \frac{b_T}{\mathcal{L}_T}\frac{\partial Q}{\partial b_T} = n_T - \frac{3Q}{\mathcal{L}_T} = \text{constant} \qquad (210)$$

$$\left.\begin{aligned}\frac{dg_T}{dt} &= \frac{\partial H}{\partial \mathcal{G}_T} = \frac{b_T}{\mathcal{G}_T}\frac{\partial Q}{b_T} + \left[\frac{1}{\mathcal{G}_S \sin j_L} + \frac{\cot a n\, j_L}{\mathcal{G}_T}\right]\frac{\partial Q}{\partial j_L} = \\ &= \frac{3m_1m_2n_Ta_S^2}{8(m_1+m_2)^2p_T^2}\left[4+e_S^2+10e_S^2\sin^2 g_S - 5\sin^2 j_L(1-e_S^2+5e^2\sin^2 g_S)\right] + \\ &+ \frac{3Gm_3a_S}{4n_Sb_Sb_T^3}\cos j_L(1-e_S^2+5e_S^2\sin g_S)\end{aligned}\right\} (211)$$

With $b_S = (m_1 + m_2)\mathcal{L}_S\mathcal{G}_S/Gm_1^2m_2^2$ and with (203), (207), (209) it is easy to verify that Q is indeed an integral of motion.

The main remaining problem is the integration of (207) and (209) with the help of the integrals of motion c, $\mathcal{L}_S$, $\mathcal{L}_T$, $\mathcal{G}_T$ and, especially, Q.

Since $\mathcal{G}_S = \mathcal{L}_S\sqrt{(1 - e_S^2)}$ the last identity of (203) can be written :

$$\cos j_L = \frac{A}{\sqrt{(1-e_S^2)}} - B\sqrt{(1-e_S^2)} \; ; \; 0 \leq j_L \leq \pi \qquad (212)$$

the two constants A and B being defined by :

$$A = (c^2 - \mathcal{G}_T^2)/2\mathcal{G}_T\mathcal{L}_S \ ; \ B = \mathcal{L}_S/2\mathcal{G}_T \quad ; \quad B > 0 \ ; \ \left.\begin{array}{l} -1/4B \leq A \leq B+1 \\ \text{if } B \leq 0.5 \text{ then } B-1 \leq A \end{array}\right\} \tag{213}$$

On the other hand, since a_S and b_T are constant, the integral Q = constant is equivalent to :

$$Z = (1 - e_S^2)(1 + \sin^2 j_L) + 5e_S^2 \sin^2 j_L \sin^2 g_S = \text{constant} \ ; \ 0 \leqq Z \leqq 5 \tag{214}$$

If we write :

$$x = 1 - e_S^2 = \mathcal{G}_S^2/\mathcal{L}_S^2 \ ; \ 0 \leqq x \leqq 1 \tag{215}$$

we obtain from (206) and (207) :

$$\frac{dx}{dt} = 2\mathcal{G}_S \cdot \mathcal{L}_S^{-2} \cdot d\mathcal{G}_S/dt = -\frac{15Gm_3}{4n_S b_T^3} x^{1/2} \cdot e_S^2 \cdot \sin^2 j_L \cdot \sin 2g_S \tag{216}$$

The right-hand side of (216) can easily be expressed in terms of x and the constants of motion, indeed we can define the polynomials $P_1(x)$ and $P_2(x)$ by the two following identities :

$$5e_S^2 \sin^2 j_L \sin^2 g_S = P_1(x) = B^2x^2 - 2x(1 + AB) + Z + A^2 \tag{217}$$

$$\left.\begin{array}{l} 5e_S^2(1 - e_S^2)\sin^2 j_L \cos^2 g_S = P_2(x) = 4B^2x^3 - x^2(3 + 8AB + 5B^2) + \\ \qquad + x(5 - Z + 4A^2 + 10AB) - 5A^2 \end{array}\right\} \tag{218}$$

and then obtain :

$$dx/dt = \pm \frac{3Gm_3}{2n_S b_T^3} \{P_1(x) \cdot P_2(x)\}^{1/2} \tag{219}$$

A, B, Z, n_S and b_T being constant, the differential equation (219) is a classical quadrature and either x moves periodically back and forth between the two nearest roots of the product $P_1(x).P_2(x)$ or it tends asymptotically to one of these two roots, if the root of interest is a multiple root.

The general discussion leads to the following in terms of A,B,Z.

A) Let us call x_a and x_b the two nearest roots of the product $P_1(x)P_2(x)$, with :

$$x_a \leqq x \leqq x_b \tag{220}$$

(217) and (218) imply :

$$\{x_a < x < x_b\} \Longrightarrow \{P_1(x) > 0 \ ; \ P_2(x) > 0\} \tag{221}$$

On the other hand :

$$P_2(0) = -5A^2 \leqq 0 \ ; \ P_1(1) + P_2(1) = 0 \ ; \ (5 - 4Z)P_1(Z) + P_2(Z) = 0 \tag{222}$$

Hence, since $0 \leqq x \leqq 1$ and $x = 1 - e_S^2 \leqq Z$:

$$0 \leqq x_a \leqq x \leqq x_b \leqq \inf(1 \ ; \ Z) \tag{223}$$

B) If x_a and x_b are both simple roots of the product $P_1(x)P_2(x)$ the motion of x is periodic, it moves back and forth from x_a to x_b and there are the three following types of periodic motions.

B.1) If x_a and x_b are two roots of $P_2(x)$, the polynomial $P_1(x)$ cannot approach zero and $\sin g_S$ has a constant sign : the motion of the angle g_S is a periodic libration about either $+\pi/2$ or $-\pi/2$.

B.2) If x_a is a root of $P_2(x)$ and x_b a root of $P_1(x)$, the angle g_S is always increasing and has a periodic circulation.

B.3) If x_a is a root of $P_1(x)$ and x_b a root of $P_2(x)$ the angle g_S is always decreasing and has a periodic circulation.

This third case is rare ; it occurs if and only if :

$$2 - (A - B)^2 < Z < (1 + 2AB)/B^2 \quad \text{and} \quad (1 + AB)/B^2 < x < 1 \tag{224}$$

which require $A > 0$, $B > 1$, $Z < 2$ and $j_L > \pi/2$.

C) We will include in these periodic cases the limit cases for which $x_a = x_b$; the parameters x, e_S, j_L and g_S are then constant (the angles g_S and j_L are sometimes undefined).

D) If $x_a < x_b$ one of these two roots (but not both) can be a multiple root of the product $P_1(x)P_2(x)$. This leads to the following four asymptotical motions.

D.1) $x_b = 1 \leqq Z$; x_b is a root of both $P_1(x)$ and $P_2(x)$.

These case occurs if and only if :

$$Z = 2 - (A - B)^2 \ ; \ B^2 \leqq 1 + AB \ ; \ 3B^2 - 8AB + 5A^2 \leqq 3 \tag{225}$$

and leads to :

$$x \to 1 \; ; \; e_S \to 0 \; ; \; \cos j_L \to A - B \; ; \; \tan g_S \to - \left[\frac{2(1 + AB - B^2)}{3 - 5A^2 + 8AB - 3B^2} \right]^{1/2} \tag{226}$$

D.2) $x_b = Z \leq 1$; x_b is a root of both $P_1(x)$ and $P_2(x)$.

This case occurs if and only if :

$$BZ - \sqrt{Z} = A \leq 0 \; ; \; 5 - 3Z \leq B(10 - 8Z)\sqrt{Z} \tag{227}$$

and leads to :

$$x \to Z \; ; \; e_S \to \sqrt{(1 - Z)} \; ; \; j_L \to \pi \; ; \; \tan g_S \to - \left[\frac{2Z(1 - B\sqrt{Z})}{B\sqrt{Z}(10 - 8Z) - 5 + 3Z} \right]^{1/2} \tag{228}$$

Conversely the corresponding plane retrograde motions, i.e. the motions with $j_L = \pi$ and $(2 + 3e_S^2)/(1 + 4e_S^2) < \mathcal{G}_S/\mathcal{G}_T \leq 2$, do not have the "out of plane" stability.

D.3) x_a or $x_b = (1 + AB)/B^2 \leq \inf\{1;Z\}$; this root is a double root of $P_1(x)$.

This case occurs if and only if :

$$ZB^2 = 1 + 2AB \; ; \; A \geq 0 \; ; \; 1 + AB \leq B^2 \tag{229}$$

and leads to :

$$x \to (1 + AB)/B^2 \; ; \; e_S \to \sqrt{(B^2 - 1 - AB)/B} \; ; \; \tan j_L \to - \sqrt{AB} \; ; \; \sin g_S \to 0 \tag{230}$$

D.4) x_a or x_b is a double or triple root of the polynomial $P_2(x)$.

This fourth and last case is exceptional and complex. If we call x_0 the multiple root, we find the following relations.

$$5/9 \leq x_0 \leq 1$$

$$x_0 \leq Z \leq 5 + \left[12x_0\left(x_0 - \sqrt{(5x_0)}\right)/\left(4\sqrt{(5x_0)} - 5\right)\right] \tag{231}$$

(hence $5/9 \leq Z \leq (95 - 12\sqrt{5})/55 = 1.23940\ldots$)

The constants A and B can be given in terms of Z and x_0 :

$$\alpha = -A/Bx_0 = 1 - \frac{x_0(10 - 8Z)}{25 - 5Z - 30x_0 + 12x_0^2} \tag{232}$$

$$\beta = Bx_0 - A = \left[\frac{(5 - Z)x_0 - 3x_0^2}{5 - 4x_0} \right]^{1/2} \tag{233}$$

and then :

$$A = -\alpha\beta/(1+\alpha) \; ; \; B = \beta/x_0(1+\alpha) \tag{234}$$

(which implies : $-0.41175.. \leqq A \leqq 0$; $0.46036.. \leqq B \leqq 1$).

In this case the asymptotical motion leads to the following :

$$x \to x_0 \; ; \; e_S \to \sqrt{(1-x_0)} \; ; \; \cos j_L \to -\left[\frac{5 - Z - 3x_0}{5 - 4x_0}\right]^{1/2} \; ; \; \cos g_S \to 0 \tag{235}$$

This last type of asymptotical motions lies at the boundary of the periodic motions of type B1 and B2 and it is because of its complexity that no simple necessary and sufficient conditions have been given for the two first types of periodic motions.

However we can write the following simple rules :

1) If $Z > 2$ or even if the polynomial $P_1(x)$ has no real root on the close segment $0 \leqq x \leqq \inf(Z;1)$ the motion is always a periodic motion of type B1 with librations of g_S about either $+\pi/2$ or $-\pi/2$.

Note that $Z > 2$ implies :

$$e_S^2 \sin^2 j_L \sin^2 g_S \geqq (Z-2)/3 > 0 \; ; \; \sin^2 j_L \sin^2 g_S \geqq Z/5 > 0.4 \tag{236}$$

The three parameters e_S, $\sin j_L$, $\sin g_S$ cannot approach zero and the latitude ψ_i of the pericenter of the inner orbit with respect to the outer orbital plane remains very large ($|\sin\psi_i| = |\sin j \sin g_i| \simeq |\sin j_L \sin g_S| \geqq \sqrt{Z/5} > \sqrt{0.4}$).

2) The constant B is always non-negative and generally small, indeed from (213) and (203) :

$$B = \mathcal{L}_S/2\mathcal{G}_T \tag{237}$$

$$\mathcal{L}_S = m_1 m_2 \left[G a_S/(m_1+m_2)\right]^{1/2} \tag{238}$$

$$\mathcal{G}_T = m_3(m_1+m_2) \cdot \left[G p_T/M\right]^{1/2} \tag{239}$$

Hence :

$$B = \frac{m_1 m_2}{2m_3(m_1+m_2)} \cdot \left(\frac{M a_S}{(m_1+m_2)p_T}\right)^{1/2} \tag{240}$$

By hypothesis the distance ratio a_S/p_T is small. Hence, if m_3 is not too small, for instance is m_3 is not the smallest mass or even if m_3 is larger than the reduced mass m equal to $m_1 m_2/(m_1+m_2)$ we will have B small and will avoid the troublesome range of values of B presented after (234).

In this case the discussion is simple : $B \leq 0.46$ implies the following analysis :

$$\left.\begin{array}{ll} \{z > 2 - (A - B)^2\} \Longrightarrow & \text{periodic motion with librations of } g_S \text{ (case B.1).} \\ \{z = 2 - (A - B)^2\} \Longrightarrow & \text{either } e_S \text{ is identical to zero (case C)} \\ & \text{or } e_S \to 0 \text{ (asymptotical case D.1).} \\ \{z < 2 - (A - B)^2\} \Longrightarrow & \text{periodic motion with ever increasing } g_S \text{ (case B.2).} \end{array}\right\} \quad (241)$$

10.2.4 A concrete picture of the wide perturbations of the three-body problem

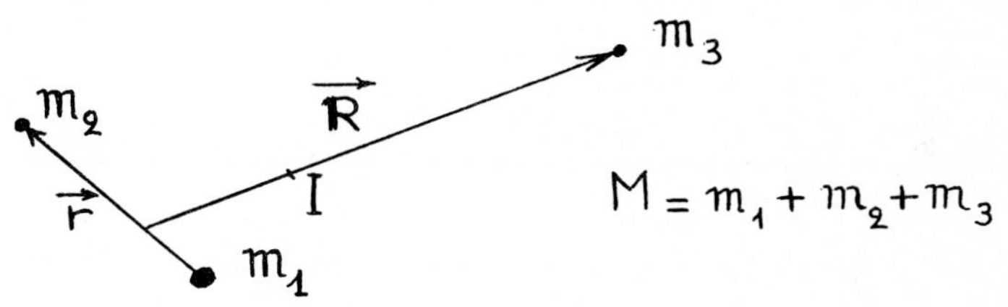

Fig. 36. The three-body problem and the Jacobi vectors $\vec{r}$ and $\vec{R}$.

The first-order study of the three-body problem when $(r/R)^3 \ll (m_1 + m_2)/M$ led in the previous Section to the following.

1) The inner orbit (orbit of $\vec{r}$) and the outer orbit (orbit of $\vec{R}$) are slowly perturbed Keplerian orbits with constant long-period semi-major axes a_S and a_T and constant long-period outer eccentricity e_T.

2) The following corresponding auxiliary long-period elements are also constant : n_S and n_T (mean angular motions), b_T (outer semi-minor axis), p_T (outer semi-latus rectum) and $\mathcal{G}_T$ (modulus of the outer angular momentum).

3) The inner inclination i_S (on the invariable plane), the outer inclination i_T and the mutual inclination j_L are related by the geometrical relations of Fig. 37 with constant c and $\mathcal{G}_T$.

4) The long-period mean anomalies ℓ_S and ℓ_T are ignorable, their derivatives are almost equal to n_S and n_T respectively, the derivative of ℓ_T is constant.

5) The angles g_T (long-period argument of pericenter of outer orbit), h_S and h_T (long-period longitude of nodes in the invariable plane) are also ignorable, their derivatives are small and $h_T = h_S + \pi$.

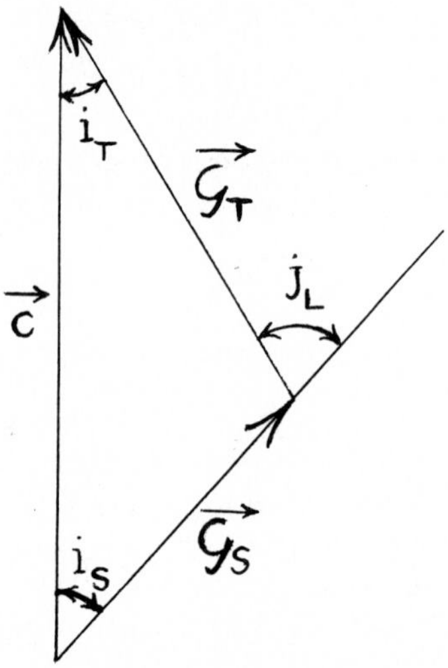

Fig. 37. Some long-period elements.

i_S = *inner inclination* ; $0 \leq i_S \leq \pi$
i_T = *outer inclination* ; $0 \leq i_T \leq \pi$
j_L = *mutual inclination* = $i_S + i_T \leq \pi$
$\vec{G}_S$ = *inner angular momentum.*
$\vec{G}_T$ = *outer angular momentum.*
$\vec{c} = \vec{G}_S + \vec{G}_T$ = *total angular momentum.*

$$c^2 = G_S^2 + G_T^2 + 2G_S G_T \cos j_L .$$

6) The remaining long-period elemnts e_S (inner eccentricity), j_L and g_S (inner argument of pericenter) are, to the first order, related by :

$$\left.\begin{aligned} de_S/dt &= K\, e_S . \sin^2 j_L . \sin 2g_S \sqrt{(1 - e_S^2)} \\ \text{where } K &= \text{constant} = 15Gm_3/8n_S b_T^3 \end{aligned}\right\} \quad (242)$$

and by the following integrals of motion :

$$Z = (1 - e_S^2)(1 + \sin^2 j_L) + 5e_S^2 \sin^2 j_L \sin^2 g_S \quad (243)$$

$$A = B(1 - e_S^2) + \cos j_L \sqrt{(1 - e_S^2)} \quad (244)$$

B being the constant defined in (240).

The remaining problem is integrable and the corresponding possible motions have been classified in the previous Section, however there are many types of motion and the results are complex.

Fortunately a very informative picture can be drawn in terms of the two following parameters only : the constant B and the angle ψ_S, i.e. the "long-period latitude of the pericenter of the inner orbit with respect to the outer orbital plane".

The difference between ψ_S and the corresponding osculating element ψ_i will remain of the first order (if the eccentricities e_i and e_S are not too small) and ψ_S will be given by :

$$\sin\psi_S = \sin j_L \sin g_S \ ; \ |\psi_S| \leqq 90° \tag{245}$$

As in (241) we will assume that $B \leqq 0.46$, which is very general since, by hypothesis, the ratio a_S/p_T is small in (240).

In these conditions the asymptotical motions of (241) will forever satisfy :

$$\psi_a \leqq |\psi_S| \leqq \psi_b \tag{246}$$

where :

$$\left.\begin{aligned} \cos\psi_a &= 0.2\left[B + \sqrt{(B^2 + 15)}\right] \\ & \qquad 29.3°\ldots \leqq \psi_a \leqq 39.2°\ldots \leqq \psi_b \leqq 46.5°\ldots \\ \cos\psi_b &= 0.2\left[-B + \sqrt{(B^2 + 15)}\right] \end{aligned}\right\} \tag{247}$$

For instance $B = 0.2$ implies $\psi_a = 35.35°\ldots$ and $\psi_b = 42.64°\ldots$ (Fig. 38).

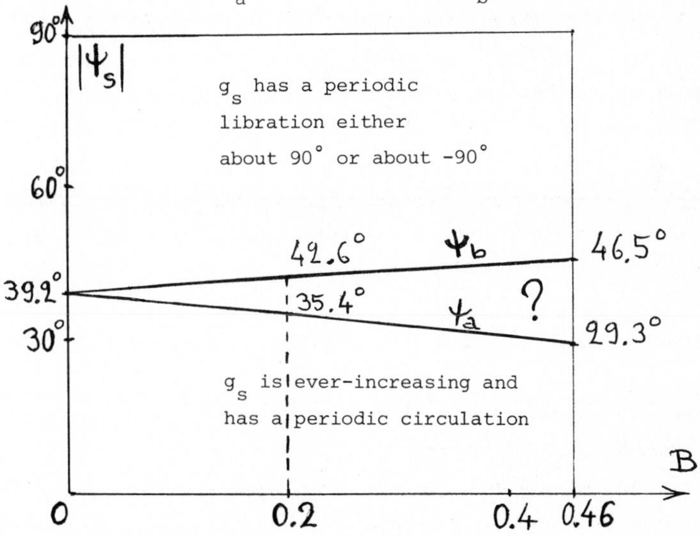

Fig. 38. Discussion of the type of motion in terms of B and the angle ψ_S (i.e. the "long-period latitude of the pericenter of the inner orbit with respect to the outer orbital plane"). $\sin\psi_S = \sin j_L . \sin g_S$ Motions of asymptotic type remain in the region of the question mark.

Hence, when $B \leq 0.46$:

1) If, at some time, $|\psi_S| < \psi_a$ the motion is of the circulation type with an ever-increasing g_S and with $|\psi_S| < \psi_b$ forever.

2) If, at some time, $|\psi_S| > \psi_b$ the motion is of the libration type with $|\psi_S| > \psi_a$ forever and with an inner pericenter always in the same hemisphere.

Notice some cases of very large perturbations. Assume for instance that initially $|\psi_S| < \psi_a$ and $j_L \sim 90°$; the motion will be of the circulation type and when g_S will reach $\pm 90°$ we will have :

$$\sin j_L = \sin|\psi_S| < \sin\psi_b \tag{248}$$

Hence j_L will either become smaller than ψ_b (when $A > 0$) or larger than $180° - \psi_b$ (when $A < 0$).

When $A = 0$ the eccentricity e_S increases to unity, there are many close approaches of the bodies m_1 and m_2 and sometimes a collision (see Section 11.7.7.6).

10.2.5 General considerations on the first-order integration

1) Let us remember that the condition of this first-order study was that $(r/R)^3$ be much smaller than $(m_1 + m_2)/M$.

On the other hand $n_S^2 a_S^3 = G(m_1 + m_2)$ and $n_T^2 a_T^3 = GM$, hence $(r/R)^3 \ll (m_1 + m_2)/M$ implies $(a_S/a_T)^3 \ll (m_1 + m_2)/M$ and $n_S \gg n_T$, i.e. the mean angular velocity of the small binary is much larger than that of the third body and its orbital period is much smaller.

2) The expressions (206)-(211),(219), (250) show that the perturbations are slow : the corresponding angular velocities are of the order of n_T^2/n_S.

3) When the evolution of x in terms of the time t is known (integration of the quadrature (219)), all the other parameters are easily obtained :
(215) gives :

$$e_S = (1 - x)^{1/2} \; ; \; \mathcal{G}_S = \mathcal{L}_S\sqrt{x} \tag{249}$$

j_L is given by (212) and g_S by (214) and (216), while ℓ_S, ℓ_T, g_T are given by the quadratures (208), (210), (211).

The osculating elements are given by the transformation (187)-(201) and the remaining ignorable parameters $\mathcal{H}_i$, $\mathcal{H}_e$, h_i, h_e are given by (180).

Hence the integration of the first-order three-body problem requires the integration of five quadratures, the quadrature (210) being very simple.

Notice that, whith the long-period longitudes of nodes h_S and h_T, the quadrature of (180) can be written as follows :

$$\left.\begin{aligned}\frac{dh_S}{dt} &= \frac{dh_T}{dt} = \frac{\partial H(\mathcal{L}_S,\mathcal{G}_S,\mathcal{L}_T,\mathcal{G}_T,\ell_S,g_S,\ell_T,g_T,c)}{\partial c} = \\ &= -\frac{c}{\mathcal{G}_S\mathcal{G}_T\sin j_L}\cdot\frac{\partial Q}{\partial j_L} \\ &= -\frac{3ca_Sn_T}{4n_Sb_S(m_1+m_2)p_T^2}\cos j_L(1-e_S^2+5e_S^2\sin^2 g_S)\end{aligned}\right\} \qquad (250)$$

where, as in (194)-(201) :

$$h_S - h_i = h_T - h_e = \frac{c}{\mathcal{G}_i\mathcal{G}_e\sin j}\frac{\partial s_1}{\partial j} \; ; \; h_T = h_S + \pi \qquad (251)$$

4) If the three-body motion of interest is not one of the exceptional asymptotical motions the first-order approximation leads to the following :

n_S, a_S, n_T, a_T, e_T, $d\ell_T/dt$ } are constant.

$d\ell_S/dt \sim n_S$ $\quad d\ell_T/dt \sim n_T$

e_S, g_S, j_L, i_S, i_T, $(i_S + i_T = j_L)$
$d\ell_S/dt$, dg_T/dt
dh_S/dt, dh_T/dt, $(h_T = h_S + \pi)$
} are periodic with a long period of the order of $2\pi M n_S(1-e_T^2)^{3/2}/m_3 n_T^2$

ℓ_S, g_T, h_S, h_T and all the osculating elements n_i, a_i, e_i, ..., g_e, h_e } are quasi-periodic.

5) If $e_S\sin j_L = 0$ the two elements e_S and j_L remain constant and the solution is simple. However when $e_S = 0$ and/or $j_L = \pi$ the motion can be unstable as already noticed with the two first types of asymptotical motions D1 and D2 (Section 10.1.3).

On the contrary all motions with $j_L = 0$ are stable and if the mutual inclination j_L is small the perturbations of e_S and j_L remain very small, the arguments of pericenter g_S and g_T are slowly increasing and the longitudes of nodes h_S and h_T are slowly decreasing... exactly as happens for the planetary orbits of the Solar System (with respect to the invariable plane and absolute axes).

For the lunar orbit we also find retrograde nodes (period 18.60 years) and a direct perigee (period 8.85 years).

For the other natural satellites we have to consider the effect of the equatorial bulge of their planet : the "inner satellites" are essentially under the influence of the equatorial bulge while the "outer satellites" are essentially governed by the three-body Sun-planet-satellite system, and it is the bulge of Uranus that stabilizes the orbits of its satellites in spite of their inclination of about 98°.

Eleven outer satellites are known : the Moon, the eight outer satellites of Jupiter (with periods between 200 and 800 days), Phoebe satellite of Saturn (period 550 days) and Nereid satellite of Neptune (period 360 days). Five of these eleven satellites have a retrograd orbit but all of them have a motion of the circulation type in spite of their generally large eccentricities and inclinations that imply large perturbations... the eccentricity of Nereid is 0.75 !

6) Finally, if the mutual inclination j_L is very large, in the vicinity of 90°, the perturbations of j_L and especially e_S become also very large even if they are slow and the three-body motion can be very unstable. It is perhaps for this reason that no natural outer satellite has an inclination between 30° and 140°.

We will see in Sections 10.4.2. and 11.7.7.6 cases leading to a collision of the two bodies m_1 and m_2 (increase of the inner eccentricity up to unity).

10.3 INTEGRATION OF THE THREE-BODY PROBLEM TO SECOND-ORDER

This question was considered by Brown (Ref. 151) and solved in the "lunar case", i.e. when $m_1 + m_2 << m_3$, by Kovalevsky (Ref. 152) who noticed the very large perturbations when the mutual inlination is in the vicinity of 90°.

In the previous Section we have neglected the error term ε in (202), the order of magnitude of which is given in (205). If we try to obtain the main terms of ε we can arrive at a simple result provided that we use a slightly different canonical transformation (the new function S_1 has the same main part but four times more terms). The new long-period elements will be then slightly different from those of the previous Section but their difference with the osculating elements will remain of first-order and will have the same main part.

We will continue to use the same notations for the new long-period elements and we will thus obtain the following improvement of the expression of the Hamiltonian.

$$H = -\frac{G^2 m^3 (m_1 + m_2)^2}{2\mathcal{L}_S^2} - \frac{G^2 \mathcal{M}^3 M^2}{2\mathcal{L}_T^2} + Q + Q_1 + Q_2 + \varepsilon_2 \quad (252)$$

(with $M = m_1+m_2+m_3$; $m = m_1m_2/(m_1+m_2)$; $\mathcal{M} = m_3(m_1+m_2)/M$).

The expression of Q is, as in (204) :

$$Q = \frac{Gmm_3a_S^2}{8b_T^3}\left[-2-3e_S^2+3\sin^2 j_L(1-e_S^2+5e_S^2\sin^2 g_S)\right] \quad (253)$$

The new terms are given by the following :

$$Q_1 = -\frac{3Gmm_3^2 n_T a_S b_S}{64Mn_S p_T^3}(3+2e_T^2)\cos j_L\left[25e_S^2+\sin^2 j_L(1-e_S^2-15e_S^2\sin^2 g_S)\right] \quad (254)$$

$$Q_2 = \frac{15Gmm_3(m_2-m_1)a_S^3 e_S e_T}{64(m_1+m_2)b_T^3 p_T}\cdot U$$

where :

$$U = (\sin g_S \sin g_T \cos j_L + \cos g_S \cos g_T)\left[4+3e_S^2-5\sin^2 j_L(1-e_S^2+7e_S^2\sin^2 g_S)\right] -$$

$$- 10(1-e_S^2)\sin^2 j_L \cos j_L \sin g_S \sin g_T \quad (255)$$

$$\varepsilon_2 = O\left\{\frac{Gmm_3a_S^4}{p_T^5}\cdot\sup\left(1 ; \frac{Ma_S}{(m_1+m_2)p_T}\right)\right\} \quad (256)$$

If Q_2 and ε_2 are considered negligible the problem can be integrated easily ; H is independent of ℓ_S, ℓ_T, g_T and there are four integrals of motion ($\mathcal{L}_S$, $\mathcal{L}_T$, $\mathcal{G}_T$ and $Q + Q_1$) leading to results qualitatively similar to those of the first order analysis (Section 10.1.).

If Q_2 is not negligible the Hamiltonian H becomes g_T dependent and $\mathcal{G}_T$ is no longer an integral of motion. The problem is no longer integrable and resonance phenomena appear, especially if in the first order analysis the increase of g_T is $2K\pi$ (K integer) during a period of e_s, j_L and g_S.

10.4 THE NUMERICAL METHODS

It is in the domain of numerical methods that the greatest progress have been made in the last twenty years and these progress seem to accelarate continuously. To the young people living with computers it seems incredible that in 1960 the computing techniques were those of the past centuries : the logarithm table and the sliderule.

In the early years of our marriage my wife gave me a beautiful and accurate sliderule that I still use when its accuracy is adequate (i.e. 0.5 %), but I feel like a dinosaur among my students who have never seen such a strange object...to-day an average analyst is able to do more computations in a single day than all of mankind in the past centuries !

Celestial Mechanics was famous for its arduous and lengthy computations requiring extreme accuracy, so it has of course received the greatest benefit from the new techniques at precisely the time when space exploration provides it with so many new and accurate data.

Whe have already seen in Section 9.4. the development given to the Copenhagen problem by the computations of Michel Hénon. It is sufficient to compare the wild behavior of the orbits of Fig. 31 or even 30 with the staid periodic orbits computed in the twenties and the thirties by the Copenhagen School (Fig. 52 and also Ref. 1 Section 9.4 pages 455-497) to understand the computer revolution. Computers have given us a new point of view, they have led to the discovery of many new phenomena (chaotic motions, strange attractors, etc...) that now appear in almost all domains of science and technology. They have emphasized the importance of that were once mathematical curiosities such as the fractals that are now noticed in many physical, biological and geological phenomena.

The fractals are those sets, such as the Cantor set, with non-integer number of dimensions. The analysis of the number of dimensions of natural fractals and strange attractors generally require numerous measurements and huge computations but will perhaps have in the future the importance of the Fourier analysis for periodic and quasi-periodic phenomena.

However the numerical methods have their limitations, their accuracy is not infinite and must be carefully checked. Integrations over too large time intervals lose their accuracy and even their significance, they must be comforted by other means of analysis.

The largest difficulty is perhaps the wide range of usual problems : it is generally impossible to look at random for solutions of a given type, and many adjacent conditions are useful and sometimes necessary.

Consider the periodic orbits of the three-body problem : no numerical integration, however accurate, can provide a periodic orbit and only a supplementary topological analysis can provide the certainty of the existence of a periodic orbit in a small neighbourhood of a given suitable computed orbit (see Section 11.4).

The number of numerical studies has now grown so far that it is impossible to have even a small idea of it. Nevertheless these studies can be classed in terms of their purposes :

1) Studies of classification (e.g. the numerous families of periodic orbits of various kinds, Ref.68-140).

2) Studies of verification (existence of this or that type of motion, confirmation or invalidation of a conjecture).

3) Studies of exploration such as those of Michel Hénon presented in Section 9.4. The study of gravitational scattering (Section 10.4.3.) has required the computation of 1.7 million close encounters between a binary and a single star ! Etc...

The studies of exploration, the method of "numerical experiment" (integration of a given arbitrary problem over very long intervals, just in order to see what happens) have so far been the most fruitful and have led to the surprises presented above : the great generality of chaotic motions, strange attractors, fractals. They have completely renewed the picture of many scientific domains and especially Celestial Mechanics.

We now present some examples of numerical studies.

10.4.1 <u>A three-body motion of the exchange type</u>

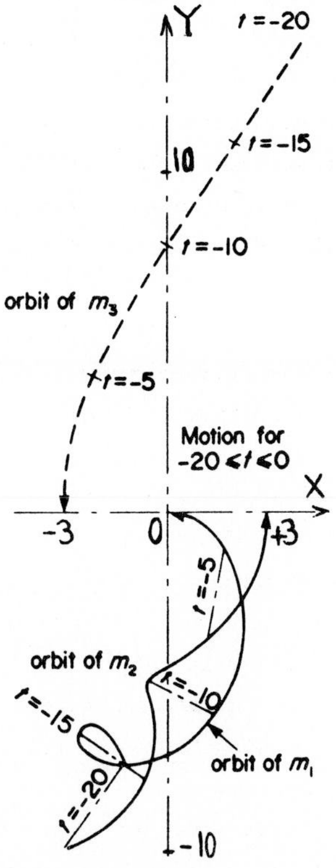

Fig. 39. The exchange motion for negative times.
Source : Szebehely, V. [153].

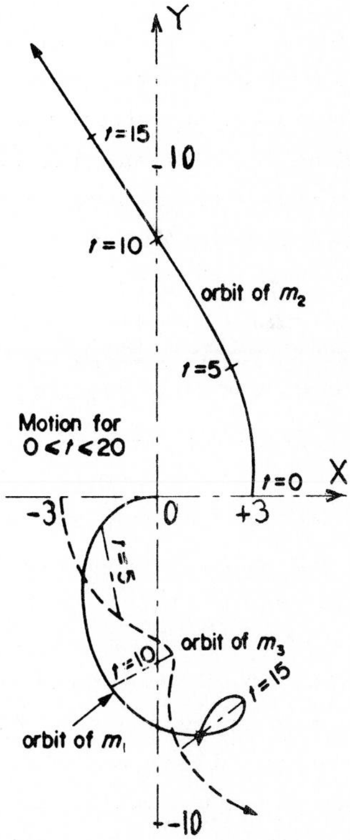

Fig. 40. The exchange motion for positive times.
Source : Szebehely, V. [153].

A symmetrical three-body solution of the exchange type has been computed by V. Szebehely (Fig. 39 and 40, and Ref. 153). The motion is planar and the conditions at the time t = 0 are the following :

$$\left.\begin{aligned} &G = 1 \;;\; m_1 = 1 \;;\; m_2 = m_3 = 2 \\ &X_1 = Y_1 = Y_2 = Y_3 = 0 \;;\; X_2 = -X_3 = 3 \\ &X'_1 = -28/123 \;;\; Y'_1 = 0 \\ &X'_2 = X'_3 = 7/123 \\ &Y'_2 = -Y'_3 = 122/123 \end{aligned}\right\} \quad (257)$$

The integrals of motion are :

$$\left.\begin{aligned} &h = \text{energy} = 0 \;;\; \vec{c} = \text{angular momentum} = (0,\ 0,\ 488/41) \\ &\text{the center of mass remains at the origin} \end{aligned}\right\} \quad (258)$$

The space-time symmetry is :

t	X_1	Y_1	X_2	Y_2	X_3	Y_3
$-t$	$-X_1$	Y_1	$-X_3$	Y_3	$-X_2$	Y_2

(259)

This symmetry allows us to compute only one half of the motion. After 30 units of time the motion has led to :

$X_1 = 5.89612$	$Y_1 = -10.57346$	$X'_1 = -0.28029$	$Y'_1 = 0.12661$	
$X_2 = -7.61014$	$Y_2 = 18.16021$	$X'_2 = -0.36598$	$Y'_2 = 0.47317$	(260)
$X_3 = 4.66208$	$Y_3 = -12.87348$	$X'_3 = 0.50612$	$Y'_3 = -0.53648$	

The masses m_1 and m_3 are close together and form a small binary (with : a = semi-major axis = 2.418 ; e = eccentricity = 0.221 ; ω_{13} = argument of pericenter = 119°), while m_2 is very far away and escapes to infinity (final velocity : $X'_{2\infty} = -0.314$; $Y'_{2\infty} = +0.387$).

The Figs. 39-40 and the space-time symmetry imply that initially the masses m_1 and m_2 were closed together in a small binary (a = 2.418 ; e = 0.221 ; ω_{12} = 61°) while m_3 was arriving from infinity $X'_{3,-\infty}$ = - 0.314 ; $Y'_{3,-\infty}$ = - 0.387). The whole motion is a beautiful exchange type motion, m_1 being with m_2 for negative times and with m_3 for positive times.

We will see in Section 11.9.5. the qualitative considerations that have led to the choice of conditions at t = 0.

10.4.2 An oscillatory motion of the second kind

The theoretical analysis of Sections 10.1.3. to 10.1.5. has shown that three-body systems with a large mutual inclination of inner to outer orbit have very large perturbations, especially on the inclination itself and on the inner eccentricity.

Let us apply this to a concrete example and consider the following initial conditions at t = 0.

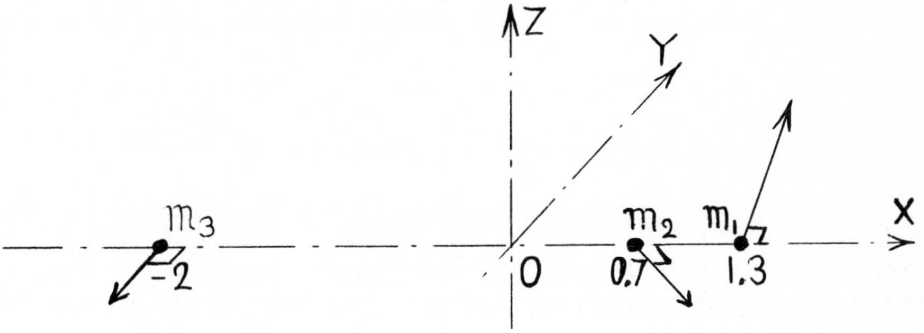

Fig. 41. The initial conditions.

G = 1 ; $m_1 = m_2 = m_3 = 1$

The three bodies are initially along the X-axis at the abscissas 1.3, 0.7 and - 2 respectively (Fig. 41).

The three initial velocities are normal to the X-axis and given by :

$$Y'_1 = 67/300 = 0.22333... ; Y'_2 = 133/300 = 0.44333... ; \qquad (261)$$

$$Y'_3 = - 2/3 = - 0.666...$$

$$Z'_1 = \{1.9637/3\}^{1/2} = 0.8090529... ; Z'_2 = - Z'_1 ; Z'_3 = 0$$

The integrals of motion are then :

$$h = -1.340067 \; ; \; \vec{c} = (0 \; ; \; -0.485432 \; ; \; 1.934) \; ; \; c = 1.993991 \tag{262}$$

The invariable plane is normal to $\vec{c}$ and we will choose Ox as the reference direction in that plane. It leads to the following osculating orbital elements at time t = 0.

1) Inner orbit :

$$\left.\begin{array}{l} a_i = 0.5 \; ; \; e_i = 0.2 \; ; \; i_i = 83.6524^\circ \; ; \; \ell_i = 180^\circ \; ; \; g_i = 0^\circ \; ; \; h_i = 0^\circ \\ \text{hence } n_i = 4 \; ; \; T_i = \text{period} = \pi/2 \; ; \; L_i = 0.5 \; ; \; G_i = 0.489898 \end{array}\right\} \tag{263}$$

2) Outer orbit :

$$\left.\begin{array}{l} a_e = 3 \; ; \; e_e = 0 \; ; \; i_e = 14.0901^\circ \; ; \; \ell_e + g_e = 0^\circ \; ; \; h_e = 180^\circ \\ \text{hence } n_e = 1/3 \; ; \; T_e = \text{period} = 6\pi \; ; \; L_e = G_e = 2 \end{array}\right\} \tag{264}$$

3) Mutual inclination = $j = i_i + i_e = 97.7425^\circ$ (265)

The motion has been integrated by Hadjidemetriou (Ref. 154) until t = 170 and the symmetry of the initial conditions allows us to extend the results to the interval $-170 \leq t \leq 0$.

The theoretical analysis gives very small variations for the parameters a_i, a_e, e_e (the corresponding long-period elements are constant) and indeed the numerical integration gives when $|t| \leq 170$:

$$0.494 < a_i < 0.501 \; ; \; 2.94 < a_e < 3.04 \; ; \; 0 \leq e_e < 0.02.$$

On the contrary the variations of the inner eccentricity e_i are very large (Fig. 42) ; e_i is initially 0.2 and goes beyond 0.999 during all of the interval $160 \leq t \leq 162$ after about 102 revolutions of the binary and 8.5 revolutions of the outer body.

When e_i is close to one the binary has very close approaches.

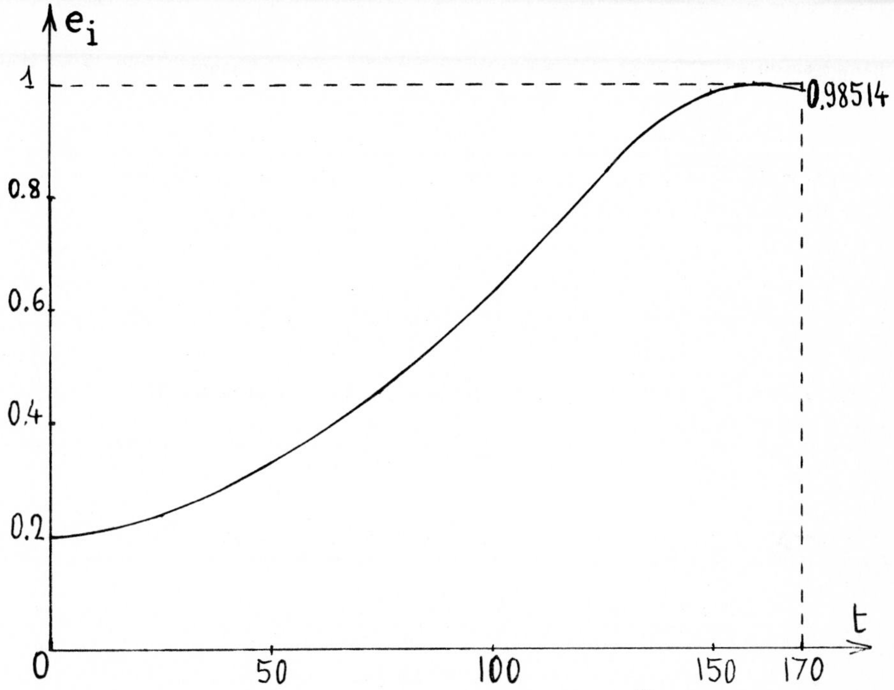

Fig. 42. Variations of the inner eccentricity e_i for $0 \leq t \leq 170$.
e_i is an even function of the time t.

The closest for $0 \leqq t \leqq 170$ occurs at t = 161.380383 after 103.5 revolutions of the binary, and leads to the following :

$$\left.\begin{array}{l} \text{Mutual distance of the binary : } r_{12} = 0.000346 \\ \text{Relative velocity of } m_1 \text{ and } m_2 \text{ : } v = 107.47 \\ \text{Inner orbit :} \\ a_i = 0.496449 \ ; \ e_i = 0.999303 \ ; \ i_i = 112.583° \ ; \ \ell_i = 0° \ ; \\ g_i = 137.866° \ ; \ h_i = 168.111° \\ \text{Outer orbit :} \\ a_e = 3.003784 \ ; \ e_e = 0.007220 \ ; \ i_e = 0.492° \ ; \ \ell_e = 334.924° \ ; \\ g_e = 76.656° \ ; \ h_e = 348.111° \\ \text{Mutual inclination } j = i_i + i_e = 113.075° \end{array}\right\} \quad (266)$$

When $|t| \leqq 170$ the mutual inclination remains in the interval (90° ; 142°), most of the variation occurs when e_i is close to one.

The long period of the three-body system of interest is the average interval of time between two successive maxima of e_i, which is about twice 161 i.e. 322 units of time.

Every maximum of e_i is very close to one and the binary undergoes an infinite number of very close approaches. It seems that this motion is of type "oscillatory motion of the second kind" with lim.inf $r_{12} = 0$ even if there is no strict collision between the point-masses m_1 and m_2 (see Section 11.7.7.6.).

10.4.3. Studies of gravitational scattering

The "gravitational scattering" is the following phenomenon.

If a binary star meets a single star it is possible that the mutual attractions of this temporary triple system lead to the disruption of the binary : all three stars escape individually to infinity (Fig. 43.A.).

By reference to quantum mechanics this first type of motion is called "ionization", and is the reverse motion of the "partial capture" (three single stars meet, one of them escapes and the other two remain in a binary).

Another common possibility is the "exchange motion" as presented in Section 10.4.1. and in Figs. 39, 40 and 43.B : after the encounter, a star of the initial binary escapes to infinity while the two other stars remain together and form a new binary.

The most common type of motion is of course the "fly-by" : after more or less perturbation the binary remains undisrupted and the incoming star escapes back to infinity, either because its passage was too far away or because it did not happen at the suitable place or time.

Gravitational scattering has been widely investigated (Ref. 156-171) and its study by Heggie, Hut and Bahcall is an excellent example of a full analysis : the qualitative analysis and the analytical approximations (Ref. 156-162) are complemented by a great number of numerical integrations (Ref. 163-171).

The qualitative analysis leads to the three above main types of motion : ionization, exchange and fly-by. These three types are not the only ones but, if we consider point-masses, all the other possible motions (collision, complete capture, oscillatory motions) have a zero probability and correspond to a set of measure zero in phase-space.

The qualitative analysis also leads to the notion of "surface of section". Indeed when the orbit of the incoming star remains too far away the motion is always of the fly-by type and the most interesting information about the importance of the other two main types of motion is the measure of the corresponding "surface of section". The analogy with quantum mechanics still goes on and, when choosing at random at infinity the initial position and velocity of the incoming star, the probabilities of "ionization" or "exchange" are directly proportional to the corresponding surfaces of section.

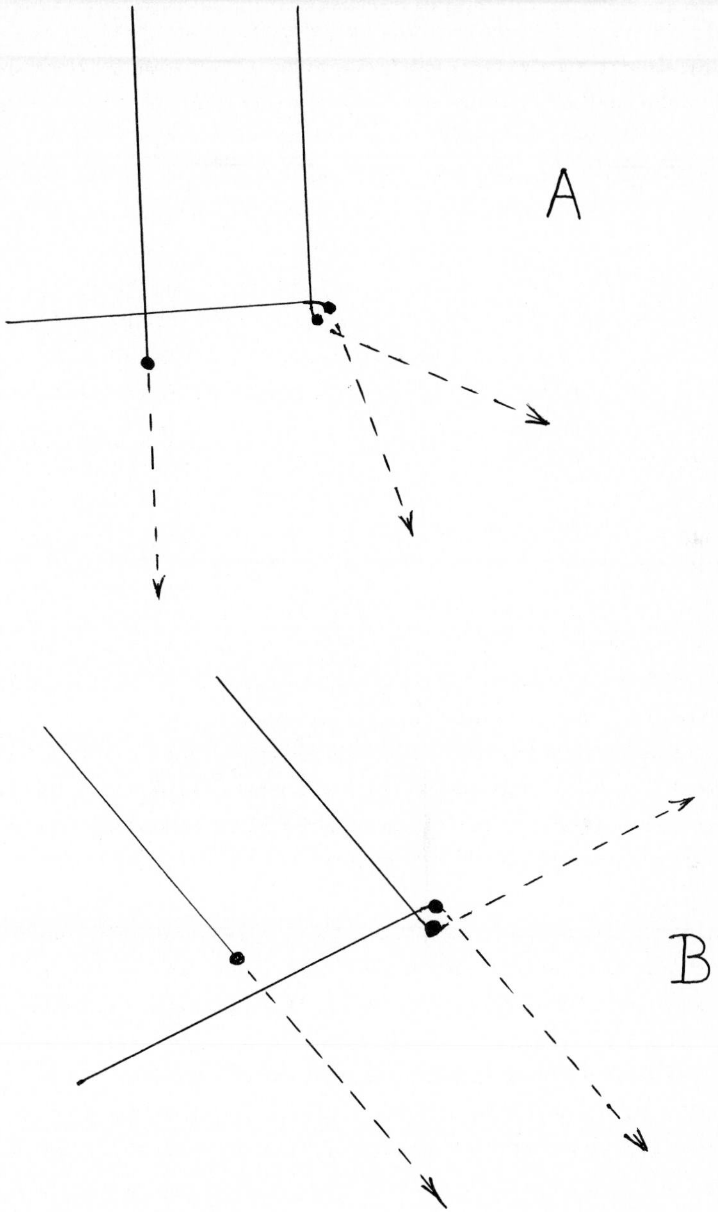

Fig. 43. Gravitational scattering for very large velocities.

A. Ionization.

B. Exchange of two equal masses.

The analytical approximations given in the Ref. 156 -160 deal especially with the case of very large incoming velocities of the third star.

In this case a substantial modification of the motion requires a very close approach between the incoming star and one star of the binary (Fig. 43), the analysis is not very different from that of the collision of billiard balls.

Let us consider the initial conditions at time minus infinity (Fig. 44).

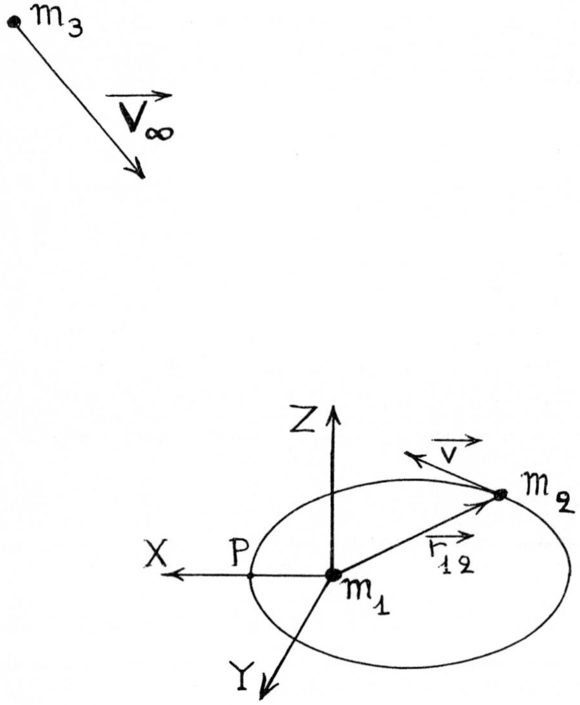

Fig. 44. The initial conditions at the time minus infinity.

The relative motion of m_2 with respect to m_1 is a Keplerian motion :

$$\left.\begin{array}{l} \text{semi-major axis} = a \\ \text{mean angular motion} = n = \left[G(m_1 + m_2)a^{-3}\right]^{1/2} \\ \text{eccentricity} = e \\ \text{semi-minor axis} = b = a\sqrt{(1 - e^2)} \\ \text{vector } m_1 \text{ to } m_2 = \vec{r}_{12} \\ \text{relative velocity} = \vec{v} = d\vec{r}_{12}/dt \end{array}\right\} \qquad (267)$$

The orbital three-hedron is $\overrightarrow{m_1,X,Y,Z}$:

$$\left.\begin{array}{l} \overrightarrow{m_1X} \text{ toward the pericenter P} \\ \overrightarrow{m_1Y} \text{ "normal-in-plane"} \\ \overrightarrow{m_1Z} \text{ "normal-out-of-plane"} \end{array}\right\} \qquad (268)$$

and finally :

$$\left.\begin{array}{l} \vec{V}_\infty = \text{velocity of arrival of } m_3 \\ \vec{V}_\infty/V_\infty = \vec{u} = \text{unit-vector of } \vec{V}_\infty \\ \vec{u} = (\alpha,\beta,\gamma) : \text{components of } \vec{u} \text{ in } \overrightarrow{m_1,X,Y,Z} \end{array}\right\} \qquad (269)$$

Notice that $\vec{V}_\infty$ is the velocity of m_3 with respect to the center of mass of the binary and for large velocities V_∞ the surface of section of ionization is divided into two parts : a part for close encouters with m_1 and a part for close encounters with m_2.

These two parts have about the same measure and the total surface of section is :

$$\Sigma_{\text{ionization}} \sim \frac{8\pi a^2 m_3^2}{(m_1 + m_2)^2 V_\infty^2} \left[\frac{2G(m_1 + m_2)}{r_{12}} - (\vec{v}\,\vec{u})^2\right] \qquad (270)$$

where :

$$\left.\begin{array}{l} G = \text{constant of the law of universal attraction} \\ r_{12}, \vec{v} : \text{Keplerian distance and velocity at the time} \\ \qquad \text{of the encounter.} \end{array}\right\} \qquad (271)$$

Surprisingly the binary is more brittle at pericenter than at apocenter.

The average value of Σ for all positions of m_1 and m_2 along their Keplerian orbits is the following :

$$\Sigma_{\text{ionization}} \text{ (time-averaged)} \sim \frac{8\pi G a m_3^2}{(m_1 + m_2)V_\infty^2} \left[2 - \frac{a\alpha^2 + b\beta^2}{a + b}\right] \qquad (272)$$

The bracket expresses the influence of the orientation and the eccentricity ; this influence is small and the bracket value is always between 1 and 2.

If we averaged over all orientations the result is independant of the eccentricity :

$$\Sigma_{\text{ionization (time and orientation-averaged)}} \sim \frac{40\pi G a m_3^2}{3(m_1 + m_2)V_\infty^2} \qquad (273)$$

These ionizations correspond to close encounters at distances of the order of $2am_3v/(m_1+m_2)V_\infty$ or less.

At large velocities V_∞ the exchanges between two masses are even more difficult than the ionizations. They can always be obtained by a suitable succession of close encounters but of course a large number of close encounters implies an extremly small surface of section.

For these very large velocities the analogy with collisions of billiards balls is very useful, a binary being two balls with zero relative velocity.

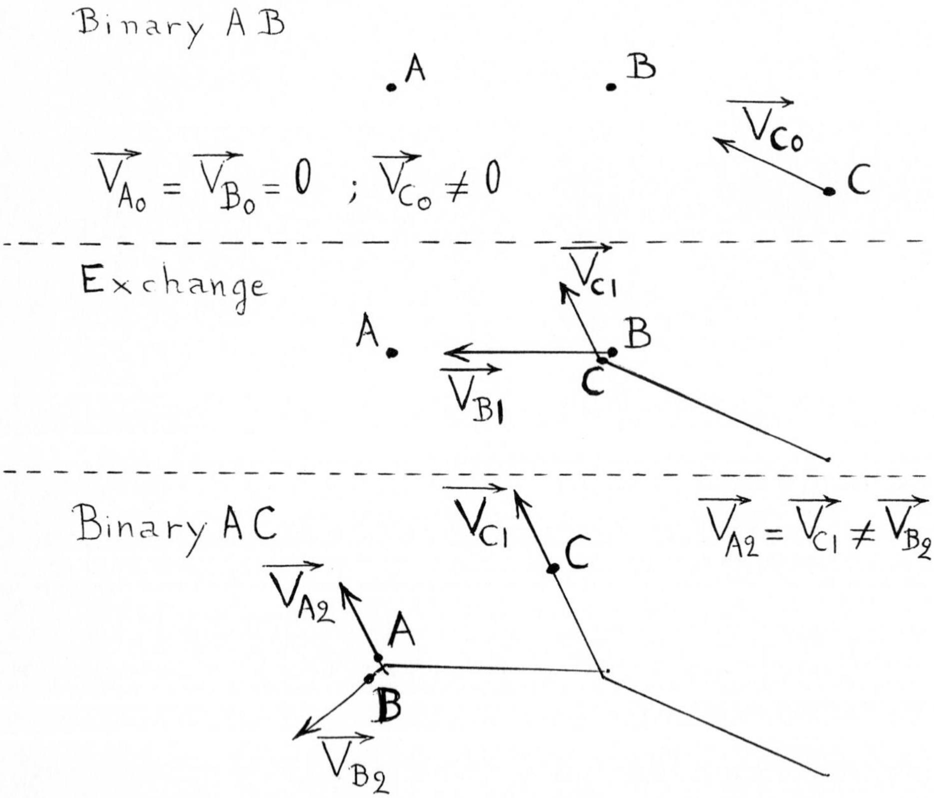

Fig. 45. Exchange between the bodies B and C with two successive close encounters.
At very large velocities the close encounters are "quasi-collisions" and the motion looks like collisions of billiards balls.

Fig. 45 shows an exchange between the bodies B and C with two sussessive "quasi-collisions" B,C and A,B. This exchange requires the two following inequalities between the masses m_A, m_B, m_C :

$$\left.\begin{aligned} & m_C \geqq m_B \\ & \frac{m_A}{m_B} \leqq 3 + \frac{m_A + m_B}{m_C} \end{aligned}\right\} \qquad (274)$$

These inequalities are given by the "billiards balls relations" of the successive velocities of the three bodies :

$$\left.\begin{aligned} & \vec{V}_{AO} = \vec{V}_{BO} = 0 \; ; \; \vec{V}_{CO} \neq 0 \\ & m_C\vec{V}_{CO} = m_B\vec{V}_{B1} + m_C\vec{V}_{C1} \\ & m_C V^2_{CO} = m_B V^2_{B1} + m_C V^2_{C1} \\ & m_B\vec{V}_{B1} = m_A\vec{V}_{A2} + m_B\vec{V}_{B2} \\ & m_B V^2_{B1} = m_A V^2_{A2} + m_B V^2_{B2} \\ & \vec{V}_{A2} = \vec{V}_{C1} \neq \vec{V}_{B2} \end{aligned}\right\} \qquad (275)$$

hence :

$$m_B m_C V^2_{B1} + m^2_C V^2_{C1} = m^2_C V^2_{CO} = (m_B\vec{V}_{B1} + m_C\vec{V}_{C1})^2 \qquad (276)$$

that is :

$$(m_C - m_B)V^2_{B1} = 2m_C\vec{V}_{B1}\vec{V}_{C1} \qquad (277)$$

and similarly :

$$(m_A + m_B)V^2_{A2} = 2m_B.\vec{V}_{A2}\vec{V}_{B1} \qquad (278)$$

The final condition $\vec{V}_{A2} = \vec{V}_{C1}$ then implies :

$$0 \leqq \frac{m_A + m_B}{2m_B} V^2_{A2} = \vec{V}_{A2}\vec{V}_{B1} = \frac{m_C - m_B}{2m_C} V^2_{B1} \qquad (279)$$

and then, indeed, $m_C \geqq m_B$ and also the second inequality of (274) since $|\vec{V}_{A2}| \cdot |\vec{V}_{B1}| \geqq \vec{V}_{A2} \cdot \vec{V}_{B1}$.

If the inequalities (274) are not satisfied it is impossible to use only the two successive close encounters B,C and A,B for an exchange of B and C. Fortunately it is always possible to obtain this exchange with a suitable succession of a sufficient number of close encounters.

For three given masses m_1, m_2, m_3 and motions with very large velocities the minimum number of close encounters that are necessary for an exchange between the masses m_2 and m_3 is a simple function of the mass ratios.

Let us put :

$$\left.\begin{aligned}
&m_i = \inf\{m_2 ; m_3\} \\
&m_S = \sup\{m_2 ; m_3\} \\
&M = m_1 + m_2 + m_3 \\
&\tan f = \{Mm_1/m_i m_S\}^{1/2} \; ; \; 0 \leqq f \leqq \pi/2 \\
&\tan g = \{Mm_i/m_1 m_S\}^{1/2} \; ; \; \frac{\pi}{2} - f \leqq g \leqq \frac{\pi - f}{2}
\end{aligned}\right\} \tag{280}$$

Let n be the minimum number of close encouters that are necessary for an exchange between m_2 and m_3. The generalization of the discussion of the above equations (274)-(279) leads to the following rules.

$$\left.\begin{aligned}
&g = \frac{\pi - f}{2} \text{ (that is } m_2 = m_3\text{) implies } n = 1 \\
&\frac{\pi - f}{2} > g \geqq \frac{f}{2} \quad \text{implies } n = 2 \\
&\frac{f}{2} > g \geqq \frac{\pi - f}{4} \quad \text{implies } n = 3 \\
&\cdots\cdots\cdots\cdots\cdots\cdots\cdots\cdots\cdots\cdots \\
&\frac{\pi - f}{2K} > g \geqq \frac{f}{2K} \quad \text{implies } n = 2K \\
&\frac{f}{2K} > g \geqq \frac{\pi - f}{2K + 2} \quad \text{implies } n = 2K + 1
\end{aligned}\right\} \tag{281}$$

The condition $g \geqq \frac{f}{2}$ is equivalent to :

$$\frac{m_1}{m_i} \leqq 2 + \frac{M}{m_S} \tag{282}$$

that is to the conditions (274) and their reverse (corresponding to the successive close encounters A,C and B,C).

The condition $g \geqq (\pi - f)/4$ is equivalent to :

$$m_1 m_s^2 + M m_i^2 \leqq 3\, m_i m_s (M + m_1) \tag{283}$$

Hence for $n \leqq 3$ it is sufficient that $m_i \geq \inf\left[\frac{m_1}{3}; \frac{m_s}{5.8}; \frac{M}{18.7}\right]$

The discussion (281) is given graphically in Fig. 46.

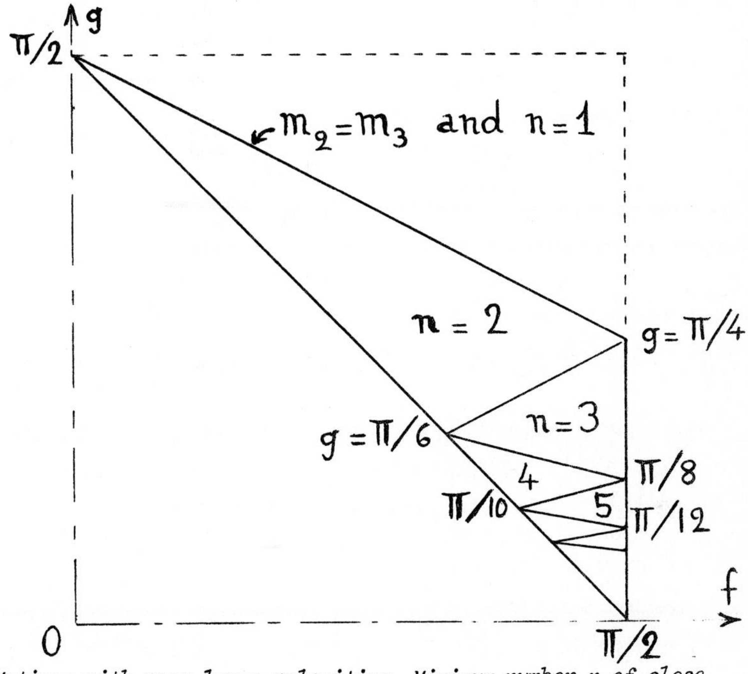

Fig. 46. Motions with very large velocities. Minimum number n of close encounters that are necessary for an exchange of the masses m_2 and m_3.

$m_i = \inf(m_2\, ;\, m_3)\ ;\ m_s = \sup(m_2\, ;\, m_3)$

$M = m_1 + m_2 + m_3$

$\tan f = \{Mm_1/m_i m_s\}^{1/2}$; $0 \leqq f \leqq \pi/2$

$\tan g = \{Mm_i/m_1 m_s\}^{1/2}$; $(\pi/2) - f \leqq g \leqq (\pi - f)/2$

If $n \geqslant 2$ the limit cases have alternately (m_i, m_s) and (m_i, m_1) close encounters with an almost collinear motion.

Exception to the rules presented in (281): if $m_1 = 0 < m_i < m_s$ the angles f and g are 0 and π/2 however n is 2 and not 1.

The "surfaces of section" of these exchange motions have been analysed only for n = 1, that is $m_2 = m_3$, in the Ref. 168 . This case is of course the simplest but it is also the most important and in the other cases the "surfaces of section" are much smaller.

For large velocities V_∞ and with the notations (267)-(271) we obtain :

$$\Sigma_{\text{exchange of equal masses } m_2 \text{ and } m_3} \sim \frac{4\pi G^2 m_3^2}{V_\infty^6}\left[\frac{2G(m_1 + m_2)}{r_{12}} - (\vec{v}\,\vec{u})^2\right] \qquad (284)$$

Since V_∞ is large this surface of section Σ_{exchange} is much smaller than $\Sigma_{\text{ionization}}$ given in (270).

The average values of Σ_{exchange} are proportional to these of $\Sigma_{\text{ionization}}$ given in (272) and (273) :

$$\Sigma_{\text{exchange (time-averaged)}} \sim \frac{4\pi G^3 m_3^2 (m_1 + m_2)}{V_\infty^6\, a}\left[2 - \frac{a\alpha^2 + b\beta^2}{a + b}\right] \qquad (285)$$

$$\Sigma_{\text{exchange (time and orientation-averaged)}} \sim \frac{20\pi G^3 m_3^2 (m_1 + m_2)}{3V_\infty^6\, a} \qquad (286)$$

These expressions of Σ_{exchange} are of course also valid for the exchange of m_1 and m_3 if $m_1 = m_3$. Furthermore if $m_1 = m_2 = m_3$ these two expressions must be added if we need the total surface of section for exchange.

As recognized in the studies of analytical approximations (Ref. 157-162, 171) the cases of slow incoming velocities of the third star are much more difficult.

The ionizations require a non-negative energy integral and they thus disappear for negative energies but on the contrary the exchanges benefit from the large "focusing effect" related to slow incoming velocities and the surface of section Σ_{exchange} increases as V_∞^{-2} when V_∞ goes to zero.

Numerical analyses have taken the succession of the powerless qualitative and analytical studies (Ref. 163-171). Some spectacular examples show the difficulty of the question, the "resonances" that appear for slow incoming velocities, the strong perturbations, the great number of loops and large elliptic ejections and returns until the final very close encounter, followed by the definitive escape of one of the three stars (Ref. 168 , Figs. 3 and 4 ; see also in this book the figures 96-102 in pages 285-290).

The major defect of numerical analyses is their particularity in providing excellent informations on the solution of interest and no information at all on the other solutions. This defect can be partially compensated by statistical methods of analysis such as the Monte-Carlo method : some initial parameters are defined (e. g. the semi-major axis and the eccentricity of the binary, the modulus of the incoming velocity $\vec{V}_\infty$, etc ...) some other parameters are chosen at random (e. g. the direction of $\vec{V}_\infty$, the position of the components of the binary along their orbit, etc ...) and the information is concentrated

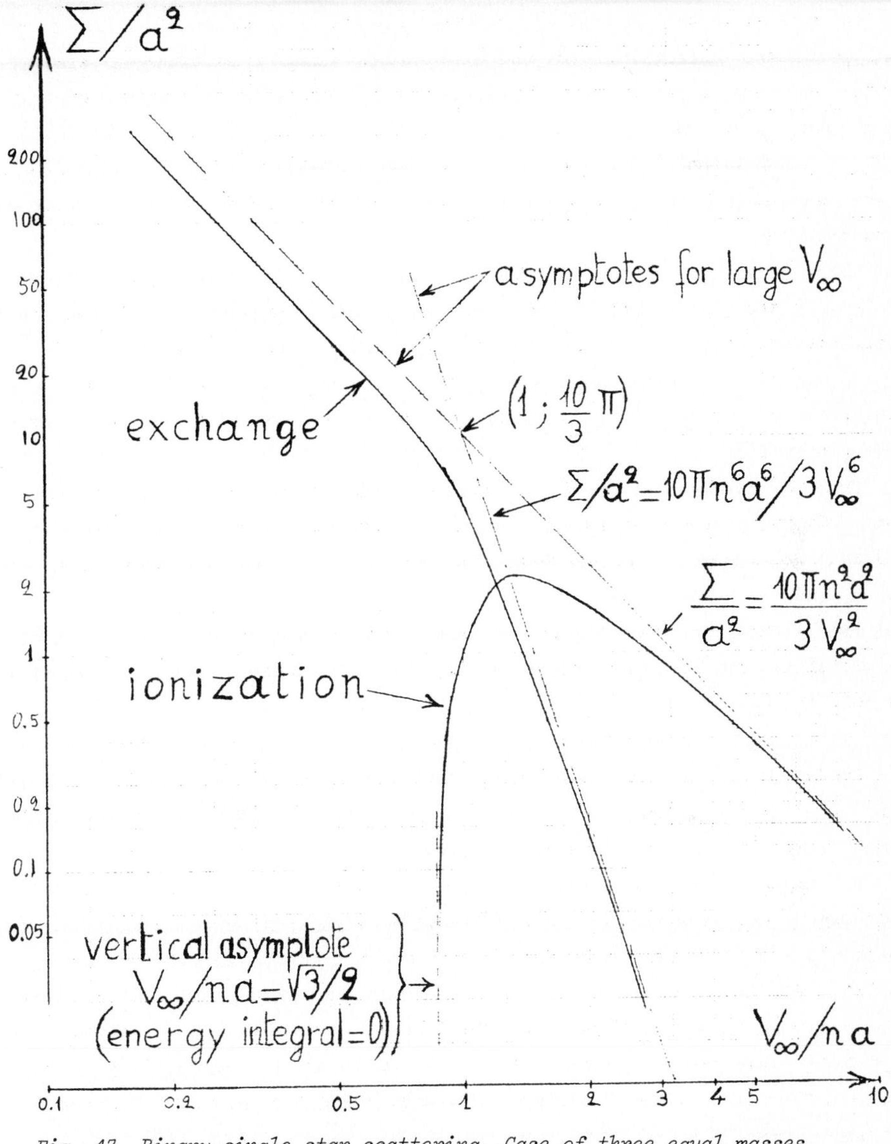

Fig. 47. Binary-single star scattering. Case of three equal masses.

The figure presents the time and direction-averaged values of the surfaces of section Σ *of ionization and exchange in terms of the modolus* V_∞ *of the incoming velocity (i. e. the velocity at infinity of the third star with respect to the center of mass of the binary).*

The mean angular motion of the binary is n and a is its semi-major axis (hence na is the root mean square of its relative velocity and $n^2a^3 = G(m_1 + m_2)$*.*

The accuracy of the figure is about 10% and to this accuracy the results seem insensitive to the eccentricity of the binary.

Source : Hut, P. and Bahcall, J.N. [168].

on some simple parameters (e. g. the average surface of section...).

Of course these statistical methods only give approximate informations but at least the uncertainty of the results is known from ordinary statistical rules and this uncertainties decreases as the number of numerical experiments increases.

In the reference 168 about 1.7 million numerical orbit calculations have been carried out for binary-single star scattering in the case of three equal masses.

Three particular values of the eccentricity of the binary were considered : 0 ; 0.7 and 0.99, and the corresponding averaged surfaces of section are presented in Fig. 47 for both ionization and exchange in terms of the ratio V_∞/na of the incoming velocity V_∞ to the root mean square na of relative velocity of the binary (n = mean angular motion and a = semi-major axis of the binary).

The two oblique asymptotes correspond of course to the relations (273) and (286) with a factor two for the second relation since we consider the total surface of section of exchange motions and not only the exchanges of m_1 and m_3 or m_2 and m_3.

The accuracy of the figure is about 10% (except in the vincinity of the asymptotes) and to this accuracy the three eccentricities considered give almost the same results.

These studies of gravitational scattering have of course great interest for a better understanding of the origin of binaries, their long-term evolution and the equilibrium between single stars, binaries and multiple systems.

The small fly-by effects have also been investigated in the Ref. 168-169 and it seems that, on average, "soft binaries become softer and hard binaries become harder".

It would be very useful to analyse the encounter of three single stars and to look for the probability of formation of new binaries. It seems that in ordinary galaxies this probability is smaller than the probability of gravitational scattering of binaries, at least when the average star density is not too large.

Another very interesting and useful investigation would be the study of gravitational scattering for unequal masses. There are certainly many mass effects and the asymptotic expression (273) seems to show that heavy binaries are more stable.

We will see in Section 11.7.7.6 an application of gravitational scattering and of the notion of surface of section : the number of indirect collisions of stars (through the formation of unstable triple systems) is about 10 to 100 times greater than the direct collision of ordinary single stars.

10.5 PERIODIC ORBITS AND NUMERICAL METHODS

We will see in Chapter 11 the qualitative aspects of the study of periodic solutions and the importance of their analysis. They have been described as "our best tool for the understanding of the three-body problem" (Henri Poincaré, ref. 13, p. 82.

The great interest of periodic solutions is that they are known for all time as soon as they are known for one period, and we will see that they are the backbone of the set of solutions. Their study by accurate modern numerical methods is easy, which explains why they are so popular. This can be seen from Fig. 48 showing the contents of the monthly "Celestial Mechanics" for March 1985.

CELESTIAL MECHANICS / *Vol. 35 No. 3 March 1985*

JAGADISH SINGH and BHOLA ISHWAR / Effect of Perturbations in the Stability of Triangular Points. In the Restricted Problem of Three Bodies with Variable Mass 201

S. FERRAZ-MELLO / Resonance in Regular Variables. I: Morphogenetic Analysis of the Orbits in the Case of a First-Order Resonance 209

S. FERRAZ-MELLO / Resonance in Regular Variables. II: Formal Solutions for Central and Non-Central First-Order Resonance 221

GERARD GÓMEZ and MIQUEL NOGUERA / Some Manifolds of Periodic Orbits in the Restricted Three-Body Problem 235

C. ZAGOURAS and V. V. MARKELLOS / Three-Dimensional Periodic Solutions Around Equilibrium Points in Hill's Problem 257

ANDREA MILANI and ANNA M. NOBILI / Resonant Structure of the Outer Solar System 269

JOHNNY H. KWOK and PAUL E. NACOZY / Periodic Orbits of the General Three-Body Problem for the Sun-Jupiter-Saturn System 289

Fig. 48. The contents list of the monthly "Celestial Mechanics" for an average month of 1985.

Three of the seven papers deal with periodic orbit.

It was traditional to discriminate between "absolute periodic orbits" for which the three bodies come back to their initial positions and velocities after a period, and "relative periodic orbits" : for which the three bodies come back to the same relative positions and velocities after a period but the absolute positions and velocities differ by a rotation about the angular momentum direction.

This subtle and artificial distinction is generally neglected to-day, and many numerical methods give only the relative positions and velocities. The essential point is that from a single period of the solution one knows the entire solution and, with the exception of their definition, no particular property distinguishes between absolute and relative periodic orbits. It is sometimes said that the relative periodic orbits are periodic in a suitable uniformly rotating set of axes.

We have already met some numerically computed orbits : in the Hill case, some orbits of Strömgren families f, g, g' (Figs 18-20, Section 8.5), in the circular restricted case the three-dimensional Halo orbits of Fig. 24 (Section 9.1) and in the Copenhagen case the periodic orbits of Figs 27-30 (Section 9.4). However, these symmetrical and very simple periodic orbits are misleading and many periodic orbits are without symmetry and/or extremely complex (especially in the general three-body case).

It is of course impossible to list all publications on periodic orbits since the early works of Darwin, Moulton, Strömgren (see for instance the references 68 - 140 and the surveys 131 - 132), but at least we can attempt a classification of the periodic orbits and their usual properties.

1) We have already just seen the difference between absolute and relative periodic orbits.

2) An essential property is the stability.

Generally, only the first-order stability is considered (see Section 10.7). Very rarely a complete analysis of the stability of the periodic solution of interest is done as presented in Section 8.8.2 for the Lagrangian solutions of the plane circular restricted three-body problem.

3) For more than half a century hardly any non-symmetrical solutions were computed. This simplifies the initial and final conditions that must satisfy one of the Poincaré space-time symmetry conditions (Figs 49 and 50), and it allows the computation of only a half-period. However, the asymmetrical periodic orbits about the Lagrangian equilibrium points L_4 and L_5 have been known from very early days.

It seems that the Ref.172 be the first dealing with arbitrary asymmetrical periodic orbits.

These asymmetrical periodic orbits are very common to-day.

4) We can also classify the periodic orbits according to the problem of interest (Fig. 51) :

4.1) The Schubart orbits are obtained in the rectilinear three-body problem (Ref.135).

4.2) Most of the earliest known periodic orbits are computed in the plane circular restricted three-body problem.

4.3) The Halo orbits about the Lagrangian equilibrium points are computed in the three-dimensional circular restricted three-body problem.

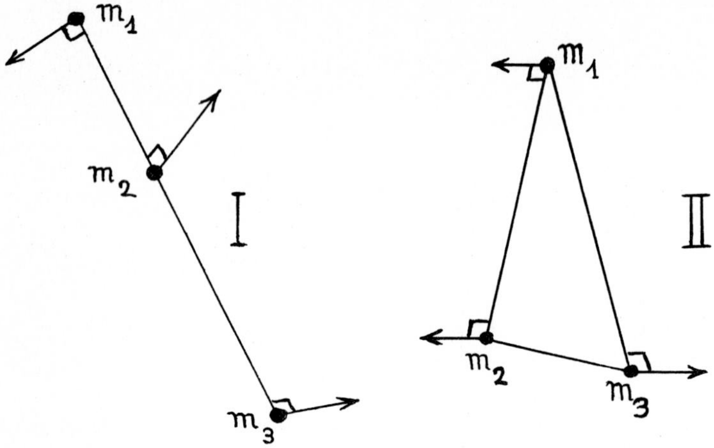

Fig. 49. Usual Poincaré space-time symmetry conditions.
If, at initial time $t = 0$, the three velocities are normal to all three radius-vectors, then for all t the vectors $\vec{r_J}(t)$ and $\vec{r_J}(-t)$ are symmetrical either with respect to the initial straight line of the three bodies (case I) or with respect to their initial plane (case II).
The three mutual distances are even functions of the time t.

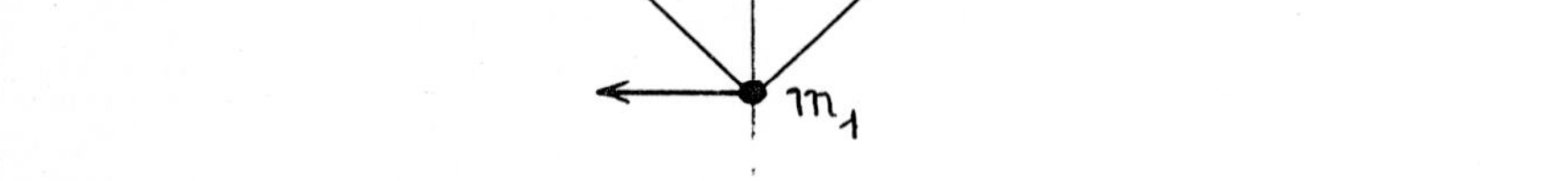

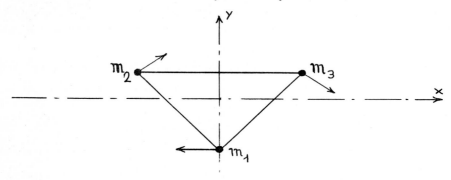

Fig. 50. Two supplementary cases of spsce-time symmetry conditions for systems with two equal masses.
Let us consider the following initial conditions at the time $t = 0$:

$m_2=m_3$; $x_1=z_1=z_2=z_3=0$; $x_3=-x_2$; $y_3=y_2$; $y'_1=0$; $x'_3=x'_2$; $y'_3=-y'_2$; $z'_3=\varepsilon z'_2$ with either $\varepsilon=+1$ (case III) or $\varepsilon=-1$; $z'_1=0$ (case IV)

These initial conditions imply the following space-time symmetries at the times t and $-t$.

t	x_1, y_1, z_1	x_2, y_2, z_2	x_3, y_3, z_3
$-t$	$-x_1, y_1, -\varepsilon z_1$	$-x_3, y_3, -\varepsilon z_3$	$-x_2, y_2, -\varepsilon z_2$

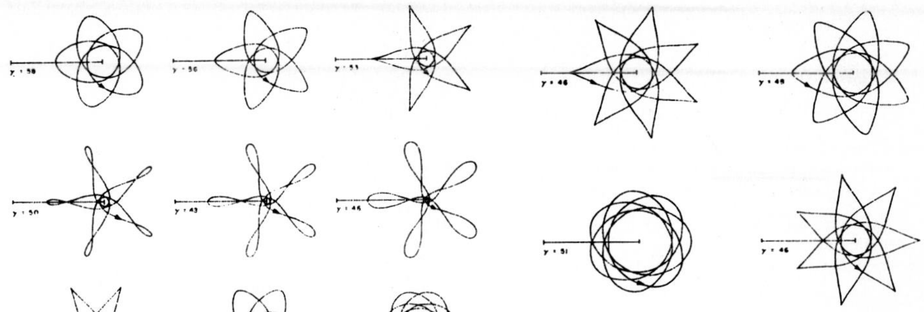

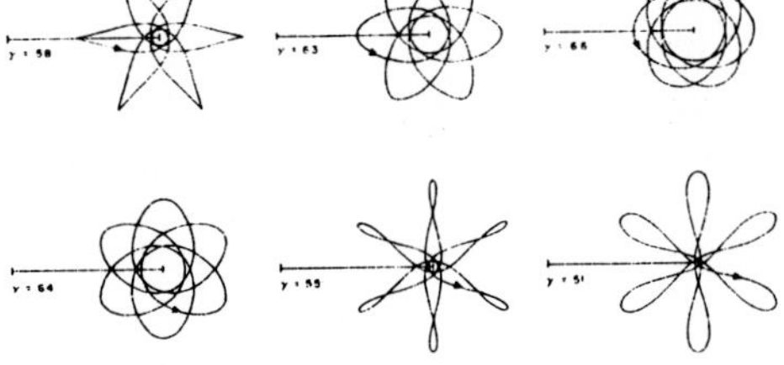

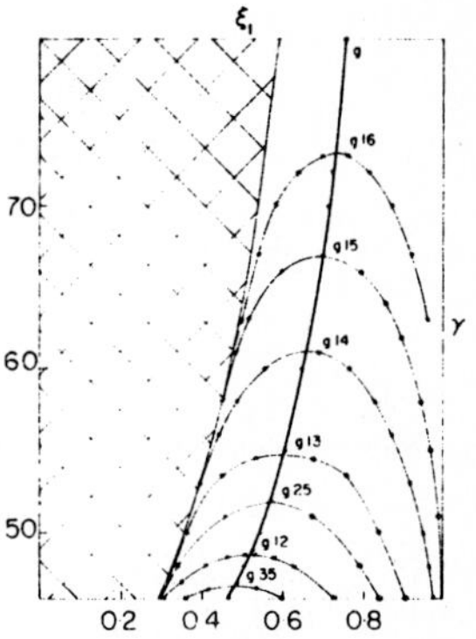

Sun-Jupiter circular restricted three-body problem - Some periodic orbits of the ref. 125 (in the rotating set of axes).

The opposite diagram shows the "branching" of the different families (with $\gamma = 2\,\Pi^2\Gamma$).

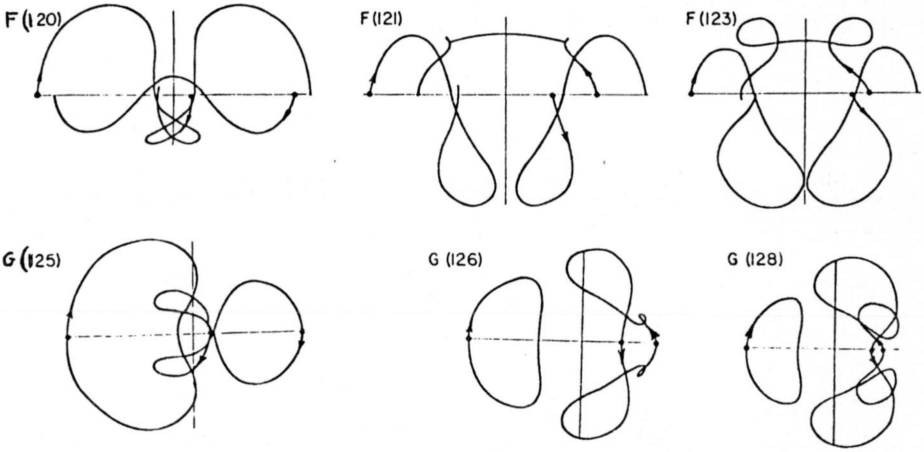

Some periodic orbits of the ref. 137.

Three equal masses - Orbits F : half a period - Orbits G : a full period.

The set of axes is uniformly rotating and corresponds to the "angle of rotation".

Fig. 51. Examples of periodic orbits.

See also: the dotted orbit of Fig. 27

the orbits of Figs. 78, 79, 84, 89-93, 114, 115, 117-119.

Source : Markellos, V.V. [125], and Davoust, E. and Broucke, R. [137].

4.4) The periodic orbits of the elliptic restricted three-body problem (either planar or three-dimensional) are always absolute periodic orbits ; their period is a multiple of the period of the primaries.

4.5) Nowadays most periodic orbits are computed in the planar general three-body problem and many in the three-dimensional general three-body problem.

5) In the qualitative study of the "planetary case" (i.e. a major mass and two small masses), Poincaré defined the first three "species" of periodic solutions (Ref 13, p.79-161) the orbits with small eccentricities and zero inclination (first species), the orbits with large eccentricities and zero inclination (second species) and the orbits with small eccentricities and large mutual inclination (third species). These three species are presented in the Fig. 52, their construction will be discussed in Section 11.4.1 on qualitative studies.

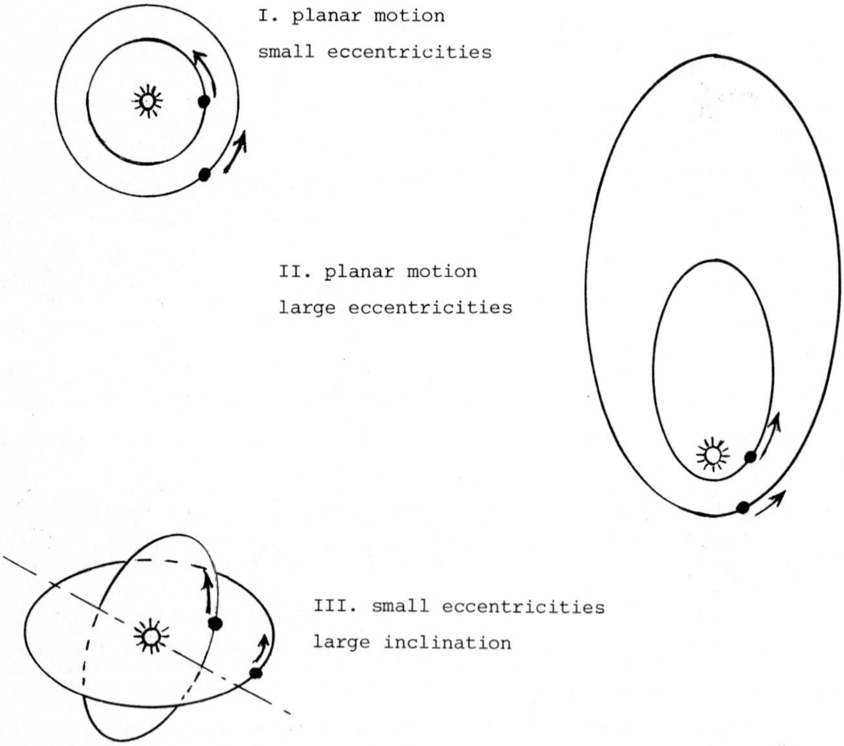

Fig. 52. The first three species of Poincaré periodic solutions. The orbits of the second species almost always have the same direction of motion.

This classification can be enlarged considerably and the number of known families of periodic solutions is now very large.

The great number and variety of known periodic orbits favour the famous Poincaré conjecture "The periodic orbits are everywhere dense in the set of bounded solutions of the three-body problem".

This conjecture can even be extended to the following : "For any three given masses, the periodic orbits are everywhere dense in the subset of bounded or oscillatory orbits and even in the subset of all orbits without hyperbolic escape neither for negative times nor for positive times".

These conjectures emphasize the interest of periodic orbits. However, notice that the orbits having at least one hyperbolic escape, in the past and/or in the future, belong to an open set without periodic orbits. We have seen in Section 5.1.4 that the orbits with a positive energy integral have hyperbolic escapes at both ends (if they have no triple collision).

10.5.1. Computation of periodic orbits. The method of analytic continuation. The utmost reduction of the three-body problem and the elimination of trivial side-effects.

Many methods have been used for the demonstration of the existence of periodic orbits, such as the application of the fixed point theorem and the methods of power series or Fourier series. However the method of analytic continuation is the most popular and allows easy computation of whole families of periodic orbits.

The principle of this method of analytic continuation is simple. Let us start from some already known periodic orbit, for instance a Lagrangian or Eulerian solution of the Section 8.1, and let us slightly modify the initial conditions and/or the three masses. Is it then possible to find a new periodic orbit in the vicinity of our initial solution ?

The trivial modifications are of course not considered (translation and/or rotation in space and/or in time ; modifications of the scales of lengths, masses and times).

The meaningful modifications can be associated with the Delaunay parameters $\mathcal{L}_i, \mathcal{G}_i, \mathcal{L}_e, \mathcal{G}_e, \ell_i, g_i, \ell_e, g_e$ of the Hamiltonian system (93) and (94) with three degrees of freedom.

Let us recall the meaning of these Delaunay parameters.

Fig. 53 gives the usual Jacobi decomposition of the three-body problem with the two vectors $\vec{r}$ and $\vec{R}$ and their time-derivatives $\vec{v}$ and $\vec{V}$.

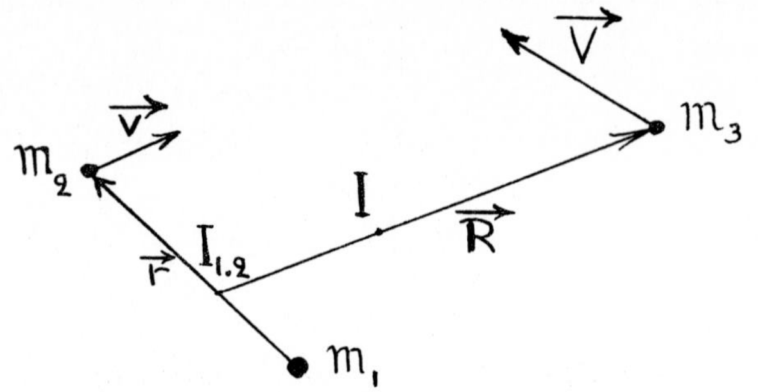

Fig. 53. The Jacobi vectors $\vec{r}$ *and* $\vec{R}$ *and their derivatives* $\vec{v}$ *and* $\vec{V}$.

$\vec{r} = \overrightarrow{r_{1.2}} = \overrightarrow{(m_1, m_2)}$

$\vec{R} = \overrightarrow{(I_{1.2}, m_3)}$

$\vec{v} = d\vec{r}/dt$; $\vec{V} = d\vec{R}/dt$

$I_{1.2}$ is the center of mass of the "binary" m_1, m_2 and I is the general center of mass.

When r/R is small ("lunar case") or when $(m_2 + m_3)/m_1$ is small ("planetary case"), the accelerations of $\vec{r}$ and $\vec{R}$ are close to the following Keplerian accelerations :

$$\left.\begin{aligned} d^2\vec{r}/dt^2 &\simeq - G(m_1+m_2)\vec{r}/r^3 \\ d^2\vec{R}/dt^2 &\simeq - G(m_1+m_2+m_3)\vec{R}/R^3 \end{aligned}\right\} \quad (287)$$

and the three-body motion is composed of two slowly perturbed Keplerian motions.

These two "osculating Keplerian motions" are interesting even if the perturbations are large ; the osculating Keplerian orbit of $\vec{r}$ is the "inner orbit" O_i (Fig. 54) and the osculating Keplerian orbit of $\vec{R}$ is the outer or "exterior orbit" O_e.

It is simple to use the "invariable plane" as reference plane and some fixed direction $\vec{\gamma}$ as reference direction in that plane. In these conditions the orbital elements of inner and exterior orbits O_i and O_e are the following

$$n_i,\ a_i,\ e_i,\ i_i,\ \ell_i,\ g_i,\ h_i \quad (288)$$

and :

$$n_e,\ a_e,\ e_e,\ i_e,\ \ell_e,\ g_e,\ h_e \quad (289)$$

(being respectively mean angular motion, semi-major axis, eccentricity, inclination, mean anomaly, argument of pericenter and longitude of node (Fig. 54).

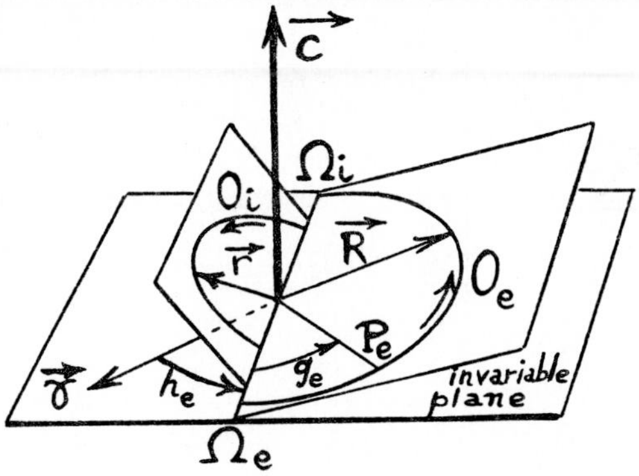

Fig. 54. The inner orbit O_i ,the outer orbit O_e and the invariable plane. The invariable plane is normal to the angular momentum $\vec{c}$.

These elements are of course related by the usual Keplerian relations :

$$n_i^2 a_i^3 = G(m_1 + m_2) \quad ; \quad n_e^2 a_e^3 = G(m_1 + m_2 + m_3) \tag{290}$$

$$a_i = \left[\frac{2}{r} - \frac{v^2}{G(m_1 + m_2)}\right]^{-1} \quad ; \quad a_e = \left[\frac{2}{R} - \frac{V^2}{G(m_1 + m_2 + m_3)}\right]^{-1} \tag{291}$$

etc...

The six parameters ℓ_i, g_i, h_i, ℓ_e, g_e, h_e are six Delaunay elements, their six conjuguate parameters L_i, $\mathcal{G}_i$, H_i, L_e, $\mathcal{G}_e$, H_e are given by the following :

$$M = m_1 + m_2 + m_3 \; ; \; m = m_1 m_2/(m_1+m_2) \; ; \; \mathcal{M} = m_3(m_1+m_2)/M \tag{292}$$

and :

$$\left.\begin{array}{lll} L_i = m\, n_i a_i^2 & ; & L_e = \mathcal{M}\, n_e a_e^2 \\ \mathcal{G}_i = L_i\sqrt{(1-e_i^2)} & ; & \mathcal{G}_e = L_e\sqrt{(1-e_e^2)} \\ H_i = \mathcal{G}_i \cos i_i & ; & H_e = \mathcal{G}_e \cos i_e \end{array}\right\} \tag{293}$$

$\mathcal{G}_i$ and $\mathcal{G}_e$ are respectively equal to the moduli of angular momenta of inner and outer orbits, that is to $|m\vec{r} \times \vec{v}|$ and $|\mathcal{M}\vec{R} \times \vec{V}|$. They are then related by the geometrical relations of Fig. 55 to the inclinations i_i and i_e and to the total angular momentum $\vec{c}$:

$$\vec{c} = m\vec{r} \times \vec{v} + \mathcal{M}\vec{R} \times \vec{V} \tag{294}$$

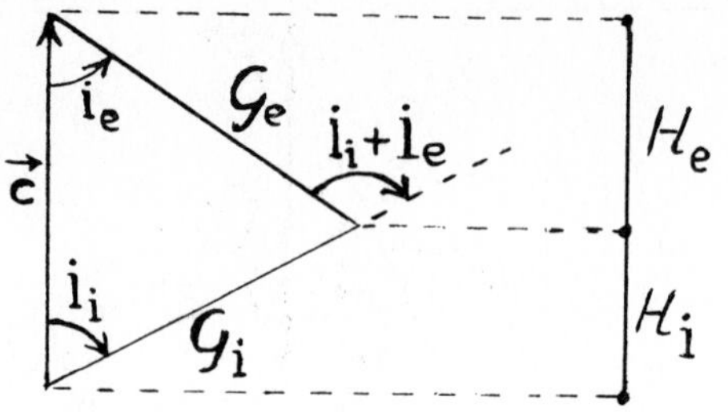

Fig. 55. The angular momenta $\mathcal{G}_i$, $\mathcal{G}_e$, $\vec{c}$ and the inclinations i_i and i_e.

The parameters $\mathcal{H}_i$ and $\mathcal{H}_e$ are then related to c, $\mathcal{G}_i$ and $\mathcal{G}_e$ by :

$$\mathcal{H}_i/\mathcal{G}_i = \cos i_i = (c^2+\mathcal{G}_i^2-\mathcal{G}_e^2)/2c\mathcal{G}_i \tag{295}$$

that is :

$$\left.\begin{aligned} \mathcal{H}_i &= (c^2+\mathcal{G}_i^2-\mathcal{G}_e^2)/2c \\ \mathcal{H}_e &= (c^2+\mathcal{G}_e^2-\mathcal{G}_i^2)/2c \\ \mathcal{H}_i + \mathcal{H}_e &= c \end{aligned}\right\} \tag{296}$$

On the other hand the longitudes of the nodes in the invariable plane, h_i and h_e (Fig. 53), are related by :

$$h_e = h_i + \pi \tag{297}$$

These two longitudes are "ignorable" and, with (296), allow the reduction of the classical Delaunay system (33) with six degrees of freedom to the following Hamiltonian system with only three degrees of freedom.

The energy integral h is identical to the Delaunay Hamiltonian H :

$$h = \frac{1}{2}(mv^2 + \mathcal{M}V^2) - G\left(\frac{m_1m_2}{r} + \frac{m_1m_3}{r_{1.3}} + \frac{m_2m_3}{r_{2.3}}\right) \tag{298}$$

Indeed the twelve Delaunay elements provide the relative positions and velocities, and then h (for given G and masses) :

$$h = h(\mathcal{L}_i, \mathcal{G}_i, \mathcal{H}_i, \mathcal{L}_e, \mathcal{G}_e, \mathcal{H}_e, \ell_i, g_i, h_i, \ell_e, g_e, h_e) \tag{299}$$

and even, with (296) and with the ignorable character of h_i and h_e :

$$h = H = h(\mathcal{L}_i, \mathcal{G}_i, \mathcal{L}_e, \mathcal{G}_e, \ell_i, g_i, \ell_e, g_e, c) \tag{300}$$

This intermediate expression leads to the Hamiltonian system (90), (91) with four degrees of freedom. However the utmost reduction of the system requires the elimination of the time, the choice of a fast-moving parameter as ℓ_i (or $\ell_i + g_i$ or ℓ_e etc ...) as parameter of description and its conjuguate as Hamiltonian. Hence we must reverse (300) and write :

$$\mathcal{L}_i = \mathcal{L}_i(h, \mathcal{G}_i, \mathcal{L}_e, \mathcal{G}_e, \ell_i, g_i, \ell_e, g_e, c) \tag{301}$$

and we then obtain the following Hamiltonian system with three degrees of freedom :

$$\left.\begin{aligned}
&\frac{d\mathcal{G}_i}{d\ell_i} = \frac{\partial\mathcal{L}_i}{\partial g_i} \;;\quad \frac{d\mathcal{L}_e}{d\ell_i} = \frac{\partial\mathcal{L}_i}{\partial\ell_e} \;;\quad \frac{d\mathcal{G}_e}{d\ell_i} = \frac{\partial\mathcal{L}_i}{\partial g_e} \\
&\frac{dg_i}{d\ell_i} = -\frac{\partial\mathcal{L}_i}{\partial\mathcal{G}_i} \;;\quad \frac{d\ell_e}{d\ell_i} = -\frac{\partial\mathcal{L}_i}{\partial\mathcal{L}_e} \;;\quad \frac{dg_e}{d\ell_i} = -\frac{\partial\mathcal{L}_i}{\partial\mathcal{G}_e}
\end{aligned}\right\} \tag{302}$$

The two integrals of motion h and c are constant, $\mathcal{L}_i$ is given by (301), $\mathcal{H}_i$ and $\mathcal{H}_e$ by (296) and the ignorable parameters h_i, h_e and the time t by :

$$\left.\begin{aligned}
&h_e = h_i + \pi \;;\; dh_i/d\ell_i = dh_e/d\ell_i = -\,\partial\mathcal{L}_i/\partial c \\
&dt/d\ell_i = \partial\mathcal{L}_i/\partial h
\end{aligned}\right\} \tag{303}$$

This Hamiltonian system (301)-(303) has many singularities (planar motions, small eccentricities, hyperbolic motions, etc...) which were considered in Section 7.

10.5.2 The method of analytic continuation for three given masses

Let us consider an arbitrary three-body periodic orbit and look for neighbouring periodic orbits with the same values of the three masses.

After one period of our periodic orbit, the six state parameters $\mathcal{G}_i$, $\mathcal{L}_e$, $\mathcal{G}_e$, g_i, ℓ_e, g_e and the description parameter ℓ_i of the system (302) return to their initial values (i.e., of course, the angular parameters ℓ_i, g_i, ℓ_e, g_e have an integer number of revolutions). Because of (301) and (296) the parameters L_i, H_i and H_e also return to their initial values.

The three remaining parameters are the ignorable parameters h_i, h_e and t, their evolution is the following :

1) During each period the increase of the time t is equal to the "period" T :

$$\left.\begin{aligned} &t_1 - t_0 = T \ : \ \text{one period} \\ &t_n - t_0 = nT : \ n \ \text{periods} \end{aligned}\right\} \quad (304)$$

2) For the longitudes of the nodes h_i and h_e, always related by $h_e = h_i + \pi$, the increase per period is a constant angle α, the "angle of rotation" :

$$\left.\begin{aligned} &h_{i1} - h_{i0} = h_{e1} - h_{e0} = \alpha \\ &h_{in} - h_{i0} = n\alpha \end{aligned}\right\} \quad (305)$$

Notice that if $\alpha = 0$ the periodic orbit of interest is an "absolute periodic orbit". If $\alpha = 2\pi p/q$ (where p and q are integers) the periodic orbit becomes an absolute periodic orbit if we consider q periods.

Let us now consider the vicinity of this periodic orbit.

Let us call ℓ_{i0} the initial value of the parameter of description and ℓ_{in} its value after n periods (with $\ell_{in} = \ell_{i0} + 2kn\pi$; k integer).

For a neighbouring orbit the six state parameters will be the following.

1) At ℓ_{i0} :

$$G_{i0}+\delta G_{i0} \ ; \ L_{e0}+\delta L_{e0} \ ; \ G_{e0}+\delta G_{e0} \ ; \ g_{i0}+\delta g_{i0} \ ; \ \ell_{e0}+\delta\ell_{e0} \ ; \ g_{e0}+\delta g_{e0} \quad (306)$$

2) At ℓ_{in} :

$$G_{i0}+\delta G_{in} \ ; \ L_{e0}+\delta L_{en} \ ; \ G_{e0}+\delta G_{en} \ ; \ g_{in}+\delta g_{in} \ ; \ \ell_{en}+\delta\ell_{en} \ ; \ g_{en}+\delta g_{en} \quad (307)$$

with :

$$\left.\begin{aligned} &g_{in} = g_{i0} + 2k_1 n\pi \ ; \ \ell_{en} = \ell_{e0} + 2k_2 n\pi \ ; \\ &g_{en} = g_{e0} + 2k_3 n\pi \ ; \ k_1, k_2, k_3 \ \text{integers} \end{aligned}\right\} \quad (308)$$

We will denote by $\vec{\delta}_0$ the small vector with the six components $\delta\mathcal{G}_{i0}$ to δg_{e0} and by $\vec{\delta}_n$ the corresponding vector $\delta\mathcal{G}_{in}$ to δg_{en}.

If we consider neighbouring orbits with the same values of the integrals h and c, the correspondence between $\vec{\delta}_0$ and $\vec{\delta}_n$ has the following form given by the Floquet theory (Ref. 173-174).

$$\left.\begin{aligned} \vec{\delta}_1 &= A\vec{\delta}_0 + O(\delta_0^{\,2}) \\ \vec{\delta}_n &= A^n\vec{\delta}_0 + O(\delta_0^{\,2}) \end{aligned}\right\} \qquad (309)$$

A is a real 6 × 6 matrix, that can be obtained either by the computation of six suitable neighbouring orbits or by the integration of the "variational equations" of the periodic solution of interest. These auxiliary computations are generally done with a system more directly related to $\vec{r}$, $\vec{v}$, $\vec{R}$, $\vec{V}$ than the system (301)-(302).

The matrix A has several names, being called either the Floquet matrix or the monodromy matrix or the transition matrix for ℓ_{i0} and ℓ_{i1} ; its eigenvalues are the "Floquet multipliers".

For a given periodic solution, the expression of A is a function of the initial ℓ_{i0}, it is also a function of the chosen parameter of description, but the "Floquet multipliers" are the same in all these cases. These Floquet multipliers will be the essential elements in the analysis of stability (Section 10.7).

The theory of Hamiltonian systems shows that A has a particular form, it is a "symplectic matrix". This means that if we denote by J the following matrix :

$$J = \begin{vmatrix} 0 & 0 & 0 & -1 & 0 & 0 \\ 0 & 0 & 0 & 0 & -1 & 0 \\ 0 & 0 & 0 & 0 & 0 & -1 \\ 1 & 0 & 0 & 0 & 0 & 0 \\ 0 & 1 & 0 & 0 & 0 & 0 \\ 0 & 0 & 1 & 0 & 0 & 0 \end{vmatrix} = \begin{vmatrix} 0, & -I \\ I, & 0 \end{vmatrix} \qquad (310)$$

and if A^T is the transposed matrix of A, we have the following relation :

$$A^T JA = J \qquad (311)$$

Real symplectic matrices have a real and reciprocal characteristic polynomial hence their determinant is + 1 and if λ is an eigenvalue, λ^{-1}, $\bar{\lambda}$ and $(\bar{\lambda})^{-1}$ are also eigenvalues.

Our purpose is to find neighbouring periodic orbits, i.e. to find $\vec{\delta}_0$ such that $\vec{\delta}_1 = \vec{\delta}_0$ or at least $\vec{\delta}_n = \vec{\delta}_0$ for some positive n. This is impossible for small $\vec{\delta}_0$ and not too large n if + 1 is not an eigenvalue of A or at least of A^n, and in general there is no periodic orbit of not too large a period and of the same masses and integrals of motion in the vicinity of a given periodic orbit.

What are the possibilities provided by the choice of neighbouring values of the integrals of motion h and c ?

A major point is the existence of the scale transformation : for any non-zero constant k, if $\vec{r}_j(t)$ with $j = \{1,2,3\}$ is a solution of the three-body problem, then $\vec{\rho}_j(t) = k^2.\vec{r}_j(t/k^3)$ is another solution (see Section 5.3). For these two slutions the angular parameters ℓ_i, g_i, h_i, ℓ_e, g_e, h_e are the same, while L_i, G_i, H_i, L_e, G_e, H_e, c must be multiplied by k and h by k^{-2}.

This trivial transformation reduces our possibilities, it is useless to modify both h and c if the product c^2h is unchanged. There remains only one real parameter instead of two for non-trivial modifications, and most people keep h constant and modify only the angular momentum c. They generally either keep $h = -1$ or $h = -(m_1m_2+m_1m_3+m_2m_3)$, and indeed three-body motions with a positive or zero energy integral h cannot be periodic (see Section 5.1.4, the Lagrange-Jacobi identity). Thus with a constant negative h and a variable c, all useful values of the product c^2h, i.e. all negative or zero values,. can be reached.

What then are, for constant energy integral h, the possibilities given by variations of the modulus c of the angular momentum ?

The equation (309) can be extended to the following :

$$\vec{\delta}_1 = A\vec{\delta}_0 + \vec{B}.\delta c + O(\delta_0^{\ 2}, (\delta c)^2) \tag{312}$$

Like the matrix A, the vector $\vec{B}$ (with six components) can be obtained either by computation of a suitable neighbouring orbit (an orbit starting at h, G_{i0}, L_{e0}, G_{e0}, ℓ_{i0}, g_{i0}, ℓ_{e0}, g_{e0}, $c + \delta c$) or by the corresponding integration of the "variational equations" of the periodic solution of interest.

The equation (312) gives also :

$$\vec{\delta}_2 = A\vec{\delta}_1 + \vec{B}.\delta c + O(\delta_1^{\ 2}, (\delta c)^2) \tag{313}$$

hence :

$$\vec{\delta}_2 = A^2\vec{\delta}_0 + (A + I)\vec{B}.\delta c + O(\delta_0^{\ 2}, (\delta c)^2) \tag{314}$$

I being the unit 6 × 6 matrix.

This leads to :

$$\vec{\delta}_n = A^n\vec{\delta}_0 + (A^{n-1} + A^{n-2} + \dots + A^2 + A + I)\vec{B}\delta c + O\left(\delta_0{}^2, (\delta c)^2\right) \qquad (315)$$

If A has no eigenvalue equal to unity, i.e. if (A - I) is invertible the equation (315) is equivalent to the following :

$$\vec{\delta}_n = A^n\vec{\delta}_0 + (A^n - I)(A - I)^{-1}\vec{B}\delta c + O\left(\delta_0{}^2, (\delta c)^2\right) \qquad (316)$$

We are looking for periodic orbits in the vicinity of our initial periodic orbit, hence let us look for $\vec{\delta}_1 = \vec{\delta}_0$ in the equation (312).

If A has no eigenvalue equal to unity we will say that the periodic orbit of interest is "regular", and neighbouring periodic orbits are given, to first order, by the following :

$$\vec{\delta}_0 = (I - A)^{-1} \cdot \vec{B}\delta c \qquad (317)$$

that is, for the differential elements :

$$d\vec{\delta}_0/dc = (I - A)^{-1} \cdot \vec{B} \qquad (318)$$

and thus, as long as A has no eigenvalue equal unity (i.e. as long as (I-A) is invertible), it is possible to build step by step a non-trivial one-parameter family of periodic orbits in terms of the modulus c of the angular momentum.

We will see in Section 10.6.2 and in the equations (370) a simpler construction of this family when the initial periodic orbit has a space-time symmetry property.

Thus, if we consider a three-body system of three given masses, each regular periodic orbit belongs to a non-trivial one-parameter family of periodic orbits. These one-parameter families are generally expressed in terms of the angular momentum c for a fixed negative value of the energy integral h (see for instance the families of pseudo-circular orbits for three equal masses in Section 10.8.1).

These general results require the following comments :

1) The Hamiltonian system (301)-(303) is not valid for planar motions. Fortunately the corresponding suitable Hamiltonian system (103)-(105) of Chapter 7 leads to exactly the same analysis and the same conclusions : each regular planar periodic orbit belongs to a non-trivial one-parameter family of planar periodic orbits with the same three masses.

2) The Hamiltonian systems (301)-(303) and (103)-(105) are not valid for the restricted three-body problem and we must therefore use another Hamiltonian system, for instance the system (171), (172) of Section 9.3. Again this leads to the same analysis and to similar conclusions in the four possible cases (planar or three dimensional problem, circular or elliptic restricted three-body problem).

In each case, if the corresponding matrix A has no eigenvalue equal to unity, the periodic orbit of interest is regular and belongs to a non-trivial one-parameter family of the case of interest, such as the periodic orbits of Figs. 18, 19, 20 (Section 8.5) and of Fig. 24 (Section 9.1).

The variation of c in equations (312)-(318) becomes a variation of the eccentricity e in the elliptic restricted case and a variation of the Jacobi integral Γ in the circular restricted case.

3) What happens when the matrix A has an eigenvalue equal to unity ? i.e. when $\det(I - A) = 0$ and the periodic orbit of interest is "singular".

The equations (317) and (318) cannot be used and, when looking for $\vec{\delta}_1 = \vec{\delta}_0$, the equation (312) gives only :

$$(I - A)\vec{\delta}_0 = \vec{B}\delta c + O\left(\delta_0^{\,2}, (\delta c)^2\right) \tag{319}$$

This equation has not necessarily a non-zero solution, and thus isolated periodic orbits are not excluded even if none of them is known.

Notice that the matrix A, being symplectic, has an even number of eigenvalues equal to unity and so it has at least a second unity eigenvalue as soon as there is a first. Hence the phenomenon of "branching" is not surprising : usually two, or several, families of periodic orbits meet at points where $\det(I - A) = 0$ and the equation (319) can have several directions of solutions there, at least for $\delta c = 0$.

An example is given by the two g and g' Strömgren families (Figs. 18 and 19, Section 8.5). These two families have a common orbit g_1 for $\Gamma = 4.499986$. This orbit g_1 is also the limit of stability for the family g, a natural phenomenon that will be explained in Section 10.7.6 on stability.

Another type of branching is given by the second equation (309) : when the matrix A^n has eigenvalues equal to unity, a second family with locally the period nT can be branched to the first family. This phenomenon is very common and leads to a very large number of families of periodic orbits.

4) As a final remark on these families of periodic orbits, let us use the information given by the "eleventh local integral" of Section 5.3.

We will use the expression (79) of this "eleventh local integral" Q_D between two neighbouring orbits O_1 and O_2 of the family of interest, and we will assume that these two periodic orbits have the same period T (after, if necessary, a slight scale transformation of one of these orbits).

Let us give subscript 1 to the elements of O_1 and subscript 2 to the elements of O_2 ; on the other hand the parameter H used in (79) is the Hamiltonian identical to the energy integral h.

Hence in (79) :

$$\mathcal{L}_i = \mathcal{L}_{i1} \; ; \; \delta\mathcal{L}_i = \mathcal{L}_{i2} - \mathcal{L}_{i1} \; ; \; \text{etc} \ldots \tag{320}$$

and :

$$\left.\begin{aligned} Q_D = {} & (\mathcal{L}_{i1}+\mathcal{L}_{i2})(\ell_{i2}-\ell_{i1}) + (\mathcal{G}_{i1}+\mathcal{G}_{i2})(g_{i2}-g_{i1}) + (\mathcal{H}_{i1}+\mathcal{H}_{i2})(h_{i2}-h_{i1}) + \\ & + (\mathcal{L}_{e1}+\mathcal{L}_{e2})(\ell_{e2}-\ell_{e1}) + (\mathcal{G}_{e1}+\mathcal{G}_{e2})(g_{e2}-g_{e1}) + (\mathcal{H}_{e1}+\mathcal{H}_{e2})(h_{e2}-h_{e1}) + \\ & + 6t(h_2-h_1) \end{aligned}\right\} \tag{321}$$

Let us consider all these elements at same initial time t_0 and, after a period, at the final time t_f equal to $t_0 + T$.

A) On both orbits the parameters $\mathcal{L}_i$, $\mathcal{G}_i$, $\mathcal{H}_i$, $\mathcal{L}_e$, $\mathcal{G}_e$, $\mathcal{H}_e$ come back to their initial values after a period.

B) The increases of ℓ_i, g_i, ℓ_e, g_e are multiples of 2π. The differences $(\ell_{i2} - \ell_{i1})$, $(g_{i2} - g_{i1})$, $(\ell_{e2} - \ell_{e1})$, $(g_{e2} - g_{e1})$ return to their initial values.

C)

$$\begin{aligned} h_{i1}(t_f) - h_{i1}(t_0) &= h_{e1}(t_f) - h_{e1}(t_0) = \alpha_1 \\ h_{i2}(t_f) - h_{i2}(t_0) &= h_{e2}(t_f) - h_{e2}(t_0) = \alpha_2 \end{aligned} \tag{322}$$

where α_1 and α_2 are the "angles of rotation" (per period) of the orbits O_1 and O_2 respectively. They have already been presented in (305).

D) The "eleventh local integral" Q_D remains almost constant :

$$Q_D(t_f) = Q_D(t_0) + (\text{third order}) \tag{323}$$

On the other hand the angular momenta c_1 and c_2 are given by the following

$$\begin{aligned} \mathcal{H}_{i1} + \mathcal{H}_{e1} &= c_1 \\ \mathcal{H}_{i2} + \mathcal{H}_{e2} &= c_2 \end{aligned} \tag{324}$$

Hence the equations (320)-(324), and the comparison of $Q_D(t_0)$ with $Q_D(t_f)$, lead to :

$$(c_1 + c_2)(\alpha_2 - \alpha_1) + 6T(h_2 - h_1) = \text{(third order)} \tag{325}$$

that is, for the differential elements :

$$c \cdot d\alpha + 3T \cdot dh = 0 \tag{326}$$

Let us recall that we have chosen the periodic orbits O_1 and O_2 of the family of interest with the same period T.

If we take account of the possibilities given by the scale transformation of Section 5.3 (a transformation in which α and T^2h^3 are constant, that is $d\alpha = 0$ and $3Tdh + 2hdT = 0$) we extend (326) into the following relation (that is also a consequence of (80)) :

$$c \cdot d\alpha + 3T \cdot dh + 2h \cdot dT = 0 \tag{327}$$

For instance, if we characterized the family of periodic orbits of interest by a non-trivial one-parameter family of periodic orbits of constant energy integral h, we would find along this family the following relation between the variations of the angle of rotation α and those of the period T :

$$c \cdot d\alpha + 2h \cdot dT = 0 \tag{328}$$

Hence, since $c \geqq 0$ and $h < 0$, the angle of rotation α and the period T would always vary in the same direction.

With an obvious suitable definition of the angle of rotation per period, α, these relations (326), (327), (328) remain valid for the particular cases of planar motions and of motions in the restricted three-body problem.

10.5.3 The method of analytic continuation and the modification of masses.

The modification of masses considerably increases the possibilities given by the method of analytic continuation for the construction of families of periodic orbits. However we must notice the following phenomenon.

There is a trivial "variation of the scale of masses" very similar to the scale transformation of Section 5.3, and a new solution of the three-body problem can be obtained by a modification of the scales of length, mass and time that maintains the value of the constant G of the law of the universal attraction (i.e. $G = 6.672 \times 10^{-11} m^3/s^2.kg$). For instance, a multiplication of the distances by the factor k and of the masses by k^3 preserves the Newtonian character of the motion.

This trivial construction of new solutions is not the purpose of the method of analytic continuation, and it effectively wastes one parameter among the possibilities of this method, leaving only two new parameters that are sometimes called the two "mass ratios".

The method of analysis is almost identical to that of the previous Section. We again find the equation (309) relating the small discrepancies $\delta\mathcal{G}_i$, $\delta\mathcal{L}_e$, $\delta\mathcal{G}_e$, δg_i, $\delta\ell_e$, δg_e before and after one or several periods when the three masses and the two integrals of motions are maintained :

$$
\begin{aligned}
&\vec{\delta} = (\delta\mathcal{G}_i, \delta\mathcal{L}_e, \delta\mathcal{G}_e, \delta g_i, \delta\ell_e, \delta g_e) \\
&\vec{\delta}_n = \vec{\delta}(\ell_{in}) \quad ; \quad \ell_{in} = \ell_{i0} + 2kn\pi \\
&\vec{\delta}_1 = A\vec{\delta}_0 + O\delta_0^{\,2} \\
&\vec{\delta}_n = A^n\vec{\delta}_0 + O(\delta_0^{\,2})
\end{aligned}
\tag{329}
$$

The "monodromy matrix" A is a real symplectic 6 × 6 matrix.

Let us now slightly modify the mass m_j (with $j = \{1,2,3\}$) we will obtain an equation very similar to the equation (312) with the same matrix A :

$$\vec{\delta}_1 = A\vec{\delta}_0 + \vec{C}_j\delta m_j + O\left(\delta_0^{\,2}, (\delta m_j)^2\right) \tag{330}$$

It of course leads to the same analysis :

1) If the periodic orbit is regular the research of new neighbouring periodic orbits, i.e. the research of $\vec{\delta}_1 = \vec{\delta}_0$, leads to :

$$\vec{\delta}_0 = (I - A)^{-1}\,\vec{C}_j\delta m_j + O\left((\delta m_j)^2\right) \quad ; \quad j = \{1,2,3\} \tag{331}$$

Thus, since the variations of the three masses introduce only two new non-trivial parameters, the initial periodic orbit belongs to a non-trivial three-parameter family of periodic orbits. One of these parameters is related to the variations of integrals of motion, and was presented in the previous Section ; the other two are related to the variations of the mass ratios.

2) The same analysis can be made for planar motions with the Hamiltonian system (103)-(105) of Chapter 7, and thus if the initial periodic orbit is planar all periodic orbits of its three parameter family are also planar. However, only two-parameter parts of these families belong to motions of the restricted three-body problem, since for these motions only one mass ratio is available.

The phenomenon of "branching" between different families of periodic orbits again exists at singular orbits (where det(I - A) = 0), and is very often accompanied by an exchange of stability. The branching also exists at periodic orbits for which $\det(I - A^n) = 0$ between families with locally periods which are multiples of each other and this allows a very large number of non-trivial three-parameter families of periodic orbits to be obtained easily.

A beautiful application can be found in the Ref. 175, where a periodic orbit of the circular restricted three-body problem is linked to periodic orbit of the elliptic three-body problem (with eccentricity e = 0.292) through periodic orbits of the general three-body problem.

This analysis can be applied to the particular case of rectilinear periodic orbits :

1) Rectilinear motions always have a zero angular momentum and then the variations of integrals of motion cannot give a non-trivial parameter.

2) On the contrary the variation of the mass ratios have the same interest as in the other cases and hence the families of rectilinear periodic orbits have two non-trivial parameters.

Michel Hénon in the Ref.176 has determined the stability of all orbits of the simplest family of symmetric rectilinear periodic orbits, the family with only two binary collisions per period (see Fig. 64 in Section 10.7.4.1).

10.6 PERIODIC ORBITS AND SYMMETRY PROPERTIES

10.6.1 The four types of space-time symmetries

Fig. 56 presents again (in terms of the Jacobi vectors $\vec{r}$, $\vec{R}$ and their derivatives $\vec{v}$, $\vec{V}$) the space-time symmetries already shown in Figs. 49 and 50.

We will see in Section 11.2.2. two other cases of space symmetries (but not space-time). However these two cases have little correlations with the periodic orbits and they will not be considered in this Section.

The conditions I, II, III, IV of Fig. 56 can be expressed in terms of the Delaunay elements of Chapter 7 and Section 10.5.1.

The conditions I correspond to :

$$\ell_i,\ g_i,\ \ell_e,\ g_e = 0 \text{ or } \pi \tag{332}$$

The conditions II correspond to :

$$\ell_i,\ \ell_e = 0 \text{ or } \pi \ ; \quad g_i,\ g_e = \pm\,\pi/2 \tag{333}$$

The conditions III correspond to :

$$m_1 = m_2 \ ; \ \ell_i, \ell_e, g_e = 0 \text{ or } \pi \ ; \ g_i = \pm \pi/2 \tag{334}$$

The conditions IV correspond to :

$$m_1 = m_2 \ ; \ \ell_i, g_i, \ell_e = 0 \text{ or } \pi \ ; \ g_e = \pm \pi/2 \tag{335}$$

Notice the existence of two limit cases with a binary collision.

For planar motions the cases I and II are identical ($\ell_i, \ell_e, h_e - h_i = 0$ or π) and so are the cases III and IV ($m_1 = m_2$; $\ell_i, \ell_e = 0$ or π ; $h_e - h_i = \pm \pi/2$). Notice that type II can also occur with $\vec{v} = \vec{V} = 0$, that is $\ell_i = \ell_e = \pi$; $e_i = e_e = 1$, $H_i = H_e = 0$.

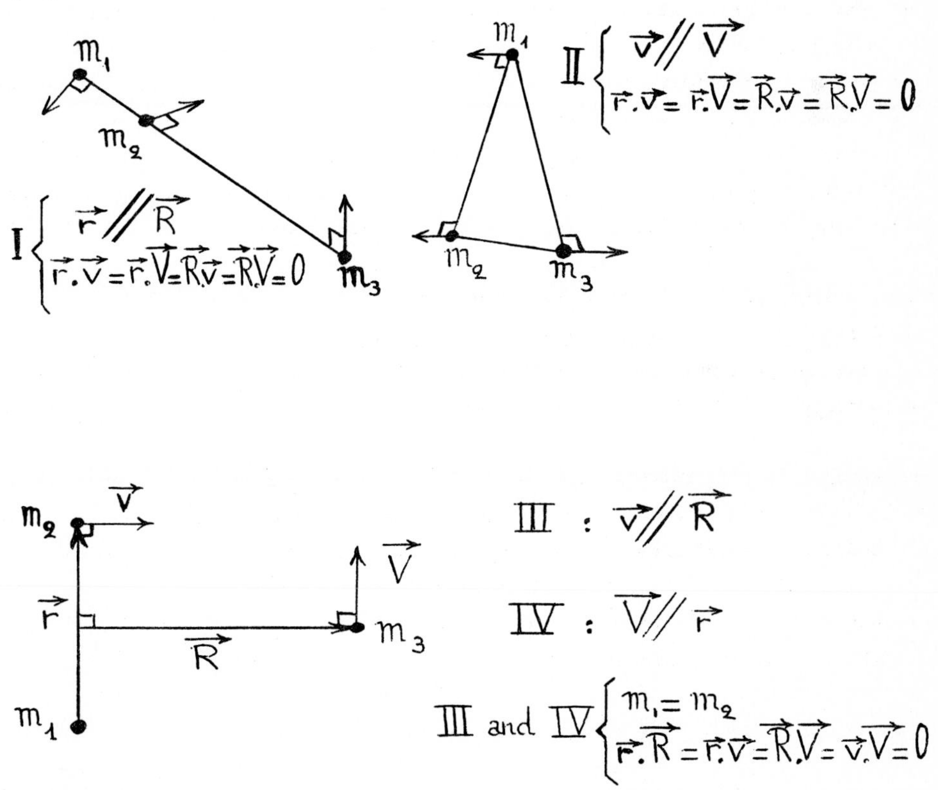

Fig. 56. The Poincaré space-time symmetry conditions (I and II) and, for systems with two equal masses, the supplementary space-time symmetry conditions III and IV.

When r/R and $(m_1 + m_2)/m_3$ are infinitely small (Hill motions) the condition $m_1 = m_2$ of cases III and IV is no longer required.

These space-time symmetries lead to the following.

A) With the exception of the very particular case B below, the momenta $\mathcal{L}_i, \mathcal{G}_i, \mathcal{H}_i, \mathcal{L}_e, \mathcal{G}_e, \mathcal{H}_e$ are even functions of the time with respect to the instant t_s of symmetry, and the angles $\ell_i, g_i, h_i, \ell_e, g_e h_e$ are odd functions of the time with respect to the conditions at t_s (Fig. 57).

$$\forall t : \mathcal{L}_i(t_s + t) = \mathcal{L}_i(t_s - t) \tag{336}$$

$$\forall t : \{h_e(t_s + t) - h_e(t_s)\} = - \{h_e(t_s - t) - h_e(t_s)\} \tag{337}$$

B) In the plane case the Delaunay elements are $\mathcal{L}_i, \mathcal{H}_i, \ell_i, h_i, \mathcal{L}_e, \mathcal{H}_e, \ell_e, h_e$ the angles h_i and h_e being the longitudes of pericenters. Then if the space-time symmetry is either of the type $\vec{v} = \vec{V} = 0$ or of the type $\vec{r} = \vec{V} = 0$ the angular momentum is zero, the elements $\mathcal{L}_i, h_i, \mathcal{L}_e, h_e$ are even functions of t while $\mathcal{H}_i, \ell_i, \mathcal{H}_e, \ell_e$ are odd functions.

Let us now consider a periodic orbit (period T) with an instant t_s of space-time symmetry, of course all instants t_s + nT with n integer will also be instants of identical space-time symmetries.

Notice that the instants t_s + (T/2) + nT will also be instants of space-time symmetries of the same type I, II, III or IV. Indeed :

A) Because of the periodicity :

$$\ell_i(t_s + \frac{T}{2}) = \ell_i(t_s - \frac{T}{2}), \quad \text{modulo } 2\pi \tag{338}$$

B) Because of the symmetry and the relation (337) :

$$\ell_i(t_s + \frac{T}{2}) - \ell_i(t_s) = \ell_i(t_s) - \ell_i(t_s - \frac{T}{2}) \tag{339}$$

Hence :

$$\ell_i(t_s + \frac{T}{2}) - \ell_i(t_s) = 0 \text{ or } \pi, \text{ modulo } 2\pi \tag{340}$$

The same is true for the angles g_i, ℓ_e, g_e, and thus we find at t_s + T/2 the symmetry conditions of the time t_s (either (332) or (333) or (334) or (335)). However the other orbital elements are generally very different at t_s and at t_s + T/2 .

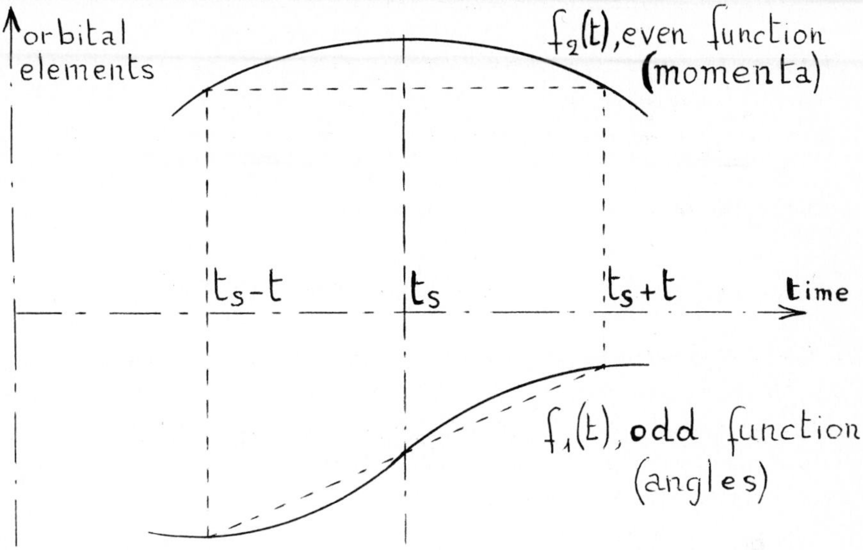

Fig.57. Space time symmetries.

When a three-body system has a space-time symmetry at the time t_s its radius-vectors and its velocity vectors have the symmety properties described in Figs.49, 50 and 56. The shape and the size of the triangle of the three bodies are the same at $t_s + t$ and $t_s - t$ (with exchange of the equal masses for symmetries of types III or IV).

The orbital elements have simple symmetry properties, they are either odd functions of the time like the function $f_1(t)$ of this figure, or even functions like the function $f_2(t)$; this requires that in case of space-time symmetries of type III or IV, the Jacobi vector $\vec{r}$ joins the symmetrical equal masses.

Usually, the angles ℓ_i, g_i, h_i, ℓ_e, g_e, h_e (true anomalies, arguments of pericenters, longitudes of nodes) are odd functions of t while the other elements (semi-major axes, excentricities, inclinations, Delaunay momenta L_i, G_i, H_i, L_e, G_e, H_e) are even functions. However, in the very particular cases when the angular momentum is zero and when the space-time symmetry is either of the type $\vec{v} = \vec{V} = 0$ or of the type $\vec{r} = \vec{V} = 0$ (case with a binary collision), the motion is planar, the angles h_i and h_e are the longitudes of pericenters, the elements H_i, ℓ_i, H_e, ℓ_e are odd functions of t while L_i, h_i, L_e, h_e are even functions.

Conversely, if t_{s0} and t_{s1} are two instants of space-time symmetry of the orbit of interest :

A) The relation (336) leads to :

$$\forall \alpha : L_i(t_{s0} + \alpha) = L_i(t_{s0} - \alpha) \quad ; \forall \beta : L_i(t_{s1} + \beta) = L_i(t_{s1} - \beta) \qquad (341)$$

Hence, with $t = t_{s0} + \alpha$ and $t_{s0} - \alpha = t_{s1} - \beta$:

$$\forall t : \mathcal{L}_i(t) = \mathcal{L}_i(2t_{s0} - t) = \mathcal{L}_i(t + 2t_{s1} - 2t_{s0}) \tag{342}$$

Thus $2(t_{s1} - t_{s0})$ is a period for $\mathcal{L}_i$ and similarly for $\mathcal{G}_i$, $\mathcal{H}_i$, $\mathcal{L}_e$, $\mathcal{G}_e$, $\mathcal{H}_e$.

B) Let us put :

$$t_{s2} = 2t_{s1} - t_{s0} \quad ; \quad t_{sn} = t_{s0} + n(t_{s1} - t_{s0}) \quad ; \quad n \in \mathbb{Z} \tag{343}$$

Hence, from (337) :

$$\ell_i(t_{s2}) = 2\ell_i(t_{s1}) - \ell_i(t_{s0}) \quad ; \quad g_i(t_{s2}) = 2g_i(t_{s1}) - g_i(t_{s0}) \tag{344}$$

and, with (332)-(335) :

$$\ell_i(t_{s0}) = 0 \text{ or } \pi \; ; \; \ell_i(ts_1) = 0 \text{ or } \pi \; ; \; \ell_i(t_{s2}) = \ell_i(t_{s0}) \text{ modulo } 2\pi \tag{345}$$

$$g_i(t_{s0}) = 0, \pm \pi/2 \text{ or } \pi \; ; \; g_i(t_{s1}) = 0, \pm \pi/2 \text{ or } \pi \; ;$$
$$g_i(t_{s2}) = g_i(t_{s0}) + 0 \text{ or } \pi \text{ modulo } 2\pi \tag{346}$$

and similarly :

$$\ell_e(t_{s2}) = \ell_e(t_{s0}) \quad ; \quad g_e(t_{s2}) = g_e(t_{s0}) + 0 \text{ or } \pi \text{ modulo } 2\pi \tag{347}$$

Thus :

B1) We find back at t_{s2} the symmetry conditions of the type met at t_{s0}.

B2) This can be extended to all t_{sn} : the t_{sn} with even n are instants of space-time symmetry of the t_{s0} type and the t_{sn} with odd n are instants of space-time symmetry of the t_{s1} type. These two types can be different ; for instance the Poincaré motions of the third species (Fig. 52) have alternately symmetries of types I and II.

B3) At t_{s2}, or at least at t_{s4}, the ten Delaunay elements $\mathcal{L}_i$, $\mathcal{G}_i$, $\mathcal{H}_i$, ℓ_i, g_i, $\mathcal{L}_e$, $\mathcal{G}_e$, $\mathcal{H}_e$, ℓ_e, g_e return to their initial t_{s0} values. Hence orbits with two instants t_{s0} and t_{s1} of space-time symmetry are periodic with a period T that divides $2(t_{s1} - t_{s0})$ or at least $4(t_{s1} - t_{s0})$.

As already noticed the ignorable parameters h_i and h_e are related by $h_e = h_i + \pi$. They are not periodic in general and their increase per period is the "angle of rotation" of the periodic orbit of interest.

Thus, finally, symmetric periodic orbits have either two or four instants of space-time symmetry per period. In former case all symmetries are of the same type and in the latter case the symmetries at t_s and $t_s + T/2$ necessarily have a type different from those at $t_s + T/4$ and $t_s + 3T/4$. In these cases with four instants of symmetry per period the true period of $\mathcal{L}_i, \mathcal{G}_i, \mathcal{H}_i, \ell_i, \mathcal{L}_e, \mathcal{G}_e, \mathcal{H}_e, \ell_e$ is only T/2 while g_i and/or g_e increase by $(2k + 1)\pi$ during this interval or time.

Let us notice the following exceptional cases :

1) In the circular Eulerian motions (Section 8.1), all instants t are instants of space-time symmetry of types I and II. In the circular Lagrangian motions with two equal masses (same Section 8.1) all instants are instants of space-time symmetry of types III and IV.

These particular motions are the only ones with non-isolated instants of space-time symmetry.

2) For systems with three equal masses the space-time symmetries of types III or IV can occur with any of the three pairs of equal masses.

This can lead to, at most, twelve instants of space-time symmetry per period (see Section 10.8.2 and Fig. 78).

10.6.2 Families of symmetric periodic orbits

Section 10.5.2 and 10.5.3 present the construction by continuity of non trivial three-parameter families of periodic orbits.

The periodic orbits are either "regular" (if the matrix A of (309) has no eigenvalue equal to unity) or "singular" in the opposite case. Each regular periodic orbit belongs to a well defined non-trivial three-parameter family of periodic orbits and these families meet at the singular periodic orbits.

If a regular periodic orbit is planar all periodic orbits of its family are planar.

Let us now demonstrate the following.

A) If a regular periodic orbit has space-time symmetries of types I and/or II, all periodic orbits of its family have the same type or types of symmetry.

B) If a regular periodic orbit has space-time symmetries of types III and/or IV, it necessarily has two equal masses, and all periodic orbits of its family that have the same equal masses and that can be obtaines by continuity also have the same type or types of symmetry.

This continuous sub-family of symmetric periodic orbits has only two non-trivial parameters, one of the mass ratios being indeed equal to unity and is no longer available.

Let us consider the matrix A of (309) as it is when ℓ_{i0} and all ℓ_{in} correspond to a space-time symmetry (hence, modulo 2π, $\ell_{i0} = \ell_{in} = 0$ or π) and let us decompose A into its four corners :

$$A = \begin{pmatrix} U, & V \\ W, & X \end{pmatrix} \tag{348}$$

Similarly, the matrix J of (310) is :

$$J = \begin{pmatrix} 0, & -I \\ I, & 0 \end{pmatrix} \tag{349}$$

The property of symplecticity, i.e. $A^T JA = J$ leads to :

$$A^{-1} = - JA^T J = \begin{pmatrix} X^T, & -V^T \\ -W^T, & U^T \end{pmatrix} \tag{350}$$

T being the symbol of transposition.

Let us also decompose the vector $\vec{\delta}$ into its two main parts :

$$\vec{\delta} = \begin{pmatrix} \vec{\alpha} \\ \vec{\beta} \end{pmatrix} \quad ; \quad \vec{\alpha} = \begin{pmatrix} \delta\mathcal{G}_i \\ \delta\mathcal{L}_e \\ \delta\mathcal{G}_e \end{pmatrix} \quad ; \quad \vec{\beta} = \begin{pmatrix} \delta g_i \\ \delta \ell_e \\ \delta g_e \end{pmatrix} \tag{351}$$

Hence $\vec{\delta}_1 = A\vec{\delta}_0$ (to first order) can be written :

$$\left.\begin{aligned} \vec{\alpha}_1 &= U\vec{\alpha}_0 + V\vec{\beta}_0 \\ \vec{\beta}_1 &= W\vec{\alpha}_0 + X\vec{\beta}_0 \end{aligned}\right\} \tag{352}$$

We also have $\vec{\delta}_{-1} = A^{-1}\vec{\delta}_0$ (to first order), that is with (350) :

$$\left.\begin{aligned} \vec{\alpha}_{-1} &= X^T\vec{\alpha}_0 - V^T\vec{\beta}_0 \\ \vec{\beta}_{-1} &= -W^T\vec{\alpha}_0 + U^T\vec{\beta}_0 \end{aligned}\right\} \tag{353}$$

The space-time symmetries of the three-body problem corresponds to the following transformation :

$$A\begin{pmatrix} \vec{\alpha}_0 \\ \vec{\beta}_0 \end{pmatrix} = \begin{pmatrix} \vec{\alpha}_1 \\ \vec{\beta}_1 \end{pmatrix} \Longrightarrow A\begin{pmatrix} \vec{\alpha}_0 \\ -\vec{\beta}_0 \end{pmatrix} = \begin{pmatrix} \vec{\alpha}_{-1} \\ -\vec{\beta}_{-1} \end{pmatrix} \tag{354}$$

Hence, with (352) and (353) :

$$U = X^T \quad ; \quad V = V^T \quad ; \quad W = W^T \quad ; \quad X = U^T \tag{355}$$

and :

$$A^{-1} = \begin{pmatrix} U, & -V \\ -W, & X \end{pmatrix} \tag{356}$$

Let us now consider $\vec{\delta}_{1/2}$, $\vec{\alpha}_{1/2}$, $\vec{\beta}_{1/2}$, the small discrepancies obtained at mid-period for $\ell_i = \ell_{i(1/2)} = (\ell_{i0}+\ell_{i1})/2$ and the corresponding matrices A' and A" :

$$\vec{\delta}_{1/2} = A'\vec{\delta}_0 + O(\delta_0^{\ 2}) \quad ; \quad \vec{\delta}_1 = A''\vec{\delta}_{1/2} + O(\delta_{1/2}^{\ 2}) \quad ; \quad A''\,A' = A \tag{357}$$

$$A' = \begin{pmatrix} U', & V' \\ W', & X' \end{pmatrix} \quad ; \quad A'' = \begin{pmatrix} U'', & V'' \\ W'', & X'' \end{pmatrix} \tag{358}$$

A' and A" are also symplectic matrices, hence, as in (350) :

$$A'^{-1} = \begin{pmatrix} X'^T, & -V'^T \\ -W'^T, & U'^T \end{pmatrix} \quad ; \quad A''^{-1} = \begin{pmatrix} X''^T, & -V''^T \\ -W''^T, & U''^T \end{pmatrix} \tag{359}$$

On the other hand $\vec{\delta}_{-1/2} = A''^{-1}\vec{\delta}_0 + O(\delta_0^{\ 2})$ and thus the space-time symmetry already met in (354) and (355) leads also to :

$$U' = X''^T \ , \ V' = W''^T \ , \ W' = W''^T \ , \ X' = U''^T \tag{360}$$

that is :

$$A'' = \begin{pmatrix} X'^T, & V'^T \\ W'^T, & U'^T \end{pmatrix} \quad ; \quad A''^{-1} = \begin{pmatrix} U', & -V' \\ -W', & X' \end{pmatrix} \tag{361}$$

Notice the similarity between A^{-1} in (356) and A''^{-1} in (361).

Our purpose is to demonstrate that if the symmetric periodic orbit of interest is regular (i.e. if det.(I - A) ≠ 0), the neighbouring orbits of its family are also symmetric.

These neighbouring periodic orbits were determined by the research of $\vec{\delta}_1 = \vec{\delta}_0$ in the equations (312) and (330) that is (to first order) :

$$\vec{\delta}_1 = A\vec{\delta}_0 + \vec{B}\delta c \tag{362}$$

and :

$$\vec{\delta}_1 = A\vec{\delta}_0 + \vec{c}_j\ \delta m_j \tag{363}$$

where the effects of δc (a small variation of the angular momentum) or of δm_j (a small variation of the mass m_j) are introduced.

When looking for $\vec{\delta}_1 = \vec{\delta}_0$ these equations (362) and (363) are equivalent to :

$$\vec{\delta}_0 = (I - A)^{-1}\,\vec{B}\delta c \tag{364}$$

or :

$$\vec{\delta}_0 = (I - A)^{-1}\,\vec{C}_j\delta m_j \tag{365}$$

and, since $\det(I - A) \neq 0$, these equations have one and only one solution.

On the other hand the investigation of neighbouring symmetric periodic orbits is usually done by Poincaré method, that is by the research of orbits with two successive instant of space-time symmetry. We have seen in the previous Section that such orbits are indeed periodic.

The space-time symmetry conditions at ℓ_{i0} and at $\ell_{i(1/2)}$ in the vicinity of the orbit of interest are simply $\vec{\beta}_0 = 0$ and $\vec{\beta}_{1/2} = 0$, hence we must study the equivalent of (362) and (363) for half a period, that is :

$$\vec{\delta}_{1/2} = A'\vec{\delta}_0 + \vec{B}'.\delta c \tag{366}$$

and :

$$\vec{\delta}_{1/2} = A'\vec{\delta}_0 + \vec{C}'_j\delta m_j \tag{367}$$

Let us assume that $\vec{\beta}_0 = 0$, we will find in these two cases :

$$\vec{B}' = \begin{pmatrix}\vec{\alpha}_c\\ \vec{\beta}_c\end{pmatrix} \quad ; \quad \vec{\beta}_{1/2} = W'\vec{\alpha}_0 + \vec{\beta}_c\ .\ \delta c \tag{368}$$

$$\vec{C}'_j = \begin{pmatrix}\vec{\alpha}_j\\ \vec{\beta}_j\end{pmatrix} \quad ; \quad \vec{\beta}_{1/2} = W'\vec{\alpha}_0 + \vec{\beta}_j\ .\ \delta m_j \tag{369}$$

If W' is invertible the conditions $\vec{\beta}_0 = 0$, $\vec{\beta}_{1/2} = 0$ lead to the following vector $\vec{\delta}_0$:

$$\vec{\delta}_0 = \begin{pmatrix}\vec{\alpha}_0\\ \vec{0}\end{pmatrix}, \text{ with } \begin{array}{l}\vec{\alpha}_0 = -W'^{-1}\vec{\beta}_c\ .\ \delta c\\ \text{or}\\ \vec{\alpha}_0 = -\,W'^{-1}\vec{\beta}_j\ .\ \delta m_j\end{array} \tag{370}$$

The corresponding orbits are periodic and are necessarily also the periodic orbits given by (364) or (365) since those equations have only one solution. Furthermore all these neighbouring periodic orbits are also symmetric, as we intended. Thus, in order to achieve our demonstration, we must verify that $\det.(I - A) \neq 0$ implies W' invertible, that is $\det.(W') \neq 0$.

The relation between $(I - A)$ and W' is simple :

$$I = A''\ .\ A''^{-1} \quad ; \quad A = A''A' \quad ; \quad I - A = A''(A''^{-1} - A') \tag{371}$$

that is, with (358) and (361) :

$$(I - A) = - 2A'' \cdot \begin{pmatrix} 0, & V' \\ W', & 0 \end{pmatrix} \tag{372}$$

The symplectic matrix A" has a determinant equal to one and thus :

$$|\det.(I - A)| = 2 \cdot |\det(V') \cdot \det(W')| \tag{373}$$

Hence finally we indeed find that $\det(I - A) \neq 0$ implies $\det.(W') \neq 0$ and then if a regular periodic orbit has space-time symmetries of types I and/or II all orbits of its non-trivial three-parameter family of periodic orbits have the same space-time symmetries, while if the space-time symmetries are of types III and/or IV they also appear in all periodic orbits of the continuous non-trivial two-parameter family that satisfies the condition of equality of the two suitable masses.

Conversely, for instance, if in a family of asymmetrical periodic orbits we find an orbit with symmetries of type I and/or II this periodic orbit is necessarily singular.

Let us notice the following remarks.

1) The demonstration of this Section assumes that the description parameter ℓ_i of the Hamiltonian system (301)-(303) of Section 10.5.1 really is a decription parameter, at least in the vicinity of ℓ_{i0} and $\ell_{i(1/2)}$. This means that the variation of ℓ_i is monotonic in terms of t, at least in some neighbourhood of the symmetry conditions.

Fortunately, if this is not the case, it is always possible to choose another suitable description parameter such as ℓ_e, g_i, g_e, $\ell_i + g_i$, etc ... and all the conclusions remain the same.

Similar considerations can be made for plane motions and for the corresponding Hamiltonian system (103)-(105) of Chapter 7.

2) The equations (370) are much simpler than the equivalent general equations (364), (365) and the numerical computations of families of symmetrical periodic orbits are much easier than those of families of asymmetrical periodic orbits. These computations require the integration of only one half or even sometimes one quarter of the period, and the initial values of the angular orbital elements are simple and given by the symmetry conditions (332)-(335).

Furthermore, notice that the dimensionality of the matrix W' is only one half of that of the matrix (I - A), and it is more rarely singular.

3) Since families of periodic orbits meet each other at branching singular orbits (i.e. orbits for which $\det.(I - A) \neq 0$), the equation (373) leads us to conclude that the families of symmetrical periodic orbits meet asymmetrical

families at orbits for which det.(V') = 0, while they meet other symmetrical families at orbits for which det.(W') = 0.

10.7 THE VICINITY AND THE STABILITY OF PERIODIC ORBITS

10.7.1 Definitions and generalities

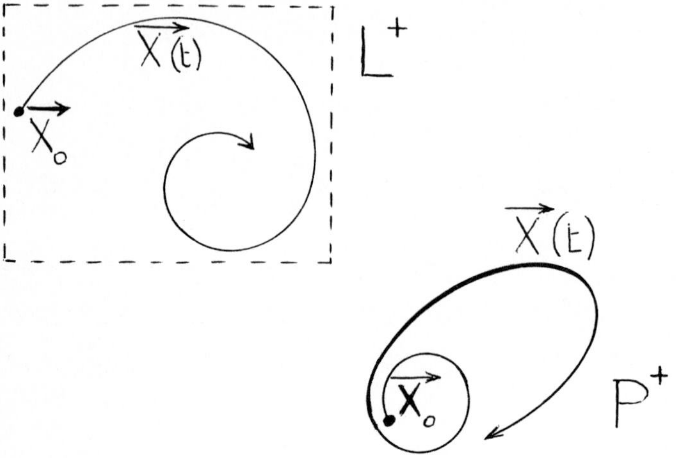

Fig. 58. The Lagrange stability L^+ *(for all future times* $\vec{X}(t)$ *remain bounded) and the Poisson stability* P^+ *(the solution* $\vec{X}(t)$ *returns an infinite number of times in any vicinity of its past states).*

The word "stability" has a great number of different meanings (Ref.177, 178) so it is always necessary to define this concept.

A trajectory X(t) is said to be "Lagrange stable" for the future (we will write L^+) iff it remains bounded for all future times (Fig. 58), and it is said Poisson stable for the future (P^+) iff it returns an infinite number of times in any vicinity of its past states.

Symmetrical definitions for the past lead to the stabilities L^- and P^-.

The Hill stability of Section 9.1 (circular restricted three-body problem) is a kind of Lagrange stability : for sufficiently large values of the Jacobi constant and suitable initial conditions the mutual distances will remain forever bounded. However, notice that the velocities are not necessarily bounded (collisions) and thus that Hill stable motions are not always bounded in phase space.

The other types of stability essentially consider the following question : "Let us consider a given solution of the problem of interest (the nominal solution), how will neighbouring solutions evolve when the time goes to infinity?"

The stability or instability are essential properties of physical laws. Stable physical law lead to deterministic phenomena : when repeating the same experiment (i.e. when doing again, almost exactly, the same experiment...) one obtains almost exactly the same results. On the contrary unstable physical laws lead to stochastic phenomena and to the analysis of statistical elements such as the pressure and temperature in the kinetic theory of gases. The other elements, for instance the velocity of a given molecule, are highly unstable, unpredictable and irreproducible...

Modern statistical mechanics is a very useful tool as useful in its own domain as the deterministic mechanics.

In Astronomy and Celestial Mechanics, the development of instabilities is slow. It is, however, sufficiently fast to explain phenomena such as the Kirkwood gaps of the asteroid belt (Fig. 59).

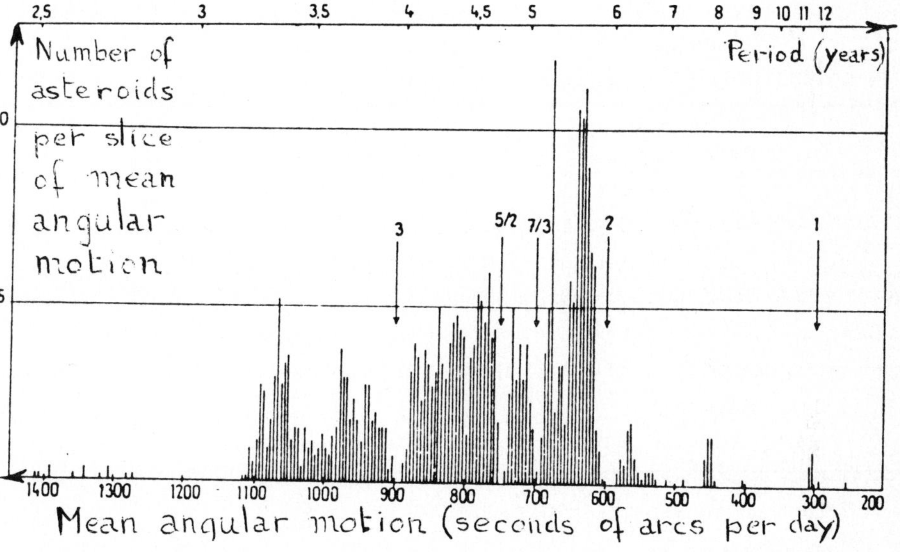

Fig. 59. Distrubution of the periods and the mean angular motions of the 1800 main asteroids.

This distrubution is irregular and shows four Kirkwood gaps. These gaps correspond to the simple ratios 2, 7/3, 5/2 and 3 between the mean angular motion of the asteroid and that of Jupiter.

On the other hand the ratio 1 corresponds to the stable orbits of the 15 Trojan asteroids already presented in section 8.3 and in Fig. 14, p. 52.

Source : Annuaire du Bureau des Longitudes, 1974.

The perturbations of Jupiter destabilize some "resonant orbits", i.e. orbits with periods in a simple ratio to the period of Jupiter.

We will henceforth use the following usual definition of stability.

Let us consider a set of differential equations with a description parameter t (generally the "time"), some state parameters $x_1, x_2, \ldots, x_n$, their "state vector" $\vec{X} = (x_1, x_2, \ldots, x_n)$ and the equations of motion :

$$\vec{dX}/dt = f(\vec{X},t) \tag{374}$$

For instance, in the three-body problem, the components $x_1, x_2, \ldots, x_n$ can be the coordinates and the veloçity components of the three bodies.

A "solution" or a "trajectory" of the system (374) is a continuous function $\vec{X}(t)$ always satisfying :

$$d\,\vec{X}(t)\,/dt = f\,[\,\vec{X}(t),t\,] \tag{375}$$

We will assume that $f(\vec{X},t)$ is continuous and continuously differentiable, at least in some vicinity of the solution of interest. This will imply the existence and the uniqueness of the solution $\vec{X}(t)$ associated with a given point $\vec{X}_0, t_0$ (these parameters $\vec{X}_0, t_0$ are called "initial conditions" even if the integration of (375) is carried out towards the past).

Notice the following.

A) The trajectory $\vec{X}(t)$ is a continuous function of its initial conditions $\vec{X}_0, t_0$.

B)These properties of existence, uniqueness and continuity do not require a continuous and continuously differentiable function $f(\vec{X},t)$. For instance it is sufficient that $f(\vec{X},t)$ be continuous in terms of $\vec{X}$ and t and locally Lipschitzian in terms of $\vec{X}$ (i.e. for any given bounded set B of the $\vec{X}$,t space the ratio $||f(\vec{X}_1,t) - f(\vec{X}_2,t)||\,/||\vec{X}_1 - \vec{X}_2||$, with $\vec{X}_1 \neq \vec{X}_2$, $(\vec{X}_1,t) \in B$ and $(\vec{X}_2,t) \in B$, is bounded).

C) In the three or n-body problem the function $f(\vec{X},t)$ is discontinuous at collisions. Fortunately, binary collisions can be regularized (Section 6 and Ref.55-58), that is suitable transformations of the $\vec{X}$,t variables lead to regular equations of motion.

However, triple and multiple collisions cannot be regularized (except for some very particular triple or multiple collisions of the rectilinear n-body problem with several infinitesimal masses, Ref.47-54), and hence this Section 10.7 will not be applicable to three-body motions with triple collisions.

Let us now consider a "nominal solution" $\vec{X}(t)$ starting at $\vec{X}_0, t_0$ and its neighbouring solutions $\vec{X}(t) + \delta\vec{X}(t)$ starting at $\vec{X}_0 + \delta\vec{X}_0, t_0$.

By definition, we will consider that the nominal solution $\vec{X}(t)$ is "stable for the future" if and only if for any positive ε there exists a positive ε_0 such that :

$$||\overrightarrow{\delta X_0}|| < \varepsilon_0 \text{ implies } ||\overrightarrow{\delta X}(t)|| < \varepsilon \text{ for all future t} \tag{376}$$

A)

$dx/dt = x$

$x = x_0 \exp(t)$

$\delta x = \delta x_0 \exp(t)$

instability for the future

stability for the past

B)

$dx/dt = -x$

$x = x_0 \exp(-t)$

$\delta x = \delta x_0 \exp(-t)$

stability for the future

instability for the past

C)

$dx/dt = 0$

$x = x_0$

$\delta x = \delta x_0$

stability for both

past and future

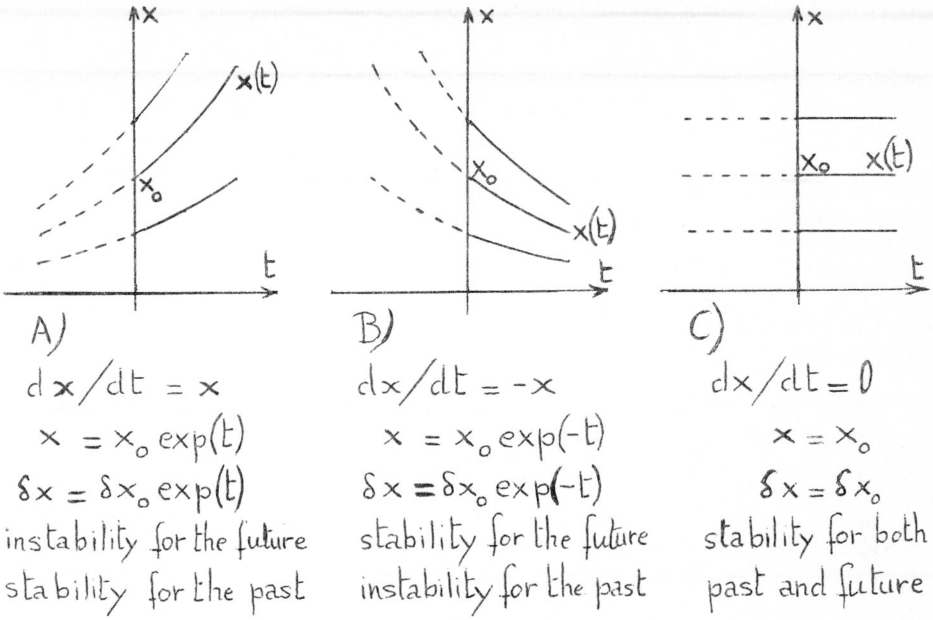

Fig. 60. Some simple examples of stable and unstable trajectories.

Because of the continuity of the set of solutions this definition of the stability for the future (or for the past) is independent of the initial time t_0 and the corresponding initial point $\vec{x}_0$. It is the "Liapunov stability".

This definition can of course be applied to any solution, but is generally reserved for periodic or quasi-periodic solutions.

10.7.2   The evolution of ignorable parameters. The orbital stability. The "in plane" stability.

We have seen in Sections 10.5.2 and 10.5.3 that regular periodic orbits belong to a non-trivial one-parameter family of periodic orbits with the same three masses and to a non-trivial three-parameter family of periodic orbits with arbitrary masses.

Let us analyse the stability of a given periodic orbit (regular or not) and let us consider an arbitrary neighbouring periodic orbit.

We can always choose a suitable fast-moving parameter (such as ℓ_i, ℓ_e, $\ell_i + g_i$, etc...) as description parameter. The two periodic orbits then have the same period with respect to the description parameter, and they remain indefinitely close to each other except for their ignorable parameters : the time t and the longitude of the nodes h_i (or h_e equal to $h_i + \pi$). Indeed during a period the increase of t is equal to T (the "period" or "temporal period") and the increase of h_i is equal to α (the "angle of rotation") ; in general T and α are not the same for the two periodic orbits of interest, and then the discrepancies δt and δh_i increase linearly with the number of periods.

Notice that the "scale transformation" of Section 5.3 gives neighbouring periodic orbits with the same angle of rotation α but with slightly different periods T, the products T^2h^3 being equal for these orbits.

Hence stríct application of the definition of stability, as given by (376) in the previous section, leads to the instability of all periodic orbits of the three-body problem (and even for all orbits...).

For a meaningful analysis of the stability of periodic orbits we must modify our point of view, and this generally leads to the following.

A) Either the ignorable parameters h_i, h_e and t are disregarded, they are "ignored". This leads to the "orbital stability" or "Poincaré stability".

B) Or only neighbouring orbits with the same three masses, the same energy and the same angular momentum are considered.

C) Many studies have a mixed point of view ; for instance, they neglect h_i, h_e and t and impose given values for the three masses. It is also customary to impose a given value for the energy integral but to let the angular momentum be arbitrary.

Most studies of the stability of plane periodic orbits neglect the neighbouring non-plane motions (they obtain the "in-plane" stability).

We will see in the next section that these different points of view lead to the same first-order analysis of the stability. However, the full analysis of stability leads to other results, and it seems that in the above case B the ignorable parameters h_i, h_e and t almost always slowly destabilize the system of interest, even if it is stable with respect to the other parameters (this last conjecture is not true in the elliptic restricted case).

10.7.3 The first-order analysis

The first-order study of the vicinity of a periodic orbit has already been partly presented in Sections 10.5.2 and 10.5.3. In the three-dimensional case the main points are the following.

1) Let us use the simplified Hamiltonian system (301), (302) or any other equivalent system with some suitable fast-moving parameter such as ℓ_e, $\ell_i + g_i$, etc...

2) If for instance ℓ_i is the description parameter, it will be ℓ_{io} initially and ℓ_{in} after n periods of the periodic orbit of interest.

3) The six state parameters $\mathcal{G}_i$, $\mathcal{L}_e$, $\mathcal{G}_e$, g_i, ℓ_e, g_e will be $\mathcal{G}_{in}$, ..., g_{en} at

ℓ_{in} for the periodic orbit of interest (with $\mathcal{G}_{in} = \mathcal{G}_{io}$; $\mathcal{L}_{en} = \mathcal{L}_{eo}$; $\mathcal{G}_{en} = \mathcal{G}_{eo}$ and, modulo 2π, $\ell_{in} = \ell_{io}$, $g_{in} = g_{io}$, $\ell_{en} = \ell_{eo}$, $g_{en} = g_{eo}$).

4) For neighbouring orbits we will obtain $\mathcal{G}_{in} + \delta\mathcal{G}_{in}$, $\mathcal{L}_{en} + \delta\mathcal{L}_{en}$, ..., $g_{en} + \delta g_{en}$ at ℓ_{in} ; the evolution of these small discrepancies is given, to first order, by the equations (309), (312)-(316) and (330), that is by the following.

$$\vec{\delta}_0 = \begin{pmatrix} \delta\mathcal{G}_{io} \\ \delta\mathcal{L}_{eo} \\ \delta\mathcal{G}_{eo} \\ \delta g_{io} \\ \delta\ell_{eo} \\ \delta g_{eo} \end{pmatrix} ; \quad \vec{\delta}_n = \begin{pmatrix} \delta\mathcal{G}_{in} \\ \delta\mathcal{L}_{en} \\ \delta\mathcal{G}_{en} \\ \delta g_{in} \\ \delta\ell_{en} \\ \delta g_{en} \end{pmatrix} ; \quad n \in \mathbb{Z} \tag{377}$$

If we consider neighbouring orbits with the same three masses and the same integrals of motion, the Floquet theory leads to the following, with the monodromy matrix or Floquet matrix A :

$$\vec{\delta}_1 = A\vec{\delta}_0 + \mathcal{O}(\delta_0^2) \tag{378}$$

$$\vec{\delta}_n = A^n\vec{\delta}_0 + \mathcal{O}(\delta_0^2) \tag{379}$$

If we consider neighbouring orbits with masses $m_j + \delta m_j$ and integrals of motion $h + \delta h$ and $c + \delta c$ we obtain :

$$\vec{\delta}_1 = A\vec{\delta}_0 + \vec{B}_c\delta c + \vec{B}_h\delta h + \sum_{j=1}^{3} \vec{C}_j\delta m_j + \mathcal{O}(\delta_0^2, \delta c^2, \delta h^2, \delta m_j^2) \tag{380}$$

which leads to :

$$\vec{\delta}_n = A^n\vec{\delta}_0 + Q_n(\vec{B}_c\delta c + \vec{B}_h\delta h + \sum_{j=1}^{3} \vec{C}_j\delta m_j) + \mathcal{O}(\delta_0^2, \delta c^2, \delta h^2, \delta m_j^2) \tag{381}$$

with :

$$\begin{aligned} &n > 0 \text{ implies } Q_n = I + A + A^2 + \ldots + A^{(n-2)} + A^{(n-1)} \\ &n < 0 \text{ implies } Q_n = -(A^{-1} + A^{-2} + \ldots + A^{n}) \end{aligned} \tag{382}$$

In both cases, if the matrix (A - I) is invertible, Q_n is given by :

$$Q_n = (A^n - I).(A - I)^{-1} \tag{383}$$

In these expressions I is the 6 × 6 identity matrix, A is a 6 × 6 real symplectic matrix and $\vec{B}_c$, $\vec{B}_h$, $\vec{C}_j$ are five column vectors with six components.

A, $\vec{B}_c$, $\vec{B}_h$, $\vec{C}_j$ are generally found either by numerical computation of suitable neighbouring orbits or by numerical integration of the variational equations of the periodic orbit of interest.

The scale transformations of the three-body problem lead to the following.

If, for small ε :

$$\vec{\delta}_0 = \varepsilon \begin{pmatrix} \mathcal{G}_{io} \\ \mathcal{L}_{eo} \\ \mathcal{G}_{eo} \\ 0 \\ 0 \\ 0 \end{pmatrix} \;;\quad \delta c = \varepsilon c \;;\quad \delta h = -\,2\varepsilon h \;;\quad \delta m_j = 0 \tag{384}$$

we obtain, to first order : $\vec{\delta}_1 = \vec{\delta}_0$.

On the other hand :

$$\vec{\delta}_0 = 0 \;;\quad \delta c = 0 \;;\quad \delta h = 5\varepsilon h \;;\quad \delta m_j = \varepsilon m_j \tag{385}$$

implies to first order $\vec{\delta}_1 = 0$.

Hence :

$$(A - I) \begin{pmatrix} \mathcal{G}_{io} \\ \mathcal{L}_{eo} \\ \mathcal{G}_{eo} \\ 0 \\ 0 \\ 0 \end{pmatrix} = 2h\vec{B}_h - c\vec{B}_c \tag{386}$$

and :

$$5h\vec{B}_h + \sum_{j=1}^{3} m_j\vec{C}_j = 0 \tag{387}$$

5) The remaining Delaunay parameters $\mathcal{L}_i$, $\mathcal{H}_i$, $\mathcal{H}_e$ will be given by (301) and (296), that is :

$$\mathcal{L}_i = \mathcal{L}_i(h,\mathcal{G}_i,\mathcal{L}_e,\mathcal{G}_e,\ell_i,g_i,\ell_e,g_e,c,m_1,m_2,m_3) \tag{388}$$

$$\left.\begin{aligned} \mathcal{H}_i &= (c^2 + \mathcal{G}_i^2 - \mathcal{G}_e^2)/2c \\ \mathcal{H}_e &= (c^2 - \mathcal{G}_i^2 + \mathcal{G}_e^2)/2c \\ \mathcal{H}_i + \mathcal{H}_e &= c \end{aligned}\right\} \quad (389)$$

6) Finally, the ignorable parameters h_i, h_e and t are given by the following.

6.1) At ℓ_{in} for the periodic orbit of interest :

$$\left.\begin{aligned} h_{in} &= h_{io} + n\alpha \\ h_{en} &= h_{eo} + n\alpha \end{aligned}\right\} \qquad \left.\begin{aligned} &h_e = h_i + \pi \\ &\alpha \text{ is the "angle of rotation" (per period)} \end{aligned}\right\} \quad (390)$$

$$t_n = t_o + nT \qquad T \text{ is the "period"} \tag{390}$$

6.2) For the neighbouring orbits, between ℓ_{in} and $\ell_{i,n+1}$:

$$\delta h_{i,n+1} - \delta h_{i,n} = \delta h_{e,n+1} - \delta h_{en} = \vec{U}\vec{\delta}_n + b_c\delta c + b_h\delta h + \sum_{j=1}^{3} c_j\delta m_j + \mathcal{O}(\delta_n^2,\delta c^2,\delta h^2,\delta m_j^2) \tag{392}$$

$$\delta t_{n+1} - \delta t_n = \vec{U}'\vec{\delta}_n + b'_c\delta c + b'_h\delta h + \sum_{j=1}^{3} c'\delta_j m_j + \mathcal{O}(\delta_n^2,\delta c^2,\delta h^2,\delta m_j^2) \tag{393}$$

Like A, $\vec{B}_c$, $\vec{B}_h$, $\vec{C}_j$ the vectors $\vec{U}$ and $\vec{U}'$ and the scalars b_c, b_h, c_j, b'_c, b'_h, c'_j can be obtained by suitable numerical computations but they are generally not considered.

The scale transformations (384) and (385) of the three-body problem imply some simple relations. Indeed $\delta t_1 - \delta t_0$ is $3\varepsilon T$ in the case (384) and $-5\varepsilon T$ in the case (385), while $\delta h_{i1} - \delta h_{io}$ is zero in both cases.

Hence :

$$\vec{U} \cdot \begin{pmatrix} \mathcal{G}_{io} \\ \mathcal{L}_{eo} \\ \mathcal{G}_{eo} \\ 0 \\ 0 \\ 0 \end{pmatrix} + cb_c - 2hb_h = 0 \quad ; \quad 5hb_h + \sum_{j=1}^{3} m_j c_j = 0 \tag{394}$$

$$\vec{U}' \cdot \begin{pmatrix} \mathcal{G}_{io} \\ \mathcal{L}_{eo} \\ \mathcal{G}_{eo} \\ 0 \\ 0 \\ 0 \end{pmatrix} + cb'_c - 2hb'_h - 3T = 0 \quad ; \quad 5hb'_h + \sum_{j=1}^{3} m_j c'_j + 5T = 0 \tag{395}$$

The "eleventh local integral" Q_D of the three-body problem was given in (80) (Section 5.3). As the scale transformation, this integral gives some simple relations between $\vec{U}$, $\vec{U}'$, b_c, b_h, b'_c, b'_h.

At ℓ_{in} the first-order expression of Q_D is :

$$Q_{Dn} = 2\{\mathcal{G}_{in}\delta g_{in} + \mathcal{L}_{en}\delta\ell_{en} + \mathcal{G}_{en}\delta g_{en} + c\delta h_{in} + 3t_n\delta h + 2h\delta t_n\} \tag{396}$$

As $\mathcal{G}_{in}$, $\mathcal{L}_{en}$, $\mathcal{G}_{en}$ are independent of n, let us put :

$$(0,0,0,\mathcal{G}_{io},\mathcal{L}_{eo},\mathcal{G}_{eo}) = \text{row vector } \vec{W} \tag{397}$$

The equality (to first order) between Q_{D0} and Q_{D1} will lead to :

$$\vec{W}(\vec{\delta}_1 - \vec{\delta}_0) + c(\delta h_{i1} - \delta h_{i0}) + 3T\delta h + 2h(\delta t_1 - \delta t_0) = 0 \tag{398}$$

that is :

$$\left.\begin{array}{l} \vec{W}\left[(A - I)\vec{\delta}_0 + \vec{B}_c\delta c + \vec{B}_h\delta h\right] + c(\vec{U}\vec{\delta}_0 + b_c\delta c + b_h\delta h) + \\ + 3T\delta h + 2h(\vec{U}'\vec{\delta}_0 + b'_c\delta c + b'_h\delta h) \end{array}\right\} = 0 \tag{399}$$

This result being valid for all $\vec{\delta}_0$, δc, δh it implies :

$$\vec{W}(A - I) + c\vec{U} + 2h\vec{U}' = 0 \tag{400}$$

$$\vec{W}\vec{B}_c + cb_c + 2hb'_c = 0 \tag{401}$$

$$\vec{W}\vec{B}_h + cb_h + 3T + 2hb'_h = 0 \tag{402}$$

The first-order long-term evolution of δh_i and δt is given by the summation of (392) and (393), that is, if $(A - I)$ is invertible :

$$\left.\begin{aligned} \delta h_{in} - \delta h_{io} &= n(b_c\delta c + b_h\delta h) + \vec{U}(A^n - I)(A - I)^{-1}\vec{\delta}_0 + \\ &+ \vec{U}\big(A^n - nA + (n-1)I\big)(A - I)^{-2}(\vec{B}_c\delta c + \vec{B}_h\delta h + \sum_{j=1}^{3}\vec{C}_j\delta m_j) + \\ &+ O(\delta_o^2, \delta c^2, \delta h^2, \delta m_j^2) \end{aligned}\right\} \tag{403}$$

and similarly :

$$\left.\begin{aligned} \delta t_n - \delta t_o &= n(b'_c\delta c + b'_h\delta h) + \vec{U}'(A^n - I)(A - I)^{-1}\vec{\delta}_0 + \\ &+ \vec{U}'\big(A^n - nA + (n-1)I\big)(A - I)^{-2}(\vec{B}_c\delta c + \vec{B}_h\delta h + \sum_{j=1}\vec{C}_j\delta m_j) + \\ &+ O(\delta_o^2, \delta c^2, \delta h^2, \delta m_j^2) \end{aligned}\right\} \tag{404}$$

We now have all the elements necessary for a long-term first-order study of neighbouring orbits.

The main equations are the equations (377)-(383) and (403), (404), where <u>the only first-order variable coefficients are n and A^n</u>.

Hence the main question is the following : "What are the evolutions of A^n and $A^n\vec{\delta}_o$ when n goes to $\pm\infty$?"

Let us assume that $\vec{\delta}_o$ is an eigenvector of A and that μ is the corresponding eigenvalue (μ is a "Floquet multiplier").

$$\begin{aligned} A\vec{\delta}_o &= \mu\vec{\delta}_o \\ A^n\vec{\delta}_o &= \mu^n\vec{\delta}_o \end{aligned} \tag{405}$$

Hence, since $A^n\vec{\delta}_o$ is equivalent to $\vec{\delta}_n$ in (379), if $|\mu| > 1$ the periodic orbit of interest is not stable, it has an "exponential instability" for the future, while if $|\mu| < 1$ it has an exponential instability for the past.

Let us recall that, the three-body system (301), (302) being Hamiltonian, the monodromic matrix A is real and "symplectic", that is :

$$A^T J A = J \tag{406}$$

with :

A^T transposed matrix of A

$$J = \begin{pmatrix} 0 & 0 & 0 & -1 & 0 & 0 \\ 0 & 0 & 0 & 0 & -1 & 0 \\ 0 & 0 & 0 & 0 & 0 & -1 \\ 1 & 0 & 0 & 0 & 0 & 0 \\ 0 & 1 & 0 & 0 & 0 & 0 \\ 0 & 0 & 1 & 0 & 0 & 0 \end{pmatrix} = \begin{pmatrix} 0 & -I \\ I & 0 \end{pmatrix} \tag{407}$$

This implies that if μ is an eigenvalue of A, μ^{-1} is also an eigenvalue, indeed :

$$\mu\vec{\delta}_o = A\vec{\delta}_o \quad ; \quad \vec{\delta}_o \text{ : eigenvector} \tag{408}$$

hence :

$$\mu\vec{\delta}_o^T = \vec{\delta}_o^T A^T \quad ; \quad \vec{\delta}_o^T \text{ : row vector} \tag{409}$$

$$\mu\vec{\delta}_o^T J A = \vec{\delta}_o^T A^T J A = \vec{\delta}_o^T J \tag{410}$$

and finally :

$$(\vec{\delta}_o^T J)A = \mu^{-1}(\vec{\delta}_o^T J) \qquad \left\{ \begin{array}{ll} (\vec{\delta}_o^T J) & \text{: row eigenvector} \\ \mu^{-1} & \text{: eigenvalue} \end{array} \right\} \tag{411}$$

Thus there are only the two following main cases :

I) If the monodromy matrix A has one or several eigenvalues with a modulus larger than one it has the same number of eigenvalues with a modulus smaller than one (Fig. 61) and the periodic orbit of interest is unstable for both past and future.

The instability is of the "exponential instability" type, as shown by (405), and the eigenvalues with a modulus smaller than one correspond to the "stable subset" of the periodic orbit of interest (i.e. the subset of trajectories asymptotic to the periodic orbit : $\vec{\delta}_n \to 0$ when $n \to \infty$). Symmetrically the eigenvalues with a modulus larger than one correspond to the "unstable subset" that is to the subset of trajectories that were asymptotic in the past to the periodic orbit : $\vec{\delta}_n \to 0$ when $n \to -\infty$.

II) In the opposite case all the eigenvalues of A have a modulus equal to one and are then on the unit circle of the complex plane (Fig. 61).

If these eigenvalues are different, the matrix A^n remains bounded for all n. If two or several eigenvalues are equal, A^n can escape to infinity, but it is a polynomial escape and no longer an exponential one. Thus the first-order approximation then leads either to the stability or to a polynomial instability.

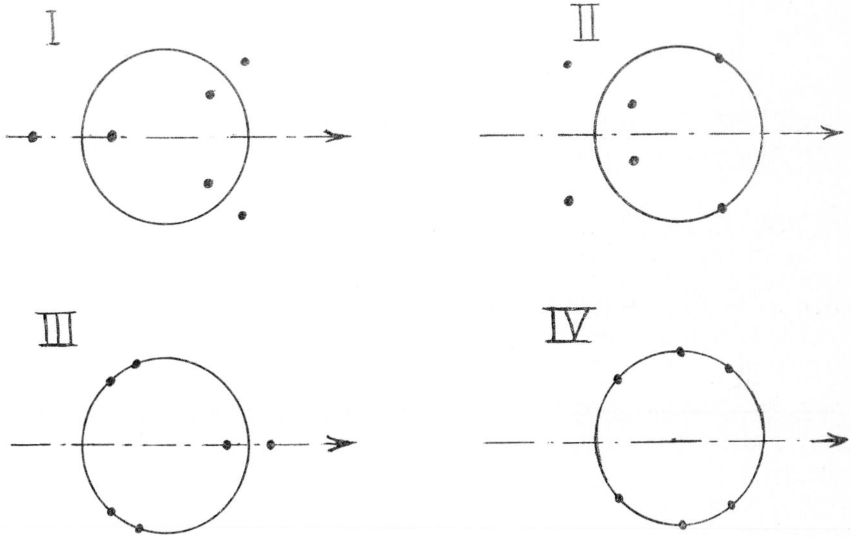

Fig. 61. The six eigenvalues of the monodromy matrix A in the complex plane and the stability of the motion.

If μ is an eigenvalue, μ^{-1}, $\bar{\mu}$ and $\bar{\mu}^{-1}$ are also eigenvalues.

Case I, II, III: some eigenvalues are not on the unit circle and the motion is exponentially unstable for both past and future.

Case IV: all eigenvalues are on the unit circle, this is the "critical case" and the first-order study is not conclusive for stability nor instability. However the eventual instabilities are slow and no longer exponential.

The eigenvalues of the monodromy matrix A are also called "Floquet multipliers".

These latter results have little interest, and the higher-order effects can either stabilize or on the contrary, destabilize the first-order results. The only certainty is the absence of exponential instability.

This case when all eigenvalues are on the unit circle is called the "critical case" ; the periodic orbit is "first-order stable" or "stable in the linear approximation". The exponential instabilities are avoided but the demonstration of either stability or instability requires the study of higher-order effects (see Section 10.7.7).

The study of the stability of Eulerian and Lagrangian motions was presented in Section 8.2.; Fig. 13 gives the results of the first-order analysis, with the zone of exponential instability and the two zones of "first-order stability", but Section 8.2.2 shows that the "first-order stability" does not imply true stability.

In Fig. 18, 19 and 20 of Section 8.5 three families of periodic orbits of the plane circular Hill problem were presented. Some of these orbits are exponentially unstable (Fig. 18) but the "stability" indicated for the other orbits is only a first-order "in-plane" stability.

These restrictions are generally implicit in most stability studies of periodic orbits.

Notice that we have presented the monodromy matrix A with the system (301) (302) and the description parameter ℓ_i. This matrix A is a function of ℓ_{io}, it depends also on the choice of the description parameter and of the state parameters. However, the eigenvalues of A, the "Floquet multipliers", express intrinsic properties of the set of neighbouring orbits and, for a given periodic orbit, they remain the same for all possible expressions of the monodromy matrix A.

10.7.4 Simple cases of the first-order analysis

10.7.4.1 Rectilinear periodic orbits

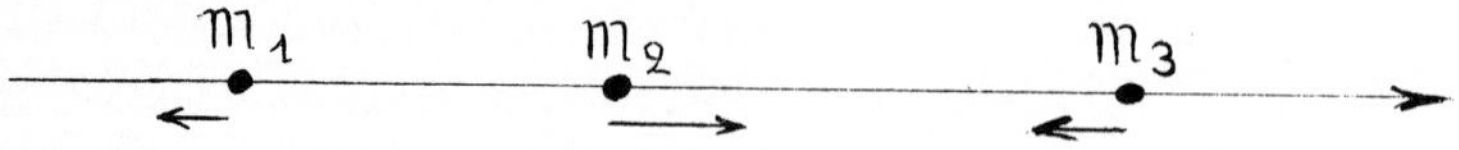

Fig. 62. A rectilinear three-body motion.
$x_1<x_2<x_3$ implies $d^2x_1/dt^2>0$ and $d^2x_3/dt^2<0$ as long as there is no collision.

In rectilinear three-body motions (Fig. 62) binary collisions are unavoidable. Fortunately these collisions can be regularized (Chapter 6 and Ref.20,55-58) and the motion remains uniquely defined as long as triple collisions are avoided.

In rectilinear three-body motions the succession of the three bodies remains constant and we will choose $x_1 \leqq x_2 \leqq x_3$ as in Fig. 62.

The rectilinear three-body problem has many periodic solutions, such as the solution of Fig. 63, and their stability can be analysed by the method of the previous section.

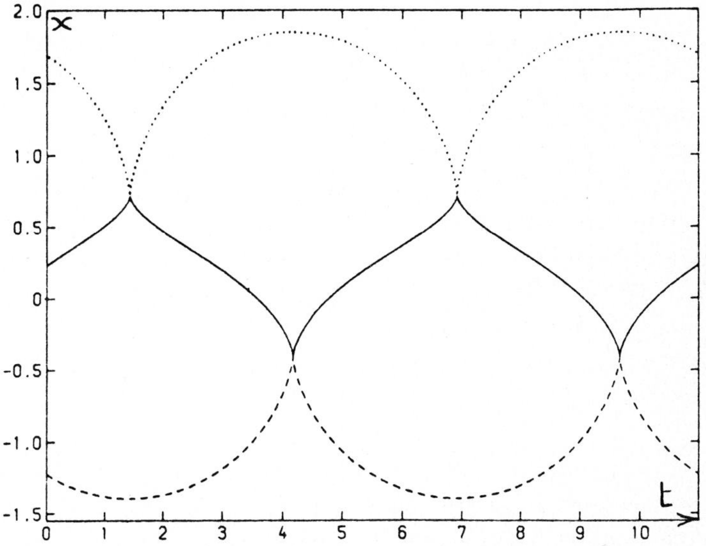

Fig. 63. (Ref.176) Abscissas $x_1(t), x_2(t), x_3(t)$ for two periods of a simple periodic orbit with two collisions per period.

The masses are $m_1 = 0.34; m_2 = 0.48; m_3 = 0.18$ and the instants of collision are instants of space-time symmetry (limit of types I and II).

Source : Hénon, M. [176].

Because of the simplicity of rectilinear motions their matrix A has the following form :

$$A = \begin{pmatrix} a_1 & 0 & 0 & b_1 & 0 & 0 \\ 0 & a_2 & 0 & 0 & b_2 & 0 \\ 0 & 0 & a_2 & 0 & 0 & b_2 \\ c_1 & 0 & 0 & d_1 & 0 & 0 \\ 0 & c_2 & 0 & 0 & d_2 & 0 \\ 0 & 0 & c_2 & 0 & 0 & d_2 \end{pmatrix} \qquad (412)$$

Hence A can be decomposed into three matrices, a matrix A_1 of longitudinal motions :

$$A_1 = \begin{pmatrix} a_1, & b_1 \\ c_1, & d_1 \end{pmatrix} \tag{413}$$

and two equal matrices of transversal motions :

$$A_2 = \begin{pmatrix} a_2, & b_2 \\ c_2, & d_2 \end{pmatrix} \tag{414}$$

The equality of the two transversal matrices is a consequence of the properties of rectilinear motions : transversal motions in the second and third dimensions are similar.

These two matrices A_1 and A_2 are "symplectic" and here this implies only $\det.(A_1) = 1$ and $\det.(A_2) = 1$.

The characteristic polynomials of A_1 and A_2 are the following :

$$\mu^2 - (a_1 + d_1)\mu + 1 = 0 \tag{415}$$

and :

$$\mu^2 - (a_2 + d_2)\mu + 1 = 0 \tag{416}$$

It is then very easy to obtain the modulus of the eigenvalues μ and thus :

$$\left.\begin{array}{l} \text{If } |a_1 + d_1| > 2 \text{ the motion is longitudinally unstable} \\ \text{If } |a_2 + d_2| > 2 \text{ the motion is transversely unstable} \\ \text{If } |a_1 + d_1| \leqq 2 \text{ and } |a_2 + d_2| \leqq 2 \text{ the motion has} \\ \text{the first-order stability (critical case).} \end{array}\right\} \tag{417}$$

This analysis was done by Michel Hénon in the Ref. 176 for all symmetric rectilinear periodic orbits with only two collisions per period (such as the orbit of Fig. 63). One and only one such orbit exists for any given three-body mass ratio and they constitute the simplest non-trivial two-parameter family of symmetric rectilinear periodic orbits.

The results of this analysis of stability are presented in terms of the three masses in the mass diagram of Fig. 64 and a detailed and accurate presentation is done in the Ref. 176 .

Notice that when $m_1 = m_3$ the periodic orbits of this family have four, and not only two, instants of space-time symmetry per period. The symmetries at quarters of period correspond to $r_{12} = r_{23}$ and $x'_1 = x'_3$, they are of types III and IV of Fig. 56 (Section 10.6.1).

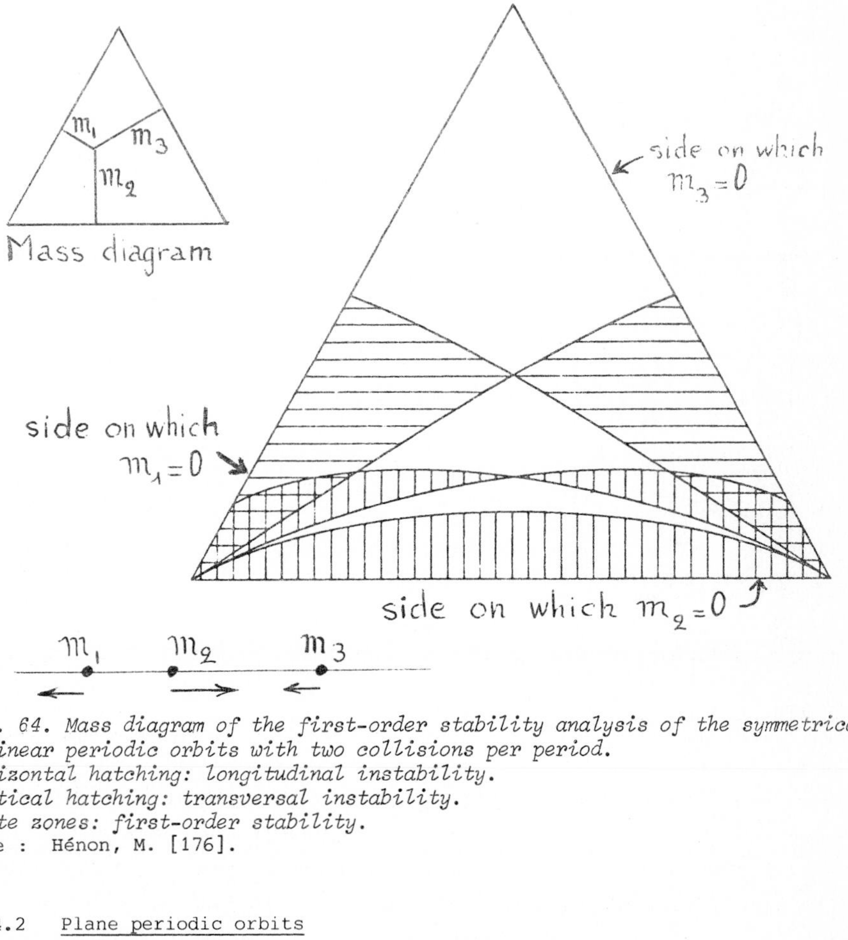

Fig. 64. Mass diagram of the first-order stability analysis of the symmetrical rectilinear periodic orbits with two collisions per period.
Horizontal hatching: longitudinal instability.
Vertical hatching: transversal instability.
White zones: first-order stability.
Source : Hénon, M. [176].

10.7.4.2 Plane periodic orbits

As emphasized by all review papers (Ref.131,132) a large majority of the studies of periodic orbits deal with plane periodic orbits.

For plane periodic orbits the analysis requires, of course, adapted orbital elements. Those of the system (301)-(302) used for three-dimensional systems are singular here, and the in-plane analysis can use the system (103)-(105) of the Chapter 7. The main point is always to extract six significant linear infinitesimals from the twelve components of $\delta\vec{r}$, $\delta\vec{R}$, $\delta\vec{v}$, $\delta\vec{V}$ after a period and this is easy since the integrals of motion h and $\vec{c}$ are not modified and since the time and the longitude of nodes are ignorable.

For plane periodic orbits, the matrix A has of course fewer symmetries than in (412) for rectilinear periodic orbits :

$$A = \begin{pmatrix} a_1 & b_1 & 0 & c_1 & d_1 & 0 \\ e_1 & f_1 & 0 & g_1 & h_1 & 0 \\ 0 & 0 & a_2 & 0 & 0 & b_2 \\ j_1 & k_1 & 0 & \ell_1 & m_1 & 0 \\ n_1 & p_1 & 0 & q_1 & r_1 & 0 \\ 0 & 0 & c_2 & 0 & 0 & d_2 \end{pmatrix} \tag{418}$$

Hence A can be decomposed into two matrices, the matrix A_1 for "in-plane motions" :

$$A_1 = \begin{pmatrix} a_1 & b_1 & c_1 & d_1 \\ e_1 & f_1 & g_1 & h_1 \\ j_1 & k_1 & \ell_1 & m_1 \\ n_1 & p_1 & q_1 & r_1 \end{pmatrix} \tag{419}$$

and the matrix A_2 for "out-of-plane motions" :

$$A_2 = \begin{pmatrix} a_2 & b_2 \\ c_2 & d_2 \end{pmatrix} \tag{420}$$

The discussion of out-of-plane stability is identical to that of transversal stability in (416) and (417) but it is generally neglected and most papers only consider the in-plane stability. However, notice that this out-of-plane stability is not obvious, and many plane retrograde motions have not the out-of-plane stability, for instance those that are limits of the asymptotical motions (228) (Section 10.2.3), (i.e. the retrograde motions with small r/R and with $(2 + 3e_S^2)/(1 + 4e_S^2) < \mathcal{G}_S/\mathcal{G}_T \leq 2$.

The matrix A_1 being symplectic, its characteristic polynomial is reciprocal :

$$\mu^4 + a\mu^3 + b\mu^2 + a\mu + 1 = 0 \tag{421}$$

with :

$$a = -(a_1 + f_1 + \ell_1 + r_1) = -\,\mathrm{trace}(A_1) \tag{422}$$

$$b = a_1f_1 + a_1\ell_1 + a_1r_1 + f_1\ell_1 + f_1r_1 + \ell_1r_1 - \\ - b_1e_1 - c_1j_1 - d_1n_1 - g_1k_1 - h_1p_1 - m_1q_1 \tag{423}$$

The discussion of the modulus of the eigenvalues μ is easy, let us put :

$$x = \mu + \frac{1}{\mu} \tag{424}$$

hence from (421) :

$$x^2 + ax + (b - 2) = 0 \tag{425}$$

For first-order stability all eigenvalues μ must have a modulus equal to one, hence the two roots x_1 and x_2 of (425) must be real and in -2 ; $+2$. This happens if and only if :

$$8|a| - 16 \leqq 4b - 8 \leqq a^2 \leqq 16 \tag{426}$$

The corresponding zone of the a,b plane is drawn in Fig. 65.

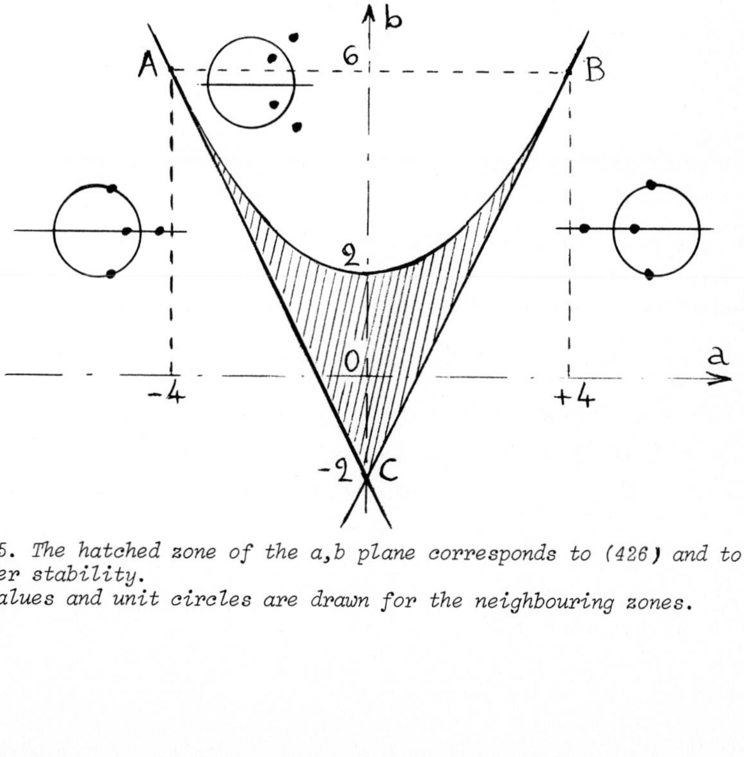

Fig. 65. The hatched zone of the a,b plane corresponds to (426) and to the first-order stability.
Eigenvalues and unit circles are drawn for the neighbouring zones.

The number of studies of plane periodic orbits is very large (Ref.131,132) and no general rules of stability or instability have been found. Orbits with a well established hierarchy of two-body motions and with small perturbations are very often stable however orbits with very large perturbations such as those of Figs. 63 and 64 can also be stable.

Nevertheless, it seems that orbits with masses of the same order of magnitude and with a triple close approach are always unstable.

Let us finally notice the importance of simple resonances. For instance, in the prograde Poincaré periodic orbits of the fisrt species (i.e. the Sun and two small planets revolving in the same direction on coplanar and almost circular orbits, Fig. 66), if the ratio of the period of revolution of the two planets is 1/3, 3/5, ..., (2n-1)/(2n+1)... the motion is unstable even if the masses of the planets are very small.

These results agree with the gaps of Fig. 59 (Section 10.7.1).

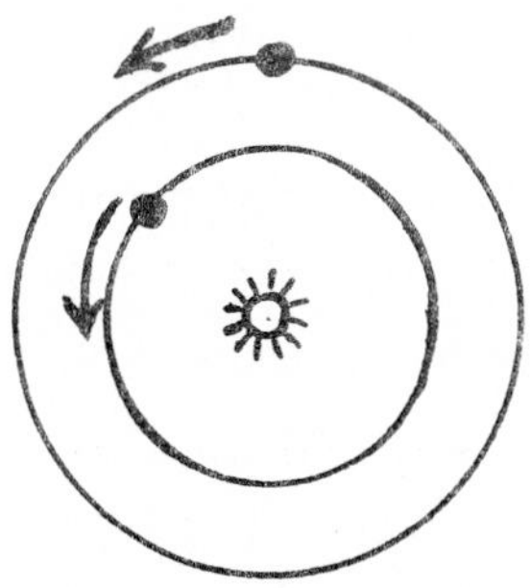

Fig. 66. Prograde Poincaré periodic orbit of the first species.

10.7.4.3 Symmetric periodic orbits

We have seen in Section 10.6.2 that for a periodic orbit with space-time symmetries, it is sufficient to analyse only half a period, and the matrix A is the following product :

$$A = A''A' \tag{427}$$

with, to the first order, for neighbouring orbits of the same masses and integrals of motion (ℓ_{io}, $\ell_{i(1/2)}$, ℓ_{i1} corresponding to space-time symmetries) :

$$\vec{\delta}_{1/2} = A'\vec{\delta}_0 \;\; ; \;\; \vec{\delta}_1 = A''\vec{\delta}_{1/2} \tag{428}$$

that is :

$$\vec{\delta}_1 = A''A'\vec{\delta}_0 = A\vec{\delta}_0 \tag{429}$$

A' can be obtained by analysis of the first half-period from ℓ_{io} to $\ell_{i(1/2)}$ and it is easily related to A" and A. Let us decomposed A' into its four corners:

$$A' = \begin{pmatrix} U' & V' \\ W' & X' \end{pmatrix} \tag{430}$$

We then obtain, as in (361) where T stands for transposition :

$$A'' = \begin{pmatrix} X'^T & V'^T \\ W'^T & U'^T \end{pmatrix} \quad ; \quad A = A''A' \tag{431}$$

Notice that $A_{1/2} = A'A''$ corresponds to the period $\ell_{-1/2}, \ell_{1/2}$ and leads, of course, to the same analysis : it has the same eigenvalues as A.

If the periodic orbit of interest has four space-time symmetries per period, as for instance the orbits of the "third species" (Figs. 52 and 118), a supplementary reduction is possible and it is sufficient to analyse one quarter of a period.

We then obtain with obvious notations and with suitable matrices a', a", a''', a^{IV} :

$$\vec{\delta}_{1/4} = a'\vec{\delta}_0 \; ; \; \vec{\delta}_{1/2} = a''\vec{\delta}_{1/4} \; ; \; \vec{\delta}_{3/4} = a'''\vec{\delta}_{1/2} \; ; \; \vec{\delta}_1 = a^{IV}\vec{\delta}_{3/4} \tag{432}$$

Hence :

$$\vec{\delta}_{1/2} = a''a'\vec{\delta}_0 = A'\vec{\delta}_0 \; ; \; a''a' = A' \; ; \; a^{IV}a''' = A'' \tag{433}$$

on the other hand :

$$a' = \begin{pmatrix} u' & v' \\ w' & x' \end{pmatrix} \; ; \; a'' = \begin{pmatrix} x'^T & v'^T \\ w'^T & u'^T \end{pmatrix} \; ; \; a''' = a' \; ; \; a^{IV} = a'' \; ; \; A'' = A' \tag{434}$$

Thus, since $A = A''A' = A'^2$, the matrix a' contains all the necessary informations.

Notice that $A = A'^2$, so the Floquet multipliers, i.e. the eigenvalues of A, are the squares of the eigenvalues of A', and it is then sufficient to apply the stability rules to the eigenvalues of A' : if all these eigenvalues have a modulus equal to one the periodic orbit of interest has first-order stability.

As a general rule, it is sufficient to analyse a symmetric periodic orbit and its vicinity between two successive symmetries.

10.7.4.4 Circular restricted case and Hill case

In the circular restricted three-body problem (Section 9.1) and in the Hill problem (Section 9.2) the existence of the Jacobi integral implies that at least two Floquet multipliers are equal to one.

This phenomenon is characteristic of the existence of a supplementary differentiable integral of motion, provided that this integral is not stationary at the periodic solution of interest.

Indeed let us assume that locally, for the given values of the masses and the usual integrals of motion, the supplementary integral is equivalent to $u + \vec{v}\vec{\delta}_0$ at t_0 and to $u + \vec{v}\vec{\delta}_1$ at t_1. Then to first order :

$$\forall\vec{\delta}_0 : u + \vec{v}.\vec{\delta}_1 = u + \vec{v}.\vec{\delta}_0 \quad ; \quad \vec{v} \neq 0 \tag{435}$$

However, to the same order :

$$\vec{\delta}_1 = A\vec{\delta}_0 \tag{436}$$

Hence :

$$\forall\vec{\delta}_0 : \vec{v}A\vec{\delta}_0 = \vec{v}.\vec{\delta}_0 \tag{437}$$

that is :

$$\vec{v}A = \vec{v} \tag{438}$$

Thus $\vec{v}$ is a row eigenvector and one is an eigenvalue of A. This of course implies that a second Floquet multiplier, at least, is equal to one since for symplectic monodromy matrices the number of Floquet multipliers equal to one is always even.

This allows a simplification of the first-order stability study of periodic solutions of the circular restricted three-body problem and of the Hill problem : it is sufficient to analyse the neighbouring solutions with the same three masses, the same energy integral, the same angular momentum and the same Jacobi integral.

This simplification leads to a smaller monodromy matrix A, with only 4 × 4 components, as the matrix A_1, of (419) in Section 10.7.4.2.

Let us remember that the Delaunay parameters are singular in the circular restricted three-body problem and in the Hill problem. It is then necessary to use other suitable Hamiltonian parameters for both the above demonstration (435)-(438) and the study of these two problems.

The following simple Hamiltonian system has already been presented in the elliptic case in (171)-(172) (Section 9.3).

Let us use as Oxy plane the orbital plane of the primaries and let us use a set of axes rotating with the primaries (Fig. 67).

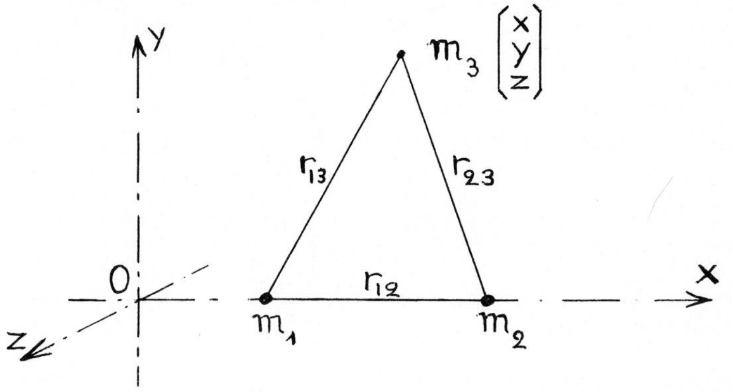

Fig. 67. The set of axes Oxyz rotating with the primaries m_1 and m_2.

Hence the primaries m_1 and m_2 have fixed positions in the Oxy plane and are usually along the x-axis (the origine O is usually some remarkable point : center of mass, primary, Lagrange equilibrium point, etc...).

If x, y , z are the coordinates of the small body m_3 and p_x, p_y, p_z their conjugate parameters, the Hamiltonian H of the motion of m_3 is the following :

$$H = \frac{1}{2}\left[(p_x+\omega y)^2+(p_y-\omega x)^2+p_z^2+\omega^2 z^2\right] -Gm_1\left(\frac{1}{r_{13}} + \frac{r_{13}^2}{2r_{12}^3}\right)-Gm_2\left(\frac{1}{r_{23}} + \frac{r_{23}^2}{2r_{12}^3}\right) \qquad (439)$$

In this equation ω is the rate of rotation of the set of axes :

$$\omega^2 r_{12}^3 = G(m_1 + m_2) \qquad (440)$$

and the distances r_{13} and r_{23} must be expressed in terms of the fixed quantities and the parameters x, y, z.

The Jacobi integral is then equal to (-2H) and the equations of motion are the following :

$$\left.\begin{array}{ll} dx/dt = \partial H/\partial p_x = p_x + \omega y & \quad p_x = dx/dt - \omega y \\ dy/dt = \partial H/\partial p_y = p_y - \omega x & \quad p_y = dy/dt + \omega x \\ dz/dt = \partial H/\partial p_z = p_z & \quad p_z = dz/dt \\ dp_x/dt = -\partial H/\partial x \;;\; dp_y/dt = -\partial H/\partial y \;;\; dp_z/dt = -\partial H/\partial z & \end{array}\right\} \qquad (441)$$

Most authors use then a suitable "surface of section", as the plane $y=0$ in Figs. 28, 29, 32, 33 of Section 9.4 (the variable y then plays a role similar to that of the description parameter ℓ_i in the previous sections).

The periodic solution of interest crosses the surface of section S at one or several points, for instance at C (Fig. 68).

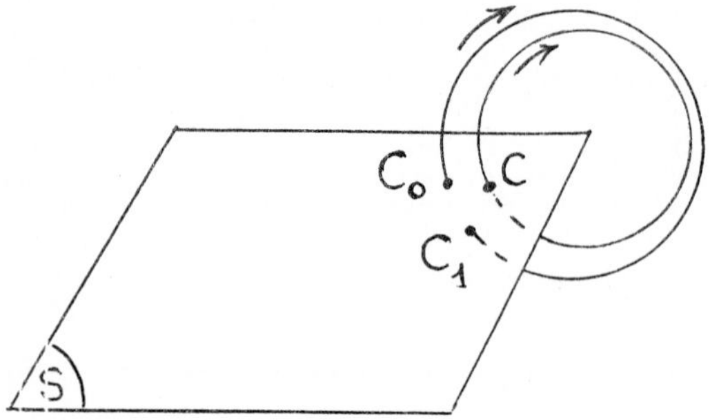

Fig. 68. The surface of section S, the periodic solution of interest crossing S at C and a neighbouring solution crossing S at C_0 and C_1.

For a neighbouring solution of the same Jacobi integral the successive crossing points in the vicinity of C are C_0, C_1, C_2, ... The small vectors $\vec{CC_0}$, $\vec{CC_1}$, ... correspond to small discrepancies δx, δz, δp_x, δp_z of the remaining parameters (if S is a y = constant surface) and they lead to the monodromy matrix A we are looking for :

$$\begin{pmatrix} \delta x_1 \\ \delta z_1 \\ \delta p_{x1} \\ \delta p_{z1} \end{pmatrix} = A \begin{pmatrix} \delta x_0 \\ \delta z_0 \\ \delta p_{x0} \\ \delta p_{z0} \end{pmatrix} + \text{second order terms} \tag{442}$$

Hence for the periodic solutions of the circular restricted three-body problem and of the Hill problem the useful monodromy matrix A is only a 4×4 matrix and its computation is much simpler than in the general case.

Notice I. Usually the Hill problem is considered as the limit of the Hamiltonian problem (439)-(441) when :

$$\left.\begin{array}{l} G = 1 \;;\; m_2 = 1 \;;\; \omega = 1 \;;\; (\text{hence } m_1 + 1 = r_{13}^3) \\ m_2 \text{ is at the origin} \\ m_1 \text{ goes to infinity along the x-axis} \end{array}\right\} \tag{443}$$

This leads to the following Hamiltonian :

$$H = \frac{1}{2}\left[(p_x + y)^2 + (p_y - x)^2 + p_z^2 + z^2 - 3x^2\right] - (x^2 + y^2 + z^2)^{-1/2} \tag{444}$$

Notice II. The demonstration (435)-(438) is very useful when, in some given dynamical system, we look for new unknown integrals of motion.

If a new everywhere differentiable integral of motion exists, the monodromy matrix A has at least one eigenvalue equal to one for all periodic solutions in the vicinity of which the seeked integral of motion is not stationary.

10.7.5 First-order stability, the general discussion

The discussion of first-order stability was presented in Section 10.7.3 and in Fig. 61 : If the "Floquet multipliers", i.e. the six eigenvalues of the monodromy matrix A, are all with a modulus equal to one the periodic orbit of interest has the first-order stability ; if not, this orbit is exponentially unstable for both past and future.

Let us simply relate this property to the characteristic polynomial of the modromy matrix A.

The matrix A being real and symplectic, its characteristic polynomial $P(\mu)$ is real and reciprocal :

$$P(\mu) = \mu^6 + a\mu^5 + b\mu^4 + c\mu^3 + b\mu^2 + a\mu + 1 \tag{445}$$

with a = - trace A, etc...

The equation $P(\mu) = 0$ can be solved in the following way.

Let us put :

$$\mu + \frac{1}{\mu} = x \tag{446}$$

Then from (445) when $P(\mu) = 0$:

$$x^3 + ax^2 + (b - 3)x + (c - 2a) = 0 \tag{447}$$

For the first-order stability all μ'_s must have a modulus equal to one which happens if and only if the three roots x_1, x_2, x_3 of (447) are real and on the close segment -2 ; +2 .

It is customary to write :

$$\left.\begin{array}{ll} -a = s & s = \text{"sum"} = x_1 + x_2 + x_3 \\ b - 3 = q & q = x_1x_2 + x_1x_3 + x_2x_3 \\ 2a - c = p & p = \text{"product"} = x_1x_2x_3 \\ x^3 - sx^2 + qx - p = 0 & \end{array}\right\} \tag{448}$$

With these notations the conditions of first-order stability are classical. The periodic orbit of interest is first-order stable if and only if :

$$4|s| - 12 \leqq q \leqq 12 \tag{449}$$

$$|4s + p| \leqq 2q + 8 \tag{450}$$

$$s^2q^2 - 4q^3 - 4s^3p + 18sqp - 27p^2 \geqq 0 \tag{451}$$

These conditions imply of course $|s| \leqq 6$; $-4 \leqq q \leqq 12$; $|p| \leqq 8$. The condition (451) is the condition of reality of the three roots x_1, x_2, x_3.

These conditions (448)-(451) generalize the conditions (417) and (426). They are not simple but at least they allow one to know immediately, without solving $P(\mu) = 0$, whether or not a given monodromy matrix A corresponds to first-order stability.

10.7.6 <u>On the evolution of first-order stability along the families of periodic orbits</u>.

As already shown in Figs. 18, 19, 20 of Section 8.5 and in Fig. 64 of Section 10.7.4.1, first-order stability can vary along a given family of periodic orbits.

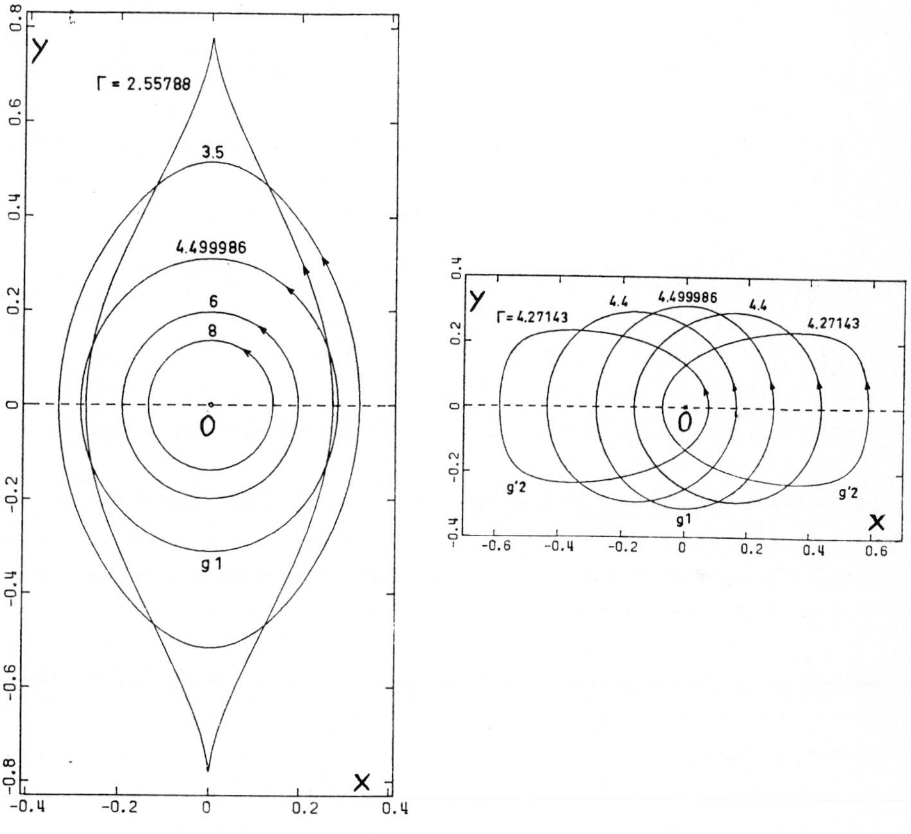

Fig. 69. The phenomenon of "exchange of stability".
These two families of periodic orbits of the plane Hill problem were already presented in Figs. 18 and 19 of Section 8.5. They are the Strömgren families g and g' given in terms of the Jacobi integral Γ.
When Γ > 4.499 986 the orbits of the family g have first-order stability.
The orbit g_1 with Γ = 4.499 986 belongs to the two families (branching orbit).
For smaller Γ the family g is no longer stable and the stability passes to the family g'. Source : Hénon, M. [104].

Another very common phenomenon is the exchange of stability presented in Fig. 69 that reproduces Figs. 18 and 19 : Two families of periodic orbits (the Strömgren families g and g') have a common orbit g_1. The orbits of the family g smaller than g_1 have first-order stability, but this stability is lost for larger orbits and passes to the family g'. The periodic orbit g_1 is both a "branching orbit" and a limit orbit for the stability.

These phenomena can easily be understood if we consider the monodromy matrix associated with a periodic orbit, its characteristic polynomial and its eigenvalues (the Floquet multipliers).

The monodromy matrix depends on the position considered as the initial position along the periodic orbit of interest but its characteristic polynomial and its Floquet multipliers are independant of the initial position and are functions only of the periodic orbit.

The Floquet multipliers vary continuously along the family of periodic orbits of interest, and the modifications of stability may happen in the three following ways presented in Fig. 70.

The instability can arise either by the passage of two Floquet multipliers at +1 (type I) or at -1 (type II), or by meeting of two pairs of Floquest multipliers, along the unit circle (type III).

These three types correspond to the three sides of the hatched zone of stability of Fig. 65 (Section 10.7.4.2).

Notice that the "branching orbits", where meet two or several families of periodic orbits with neighbouring periods, are also "singular periodic orbits", i.e. orbits with some Floquet multipliers equal to one (see Section 10.5.2 and 10.5.3). Thus the phenomenon of "exchange of stability" is natural for transitions of type I, and the branching orbit is then also a limit orbit for the stability.

However, these two properties can be independant :

A) In transitions of type II the limit Floquet multipliers are -1, i.e. the square root of 1, and a branching can exist but with a period doubling.

B) In transitions of type III there is generally no branching.

C) Conversely, in the vicinity of the branching of type IV (Fig. 70), all periodic orbits are unstable.

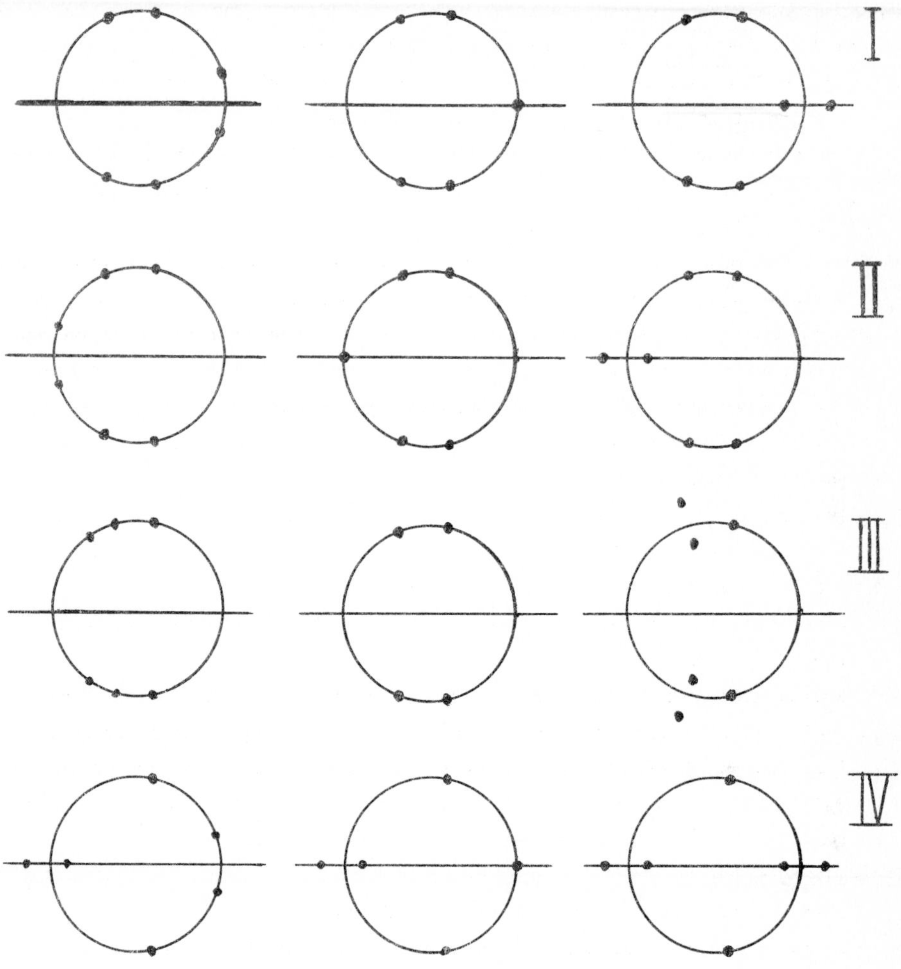

Fig. 70. The three types of apparition of instability.
μ, μ^{-1}, $\bar{\mu}$ and $\bar{\mu}^{-1}$ *are at the same time Floquet multipliers and first-order stability occurs iff all Floquet multipliers are on the unit circle.*

The multipliers can leave the unit circle either by the passage of two multipliers at +1 (type I) or at -1 (type II) or by the meeting of two pairs along the unit circle (type III).

In the type IV, two multipliers pass at +1 (branching orbit) but the orbit remain unstable.

10.7.7 Elements of the all-order stability analysis. The near-resonance theorem.

10.7.7.1 Analytic autonomous differential systems. The vicinity of a point of equilibrium.

We will assume that the equilibrium point of interest is at the origin and we will write the equations of motion as :

$$dx_K/dt = f_K(x_1, x_2, \ldots, x_n) \; ; \; K = \{1, 2, \ldots, n\} \tag{452}$$

The functions f_K can be developed into power series of $x_1, x_2, \ldots, x_n$ without constant term. These series are converging at least in a neighbourhood of the origin.

Let $\lambda_1, \lambda_2, \ldots, \lambda_n$ denote the eigenvalues of the matrix F corresponding to the linear parts of the functions f_K. The real part of the λ_j are the "Liapunov characteristic exponents" also called "Liapunov exponents".

If all λ_j are different the matrix F can be diagonalized and then a preliminary linear substitution leads to the form :

$$dx_K/dt = \lambda_K x_K + g_K(x_1, x_2, \ldots, x_n) \; ; \; K = \{1, 2, \ldots, n\} \tag{453}$$

The power series g_K begin with quadratic terms.

Even if some eigenvalues are multiple, the classical linear analysis shows that a suitable linear substitution can always lead to the following form :

$$\left.\begin{array}{l} dx_K/dt = \lambda_K x_K + \varepsilon_K x_{K+1} + g_K(x_1, x_2, \ldots, x_n) \\ \text{with : } \varepsilon_n = 0 \; ; \; \lambda_{K+1} \neq \lambda_K \text{ implies } \varepsilon_K = 0 \\ \qquad \lambda_{K+1} = \lambda_K \text{ implies } \varepsilon_K \text{ either 0 or 1} \end{array}\right\} \; K = \{1, 2, \ldots, n\} \tag{454}$$

Let us assume the indices to be chosen so that :

$$\mathcal{R}e(\lambda_1) \leqq \mathcal{R}e(\lambda_2) \leqq \mathcal{R}e(\lambda_3) \leqq \ldots \leqq \mathcal{R}e(\lambda_n) \tag{455}$$

and let us define the integers n_-, n_0 and n_+ by :

$$\left.\begin{array}{l} \mathcal{R}e(\lambda_{n_-}) < 0 \; ; \mathcal{R}e(\lambda_{(1+n_-)}) = 0 \; ; \mathcal{R}e(\lambda_{(n_0+n_-)}) = 0 \\ \mathcal{R}e(\lambda_{(1+n_0+n_-)}) > 0 \; ; \; n_- + n_0 + n_+ = n \end{array}\right\} \tag{456}$$

The corresponding vectors $\vec{x}_-$, $\vec{x}_0$, $\vec{x}_+$ are defined by :

$$\left.\begin{array}{l} \vec{x}_- = (x_1, x_2, \ldots, x_{n_-}) \\ \vec{x}_0 = (x_{(1+n_-)}, \ldots, x_{(n_0+n_-)}) \\ \vec{x}_+ = (x_{(1+n_0+n_-)}, \ldots, x_n) \end{array}\right\} \tag{457}$$

The manifold $(\vec{x}_-, 0, 0)$ has n_- dimensions and is a first-order approximation the "stable subset" while the manifold $(0, 0, \vec{x}_+)$ has n_+ dimensions and is a first-order approximation of the "unstable subset".

Let us now demonstrate the "near resonance theorem", i.e. the following generalization of the Liapunov and Hartman theorems of Ref. 179.

Even if there are resonances and/or multiple eigenvalues there exists an analytic transformation :

$$\vec{x} = (x_1, x_2, \ldots, x_n) \longrightarrow \vec{u} = (u_1, u_2, \ldots, u_n) \quad (458)$$

leading to the following properties :

$$\left.\begin{array}{l} u_K - x_K = \text{second order} \; ; \; K = \{1, 2, \ldots, n\} \\ \vec{u}_- - \vec{x}_- \; ; \; \vec{u}_0 - \vec{x}_0 \; ; \; \vec{u}_+ - \vec{x}_+ = \text{second order} \end{array}\right\} \quad (459)$$

$$du_K/dt = \lambda_K u_K + \varepsilon_K u_{K+1} + j_K(u_1, u_2, \ldots, u_n) \; ; \; K = \{1, 2, \ldots, n\} \quad (460)$$

The λ_K and ε_K are those of (454) while the power series j_K begin with quadratic terms, are convergent in a suitable neighbourhood of the origin and verify the following :

The power series $j_K(\vec{u})$ have only "sufficiently resonant" terms, i.e. terms $J_K u_1^{\alpha_1} \cdot u_2^{\alpha_2} \cdot \ldots \cdot u_n^{\alpha_n}$ for which :

$$\left|\lambda_K - \sum_{m=1}^{n} \alpha_m \lambda_m\right| \leqq \varepsilon(\alpha_1 + \alpha_2 + \ldots + \alpha_n) \quad (461)$$

ε being an arbitrary positive constant.

Because of the relations (455), (456) the power series $j_K(\vec{u}_-,0,0)$ and $j_K(0,0,\vec{u}_+)$ will have only a finite number of terms if ε is sufficiently small. Furthermore these terms will be strictly resonant and even :

$$\left.\begin{array}{ll} K \geqq n_- & \text{implies } j_K(\vec{u}_-,0,0) = 0 \\ K \leqq 1 + n_0 + n_- & \text{implies } j_K(0,0,\vec{u}_+) = 0 \end{array}\right\} \quad (462)$$

In these conditions the manifold $(\vec{u}_-,0,0)$ will be exactly the "stable subset" while the manifold $(0,0,\vec{u}_+)$ will be the "unstable subset" and the solutions in these two sets will be very simple.

Furthermore in a sufficiently small neighbourhood of the origin all bounded solutions will belong to a "central subset" or "critical subset" with n_0 dimensions and in the vicinity of the subset $(0,\vec{u}_0,0)$. These bounded solutions have their own stable and unstable subsets with at least n_- and n_+ dimensions.

We will prove this theorem only under the restriction that all ε_K of (454) are zero (e.g. all eigenvalues λ_K are distinct from one another). For the case of multiple eigenvalues and ε_K equal to one the demonstration is more intricate but causes no real conceptual difficulties (see Appendix 3).

Let us define the analytic transformation $(\vec{x}) \longrightarrow (\vec{u})$ by the following :

$$x_K = u_K + \phi_K(u_1, u_2, \ldots, u_n) \; ; \; K = \{1, 2, \ldots, n\} \quad (463)$$

The power series ϕ_K will begin with quadratic terms and we obtain :

$$du_K/dt = \lambda_K u_K + j_K(u_1, u_2, \ldots, u_n) \; ; \; K = \{1, 2, \ldots, n\} \qquad (464)$$

As $\phi_K(\vec{u})$ and $g_K(\vec{x})$ the power series $j_K(\vec{u})$ are without constant and linear terms. Our purpose is that these series $j_K(\vec{u})$ contain only "sufficiently resonant" terms, as defined by (461).

The relations between the power series $\phi_K(\vec{u})$ and $j_K(\vec{u})$ are given by :

$$dx_K/dt = \lambda_K x_K + g_K(\vec{x}) = \lambda_K u_K + \lambda_K \phi_K(\vec{u}) + g_K(\vec{x}) \qquad (465)$$

and :

$$dx_K/dt = d\left[u_K + \phi_K(\vec{u})\right]/dt = \lambda_K u_K + j_K(\vec{u}) + \sum_{m=1}^{n} \frac{\partial \phi_K}{\partial u_m} \left(\lambda_m u_m + j_m(\vec{u})\right) \qquad (466)$$

hence :

$$\lambda_K \phi_K(\vec{u}) + g_K(\vec{x}) = j_K(\vec{u}) + \sum_{m=1}^{n} \frac{\partial \phi_K}{\partial u_m} \left(\lambda_m u_m + j_m(\vec{u})\right) \qquad (467)$$

that is for $K = \{1, 2, \ldots, n\}$:

$$j_K(u_1, u_2, \ldots, u_n) + \sum_{m=1}^{n} \frac{\partial \phi_K}{\partial u_m} \lambda_m u_m - \lambda_K \phi_K(\vec{u}) = g_K(\vec{x}) - \sum_{m=1}^{n} \frac{\partial \phi_K}{\partial u_m} j_m(\vec{u}) \qquad (468)$$

Using relations (463) the power series $g_K(\vec{x})$ can be developed in terms of $u_1, u_2, \ldots, u_n$ and these equations (463) and (468) allow a formal development of the power series $j_K(\vec{u})$ and $\phi_K(\vec{u})$. Indeed if these series are known for all K up to the degree (p-1) the right-hand member of (468) is known in terms of $u_1, u_2, \ldots, u_n$ up to the degree p and this allows a single choice of the terms of degree p of $j_K(\vec{u})$ and $\phi_K(\vec{u})$, using the following convention.

Assume that we consider the term in $u_1^{\alpha_1} \cdot u_2^{\alpha_2} \ldots u_n^{\alpha_n}$ and that for this term the right-hand member of (468) is $B_K u_1^{\alpha_1} \cdot u_2^{\alpha_2} \ldots u_n^{\alpha_n}$.

Let us denote by $J_K u^{\alpha_1} \ldots u^{\alpha_n}$ and $F_K u^{\alpha_1} \ldots u^{\alpha_n}$ the corresponding terms of $j_K(\vec{u})$ and $\phi_K(\vec{u})$.

The equation (468) gives :

$$J_K + \left[\sum_{m=1}^{n} (\alpha_m \lambda_m) - \lambda_K\right] F_K = B_K \qquad (469)$$

By convention we will choose $J_K = 0$ and then $F_K = B_K \Big/ \left[\sum_{m=1}^{n} (\alpha_m \lambda_m) - \lambda_K\right]$ if the near-resonance condition (461) is not satisfied. If that condition is satisfied we will choose $F_K = 0$ and $J_K = B_K$.

Hence the power series $j_K(\vec{u})$ and $\phi_K(\vec{u})$ are uniquely defined for any K and the only remaining question is that of the convergence of these power series.

As a definition we will say that a power series $a(\vec{x})$ or $a(\vec{u})$ is "majorized" by the power series $b(\vec{x})$ or $b(\vec{u})$ if for all terms the coefficient of the power series b is real, non negative and larger than or equal to the modulus of the coefficient of the corresponding term of the series a. This implies of course that the radius of convergence of the series b is smaller than or equal to that of the series a.

Let us also define the formal power series $h_K(\vec{u})$ without constant term by :

$$j_K(\vec{u}) = \sum_{m=1}^{n} \frac{\partial h_K(\vec{u})}{\partial u_m} \cdot u_m \quad ; \quad K = \{1, 2, \ldots, n\} \tag{470}$$

For all K the power series $h_K(\vec{u})$ and $j_K(\vec{u})$ have the same radius of convergence.

On the other hand the power series $g_K(\vec{x})$ are convergent at least in some neighbourhood of the origin and hence exists a positive constant c_1 such that :

$$\left.\begin{array}{l} \text{All power series } g_K(\vec{x}) \text{ are majorized by the power series of :} \\ \\ c_1X^2/(1 - c_1X) \text{ with } X = x_1 + x_2 + \ldots + x_n \end{array}\right\} \tag{471}$$

Thus, with ε being the positive constant defined in (461), the power series $h_K(\vec{u}) + \varepsilon\phi_K(\vec{u})$ are majorized by the power series $\varepsilon\psi_K(\vec{u})$ defined by :

$$\left.\begin{array}{l} X = x_1 + x_2 + \ldots + x_n \\ \\ x_K = u_K + \psi_K(\vec{u}) \quad ; \quad K = \{1, 2, \ldots, n\} \\ \varepsilon \displaystyle\sum_{m=1}^{n} \frac{\partial\psi_K(\vec{u})}{\partial u_m} \cdot u_m = \frac{c_1X^2}{1 - c_1X} + \varepsilon \sum_{m=1}^{n} \sum_{p=1}^{n} \frac{\partial\psi_K}{\partial u_m} \cdot \frac{\partial\psi_m}{\partial u_p} \cdot u_p \end{array}\right\} \tag{472}$$

As $j_K(\vec{u})$ and $\phi_K(\vec{u})$ uniquely defined by (461)-(469) the power series $\psi_K(\vec{u})$ are uniquely defined by (472) and the comparison of their coefficients indeed shows that for all K the power series $\varepsilon\psi_K(\vec{u})$ majorize $h_K(\vec{u}) + \varepsilon\phi_K(\vec{u})$. They also majorize individually $h_K(\vec{u})$ and $\varepsilon\phi_K(\vec{u})$.

On the other hand (472) is the same for all K and then the power series $\psi_1(\vec{u}), \psi_2(\vec{u}), \ldots, \psi_n(\vec{u})$ are identical.

It is of course sufficient to study the convergence of the power series $\psi_K(\vec{u})$ for $u_1 = u_2 = \ldots = u_n = u$ that is to study the convergence of the power series $\psi(u)$ defined by the following :

$$\left.\begin{array}{l} x_K = u + \psi(u) = x_1 = x_2 = \ldots = x_n \\ \\ X = nx_K = n(u + \psi(u)) \\ \\ \dfrac{d\psi}{du} \cdot u = \dfrac{c_1n^2(u + \psi)^2}{\varepsilon[1 - c_1n(u + \psi)]} + u\left(\dfrac{d\psi}{du}\right)^2 \end{array}\right\} \tag{473}$$

$\psi(u)$ is majorized by $u \cdot \frac{d\psi}{du}$ and they are then both majorized by the power series F(u) given by :

$$F = \frac{c_1 n^2 (u + F)^2}{\varepsilon[1 - c_1 n(u + F)]} + \frac{F^2}{u} \tag{474}$$

F is an algebraic function of u with a positive radius of convergence and this completes the proof of convergence.

The power series F is majorized by this of $0.25u\left[1 - \sqrt{1 - 4c_1 nu(1 + 4n\varepsilon^{-1})}\right]$ with a radius of convergence equal to $\varepsilon/4c_1 n(4n + \varepsilon)$.

Notice that when ε goes to zero the radius of convergence of the power series F(u), $\psi(u)$ and $\psi_K(\vec{u})$ also goes to zero. Unfortunately this is generally also the case for the power series $j_K(\vec{u})$ and $\phi_K(\vec{u})$ which forbids choice of $\varepsilon = 0$ in (461).

The consequences of the analytic transformation $(\vec{x}) \to (\vec{u})$ exposed in (458)-(462) and given in (463)-(469) are the following.

The relations (462) imply that the manifolds $(\vec{u}_-,0,0)$ and $(0,0,\vec{u}_+)$ are invariant, they are the stable and unstable subsets.

In these two subsets only strictly resonant j_K terms appear, which implies that the solutions have the following form with polynomials $P_K(t)$:

$$u_K = P_K(t) \cdot \exp(\lambda_K t) \;; \quad \text{with :} \quad \begin{array}{ll} \text{either} & K = \{1, 2, \ldots, n_-\} \\ \text{or} & K = \{1 + n_0 + n_-, \ldots, n\} \end{array} \tag{475}$$

The polynomials $P_K(t)$ have at most the degree $1 + (n_- - 2)\mathcal{Re}(\lambda_1)/\mathcal{Re}(\lambda_{n_-})$ in the stable subset and the degree $1 + (n_+ - 2)\mathcal{Re}(\lambda_n)/\mathcal{Re}(\lambda_{(1+n_0+n_-)})$ in the unstable subset, and again we find the classical result.

A) If at least one eigenvalue has a positive real part the equilibrium point of interest is unstable (exponential instability).

B) If all eigenvalues λ_K have a negative real part the vector $\vec{u}_-$ is $\vec{u}$ and the equilibrium point of interest is stable (asymptotic and even exponential stability).

Let us now consider the bounded trajectories remaining in a small vicinity of the point of equilibrium and let us demonstrate that they will all remain in a manifold with at most n_0 dimensions : the "central subset".

Let us consider two solutions $\vec{x}(t)$ and $\vec{y}(t)$ of the equations of motion (454) and let us define the following real function $A(\vec{x},\vec{y})$ by :

$$A = \sum_{K=1}^{n} a_K (y_K - x_K)(\bar{y}_K - \bar{x}_K) \tag{476}$$

The real constant a_K and the real constant Q will be chosen so that for $||\vec{x}||$ and $||\vec{y}||$ sufficiently small :

$$dA/dt \geqq QA \tag{477}$$

Let us put :

$$y_K - x_K = \delta x_K \quad ; \quad K = \{1, 2, \ldots, n\} \tag{478}$$

hence, with the equations of motion (454) :

$$(dA/dt) - QA = \sum_{K=1}^{n} a_K \left[(\lambda_K + \overline{\lambda}_K - Q)\delta x_K \cdot \overline{\delta x}_K + \varepsilon_K(\delta x_K \overline{\delta x}_{K+1} + \overline{\delta x}_K \delta x_{K+1}) + \right. \\ \left. + \left[g_K(\vec{y}) - g_K(\vec{x})\right]\overline{\delta x}_K + \left[\overline{g}_K(\vec{y}) - \overline{g}_K(\vec{x})\right]\delta x_K \right] \tag{479}$$

The last terms with g_K are small with respect to $(\delta x)^2$ and if the other terms, i.e. the quadratic terms, give a positive definite function (in terms of the quantities δx_K's) the difference dA/dt - QA will be non-negative for sufficiently small $||\vec{x}||$ and $||\vec{y}||$.

This condition of positive definiteness is satisfied if for instance :

$$\left.\begin{array}{l} \text{I) For all K up to n : } Q \neq \lambda_K + \overline{\lambda_K} \\ \text{II) All } a_K \text{ have the sign of } (\lambda_K + \overline{\lambda_K} - Q) \\ \text{III) For the } \varepsilon_K\text{'s equal to one : } (a_{K+1})/a_K \geqq 4/(\lambda_K + \overline{\lambda_K} - Q)^2 \end{array}\right\} \tag{480}$$

It is then very easy to build suitable $A(\vec{x},\vec{y})$ functions and we will consider two of them called B and C.

$$\left.\begin{array}{l} B = \sum_{K=1}^{n} b_K(y_K - x_K)(\overline{y_K} - \overline{x_K}) \\ \text{with : } Q_B = \mathcal{R}e(\lambda_{n_-}) < 0 \\ \text{hence : } K \leqq n_- \text{ implies } \mathcal{R}e(\lambda_K) < 0 \; ; \; b_K < 0 \\ \qquad\quad K \geqq 1 + n_- \text{ implies } \mathcal{R}e(\lambda_K) \geqq 0 \; ; \; b_K > 0 \end{array}\right\} \tag{481}$$

and :

$$\left.\begin{aligned} & C = \sum_{K=1}^{n} c_K (y_K - x_K)(\overline{y_K} - \overline{x_K}) \\ & \text{with : } Q_C = \mathcal{Re}\left(\lambda_{(1+n_0+n_-)}\right) > 0 \\ & \text{hence : } K \leq n_0 + n_- \quad \text{implies } \mathcal{Re}(\lambda_K) \leq 0 \ ; \quad c_K < 0 \\ & \qquad\quad K \geq 1 + n_0 + n_- \quad \text{implies } \mathcal{Re}(\lambda_K) > 0 \ ; \quad c_K > 0 \end{aligned}\right\} \tag{482}$$

The conditions $dB/dt \geq Q_B B$ and $dC/dt \geq Q_C C$ are satisfied within a neighbourhood N of the origin ; these conditions lead to :

$$t_2 \geq t_1 \quad \text{implies} \left\{\begin{aligned} & B(t_2) \geq B(t_1) \cdot \exp\{Q_B(t_2 - t_1)\} \\ & C(t_2) \geq C(t_1) \cdot \exp\{Q_C(t_2 - t_1)\} \end{aligned}\right\} \tag{483}$$

Hence, since $Q_C > 0$, if $C(t_1)$ is positive the function C will increase exponentially for increasing time t and symmetrically if $B(t_2)$ is negative the function B will decrease and escape exponentially for decreasing time t.

Let us consider now a bounded orbit $\vec{x}(t)$ remaining in the neighbourhood N, we find immediately in any of its vicinities orbits escaping exponentially in the future (with positive initial C) or in the past (with negative initial B). Furthermore if $\vec{y}(t)$ is another bounded orbit of the neighbourhood N we necessarily have at any time :

$$C\left(\vec{x}(t),\vec{y}(t)\right) \leq 0 \leq B\left(\vec{x}(t),\vec{y}(t)\right) \tag{484}$$

Hence, the vectors $\vec{x}_-$, $\vec{x}_0$, $\vec{x}_+$ being defined in (457), we find that at any time between two bounded orbits of the neighbourhood N :

$$\begin{aligned} & c_m \delta x_-^2 + c_0 \delta x_0^2 \geq c_p \delta x_+^2 \\ & b_m \delta x_-^2 \leq b_0 \delta x_0^2 + b_p \delta x_+^2 \end{aligned} \tag{485}$$

the six positive constants c_m, c_o, c_p, b_m, b_o, b_p being obvious functions of the b_K and c_K of (481), (482) :

$$\left.\begin{aligned} c_m &= -\left(\inf_{K=1}^{n_-} c_K\right) ; \quad c_o = -\left(\inf_{K=1+n_-}^{n_0+n_-} c_K\right) ; \quad c_p = \inf_{K=1+n_0+n_-}^{n} c_K \\ b_m &= -\left(\sup_{K=1}^{n_-} b_K\right) ; \quad b_o = \sup_{K=1+n_-}^{n_0+n_-} b_K ; \quad b_p = \sup_{K=1+n_0+n_-}^{n} b_K \end{aligned}\right\} \quad (486)$$

If $c_p b_m > c_m b_p$, as may easily be chosen, we can deduce from (485) :

$$||\delta\vec{x}_-|| \leq K_m||\delta\vec{x}_0|| \quad ; \quad ||\delta\vec{x}_+|| \leq K_p||\delta\vec{x}_0|| \qquad (487)$$

with :

$$K_m = \{(c_p b_o + c_o b_p)/(c_p b_m - c_m b_p)\}^{1/2} \; ; \; K_p = \{(c_m b_o + c_o b_m)/(c_p b_m - c_m b_p)\}^{1/2} \qquad (488)$$

The constants K_m and K_p can be chosen very small if the neighbourhood N itself is very small.

(487) implies that the "central subset" containing the neighbouring bounded orbits cannot have two different points with the same $\vec{x}_0$. It can be given in terms of $\vec{x}_0$ by a locally Lipschitz function $\vec{x}(\vec{x}_0)$. This manifold has at most n_0 dimensions and at the origin is tangent to the subspace $(0,\vec{x}_0,0)$.

Simple examples show that, unlike the stable and the unstable subsets, the central subset can be non-analytic in the vicinity of the equilibrium point even for analytic equations of motion (452).

Finally let us complete the picture of the set of solutions in the vicinity of the equilibrium point and let us demonstrate that the bounded solutions remaining sufficiently close to the equilibrium point have, at any time, their own stable and unstable subsets with at least n_- and n_+ dimensions.

Let us consider again the function $C(\vec{x},\vec{y})$ defined in (480), (482) and denote by $C_+(\vec{x},\vec{y})$ the function containing only the positive terms of C :

$$C_+(\vec{x},\vec{y}) = \sum_{K=1+n_0+n_-}^{n} c_K(y_K - x_K)(\overline{y_K} - \overline{x_K}) \qquad (489)$$

With $y_K - x_K = \delta x_K$, the derivative of C_+ is given by the following expression similar to (479) :

$$(dC_+/dt) - Q_C C_+ = \sum_{K=1+n_0+n_-}^{n} c_K \left[\begin{aligned} &(\lambda_K+\overline{\lambda_K}-Q_C)\delta x_K\overline{\delta x_K} + \varepsilon_K(\delta x_K\overline{\delta x_{K+1}} + \overline{\delta x_K}\delta x_{K+1}) \\ &+ \left[g_K(\vec{y})-g_K(\vec{y})\right]\overline{\delta x_K} + \left[\overline{g_K(\vec{y})} - \overline{g_K(\vec{x})}\right]\delta x_K \end{aligned}\right] \qquad (490)$$

As for the function C the $(\lambda_K + \overline{\lambda_K} - \varrho_C)$ and ε_K terms give a positive definite quadratic function F_+ but for $\vec{\delta x}_+$ terms only, i.e. terms from $K = 1+n_0+n_-$ to $K = n$.

Fortunately in the cone defined by $C(\vec{x},\vec{y}) \geqq 0$ the function F_+ will be larger than or equal to a suitable positive definite function $F(\vec{\delta x})$, and thus, in a sufficiently small neighbourhood $\mathcal{N}'$ of the origin we will have :

$$\left.\begin{array}{l} \varrho_C = \mathcal{R}e\{\lambda_{(1+n_0+n_-)}\} > 0 \ ; \quad C_+ \geqq 0 \ ; \quad C_+ \geqq C \\ dC/dt \geqq \varrho_C C \\ C \geqq 0 \ \text{ implies } \ dC_+/dt \geqq \varrho_C C_+ \end{array}\right\} \tag{491}$$

Notice that if $C \geqq 0$ the function C_+ is equivalent to $||\vec{y} - \vec{x}||^2$ and can be used as the square of a distance.

Within the neighbourhood $\mathcal{N}'$, consider a bounded trajectory $\vec{x}(t)$ and let us try to build the corresponding unstable subset as it is at the time t_1, at least in the vicinity of $\vec{x}(t_1)$.

The neighbouring trajectories $\vec{y}(t)$ will be interesting if, with $t_0 < t_1$:

$$\left.\begin{array}{l} C\left(\vec{x}(t_0),\vec{y}(t_0)\right) \geqq 0 \\ C_+\left(\vec{x}(t_1),\vec{y}(t_1)\right) \leqq \delta \end{array}\right\} \tag{492}$$

Because of (491) we have in the time interval $\left[t_0,t_1\right]$:

$$0 \leqq C\left(\vec{x}(t),\vec{y}(t)\right) \leqq C_+\left(\vec{x}(t),\vec{y}(t)\right) \leqq \delta\exp\{\varrho_C(t - t_1)\} \leqq \delta \tag{493}$$

and we will choose the small quantity δ such that during the whole time interval $\left[t_0,t_1\right]$ the trajectories $\vec{y}(t)$ remain in the neighbourhood $\mathcal{N}'$.

Let us now define at the time t_1 the subset $\mathcal{U}_{t_0}$ to be all the possible $\vec{y}(t_1)$ satisfying (492).

For various t_0 the subsets $\mathcal{U}_{t_0}$ have the following properties :

I) $\mathcal{U}_{t_0}$ contains at least the $\vec{y}(t_1)$'s corresponding to the following $\vec{y}(t_0)$:

$$\vec{y}(t_0) - \vec{x}(t_0) = (0,0,\overrightarrow{\delta x_{0+}}) \ ; \quad ||\overrightarrow{\delta x_{0+}}|| \text{ sufficiently small} \tag{494}$$

hence $\mathcal{U}_{t_0}$ contains a manifold of n_+ dimensions that extends from $\vec{y}_1 = \vec{x}_1$, i.e. $C_+(\vec{x}_1,\vec{y}_1) = 0$, to $C_+(\vec{x}_1,\vec{y}_1) = \delta$ into all directions.

II) $t'_0 < t_0 < t_1$ implies $\mathfrak{U}_{t'_0} \subset \mathfrak{U}_{t_0}$ (495)

The subsets $\mathfrak{U}_{t_0}$ are closed and hence the limit subset $\mathfrak{U}$ equal to $\lim_{t_0 \to -\infty} \mathfrak{U}_{t_0}$ is is defined $\left(\mathfrak{U} = \bigcap_{N=1}^{\infty} \mathfrak{U}_{(t_1 - N)}\right)$, it is also closed and it contains at least an n_+ dimensional manifold that corresponds to trajectories $\vec{y}(t)$ verifying :

$$t \leqq t_1 \text{ implies } 0 \leqq C\left(\vec{x}(t),\vec{y}(t)\right) \leqq c_+\left(\vec{x}(t),\vec{y}(t)\right) \leqq \delta\exp\{Q_C(t - t_1)\} \leqq \delta \quad (496)$$

These trajectories $\vec{y}(t)$ are then exponentially asymptotic to the bounded solution $\vec{x}(t)$ of interest and the subset $\mathfrak{U}$ is either the unstable subset of $\vec{x}(t)$ at the time t_1 or at least a part of this subset.

A symmetrical construction gives the stable subset or at least the part corresponding to the subset $\mathfrak{U}$.

Very simple examples show that the unstable subset of a neighbouring bounded solution can have more than n_+ dimensions. Consider for instance the following differential system :

$$\begin{aligned} dX/dt &= X \\ dY/dt &= 0 \\ dZ/dt &= YZ \end{aligned} \quad (497)$$

The origin is a point of equilibrium with $n_- = 0$; $n_0 = 2$ and $n_+ = 1$ (indeed $\lambda_1 = \lambda_2 = 0$; $\lambda_3 = 1$).

The solution of (497) are simple :

$$X = X_0\exp(t) \quad ; \quad Y = Y_0 \quad ; \quad Z = Z_0\exp(Y_0 t) \quad (498)$$

The unstable subset of the origin is the X-axis and the central subset is the Y-Z plane in which only the points of the Y and Z axes correspond to bounded solutions (and even to points of equilibrium).

The points of the Z-axis have their own one-dimensional unstable subset parallel to the X-axis but :

I) The points of the positive Y semi-axis have each a two-dimensional unstable subset parallel to the X-Z plane.

II) The points of the negative Y semi-axis have each a one-dimensional unstable subset parallel to the X-axis and a one-dimensional stable subset parallel to the Z-axis.

Thus we know many informations about the picture of the set of solutions in the vicinity of an equilibrium point : we have seen successively the stable and the unstable subsets with their asymptotic exponential motions, the central subset with all neighbouring bounded motions (but sometimes also with unbounded motions) and finally the stable and unstable subsets of these bounded motions.

Nevertheless, in general, most neighbouring orbits are neither bounded nor asymptotic and they escape on both sides.

We will see in Section 10.7.7.3 a more detailed analysis of the motions in the central subset.

10.7.7.2 Analytic differential systems.
The vicinity of a periodic solution.

This Section is the natural generalization of the previous one and leads to almost the same results.

Periodic solutions of the differential systems are essentially met in the two-following cases :

I) In the case of autonomous systems.

$$dx_K/dt = f_K(x_1, x_2, \ldots, x_n) \quad ; \quad K = \{1, 2, \ldots, n\} \tag{499}$$

Such systems were met in the Chapters 4, 7, 9 and 10 and especially in (452), the three-body problem is of this type.

II) In the case of periodic systems.

$$\left.\begin{aligned} dx_K/dt &= f_K(x_1, x_2, \ldots, x_n, t) \\ f_K(x_1, x_2, \ldots, x_n t) &\equiv f_K(x_1, x_2, \ldots, x_n, t+P) \end{aligned}\right\} K = \{1, 2, \ldots, n\} \tag{500}$$

An example of a periodic system was met in the equations (171), (172) of the elliptic restricted three-body problem with the period $P = 2\pi$.

A periodic solution is a solution $\vec{x}(t)$ of (499) or (500) with :

$$\vec{x}(t + T) \equiv \vec{x}(t) \tag{501}$$

The period T is arbitrary in the first case and is a multiple of P in the second case.

For studying the vicinity of the periodic solution of interest let us choose this solution as origin of coordinates. The equations of motion will become the following with new functions f_K :

$$\left.\begin{aligned} &dx_K/dt = f_K(x_1, x_2, \ldots, x_n, t) \\ &f_K(x_1, x_2, \ldots, x_n, t) \equiv f_K(x_1, x_2, \ldots, x_n, t+T) \\ &f_K(0, 0, 0, \ldots, 0, t) \equiv 0 \end{aligned}\right\} \quad K = \{1,2,\ldots,n\} \tag{502}$$

The first-order study is the classical Floquet theory (Ref. 173, 174), that was touched on in Section 10.5.2.

The first-order equation is :

$$d\vec{x}/dt = M(t)\vec{x} \tag{503}$$

The Jacobian matrix M(t) is periodic in terms of t :

$$M(t) \equiv M(t + T) \tag{504}$$

Integration of (503) leads to :

$$\vec{x}(t) = C(t) \cdot \vec{x}(0) \tag{505}$$

the matrix C(t) being defined by :

$$C(0) = I \quad ; \quad dC/dt = M(t) \cdot C(t) \tag{506}$$

Let us call A the matrix C(T) it is the "Floquet matrix" or "monodromy matrix" or "transition matrix" and its eigenvalues $\mu_1, \mu_2, \ldots, \mu_n$ are the "Floquet multipliers".

(504) and (506) imply :

$$\left.\begin{aligned} &C(t + T) = C(t) \cdot A \\ &C(mT) = A^m \quad ; \quad m \text{ integer} \\ &\det C(t) = \exp\{\textstyle\int_0^t (\text{trace } M(u)) \cdot du\} \neq 0 \end{aligned}\right\} \tag{507}$$

and if $\vec{V}_j$ is an eigenvector of A :

$$A\vec{V}_j = \mu_j\vec{V}_j \quad ; \quad \mu_j \neq 0 \tag{508}$$

that is, to first-order :

$$\vec{x}(0)//\vec{V}_j \quad \text{implies} \left\{ \begin{aligned} \vec{x}(T) &= \mu_j \vec{x}(0) \\ \vec{x}(mT) &= \mu_j^m \vec{x}(0) \end{aligned} \right\} \tag{509}$$

These results lead to the usual rules of exponential stability or instability, the critical case being that in which the largest modulus of the μ_j's is equal to one.

The usual first simplification of the equations of motion (502) is a linear transformation $\vec{x}(t) \to \vec{y}(t)$, periodic in terms of t and leading to a constant matrix N in the first-order equation :

$$d\vec{y}/dt = N\,\vec{y}(t) \tag{510}$$

(510) implies of course :

$$\vec{y}(t) = \exp\{Nt\}\vec{y}(0)$$

and, with $\vec{y}(0) = \vec{x}(0)$, the linear transformation $\vec{x}(t) \to \vec{y}(t)$ will be :

$$\vec{y}(t) = \exp\{Nt\} \cdot C(t)^{-1} \cdot \vec{x}(t) \tag{511}$$

with A = C(T) this linear transformation will have the period T if and only if :

$$\exp(NT) = A \tag{512}$$

This equation (512) gives many possible N, simple functions of $I = A^0$, A, A^2, ..., $A^{(n-1)}$ and the eigenvalues $\mu_1, \mu_2, \ldots, \mu_n$.

$$N = \frac{1}{T} \sum_{j=0}^{n-1} a_j A^j \tag{513}$$

with, if all μ_j are different :

$$\begin{pmatrix} a_0 \\ a_1 \\ \vdots \\ a_{n-1} \end{pmatrix} = \begin{pmatrix} 1 & \mu_1 & \mu_1^2 & \cdots & \mu_1^{(n-1)} \\ 1 & \mu_2 & \mu_2^2 & \cdots & \mu_2^{(n-1)} \\ \vdots & \vdots & \vdots & & \vdots \\ 1 & \mu_n & \mu_n^2 & & \mu_n^{(n-1)} \end{pmatrix}^{(-1)} \cdot \begin{pmatrix} \text{Log}\,\mu_1 \\ \text{Log}\,\mu_2 \\ \vdots \\ \text{Log}\,\mu_n \end{pmatrix} \tag{514}$$

If some Floquet multipliers μ_j are multiple, the matrix N can be obtained by continuity and remains bounded. For instance if all eigenvalues are equal to μ_1 :

$$N = \frac{1}{T} \{ I\left(\mathrm{Log}\ \mu_1 - \frac{1}{2} - \frac{1}{3} - \dots - \frac{1}{(n-1)}\right) - \sum_{j=1}^{n-1} \binom{n-1}{j} \frac{(-1)^j A^j}{j\mu_1^j} \} \tag{515}$$

The eigenvalues μ_j are all different from zero and have then bounded logarithms, but notice that a non-zero number $\mu = \rho \cdot \exp(i\theta)$ has many possible logarithms equal to $\mathrm{Log}\ \rho + i\theta + 2Ki\pi$ with $K \in \mathbb{Z}$ and this leads to many possible N for a given A.

The matrices A and N have the same eigenvectors, the eigenvalues λ_j of N are given by $T\lambda_j = \mathrm{Log}\ \mu_j$, their real parts are the "Liapunov exponents".

Let us now go beyond the first-order analysis.

With a suitable set of axes the matrix N of (510) can be diagonalized or at least "Jordanized" and the equations of motion become very similar to these of the autonomous case (454) :

$$\left. \begin{array}{l} dx_K/dt = \lambda_K x_K + \varepsilon_K x_{K+1} + g_K(x_1, x_2, \dots, x_n, t) \\ \\ \text{where : } \lambda_1, \lambda_2, \dots, \lambda_n \text{ are the eigenvalues of N} \\ \\ \qquad \varepsilon_n = 0 \ ; \ \lambda_K \neq \lambda_{K+1} \text{ implies } \varepsilon_K = 0 \\ \\ \lambda_K = \lambda_{K+1} \text{ implies } \varepsilon_K = \text{either 0 or 1} \end{array} \right\} K = \{1, 2, \dots, n\} \tag{516}$$

The only difference is that the power series g_K are now periodic in terms of t.

(*) Denote by $M^*(t)$ the adjoint matrix of $M(t)$, its elements m^*_{ij} are given by transposed conjugate rule :

$$m^*_{ij} = \overline{m_{ji}}$$

The n eigenvalues of the product M^*M are real and non-negative. Let us call $\Lambda(t)$ the largest of these eigenvalues, it always verifies :

$$\Lambda \leq \sum_{i=1}^{n} \sum_{j=1}^{n} m_{ij}\overline{m_{ij}} \leq n\Lambda$$

It is always possible to choose the matrix N given in (512)-(514) and its eigenvalues λ_j such that for all j :

$$|\lambda_j| \leq \frac{1}{T} \int_0^T dt \cdot \sqrt{\Lambda(t)} = \text{mean value of } \sqrt{\Lambda}$$

If $\int_0^T dt\sqrt{\Lambda(t)} < \pi$ only one such choice of N is possible and furthermore if the monodromy matrix A is real, the corresponding matrix N, with $A = \exp(NT)$, is then also real.

Since the Liapunov characteristic exponents are the real parts of the eigenvalues λ_j, the mean value of $\sqrt{\Lambda(t)}$ also gives upper and lower bounds of these Liapunov exponents. However the mean values of the largest and the smallest eigenvalues of $\{(M + M^*)/2\}$ are better and classical upper and lower bounds.

The power series g_K have neither constant term nor first-order term and the analyticity implies that three positive constants a, b, c exist which satisfy :

$$c < 1 \tag{517}$$

and :

$$\left.\begin{array}{l} g_K(x_1,x_2,\ldots x_n,\ t) = \sum\limits_{(\vec{\alpha},m)} G_K(\vec{\alpha},m) \bullet x_1^{\alpha_1} x_2^{\alpha_2} \ldots x_n^{\alpha_n} \bullet \exp\left(\frac{2\pi m i}{T}\ t\right) \\ \text{with : } \begin{cases} \alpha_1, \alpha_2, \ldots, \alpha_n \in \mathbb{N} \ ; \ \alpha_1 + \alpha_2 + \ldots + \alpha_n \geq 2 \\ m \in \mathbb{Z} \end{cases} \\ \text{and : } \\ |G_K(\vec{\alpha},m)| \leq a \bullet b^{(\alpha_1+\alpha_2+\ldots+\alpha_n)} \bullet c^{|m|} \end{array}\right\} \tag{518}$$

We can now generalize the near resonance theorem of the previous Section to periodic solutions.

> There exists an analytic transformation :
>
> $$\vec{u} = \vec{u}(\vec{x},t) \tag{519}$$
>
> having the following properties :
>
> $$\text{I)} \quad \vec{u} - \vec{x} = O(x^2) \tag{520}$$
>
> II) The transformation is periodic in terms of t :
>
> $$\vec{u}(\vec{x},t) \equiv \vec{u}(\vec{x},t + T) \tag{521}$$
>
> III) The equations of motion of $\vec{u}$ have much less terms :
>
> $$du_K/dt = \lambda_K u_K + \varepsilon_K u_{K+1} + j_K(u_1,u_2,\ldots u_n,\ t) \quad ; \quad K = \{1,2,\ldots,n\} \tag{522}$$
>
> The λ_K's and ε_K's are those of (516) while the power series j_K have neither constant nor linear terms, are periodic in terms of t, are convergent within a small neighbourhood of the origin and satisfy the following :
>
> The power series $j_K(\vec{u},t)$ have only "sufficiently resonant" terms, i.e. terms :
>
> $$\left.\begin{array}{l} J_K(\vec{\alpha},m) \bullet u_1^{\alpha_1} u_2^{\alpha_2} \ldots u_n^{\alpha_n} \bullet \exp\left(\frac{2\pi m i}{T}\ t\right) \quad \text{for which :} \\ \left|\lambda_K - \sum\limits_{q=1}^{n} \alpha_q \lambda_q - \frac{2\pi m i}{T}\right| \leq \varepsilon(\alpha_1 + \alpha_2 + \ldots + \alpha_n) \end{array}\right\} \tag{523}$$
>
> ε being an arbitrary positive constant.

The demonstration of this theorem is only a small developement of the corresponding demonstration for equilibrium points and we will only present the main case in which all ε_K are zero (see Appendix 3 if some ε_K are one).

Let us define the analytic transformation $\vec{x} \rightarrow \vec{u}$ by the following :

$$\left.\begin{array}{l} x_K = u_K + \phi_K(u_1, u_2, \ldots, u_n, t) \quad ; \quad K = \{1, 2, \ldots, n\} \\ \text{with :} \\ \phi_K = O(u^2) \quad ; \quad \phi_K(\vec{u},t) \equiv \phi_K(\vec{u}, t + T) \end{array}\right\} \quad (524)$$

The relations between the power series ϕ_K and j_K are given by :

$$dx_K/dt = \lambda_K x_K + g_K(\vec{x},t) = \lambda_K u_K + \lambda_K \phi_K(\vec{u},t) + g_K(\vec{x},t) \quad (525)$$

and :

$$dx_K/dt = d\left[u_K + \phi_K\right]/dt = \lambda_K u_K + j_K(\vec{u},t) + \sum_{q=1}^{n} \frac{\partial \phi_K}{\partial u_q}(\lambda_q u_q + j_q) + \frac{\partial \phi_K}{\partial t} \quad (526)$$

Hence :

$$\lambda_K \phi_K(\vec{u},t) + g_K(\vec{x},t) = j_K(\vec{u},t) + \sum_{q=1}^{n} \frac{\partial \phi_K(\vec{u},t)}{\partial u_q}\left(\lambda_q u_q + j_q(\vec{u},t)\right) + \frac{\partial \phi_K(\vec{u},t)}{\partial t} \quad (527)$$

that is for $K = 1, 2, \ldots, n$:

$$\left.\begin{array}{l} j_K(\vec{u},t) + \displaystyle\sum_{q=1}^{n} \frac{\partial \phi_K(\vec{u},t)}{\partial u_q} \lambda_q u_q + \frac{\partial \phi_K(\vec{u},t)}{\partial t} - \lambda_K \phi_K(\vec{u},t) = \\ = g_K(\vec{x},t) - \displaystyle\sum_{q=1}^{n} \frac{\partial \phi_K(\vec{u},t)}{\partial u_q} j_q(\vec{u},t) \end{array}\right\} \quad (528)$$

Equations (524) and (528) allow a formal development of the power series $j_K(\vec{u},t)$ and $\phi_K(\vec{u},t)$ that have neither constant term nor linear term. Indeed assume that these series j_K and ϕ_K are known for all K up to the degree $(p - 1)$, the right-hand member of (528) is then known in terms of $(\vec{u},t)$ up to the degree p and this allow a single choice of the terms of degree p of j_K and ϕ_K according to the following convention.

Assume that we consider the term in $u_1^{\alpha_1} u_2^{\alpha_2} \ldots u_n^{\alpha_n} \cdot \exp(\frac{2\pi mi}{T} t)$ with $\alpha_1 + \alpha_2 + \ldots + \alpha_n = p$, and that for this term the right-hand member of (528) is : $B_K u_1^{\alpha_1} u_2^{\alpha_2} \ldots u_n^{\alpha_n} \cdot \exp(\frac{2\pi mi}{T} t)$

Denote by $J_K u_1^{\alpha_1} u_2^{\alpha_2} \ldots u_n^{\alpha_n} \cdot \exp(\frac{2\pi mi}{T} t)$ and $F_K u_1^{\alpha_1} u_2^{\alpha_2} \; u_n^{\alpha_n} \times$

$\times \left(\exp(\frac{2\pi mi}{T} t)\right)$ the corresponding terms of $j_K(\vec{u},t)$ and $\phi_K(\vec{u},t)$.

The equation (528) gives :

$$J_K + \left(\sum_{q=1}^{n} \alpha_q \lambda_q + \frac{2\pi mi}{T} - \lambda_K\right) F_K = B_K \quad (529)$$

and, by convention, we will choose :

$$\left.\begin{array}{l} J_K = B_K \ ; \ F_K = 0, \text{ if the near resonance condition (523)} \\ \qquad\qquad \text{is satisfied} \\ \\ J_K = 0 \ ; \ F_K = B_K \Big/ \left(\sum_{q=1}^{n} \alpha_q \lambda_q + \dfrac{2\pi m i}{T} - \lambda_K \right), \text{ if (523) is not satisfied} \end{array}\right\} \quad (530)$$

Hence for any K the power series $j_K(\vec{u},t)$ and $\phi_K(\vec{u},t)$ are periodic and uniquely defined and the only remaining question is that of the convergence of these power series.

We will again use the definition of majorization applied to all terms of the type $u_1^{\alpha_1} u_2^{\alpha_2} \ldots u_n^{\alpha_n} \cdot \exp\left\{\frac{2\pi mi}{T} t\right\}$ or $x_1^{\alpha_1} x_2^{\alpha_2} \ldots x_n^{\alpha_n} \cdot \exp\left\{\frac{2\pi mi}{T} t\right\}$. The power series $b(\vec{u},t)$ (or $b(\vec{x},t)$) majorizes the power series $a(\vec{u},t)$ (or $a(\vec{x},t)$) if for all terms its coefficients are real, non-negative and larger than or equal to the modulus of the corresponding terms of the series a. This implies of course that the radius of convergence of the power series b is smaller than or equal to that of the series a.

Let us also define the formal power series $h_K(\vec{u},t)$ without constant term by :

$$j_K(\vec{u},t) = \sum_{q=1}^{n} \frac{\partial h_K(\vec{u},t)}{\partial u_q} \cdot u_q \quad ; \quad K = \{1, 2, \ldots, n\} \tag{531}$$

For all K the power series $h_K(\vec{u},t)$ and $j_K(\vec{u},t)$ have the same radius of convergence and $H_K(\alpha_1, \alpha_2, \ldots, \alpha_n, m) = J_K(\alpha_1, \alpha_2, \ldots, \alpha_n, m)/(\alpha_1 + \alpha_2 + \ldots + \alpha_n)$.

The conditions (517), (518) and the corresponding constants a, b, c give a simple "power-Fourier series" $f(\vec{x},t)$ that majorizes all power-Fourier series $g_K(\vec{x},t)$:

$$f(\vec{x},t) = a\left[b^2X^2 + b^3X^3 + b^4X^4 + \ldots\right] \cdot \left[1 + \sum_{m=1}^{\infty} c^m \left(\exp\{m\omega_0 it\} + \exp\{-m\omega_0 it\}\right)\right] \tag{532}$$

with

$$X = x_1 + x_2 + x_3 + \ldots + x_n \quad ; \quad \omega_0 = 2\pi/T$$

The demonstration remains valid if we substitute to the Fourier series of (532) any converging Fourier series with positive coefficients and this will allow to extend the near resonance theorem to dynamical systems "developable into converging power-Fourier series", even if these systems are not analytic (the transformation $\vec{x} \to \vec{u}$ is then no more analytic but it is still developable into converging power-Fourier series).

Hence we will write with a suitable function $f(\vec{x},t)$ and with the symbol $\succ$ of majorization :

$$\left.\begin{aligned}
&K = \{1, 2, \ldots, n\} \text{ implies } f(\vec{x},t) \succ g_K(\vec{x},t) \\
&\text{with :} \\
&f(\vec{x},t) = g(t)X^2/(1 - bX) \\
&g(t) = a_0 + \sum_{m=1}^{\infty} a_m \left[\exp\{m\omega_0 it\} + \exp\{- m\omega_0 it\}\right] \\
&\text{all } a_m \geq 0 \ ; \ \sum_{m=0}^{\infty} a_m < \infty \ ; \ \omega_0 = 2\pi/T \\
&X = x_1 + x_2 + \ldots + x_n \ ; \ X^2/(1 - bX) = X^2 + bX^3 + b^2X^4 + \ldots
\end{aligned}\right\} \tag{533}$$

In these conditions, with the constant ε of (523) and the rules (530), the power-Fourier series $h_K(\vec{u},t)$, $\varepsilon\phi_K(\vec{u},t)$ and $h_K + \varepsilon\phi_K$ are majorized by the power-Fourier series $\varepsilon\psi\ (\vec{u},t)$ defined by :

$$\left.\begin{aligned}
&X = x_1 + x_2 + \ldots + x_n \\
&x_K = u_K + \psi(\vec{u},t) \\
&\varepsilon \sum_{q=1}^{n} \frac{\partial\psi}{\partial u_q} u_q = g(t)X^2/(1 - bX) + \varepsilon \sum_{q=1}^{n} \frac{\partial\psi}{\partial u_q} \cdot \sum_{p=1}^{n} \frac{\partial\psi}{\partial u_p} u_p
\end{aligned}\right\} \tag{534}$$

Let us put :

$$U = u_1 + u_2 + \ldots + u_n \ ; \ X = U + n\psi \tag{534.A}$$

The function $\psi(\vec{u},t)$ is a function of U and t only, it is defined by :

$$\left.\begin{aligned}
&\psi = \psi(U,t) \\
&\varepsilon U \frac{\partial\psi}{\partial U} = g(t) \cdot \frac{(U + n\psi)^2}{1 - b(U + n\psi)} + n\varepsilon U\left(\frac{\partial\psi}{\partial U}\right)^2 \\
&\varepsilon(U,t) = \frac{1}{2\varepsilon} g(t) \cdot U^2 + \left[\frac{bg(t)}{3\varepsilon} + \frac{2n\left(g(t)\right)^2}{3\varepsilon^2}\right]U^3 + O(U^4)
\end{aligned}\right\} \tag{534.B}$$

$\psi(U,t)$ is majorized by $\frac{U}{2}\frac{\partial\psi}{\partial U}$ and they are both majorized by $S(U,t)$ given by :

$$2\varepsilon S(U,t) = g(t) \cdot \frac{(U + nS)^2}{1 - b(U + nS)} + 4n\varepsilon \frac{S^2}{U} \tag{535}$$

$S(U,t)$ is majorized by $R(U,t)$ given by :

$$2\varepsilon R(U,t) = g(t) \cdot \frac{(U + nR)^2}{1 - b(U + nR)} + \frac{4n\varepsilon R^2}{U\left[1 - b(U + nR)\right]} \tag{535.A}$$

This algebraic equation of degree two gives :

$$R = \frac{\varepsilon(1 - bU) - nUg(t) - \sqrt{[\varepsilon^2(1 - bU)^2 - 6n\varepsilon Ug(t)]}}{n^2 g(t) + 2n\varepsilon b + 4\varepsilon nU^{-1}} \qquad (535.B)$$

The development of R in power-Fourier series is easy and we can verify that it is converging for small U. The radius of convergence corresponds to $|U| = \varepsilon / \left[b\varepsilon + 3ng(0) + \sqrt{[6b\varepsilon ng(0) + 9n^2 g^2(0)]} \right]$ that is positive for all positive ε but goes to zero with ε.

Hence from majorizing series to majorizing series we have demonstrated the near-resonance theorem at least in the general case when all coefficients ε_K of (516) are zero (full demonstration in Appendix 3).

We can of course write the following summary of the demonstration :

$$R(U,t) \succ S(U,t) \succ \frac{U}{2}\frac{\partial\psi}{\partial U} \succ \left\{ \begin{array}{l} \succ \frac{1}{2\varepsilon} j_K(\vec{u},t) \\ \succ \psi(U,t) \left\{ \begin{array}{l} \succ \psi_K(\vec{u},t) \\ \succ \frac{1}{\varepsilon} h_K(\vec{u},t) \end{array} \right. \end{array} \right\} K = \{1,\dots,n\} \qquad (536)$$

These bounds on the functions $j_K(\vec{u},t)$ and $\phi_K(\vec{u},t)$ have been use in the Ref. 180 in order to obtain limits on the slow phenomenon called "Arnold diffusion".

Hence the near-resonance theorem is demonstrated even for periodic solutions and this has many consequences.

Let us again assume the indices to be chosen so that the eigenvalues λ_j verify :

$$\mathcal{Re}(\lambda_1) \leq \mathcal{Re}(\lambda_2) \leq \dots \leq \mathcal{Re}(\lambda_n) \qquad (537)$$

and let us define the integers n_-, n_0, n_+, by :

$$\left. \begin{array}{l} \mathcal{Re}(\lambda_{n_-}) < 0 \;\; ; \;\; \mathcal{Re}(\lambda_{(1+n_-)}) = 0 \;\; ; \;\; \mathcal{Re}(\lambda_{(n_0+n_-)}) = 0 \\ \mathcal{Re}(\lambda_{(1+n_0+n_-)}) > 0 \;\; ; \;\; n_- + n_0 + n_+ = n \end{array} \right\} \qquad (538)$$

The corresponding vectors $\vec{u}_-$, $\vec{u}_0$, $\vec{u}_+$ are :

$$\left.\begin{aligned} \vec{u}_- &= (u_1, u_2, \ldots, u_{n_-}) \\ \vec{u}_0 &= \left(u_{(1+n_-)}, \ldots, u_{(n_0+n_-)}\right) \\ \vec{u}_+ &= \left(u_{(1+n_0+n_-)}, \ldots, u_n\right) \end{aligned}\right\} \quad (539)$$

and for sufficiently small ε the equations (522), (523) show that the manifolds $(\vec{u}_-,0,0)$ and $(0,0,\vec{u}_+)$ are invariant. They are the stable and unstable subsets with n_- and n_+ dimensions and with exponential asymptotical motions verifying for all K :

$$u_K = P_K(t) \cdot \exp(\lambda_K t) \quad (540)$$

The polynomials $P_K(t)$ are of course zero for the $\vec{u}_0$ components, they have at most the degree $1 + (n_- - 2)\dfrac{\mathcal{R}e(\lambda_1)}{\mathcal{R}e(\lambda_{n_-})}$ in the stable subset and at most the degree

$1 + (n_+ - 2)\mathcal{R}e(\lambda_{n_-})/\mathcal{R}e\left(\lambda_{(1+n_0+n_-)}\right)$ in the unstable subset.

Notice the agreement of these results with the usual stability rules for the future :

$$\left.\begin{aligned} &\text{If } \mathcal{R}e(\lambda_n) > 0 \text{ the motion is exponentially unstable} \\ &\text{If } \mathcal{R}e(\lambda_n) < 0 \text{ the motion is exponentially stable} \end{aligned}\right\} \quad (541)$$

By demonstrations identical to those of the previous section we now obtain similar results :

I) The bounded solutions remaining forever in a sufficiently small vicinity of the periodic solution of interest also remain in a "central subset", a manifold with at most n_0 dimensions at any time.

This manifold is periodic in terms of t and is tangent to the subspace $(0,\vec{u}_0,0)$ at the origin.

II) These neigbouring bounded solutions have their own stable and unstable subsets with exponential asymptotical motions. These subsets have at least n_- and n_+ dimensions at any time.

III) If $\mathcal{R}e(\lambda_1)\cdot\mathcal{R}e(\lambda_n) < 0$ most neighbouring orbits are neither bounded nor asymptotic and escape on both sides.

We will see in the next Section a more detailed analysis of the motions in the central subset.

10.7.7.3 Motions in the central subset. Motions in the critical case. The critical Hamiltonian case.

The near-resonance theorem (519)-(523) allows simple high-order analyses.

Let us attempt an Nth-order analysis and let us neglect all higher order terms.

The number of possible $\vec{\alpha}$ vectors for a degree $\alpha_1 + \alpha_2 + \ldots + \alpha_n$ less than or equal to N is finite and hence it is always possible to choose a sufficiently small ε in (523) such that all "sufficiently resonant terms" be strictly resonant :

$$\left.\begin{aligned} & du_K/dt = \lambda_K u_K + \varepsilon_K u_{K+1} + j_K(u_1, u_2, \ldots u_n, t) \quad ; \\ & K = \{1, 2, \ldots, n\} \end{aligned}\right\} \tag{542}$$

$$\left.\begin{aligned} & \text{with :} \qquad \vec{\alpha} = (\alpha_1, \alpha_2, \ldots, \alpha_n) \\ & j_K = \sum_{(\vec{\alpha}, m)} J_K(\vec{\alpha}, m) . u_1^{\alpha_1} u_2^{\alpha_2} \ldots u_n^{\alpha_n} \exp\left(\frac{2\pi m i}{T} t\right) + O(u_\ell^{(N+1)}) \\ & \alpha_1, \alpha_2, \ldots, \alpha_n \in \mathbb{N} \quad ; \quad 2 \leq \alpha_1 + \alpha_2 + \ldots + \alpha_n \leq N \\ & m \in \mathbb{Z} \end{aligned}\right\} \tag{543}$$

and for all non-zero j_K-terms up to the order N :

$$\lambda_K = \sum_{q=1}^{n} \alpha_q \lambda_q + \frac{2\pi}{T} mi \quad ; \quad K = \{1, 2, \ldots, n\} \tag{544}$$

We will consider the most critical cases with all $\mathcal{R}e(\lambda_K) = 0$ and we will denote by ω_K the corresponding pulsations :

$$\lambda_K = i\omega_K \quad ; \quad K = \{1, 2, \ldots, n\} \tag{545}$$

The pulsation of the periodic orbit can be called ω_0 :

$$\omega_0 = 2\pi/T \tag{546}$$

and the equations (544) are equivalent to :

$$\omega_K = m\omega_0 + \sum_{q=1}^{n} \alpha_q \omega_q \quad ; \quad K = \{1, 2, \ldots, n\} \tag{547}$$

Let us define the new variables v_K by the following relations :

$$u_K = v_K \exp(i\omega_K t) \quad ; \quad K = \{1, 2, \ldots, n\} \tag{548}$$

The equations of motion (542), (543) become :

$$dv_K/dt = \varepsilon_K v_{K+1} + \sum_{(\vec{\alpha},m)} J_K(\vec{\alpha},m) v_1^{\alpha_1} . v_2^{\alpha_2} \ldots v_n^{\alpha_n} + O(v_\ell^{(N+1)}) \tag{549}$$

These equations are <u>autonomous</u> up to the order N. Furthermore if the (n+1) pulsations ω_0 and ω_K are rationally independent, or even if they never satisfy (547) with $2 \leqq \alpha_1 + \alpha_2 + \ldots + \alpha_n \leqq N$ the equations (549) become simply :

$$dv_K/dt = \varepsilon_K v_{K+1} + O(v_\ell^{(N+1)}) \quad ; \quad K = \{1, 2, \ldots, n\} \tag{550}$$

A usual case is that in which the initial system (i.e. either (452) or (499) or (500)) is real. The Floquet multipliers μ_K are then either real or pair of complex **conjugates** and so are the eigenvalues λ_K (except for real negative μ_K). The pulsations ω_K are pairs of opposite numbers.

Two opposite ω_K are of course not rationally independent but if n is even it is possible that a critical case has rationally independent ω_0 and pairs of ω_K (at least for $2 \leqq \alpha_1+\alpha_2+\ldots+\alpha_n \leqq N$); the equations of motion (549) can be reduced to :

$$\left.\begin{aligned} v_{n-K} &= \overline{v_K} \quad ; \quad w_K = v_K \overline{v_K} \geqq 0 \\ dv_K/dt &= v_K P_K(w_1, w_2, \ldots, w_{n/2}) + O(v_\ell^{(N+1)}) \end{aligned}\right\} K = (1,2,\ldots,\tfrac{n}{2}) \tag{551}$$

The $P_K(w_1, w_2, \ldots, w_{n/2})$ are polynomials without constant term and of degree less than N/2.

These equations (551) imply the following n/2 <u>real equations</u>

$$dw_K/dt = w_K \left[P_K(\vec{w}) + \overline{P_K(\vec{w})} \right] + O(v_\ell^{(N+2)}) \quad ; \quad K = \{1,2,\ldots,\tfrac{n}{2}\} \tag{552}$$

These n/2 real equations govern the analysis of the stability and we will see in Section 10.7.7.4 that, in the critical Hamiltonian case without resonance, all polynomials $P_K(\vec{w}) + \overline{P_K(\vec{w})}$ are identically zero (this is the "generalized Birkhoff differential rotation").

In the non Hamiltonian cases the lower-order terms of equations (552) can lead to a stability (all $dw_K/dt < 0$) or an instability (at least one $dw_K/dt \geq qw_K^m$; with $q > 0$). However these motions are no longer exponential, they are either a "power-stability" or a "power-instability".

There are even some strange cases with stability for a large part of phase space and instability for another part, as in the following example :

$$\left.\begin{aligned} dw_1/dt &= - w_1^2 \\ dw_2/dt &= - 2w_1w_2 + w_2^2 \end{aligned}\right\} \quad (553)$$

Hence with initial values $w_{1.0}$, $w_{2.0}$ at $t = 0$:

$$w_1 = w_{1.0}/(1+w_{1.0}t) \;;\; w_2 = w_{2.0}/(1+w_{1.0}t)(1+(w_{1.0}-w_{2.0})t) \quad (554)$$

The future motion is "stable" if $w_{1.0} \geq w_{2.0}$ (indeed then w_1 and w_2 are both decreasing and go to zero), but it is "unstable" if $w_{1.0} < w_{2.0}$ with then w_2 going to plus infinity.

The importance of the critical cases come from the Hamiltonian problems where these critical cases are very common and not just limiting cases. Indeed the characteristic equation of eigenvalues λ_K then has only real even order terms, for instance :

$$a \neq 0 \quad ; \quad a\lambda^4 + b\lambda^2 + c = 0$$

and the conditions of critical motion (all λ_K purely imaginary, i.e. $ab \geq 0$; $b^2 \geq 4ac \geq 0$) include a sizeable range of possible a, b, c.

The same discussion is presented in Sections 10.7.4.2 and 10.7.5 in terms of the Floquet multipliers μ_K.

We will now consider these critical Hamiltonian cases at least when all ε_K are zero in (516) and (542), for instance if all eigenvalues λ_K are different.

The Hamiltonian equations, such as (33) or (171)-(172), are usually written :

$$\left.\begin{aligned}&\vec{p} = (p_1, p_2, \ldots, p_r) \quad ; \quad \vec{q} = (q_1, q_2, \ldots, q_r)\\&H = \text{Hamiltonian function} = H(\vec{p},\vec{q},t)\\&d\vec{p}/dt = -\,\partial H/\partial\vec{q} \quad ; \quad d\vec{q}/dt = \partial H/\partial\vec{p}\end{aligned}\right\} \quad (555)$$

r is the "number of degree of freedom" and 2r is the order of the differential system (555).

H will be either autonomous $(H = H(\vec{p},\vec{q}))$ or periodic in terms of t, $(H(\vec{p},\vec{q},t) \equiv H(\vec{p},\vec{q}, t+P))$ and our periodic solution $\vec{p}(t)$, $\vec{q}(t)$ of interest will have the period T, a multiple of P in the latter case.

The interest of the critical Hamiltonian case, without ε_K, is that the above general simplifications of this Section 10.7.7 can be expressed in an Hamiltonian form by usual canonical transformations.

A. Reduction to the origin.

The first simplification is a translation that will bring the periodic solution $\vec{p}(t)$, $\vec{q}(t)$ of interest to the origin with the following generatrix function :

$$\left.\begin{aligned}S &= S(\vec{p},\vec{q}_a,t,H_a)\\&= tH_a + \vec{p}.\vec{q}(t) + \vec{p}\vec{q}_a - \vec{p}(t).\vec{q}_a + F(t)\end{aligned}\right\} \quad (556)$$

where :

$$dF(t)/dt = H\left[\vec{p}(t),\vec{q}(t),t\right] - \vec{p}(t).d\vec{q}(t)/dt \qquad (557)$$

The corresponding canonical transformation is :

$$\left.\begin{aligned}&\vec{p}_a = -\,\partial S/\partial\vec{q}_a = \vec{p} - \vec{p}(t)\\&\vec{q} = \partial S/\partial\vec{p} = \vec{q}_a + \vec{q}(t)\\&t_a = \partial S/\partial H_a = t\\&H = \partial S/\partial t = H_a + \left[\vec{p} - \vec{p}(t)\right] . \frac{d\vec{q}(t)}{dt} - \vec{q}_a\,\frac{d\vec{p}(t)}{dt} + H\left[\vec{p}(t),\vec{q}(t),t\right]\end{aligned}\right\} \quad (558)$$

We indeed find $\vec{p}_a = 0$, $\vec{q}_a = 0$ along the periodic solution of interest and the first simplification is the following translation :

$$\left.\begin{aligned} &t_a = t \text{ : same parameter of description} \\ &\vec{p}_a = \vec{p} - \vec{p}(t) \quad ; \quad \vec{q}_a = \vec{q} - \vec{q}(t) \\ &H_a = H_a(\vec{p}_a,\vec{q}_a,t) = \text{new Hamiltonian function} = \\ &\quad = H(\vec{p},\vec{q},t) - \vec{p}_a \cdot \frac{d\vec{q}(t)}{dt} + \vec{q}_a \frac{d\vec{p}(t)}{dt} - H\left[\vec{p}(t),\vec{q}(t),t\right] \end{aligned}\right\} \quad (559)$$

Notice that $d\vec{q}/dt = \partial H/\partial\vec{p}$ and $d\vec{p}/dt = -\partial H/\partial\vec{q}$. Thus the new Hamiltonian function H_a has neither zero-order term nor first-order term but its higher-order terms are those of $H(\vec{p},\vec{q},t)$ in the vicinity of the periodic solution of interest.

Thus, with the new variables, the periodic solution of interest remains at the origin and the new Hamiltonian function has the period T :

$$H_a(\vec{p}_a,\vec{q}_a,t) \equiv H_a(\vec{p}_a,\vec{q}_a, t+T) \tag{560}$$

We will put :

$$H_a(\vec{p}_a\cdot\vec{q}_a\cdot t) = \sum_{n=2}^{\infty} H_{an}(\vec{p}_a,\vec{q}_a t) \tag{561}$$

H_{an} will gather the nth-order terms of H_a and will be a homogeneous polynomial of degree n in terms of the components of $\vec{p}_a$ and $\vec{q}_a$ with periodic coefficients in terms of t.

The analyticity implies that the series of (561) is converging for all t at least in a sufficiently small neighbourhood of the origin.

The first-order study is the study of the influence of H_{a2} alone. This leads to the well-known Floquet theory (Ref.173;174) with the monodromy matrix A and its eigenvalues the Floquet multipliers :

$$\begin{bmatrix} \vec{p}_a(T) \\ \vec{q}_a(T) \end{bmatrix} = A \cdot \begin{bmatrix} \vec{p}_a(0) \\ \vec{q}_a(0) \end{bmatrix} + \text{higher-order terms} \tag{562}$$

and, when $n \in \mathbb{Z}$:

$$\begin{bmatrix} \vec{p}_a(nT) \\ \vec{q}_a(nT) \end{bmatrix} = A^n \cdot \begin{bmatrix} \vec{p}_a(0) \\ \vec{q}_a(0) \end{bmatrix} + \text{higher-order terms} \tag{563}$$

Hence if μ is an eigenvalue of A and $\vec{V}$ a corresponding eigenvector :

$$\begin{bmatrix} \vec{p}_a(0) \\ \vec{q}_a(0) \end{bmatrix} = \vec{V} \text{ implies : } \begin{bmatrix} \vec{p}_a(nT) \\ \vec{q}_a(nT) \end{bmatrix} = \mu^n \vec{V} + \text{higher-order terms} \qquad (564)$$

In the Hamiltonian case the monodromy matrix A is "symplectic", its determinant is equal to one and if μ is an eigenvalue then μ^{-1} is also an eigenvalue with the same order of multiplicity.

This leads to the usual results of first- order analyses of Hamiltonian problems : If some eigenvalues μ_K have moduli different from unity half of these moduli are larger and half are smaller. The periodic solution of interest has an exponential instability for both the past and the future with corresponding stable and unstable subsets.

Stability can arise only if all eigenvalues have a unity modulus, i.e. in the "critical case" which happens frequently, however this necessary condition is insufficient and upper-order effects can destroy the stability.

The analysis can be easily extended if the monodromy matrix A has a full set of independent eigenvectors (for instance, if all Floquet multipliers μ_K are different) and we thus reach the second simplification.

B. Simplification of the second-order terms.

This second simplification is a periodic linear and real canonical transformation similar to the Floquet simplification (502)-(516). It leads from $(\vec{p}_a, \vec{q}_a, H_a)$ to $(\vec{p}_b, \vec{q}_b, H_b)$ with an autonomous second-order term :

$$H_{b2} = \sum_{K=1}^{r} \omega_K \frac{p_{bK}^2 + q_{bK}^2}{2} \qquad (565)$$

The r pulsations ω_K are related to the r pairs of conjugate Floquet multipliers μ_K, $\overline{\mu_K}$ by :

$$\mu_K = \exp(i\omega_K T) \quad ; \quad \overline{\mu_K} = \exp(-\, i\omega_K T) \qquad (566)$$

The generatrix function S of the linear canonical transformation $(\vec{p}_a, \vec{q}_a) \to (\vec{p}_b, \vec{q}_b)$ has the following form, in which $\vec{p}_a$ and $\vec{q}_b$ are column vectors and the superscripts $\mathbf{T}$ to the right are the symbol of transposition* :

$$S = t_a H_b + \frac{1}{2}\vec{p}_a^{\mathbf{T}} M_1 \vec{p}_a + \vec{p}_a^{\mathbf{T}} M_2 \vec{q}_b + \frac{1}{2}\vec{q}_b^{\mathbf{T}} M_3 \vec{q}_b \tag{567}$$

M_1, M_2, M_3 are three suitable real $r \times r$ matrices which are only functions of the time t_a and have the period T, (M_1 and M_3 are symmetrical).

The canonical transformation is then :

$$\left.\begin{aligned}
t_b &= \partial S/\partial H_b = t_a = t \text{ ; we always keep the time } t. \\
\vec{p}_b &= \partial S/\partial \vec{q}_b = M_2^{\mathbf{T}}(t)\vec{p}_a + M_3(t)\vec{q}_b \\
\vec{q}_a &= \partial S/\partial \vec{p}_a = M_1(t)\vec{p}_a + M_2(t)\vec{q}_b \\
H_a &= \partial S/\partial t_a = H_b + \frac{1}{2}\vec{p}_a^{\mathbf{T}} M_1'(t)\vec{p}_a + \vec{p}_a^{\mathbf{T}} M_2'(t)\vec{q}_b + \frac{1}{2}\vec{q}_b^{\mathbf{T}} M_3'(t)\vec{q}_b
\end{aligned}\right\} \tag{568}$$

The difference $H_b - H_a$ has only second-order terms, hence the decomposition in n^{th}-order terms :

$$H_b(\vec{p}_b,\vec{q}_b,t) = \sum_{n=2}^{\infty} H_{bn}(\vec{p}_b,\vec{q}_b,t) \tag{569}$$

implies for $n \geq 3$:

$$H_{bn}(\vec{p}_b,\vec{q}_b,t) \equiv H_{an}(\vec{p}_a,\vec{q}_a,t) \tag{570}$$

An important question is to determine the sign of the pulsations ω_K in (566), that is to determine which is μ_K and which is $\overline{\mu_K}$.

This of course can be done by determination of the three matrices $M_1(t)$, $M_2(t)$, $M_3(t)$ and the transformation (568). Fortunately a simpler method is available.

Let us call $\vec{B}_K + i\vec{C}_K$ the eigenvector corresponding to the eigenvalue μ_k in the linear transformation (562). $\vec{B}_K$ and $\vec{C}_K$ are two real column vectors with 2r components.

The eigenvector corresponding to $\overline{\mu_K}$ will be :

$$\begin{pmatrix}\vec{p}_a \\ \vec{q}_a\end{pmatrix}_K = \vec{B}_K - i\vec{C}_K \tag{571}$$

* This transposition symbol $\mathbf{T}$ sometimes is written to the left because of the english expression "transposed vector", "transposed matrix", however usually the subscripts and superscripts (indices, exponents, primes, dots, etc...) are written to the right and not to the left.

μ_K and $\bar{\mu}_K$ are determined as follows. The $2r \times 2r$ real matrix :

$$(\vec{C}_1, \vec{C}_2, \ldots, \vec{C}_r, \vec{B}_1, \vec{B}_2, \ldots, \vec{B}_r) \tag{572}$$

must be symplectic (after, if necessary, normalization of the eigenvectors $\vec{B}_K + i\vec{C}_K$).

In the first-order evolution* the eigenvectors that are initially $\vec{B}_K + i\vec{C}_K$ and $\vec{B}_K - i\vec{C}_K$ become $\vec{B}_K(t) + i\vec{C}_K(t)$ and $\vec{B}_K(t) - i\vec{C}_K(t)$ with of course for any t :

$$\left.\begin{aligned} \vec{B}_K(t + T) + i\vec{C}_K(t + T) &= \mu_K\left[\vec{B}_K(t) + i\vec{C}_K(t)\right] \\ \vec{B}_K(t + T) - i\vec{C}_K(t + T) &= \bar{\mu}_K\left[\vec{B}_K(t) - i\vec{C}_K(t)\right] \end{aligned}\right\} \tag{573}$$

and the linear periodic transformation (568) can be written :

$$\begin{bmatrix} \vec{p}_a \\ \vec{q}_a \end{bmatrix} = \sum_{K=1}^{r} (q_{bK}\cos\omega_K t - p_{bK}\sin\omega_K t)\vec{B}_K(t) + (q_{bK}\sin\omega_K t + p_{bK}\cos\omega_K t)\vec{C}_K(t)$$

It is easy to verify that this real linear transformation has the period T, as then the Hamiltonian $H_b(\vec{p}_b,\vec{q}_b,t)$ itself :

$$H_b(\vec{p}_b,\vec{q}_b,t) \equiv H_b(\vec{p}_b,\vec{q}_b,t+T) \tag{574}$$

C. Application of the near-resonance theorem.

This third simplification is a real periodic but non linear canonical transformation identical to the transformation (519) of the near resonance theorem.

The corresponding generatrix function S has the following form :

$$S = t_b H_c + \vec{p}_b\vec{q}_c + S_1(\vec{p}_b,\vec{q}_c,t_b) \tag{575}$$

The function S_1 is periodic in terms of t_b and only contains third and higher-order terms, hence the differences $\vec{p}_c - \vec{p}_b$, $\vec{q}_c - \vec{q}_b$ are at most of the second order :

---

* This first-order evolution corresponds then to the homogeneous Hamiltonian $H_{a2}(\vec{p_a}, \vec{q_a}, t)$.

$$\left.\begin{aligned} t_c &= \partial S/\partial H_c = t_b = t \\ \vec{p}_c &= \partial S/\partial \vec{q}_c = \vec{p}_b + \partial S_1/\partial \vec{q}_c \\ \vec{q}_b &= \partial S/\partial \vec{p}_b = \vec{q}_c + \partial S_1/\partial \vec{p}_b \\ H_b &= \partial S/\partial t_b = H_c + \partial S_1/\partial t_b \end{aligned}\right\} \quad (576)$$

The variables u_K of the near-resonance theorem do not correspond to the components p_{cK} and q_{cK} but to $z_{cK} = p_{cK} + iq_{cK}$ and $\bar{z}_{cK} = p_{cK} - iq_{cK}$ since $z'_{cK} = i\omega_K z_{cK}$ and $\bar{z}'_{cK} = -i\omega_K \bar{z}_{cK}$ to the first order.

The near-resonance theorem specifies that for any positive ε, there exists a real series $S_1(\vec{p}_b, \vec{q}_c, t)$ converging in a small neighbourhood of the origin and such that the corresponding transformation (576) leads to a real Hamiltonian $H_c(\vec{p}_c, \vec{q}_c, t)$ containing only "sufficiently resonant terms" when written with the z_{cK} and $\bar{z}_{cK}$:

$$\left.\begin{aligned} &\left.\begin{aligned} z_{cK} &= p_{cK} + iq_{cK} \\ \bar{z}_{cK} &= p_{cK} - iq_{cK} \end{aligned}\right\} K = \{1, 2, \ldots, r\} \\ &H_c(\vec{p}_c, \vec{q}_c, t) = \sum_{K=1}^{r} \omega_K \frac{p_{cK}^2 + q_{cK}^2}{2} + \\ &\qquad + \Sigma J(\vec{\alpha}, \vec{\beta}, m) z_{c_1}^{\alpha_1} \bar{z}_{c_1}^{\beta_1} \ldots z_{c_r}^{\alpha_r} \bar{z}_{c_r}^{\beta_r} \cdot \exp\{\frac{2\pi mi}{T} t\} \end{aligned}\right\} \quad (577)$$

with for all non-zero $J(\vec{\alpha}, \vec{\beta}, m)$ factors :

$$\left.\begin{aligned} &\vec{\alpha} = (\alpha_1, \alpha_2, \ldots, \alpha_r) \; ; \; \vec{\beta} = (\beta_1, \ldots, \beta_r) \; ; \text{ all } \alpha_K, \beta_K \in \mathbb{N} \; ; \; m \in \mathbb{Z} \\ &\sum_{K=1}^{r} (\alpha_K + \beta_K) \geq 3 \\ &\left| \sum_{K=1}^{r} (\alpha_K - \beta_K)\omega_K + \frac{2\pi m}{T} \right| \leq \varepsilon \cdot \sum_{K=1}^{r} (\alpha_K + \beta_K) \end{aligned}\right\} \quad (578)$$

Because the Hamiltonian $H_c(\vec{p}_c, \vec{q}_c, t)$ is real the complex constants $J(\vec{\alpha}, \vec{\beta}, m)$ satisfy the following relations :

$$J(\vec{\beta}, \vec{\alpha}, -m) = \overline{J(\vec{\alpha}, \vec{\beta}, m)} \quad (579)$$

The Hamiltonian $H_c(\vec{p}_c,\vec{q}_c,t)$ can be written in terms of $\vec{z}_c,\overline{\vec{z}_c},t$ only :

$$\left.\begin{aligned} H_c(\vec{p}_c,\vec{q}_c,t) &= H_c(\vec{z}_c,\overline{\vec{z}_c},t) = \\ &= \frac{1}{2}\sum_{K=1}^{r} \omega_K z_{cK}\overline{z}_{cK} + \sum_{(\vec{\alpha},\vec{\beta},m)} J(\vec{\alpha},\vec{\beta},m) \times \\ &\times z_1^{\alpha_1}\overline{z}_1^{\beta_1} \ldots z_r^{\alpha_r}\overline{z}_r^{\beta_r} \exp\{\frac{2\pi mi}{T} t\} \end{aligned}\right\} \quad (580)$$

and the Hamiltonian equations $dp_{cK}/dt = - \partial H_c/\partial q_{cK}$; $dq_{cK}/dt = = \partial H_c/\partial p_{cK}$ are equivalent to the following :

$$\frac{dz_{cK}}{dt} = 2i \frac{\partial H_c(\vec{z}_c,\overline{\vec{z}_c},t)}{\partial \overline{z}_{cK}} \quad ; \quad \frac{d\overline{z}_{cK}}{dt} = - 2i \frac{\partial H_c(\vec{z}_c,\overline{\vec{z}_c},t)}{\partial z_{cK}} \quad (581)$$

These partial derivatives are of course the usual formal partial derivatives corresponding to :

$$\frac{\partial(z_{c1}^{\alpha_1}\overline{z}_{c1}^{\beta_1})}{\partial z_{c1}} = \alpha_1 z_{c1}^{(\alpha_1-1)}\overline{z}_{c1}^{\beta_1} \quad ; \quad \frac{\partial(z_{c1}^{\alpha_1}\overline{z}_{c1}^{\beta_1})}{\partial \overline{z}_{c1}} = \beta_1 z_{c1}^{\alpha_1}\overline{z}_{c1}^{(\beta_1-1)} \quad (582)$$

These equations (577) - (582) allow an easy N^{th}-order study, indeed let us put :

$$\omega_o = 2\pi/T = \text{pulsation of the periodic orbit of interest} \quad (583)$$

For all non-zero $J(\vec{\alpha},\vec{\beta},m)$ coefficients :

$$\left.\begin{aligned} &\text{All } \alpha_K,\ \beta_K \in \mathbb{N} \quad ; \quad m \in \mathbb{Z} \\ &\quad D = \sum_{K=1}^{r} (\alpha_K + \beta_K) \geq 3 \\ &\quad \left| m\omega_o + \sum_{K=1}^{r} (\alpha_K - \beta_K)\omega_K \right| \leq \varepsilon D \end{aligned}\right\} \quad (584)$$

D is the degree of the term of interest and for $D \leq N$ only a finite nmber of terms satisfy (584), hence <u>for a sufficiently small ε all terms up to the Nth-order are strictly resonant</u>. These terms satisfy :

$$\left.\begin{aligned} &D = \sum_{K=1}^{r} (\alpha_K + \beta_K) \leq N \\ &m\omega_o + \sum_{k=1}^{r} (\alpha_K - \beta_K)\omega_K = 0 \end{aligned}\right\} \quad (585)$$

We will call p_{NK}, q_{NK}, z_{NK}, $\overline{z}_{NK}$ the corresponding parameters and H_N the corresponding Hamiltonian expressed either in terms of p_{NK}, q_{NK} or of z_{NK}, $\overline{z}_{NK}$ as in (580), with equations of motion similar to (581).

It is of course possible to improve the Hamiltonian step by step and to go from H_3 to H_4, then to H_5 and so on ... These successive improvements lead to the following conclusions :

For $N_2 > N$:
The differences $\vec{p}_N - \vec{p}_{N_2}$, $\vec{q}_N - \vec{q}_{N_2}$, $\vec{z}_N - \vec{z}_{N_2}$ are of Nth-order

The differences $H_N - H_{N_2}$ are of $(N + 1)^{th}$ - order. (586)

For a degree D smaller than or equal to N the coefficients $J(\vec{\alpha},\vec{\beta},m)$ are independent of N, they are the generalization to $r > 2$ of the "invariant Birkhoff coefficients".

D. The final simplification.

Let us define the real parameters P_{NK}, Q_{NK} and the corresponding complex parameters Z_{NK} by the following :

$$Z_{NK} = z_{NK} \cdot \exp\{- i\omega_K t\} \tag{587}$$

$$\left.\begin{aligned} P_{NK} + iQ_{NK} = Z_{NK} \quad ; \quad P_{NK} &= p_{NK}\cos\omega_K t + q_{NK}\sin\omega_K t \\ Q_{NK} &= - p_{NK}\sin\omega_K t + q_{NK}\cos\omega_K t \end{aligned}\right\} \tag{588}$$

This transformation is a usual canonical transformation and the corresponding generatrix function S and Hamiltonian $\mathcal{H}_N$ are the real functions given by the following :

$$\begin{aligned} S &= S(\vec{p}_N,\vec{Q}_N,t,\mathcal{H}_N) = \\ &= t\mathcal{H}_N + \sum_{K=1}^{r} \frac{p_{NK}^2 + Q_{NK}^2}{2} \tan\omega_K t + \frac{p_{NK}Q_{NK}}{\cos\omega_K t} \end{aligned} \tag{589}$$

The parameter of description remains the time t and the relation between H_N and $\mathcal{H}_N$ is :

$$H_N = \partial S/\partial t = \mathcal{H}_N + \sum_{K=1}^{r} \frac{p_{NK}^2 + Q_{NK}^2 + 2p_{NK}Q_{NK}\sin\omega_K t}{2\cos\omega^2{}_K t} = \\ = \mathcal{H}_N + \sum_{K=1}^{r} \omega_K \frac{p_{NK}^2 + q_{NK}^2}{2} \qquad (590)$$

The Hamiltonians $\mathcal{H}_N$ have no second-order terms and their higher-order terms are easy :

$$\mathcal{H}_N = \sum_{(\vec{\alpha},\vec{\beta},m)} J(\vec{\alpha},\vec{\beta},m) \cdot Z_{N1}^{\alpha_1}.\bar{Z}_{N1}^{\beta_1} \dots Z_{Nr}^{\alpha_r}.\bar{Z}_{Nr}^{\beta_r} \times \\ \times \exp\{i.L(\vec{\alpha},\vec{\beta},m).t\} \qquad (591)$$

with :

$$L(\vec{\alpha},\vec{\beta},m) = m\omega_o + \sum_{K=1}^{r} (\alpha_K - \beta_K)\omega_K \qquad (592)$$

Hence, with $D = \sum_{k=1}^{r} (\alpha_K + \beta_K)$ = degree of the term of interest and with (584), (585) :

$$D \leq N \text{ implies } \begin{cases} L = 0 \\ J(\vec{\alpha},\vec{\beta},m) \text{ is independent of } N \end{cases} \\ D > N \text{ implies } \quad |L| \leq \varepsilon(N) \cdot D \qquad (593)$$

Thus the Hamiltonians $\mathcal{H}_N$ have neither zero-order nor first-order nor second-order terms, they are autonomous up to the Nth-order but the higher-order terms have no longer the period T in terms of t.

When N goes to infinity the sequences $\vec{p}_N, \vec{q}_N, H_N$ and $\vec{P}_N, \vec{Q}_N, \mathcal{H}_N$ are generally diverging, which forbids use of their limits.

We will see in the next two Sections that the main part of $\mathcal{H}_N$ (i.e. the lower-order terms) is the essential element in the stability analyses.

10.7.7.4 Critical Hamiltonian Case.
The Nth-order study. The quasi-integrals.
Generalization of the "Birkhoff differential rotations".

The Hamiltonian $\mathcal{H}_N$ of the previous Section is autonomous up to the Nth-order and hence its derivative is of $(N + 1)^{th}$-order.

The difference $\mathcal{H}_{N+1} - \mathcal{H}_N$ also is of $(N + 1)^{th}$-order and thus there are two main cases :

A) Either for large N, for $N \geq M$, the Hamiltonians $\mathcal{H}_N$ have a main part of order M. We will denote by h_M this main part, it is an autonomous, real and homogenous polynomial of degree M. It is also the main part of all $\mathcal{H}_N$ with $N \geq M$.

B) Or for all N the Hamiltonians $\mathcal{H}_N$ have a main part of order larger than M.

In most problems M is equal to 4 (see Section 10.7.8).

A sequence as $\mathcal{H}_M, \mathcal{H}_{M+1}, \mathcal{H}_{M+2}, \ldots$ can be called a "quasi-integral". All terms have the same main part and have both derivatives and successive differences of larger and larger orders, they have "no secular variations to all orders".

The convergence or the divergence of this sequence gives the difference between the usual integrals and the quasi-integrals. In the latter case the terms of the quasi-integrals have slow variations but are not constant and cannot lead to strict stability conclusions.

In the above case B the quasi-integral $\mathcal{H}_M, \mathcal{H}_{M+1}, \ldots$ disappears but fortunately the sequences $\vec{P}_3, \vec{P}_4, \vec{P}_5, \ldots ; \vec{Q}_3, \vec{Q}_4, \vec{Q}_5, \ldots ;$ $\vec{Z}_3, \vec{Z}_4, \vec{Z}_5, \ldots$ are then quasi-integrals.

It is very interesting to analyse the case of rationally independent pulsations $\omega_0, \omega_1, \ldots, \omega_r$. The conditions (591)-(593) then imply $m = 0$ and $\alpha_K = \beta_K$ for all K and for all terms of $\mathcal{H}_N$ and H_N up to the Nth-order : these terms are functions of the products $Z_{NK}\overline{Z}_{NK}$ or $z_{NK}\overline{z}_{NK}$ only.

Let us denote A_{NK} these products, we obtain :

$$A_{NK} = Z_{NK}\overline{Z}_{NK} = P_{NK}^2 + Q_{NK}^2 = z_{NK}\overline{z}_{NK} = p_{NK}^2 + q_{NK}^2 \tag{594}$$

$$H_N = H_{NN}(A_{N1}, A_{N2}, \ldots, A_{Nr}) + \{(N+1)^{th}\text{-order}\} \tag{595}$$

$$\left.\begin{aligned} dp_{NK}/dt &= - \partial H_N/\partial q_{NK} = - 2q_{NK}.\partial H_{NN}/\partial A_{NK} + \{N^{th}\text{-order}\} \\ dq_{NK}/dt &= 2p_{NK}.\partial H_{NN}/\partial A_{NK} + \{N^{th}\text{-order}\} \;;\; dA_{NK}/dt = \{(N+1)^{th}\text{-order}\} \end{aligned}\right\} \tag{596}$$

These properties are true for all K from 1 to r and all N, hence the r sequences A_{3K}, A_{4K}, A_{5K}, ... are sequences of quasi-integrals.

If we neglect the terms of order N+1 and above we find that the r two-dimensional vectors $(\overrightarrow{p_{NK}, q_{NK}})$ with K = {1, 2, ..., r} have constant moduli $(\sqrt{A_{NK}})$ and constant rates of rotation in their planes $(2\partial H_{NN}/\partial A_{NK})$. These rates of rotation are only function of the A_{N1}, A_{N2}, ..., A_{Nr} and the picture of motion can be considered as the generalization to r degrees of freedom of the "Birkhoff differential rotation" (Ref.181, 115).

Thus, even for large N, our N^{th}-order picture of the motion is that of quasi-periodic solutions surrounding the periodic solution of interest. These quasi-periodic solutions are dense on tori and the number of dimensions of these tori depends on the number of non-zero vectors $(\overrightarrow{p_{NK}, q_{NK}})$.

The motion has the "all-order stability", however a complete analysis shows that sometimes many tori are destroyed by the resonances and correspond to either chaotic or diverging motions. Neverteless, according to the Kolmogorov-Arnold-Moser rules (Ref. 182), most of the sufficiently small tori usually survive.

The all-order stability is insufficient for true stability as we will see in the next Section.

We must now consider the cases in which the (r+1) pulsations ω_0, ω_1, ..., ω_r are not rationally independent.

If the rational dependences imply large integers, the r sums $p_{NK}^2 + q_{NK}^2$ will have only high-order derivatives and will be almost constant. The motion will not be very different from that of the above case, although generally some of the quasi-integrals will disappear.

The following simple rules give several useful quasi-integrals similar to $p_{NK}^2 + q_{NK}^2$.

Let us put :

$$\vec{\omega} = (\omega_0, \omega_1, \omega_2, \ldots, \omega_r) = \text{pulsation-vector} \tag{597}$$

Let us assume that there are s independent rational relations among the pulsations ω_K, these s relations being written :

$$\vec{a}_j \cdot \vec{\omega} = 0 \quad ; \quad j = \{1, 2, \ldots, s\} \tag{598}$$

with :

$$\vec{a}_j = (a_{0j}, a_{1j}, \ldots, a_{rj}) \tag{599}$$

All a_{Kj} are integers and the s vectors $\vec{a}_j$ are a basis of a s-dimensional space in the r-dimensional space of vectors normal to $\vec{\omega}$.

In these conditions there are (r - s) simple independent quasi-integrals, these quasi-integrals are given by :

$$I_N = \sum_{k=1}^{r} b_K(p_{NK}^2 + q_{NK}^2) \tag{600}$$

with :

$$\vec{b} = (0, b_1, b_2, \ldots, b_r) \; ; \; \forall \; j \in \{1, 2, \ldots, s\} : \vec{a}_j \vec{b} = 0 \tag{601}$$

If all coefficients b_1 to b_r are positive the quasi-integral (600) gives the "all-order stability". This second, and broader case, of "all-order stability" happens iff the system has no "destabilizing resonances"* (Ref. 180), i,e. no vector $\vec{a}$ such that :

$$\vec{a}\,\vec{\omega} = 0 \; ; \; \vec{a} \neq 0 \; ; \; a_0 \in \mathbb{Z} \; ; \; a_1, a_2, \ldots, a_r \in \mathbb{N} \tag{601.A}$$

The verification of $dI_N/dt = \{(N+1)^{th}\text{-order}\}$ is the following :

$$I_N = \sum_{K=1}^{r} b_K \, Z_{NK}\overline{Z_{NK}} \tag{602}$$

$\mathcal{H}_N = \mathcal{H}_N(\vec{Z}_N, \overline{\vec{Z}_N}, t)$ is given in (591)-(593)

$$dZ_{NK}/dt = 2i\partial\mathcal{H}_N/\partial\overline{Z}_{NK} \quad ; \quad d\overline{Z}_{NK}/dt = -2i\partial\mathcal{H}_N/\partial Z_{NK} \tag{603}$$

Hence :

$$\left.\begin{aligned} dI_N/dt = 2i \; . \; \Sigma\{J(\vec{\alpha},\vec{\beta},m) Z_{N1}^{\alpha_1} \overline{Z}_{N1}^{\beta_1} \ldots Z_{Nr}^{\alpha_r} \overline{Z}_{Nr}^{\beta_r} \times \\ \times \exp\left[i \; . \; L(\vec{\alpha},\vec{\beta},m) \; . \; t\right] . \sum_{K=1}^{r} b_K(\beta_K - \alpha_K)\} \end{aligned}\right\} \tag{604}$$

Relations (592) and (593) give that, up to the Nth-order :

$$\sum_{K=1}^{r} (\alpha_K + \beta_K) = D \leq N \tag{605}$$

$$m\omega_0 + \sum_{K=1}^{r} (\alpha_K - \beta_K)\omega_K = 0 \tag{606}$$

---

* Also called "positive resonances". They emphasize the importance of the signs of pulsations ω_K.

Let us call $\vec{W}$ the vector $(m, \alpha_1-\beta_1, \alpha_2-\beta_2, \ldots, \alpha_r-\beta_r)$, the equation (606) is equivalent to :

$$\vec{W} \cdot \vec{\omega} = 0 \qquad (607)$$

The (r+1) components of $\vec{W}$ are integers, hence $\vec{W}$ belongs to the s-dimensional space defined by the $\vec{a}_j$ vectors of (598), (599).

With (601) all this imply :

$$\left.\begin{aligned} &\vec{W} \cdot \vec{b} = 0 \\ &\text{that is :} \\ &\sum_{k=1}^{r} b_K(\alpha_K - \beta_K) = 0 \end{aligned}\right\} \qquad (608)$$

Hence, up to the Nth-order, all terms of dI_N/dt given in (604) are zero and the sequence of I_N for increasing N is the sequence of a quasi-integral.

Notice the following.

Remark I :

If the periodic solution of interest is motionless at a point of equilibrium of an autonomous Hamiltonian problem, we obtain :

A) The Hamiltonians H, H_a, H_b, H_c, H_N of the above analysis are autonomous and are thus integrals of motion (they are different expressions of the same integral of motion).

B) The "period" T and the corresponding pulsation $\omega_0 = 2\pi/T$ are arbitrary. Hence the a_{0j} are zero in (598)-(599) and the vector $\vec{b}$ can be chosen equal to $(0, \omega_1, \omega_2, \ldots, \omega_r)$ in (597)-(601). As a consequence the quantities J_N :

$$J_N = \sum_{K=1}^{r} \omega_K(p_{NK}^2 + q_{NK}^2) \qquad (609)$$

are a sequence of quasi-integrals.

C) This latter property agrees with the other results, since the H_N are integrals of motion, the $\mathcal{H}_N$ are quasi-integrals and :

$$2(H_N - \mathcal{H}_N) = J_N = \sum_{K=1}^{r} \omega_K(p_{NK}^2 + q_{NK}^2) \qquad (610)$$

The main part of $2H_a$ (the second-order part) is equivalent to the J_N and gives directly the pulsations ω_K, with their signs, by the usual first-order analysis of autonomous Hamiltonians problems.

Notice that if the point of equilibrium is at the origin we have $\vec{p}_a \equiv \vec{p}$; $\vec{q}_a \equiv \vec{q}$ and $H_a - H$ = constant.

Remark II :

We have seen from the equations (594)-(596) that when the pulsations $\omega_0, \omega_1, \ldots, \omega_r$ have no rational relation the motion is "integrable to all orders" and leads to a "generalized Birkhoff differential rotation".

When the pulsations $\omega_0, \omega_1, \ldots, \omega_r$ have only one rational relation the motion will remain "integrable to all orders", even if it is not truly integrable.

Let us write the rational relation with an irreducible set of integers $a_0, a_1, \ldots, a_r$:

$$a_0\omega_0 + a_1\omega_1 + a_2\omega_2 + \ldots + a_r\omega_r = 0 \tag{611}$$

The quasi-integrals I_N of (600) and (602) can be written :

$$I_{Njk} = a_j Z_{NK}\bar{Z}_{NK} - a_K Z_{Nj}\bar{Z}_{Nj} \tag{612}$$

These I_{Njk} only give (r-1) independent quasi-integrals.

An equivalent and more understandable method is to write with the suitable variable $F_N(t)$ and the initial values $Z_{NKo}\bar{Z}_{Nko} = A_{Nko}$:

$$\left.\begin{aligned} &Z_{NK}\bar{Z}_{NK} = A_{NKo} + 2a_K \cdot F_N(t) + \{(N+1)^{th}\text{-order}\} ; \\ &K = \{1, 2, \ldots, r\} \end{aligned}\right\} \tag{613}$$

On the other hand let us consider again the expression of the Hamiltonian $\mathcal{H}_N$:

$$\mathcal{H}_N = \Sigma J(\vec{\alpha}, \vec{\beta}, m) Z_{N1}^{\alpha_1}\bar{Z}_{N1}^{\beta_1} \ldots Z_{Nr}^{\alpha_r}\bar{Z}_{Nr}^{\beta_r} + \{(N+1)^{th}\text{-order}\} \tag{614}$$

with for all monomials up to the Nth-order :

$$m\omega_0 + \sum_{K=1}^{r} (\alpha_K - \beta_K)\omega_K = 0 \tag{615}$$

that is, because of (611) :

$$\frac{m}{a_0} = \frac{\alpha_1 - \beta_1}{a_1} = \frac{\alpha_2 - \beta_2}{a_2} = \ldots = \frac{\alpha_r - \beta_r}{a_r} = \text{integer} \tag{616}$$

The monomials of (614) are then products of the coefficient $J(\vec{\alpha},\vec{\beta},m)$ by powers of the products $(Z_{NK}\bar{Z}_{NK})$ and $(Z_{N1}^{a_1}Z_{N2}^{a_2}\ \ldots\ Z_{Nr}^{a_r})$.

Hence let us define the angle θ_N by :

$$\left.\begin{aligned} &\theta_N = \sum_{K=1}^{r} a_K\theta_{NK} = \sum_{K=1}^{r} a_K \operatorname{Arg} Z_{NK} \\ &(\text{where } Z_{NK} = R_{NK}\ .\ \exp\{i\theta_{NK}\}) \end{aligned}\right\} \quad (617)$$

We can now, using (613), write up to the Nth-order :

$$\mathcal{H}_N = \mathcal{H}_N(A_{N10},A_{N20},\ldots,A_{Nr0},F_N,\theta_N) \qquad (618)$$

Up to the Nth-order the A_{NK0} are constant, like $\mathcal{H}_N$ itself, and the equations of motion of F_N and θ_N are the Hamiltonian equations :

$$dF_N/dt = -\ \partial\mathcal{H}_N/\partial\theta_N \quad ; \quad d\theta_N/dt = +\ \partial\mathcal{H}_N/\partial F_N \qquad (619)$$

This autonomous Hamiltonian problem has only one degree of freedom, which is a usual case of integrability. Thus when the pulsations $\omega_0,\omega_1,\ldots,\omega_r$ have at most one rational relation the neighbouring motions are "integrable to all orders".

Notice that if two integers a_K have opposite signs the variable F_N can only have bounded variations in (613), because $Z_{NK}\bar{Z}_{NK} \geq 0$ for all K, and the system has the "all-order stability".

We thus arrive at the following rule of signs :

If the pulsations $\omega_0,\omega_1,\ldots,\omega_r$ of the analytic Hamiltonian system of interest have one and only one rational relation, the relation $a_0\omega_0 + a_1\omega_1 + \ldots + a_r\omega_r = 0$, and if :

$$\inf_{K=1}^{r} a_K < 0 < \sup_{K=1}^{r} a_K$$

the system has the "all-order stability" in the vicinity of the periodic solution of interest. (620)

It can be demonstrated that when all a_K are positive or zero the system again has the "all-order stability" if for some N and for all $A_{NKo} = 0$ the function $\mathcal{H}_N(A_{N10},\ldots,A_{Nr0},F_N,\theta_N)$ has either a strict maximum or a strict minimum at F = 0. If, on the contrary, $\mathcal{H}_N$ has in the same conditions positive and negative values of Nth-order in the vicinity of F = 0, the system has not the all-order stability.

Remark III :

If there are s independent rational relations among the pulsation $\omega_0,\omega_1,\ldots,\omega_r$ a similar simplification can be done.

The s relations are the following :

$$\left.\begin{aligned} &\vec{a}_j \cdot \vec{\omega} = 0 \quad ; \quad j = (1, 2, \ldots, s) \\ &(\vec{a}_j = (a_{0j}, a_{1j}, \ldots, a_{rj}), \text{ all integers}) \end{aligned}\right\} \tag{621}$$

The relations (613) become the following with the r initial values A_{NKo} and the s variables F_{Nj} :

$$Z_{NK}\bar{Z}_{NK} = A_{NKo} + 2\sum_{j=1}^{s} a_{Kj}F_{Nj} + \{(N+1)^{th}\text{-order}\} \tag{622}$$

While the s angles θ_{Nj} are defined by :

$$\theta_{Nj} = \sum_{K=1}^{r} a_{Kj} \text{ Arg } Z_{NK} \tag{623}$$

The equations (619), (620) are then, up to the Nth-order, generalized into :

$$\left.\begin{aligned} &\mathcal{H}_N = \mathcal{H}_N(A_{N10},A_{N20},\ldots,A_{Nr0},F_{N1},\ldots,F_{Ns}\theta_{N1},\ldots,\theta_{Ns}) \\ &dF_{Nj}/dt = -\,\partial\mathcal{H}_N/\partial\theta_{Nj} \;;\; d\theta_{Nj}/dt = \partial\mathcal{H}_N/\partial F_{Nj} \;;\; j = \{1,2,\ldots,s\} \end{aligned}\right\} \tag{624}$$

However when s ≥ 2 this autonomous Hamiltonian system with s degrees of freedom is generally no longer integrable.

10.7.7.5 The six main types of stability and instability.

Analytical systems have three main types of instability and also three main types of stability in the vicinity of a periodic solution :

The exponential instability	
The power-m instability	
The Arnold diffusion The non-asymptotic stability	} "all-order stability"
The power-m stability The exponential stability	} asymptotical stabilities impossible for conservative systems.

For non-critical problems we have met the exponential instability and the exponential stability according to the sign of the largest "Liapunov characteristic exponent" $\mathcal{R}e(\lambda_n)$ or to the modulus of the largest Floquet multiplier μ_n
(with $\mu_n = \exp(\lambda_n T)$).

In Hamiltonian problems, the Liapunov characteristic exponents are pairs of opposites, and then the exponential stability is impossible.

We can define a suitable "distance" δ between the periodic solution of interest and a neighbouring solution. The derivative δ' of δ will, for almost orbits, rapidly become equivalent to $\mathcal{R}e(\lambda_n)\,\delta$, with the corresponding exponential behavior.

In the critical cases the largest Liapunov characteristic exponent $\mathcal{R}e(\lambda_n)$ is zero, these exponents are even all zero for critical Hamiltonian problems, and the higher-order studies are necessary.

We have already come across the "all-order stability", for instance for the "generalized Birkhoff rotation" of Section 10.7.7.4.

This "all-order stability" leads to the following with a suitable "distance" δ between the periodic solution of interest and a neighbouring solution.

$$\forall N : d\delta/dt = o(\delta^N) \tag{625}$$

For instance $d\delta/dt = \exp\{-1/\delta\}$ or $d\delta/dt = O\{\exp(-1/\sqrt{\delta})\}$ as in the Arnold analysis of Ref. 182.

The "all-order stability" does not imply the stability it allows also the very slow "Arnold diffusion".

We have the certainty of the "all-order stability" if the integrals of motion and the quasi-integrals do not allow an escape, for instance the quasi-integrals $\mathcal{H}_N$ of (590) and the (r-s) quasi-integrals I_N of (600).

A simple case of "all-order stability" is that when all b_K can be chosen positive in (600) ; in this case the quasi-integral I_N :

$$I_N = \sum_{K=1}^{r} b_K(p_{NK}^2 + q_{NK}^2) \tag{626}$$

is sufficient to establish the "all-order stability".

This example emphasizes the importance of the signs of the pulsations ω_K that govern those of coefficients b_K. These signs are defined in (565)-(572).

If a critical system has not the "all-order stability" it can have a "power-m instability" with a derivative δ' of the order of δ^m.

Let us consider for instance the following example usign the notations of the two previous Sections.

Let us assume that for large N the Hamiltonians $\mathcal{H}_N$ are equal to their main part, i.e. the homogeneous polynomial h_M of degree M larger than 2.

The homogeneous polynomial h_M is independent of N and t, and we can have a "radial escape" along the Q_1 axis if h_M has the following form :

$$h_M = h_M(\vec{P},\vec{Q}) = KP_1Q_1^{M-1} + \text{terms with } Q_1 \text{ at the power (M-2) at most.} \tag{627}$$

Hence, along the Q_1 axis, all derivatives are zero with the only exception of dQ_1/dt :

$$dQ_1/dt = \partial h_M/\partial P_1 = KQ_1^{(M-1)} \tag{628}$$

If the coefficient K is positive the equation (628) leads to the following escape motions :

$$\left.\begin{array}{l} \vec{P} \equiv 0 \quad ; \quad Q_2, Q_3, \ldots, Q_r \equiv 0 \\ Q_1 = \{K(M-2)(t_f - t)\}^{1/(2-M)} \end{array}\right\} \tag{629}$$

These "radial escapes" (628), (629) are a prototype of a "power-m instability" with $m = M - 1$. They are asymptotic to the origin in the past.

We will call "main problem" the problem of Hamiltonian motions with the autonomous homogeneous Hamiltonian h_M of degree M.

It can be shown that, if the "main problem" has a power-(M-1) escaping solution asymptotic to the origin in the past, the general problem also has solutions of the same type and has then a power-(M-1) instability. The "radial escapes", of arbitrary direction, are the simplest examples of these solutions ; they are very common because of the homogeneity of h_M.

We thus arrive at the following types of stability or instability with a suitable "distance" δ between the periodic solution of interest and the neighbouring solutions :

A) The exponential instability

$$\mathcal{R}e(\lambda_n) > 0 \quad ; \quad d\delta/dt = O\{\delta\} \tag{630}$$

B) The power-m instability

$$d\delta/dt = O\{\delta^m\} \tag{631}$$

C) The "all-order stability"

$$\forall\ N : d\delta/dt = o\{\delta^N\} \tag{632}$$

This third case give either the very slow "Arnold diffusion" or the non-asymptotic stability.

The minimum duration necessary to move from the infinitesimal distance $\delta = \varepsilon$ to some finite distance $\delta = A$ is of the order of $\mathrm{Log}(1/\varepsilon)$ in the first case, of the order of $\varepsilon^{(1-m)}$ in the second case and extremly long in the Arnold diffusion which is physically not very different from the true stability : What would be the meaning of an instability requiring millions of years to appear ?

For non-Hamiltonian problems we must add the possibilities of asymptotic stability* :

---

* An asymptotic stability slower than all "power-m stabilities" is not impossible, even if no example is known.

D) The exponential stability :

$$\mathcal{R}e(\lambda_n) < 0 \quad ; \quad \exists K > 0 / \; d\delta/dt \leq -K\delta \tag{633}$$

E) The "power-m stability" as in (551), (552) :

$$d\delta/dt = O\{\delta^m\} \text{ and } \exists K > 0 / \; d\delta/dt \leq -K\delta^m \tag{634}$$

The first-order study is sufficient for the determination of the exponential stability or instability. The mth-order study is generally sufficient for the determination of the power-m stability or instability, especially when a radial escape exists, but it seems very difficult to discriminate between the Arnold diffusion and the non-asymptotic stability even when the quasi-integrals imply the all-order stability.

A full analysis of this question remains to be done :

I) Determination of general criteria for a power-m stability or instability.

II) Study of the critical cases with multiple eigenvalues λ_K or multiple Floquet multipliers μ_K (cases with some $\varepsilon_k = 1$ in (516) and (542)).

III) Research of general criteria for the Arnold diffusion and the true stability.

10.7.7.6 A lower bound of m for a "power-m instability".

Let us consider an analytic Hamiltonian system with a periodic solution of period T and let us assume that the first-order stability analysis leads to a critical stability.

The system has r degrees of freedom and the first-order stability analysis of the periodic solution leads to (r+1) pulsations $\omega_0, \omega_1, \ldots, \omega_r$ with $\omega_0 = 2\pi/T$ and $\vec{\omega} = (\omega_0, \omega_1, \ldots, \omega_r)$.

Let us look for $\vec{a}_j$ vectors satisfying the following for a given integer w :

$$\left.\begin{array}{l} \text{A) The components } a_{oj}, a_{1j}, \ldots, a_{rj} \text{ of } \vec{a}_j \text{ are integers.} \\[2ex] \text{B) } \vec{a}_j \cdot \vec{\omega} = 0 \\[2ex] \text{C) } |a_{1j}| + |a_{2j}| + \ldots + |a_{rj}| \leq w \end{array}\right\} \tag{635}$$

Let us now look for a vector $\vec{b}$ such that :

$$\left.\begin{array}{l} \text{A) } \vec{b} = (0,\ b_1,\ b_2,\ b_3,\ \ldots,\ b_r) \\ \text{B) The components } b_1,\ b_2,\ \ldots,\ b_r \text{ are positive*} \\ \text{C) } \vec{a}_j\ .\ \vec{b} = 0 \text{ for all } \vec{a}_j \text{ vectors of (635)} \end{array}\right\} \quad (636)$$

If such a vector $\vec{b}$ exists then for large N the sum S_N :

$$S_N = \sum_{K=1}^{r} b_K(p_{NK}^2 + q_{NK}^2) \qquad (637)$$

has a derivative of order at least (w+1). It square root will be an excellent "distance" δ with a derivative of order at least w and then w will be a lower bound of m for a "power-m instability".

Notice the following.

I) A vector $\vec{b}$ is impossible if and only if a suitable linear composition of $\vec{a}_j$ vectors gives a non-zero vector $\vec{a} = (a_0, a_1, \ldots, a_r)$ with positive or zero $a_1, a_2, \ldots, a_r$.

II) The signs of the pulsations ω_K, as defined in (565)-(572), are of course an essential element of the above analysis.

III) This Section is based on the analysis of Section 10.7.7.4 that requires non-multiple eigenvalues λ_K equal to $\pm\, i\omega_K$. However it seems that the above results are true even if there are equal or opposite pulsations.

10.7.8 Two conjectures on the stability or instability of periodic solutions of analytic Hamiltonian systems.

The first conjecture is an extension of the Kolmogorov-Arnold-Moser results on generic mappings on surfaces (Ref. **182**).

"If the integrals of motion allow the system to be reduced either to a Hamiltonian system with one degree of freedom, or to an autonomous Hamiltonian system with two degrees of freedom, if furthermore the periodic solution of interest has the

---

* The word "positive" is used here in its usual meaning equiva--lent to "strictly positive" for the so-called "modern mathemati--cians". Otherwise $\vec{b} = 0$ would be a solution.

"all-order stability", the Arnold diffusion is impossible and the solution of interest is stable"*.

The analyticity of the Hamiltonian system of interest is fundamental and all known counter-examples are non-analytic.

The second conjecture is a relation between the type of stability or instability when $t \to +\infty$ and when $t \to -\infty$, we will call it the past-future symmetry conjecture.

For Hamiltonian systems we have met four types :

The exponential instability

The power-m instability

The Arnold diffusion

The non-asymptotic stability.

The past-future symmetry conjecture is then the following : "A periodic solution of an analytic Hamiltonian system has the same type of stability or instability for the future and for the past". However a different m can apply to the power-m instability in the two cases.

This conjecture is obvious for the exponential instability, indeed for Hamiltonian systems the Liapunov characteristic exponents are pairs of opposite numbers.

The conjecture is also easy to demonstrate in the case of stability.

Let us use the parameters $\vec{p}_a$, $\vec{q}_a$ of (559) and let us define in the $(\vec{p}_a,\vec{q}_a,t)$ space the open domain D_ε of points at less than ε from the periodic orbit of interest.

$$\{\vec{p}_a,\vec{q}_a,t\} \in D_\varepsilon \iff p_a^2 + q_a^2 < \varepsilon^2 \tag{638}$$

Let us call $D_{\varepsilon p}$ and $D_{\varepsilon f}$ the open sets of the $(\vec{p}_a,\vec{q}_a,t)$ space that correspond to the past and to the future of D_ε along the trajectories of the Hamiltonian system $H_a(\vec{p}_a,\vec{q}_a,t)$:

$$\{\vec{p}_a,\vec{q}_a,t\} \in D_{\varepsilon f} \iff \left\{ \begin{array}{l} \exists\, (\vec{p}_{a0},\vec{q}_{a0},t_0) \text{ such that :} \\ 1°)\ t_0 < t \ ;\ 2°)\ p_{a0}^2 + q_{a0}^2 < \varepsilon^2 \\ 3°) \text{ The equations of motion lead} \\ \qquad \text{from } (\vec{p}_{a0},\vec{q}_{a0},t_0) \text{ to } (\vec{p}_a,\vec{q}_a,t) \end{array} \right\} \tag{639}$$

---

* The eventual decomposability of the system of interest can of course help to reduce it to Hamiltonian systems with one or two degrees of freedom.

Because of the periodicity T the domains D_ε, $D_{\varepsilon p}$, $D_{\varepsilon f}$ are time-periodic with the period T. The time t can be considered as an angle-variable.

Because of the stability the domain $D_{\varepsilon f}$ is bounded, at least for sufficiently small ε :

$$(\vec{p}_a,\vec{q}_a,t) \in D_{\varepsilon f} \Longrightarrow p_a^2 + q_a^2 < \delta(\varepsilon) \;;\; \delta \text{ goes to zero with } \varepsilon \tag{640}$$

The main proposition is the following : "a point B of $D_{\varepsilon p}$ cannot be out of the closure of $D_{\varepsilon f}$".

Let us demonstrate that proposition by contradiction ; thus let us assume the following :

$$\left.\begin{array}{l} \text{B is the point } (\vec{p}_{aB},\vec{q}_{aB},t_B) \;;\; B \in D_{\varepsilon p} - \overline{D}_{\varepsilon f} \\ \text{Future of B : } \{\vec{p}_a(t),\vec{q}_a(t),t\} \text{ with } t > t_B \\ \text{B belongs to } D_{\varepsilon p}\text{, hence there exists } t_c \text{ larger than } t_B \\ \text{and such that :} \\ \{\vec{p}_a(t_c),\vec{q}_a(t_c),t_c\} \in D_\varepsilon \\ t > t_c \Longrightarrow \{\vec{p}_a(t),\vec{q}_a(t),t\} \in D_{\varepsilon f} \end{array}\right\} \tag{641}$$

$D_{\varepsilon p}$ and D_ε are open and hence, for the same instants t_B and t_c, a small open neighbourhood $\Delta(t_B)$ of B in the $t = t_B$ space has the same properties :

$$\left.\begin{array}{l} \Delta(t_B) \subset D_{\varepsilon p} \quad ; \quad \Delta(t_B) \cap \overline{D}_{\varepsilon f} = \emptyset \\ \Delta(t_c) \subset D_\varepsilon \\ t > t_c \Longrightarrow \Delta(t) \subset D_{\varepsilon f} \end{array}\right\} \tag{642}$$

Let us call t_1 an instant such that :
$t_1 > t_c$; $t_1 = t_B + nT$; n integer
and let us call t_K the instants defined by :
$t_K = t_B + KnT$; K positive integer.

Let us consider the time t as an angle variable of period T, hence all instants t_B, t_1, t_2, ..., t_K, ... correspond to the same instant $t = t_B$ and all domains $\Delta(t_B)$, $\Delta(t_1)$, ..., $\Delta(t_K)$, ... belong to the same $(\vec{p}_a, \vec{q}_a)$ space at $t = t_B$.

These domains have the following properties :

A) They all have the same positive volume in their $(\vec{p}_a, \vec{q}_a)$ space since the Hamiltonian motions are volume preserving.

B) For any positive integer K the sets $\Delta(t_B)$ and $\Delta(t_K)$ are disjoint since $\Delta(t_B) \cap \overline{D}_{\varepsilon f}$ is empty and $\Delta(t_K) \subset D_{\varepsilon f} \subset \overline{D}_{\varepsilon f}$.

C) For any pair of different positive integers K and K' the sets $\Delta(t_K)$ and $\Delta(t_{K'})$ are disjoint. Indeed, if $K < K'$, the common part $\Delta(t_K) \cap \Delta(t_{K'})$ would correspond after the interval of time KnT to the common part $\Delta(t_B) \cap \Delta(t_{(K'-K)})$ and this intersection is empty.

Hence, in the $(\vec{p}_a, \vec{q}_a)$ space of equation $t = t_B$ or $t = t_K$, the denumerable sequence of disjoint $\Delta(t_K)$ sets cannot remain in the bounded domain $D_{\varepsilon f}$ since these disjoint sets all have the same positive volume. We have thus reached a contradiction and it is impossible that B belongs to $D_{\varepsilon p} - \overline{D}_{\varepsilon f}$.

This implies :

$$D_{\varepsilon p} \subset \overline{D}_{\varepsilon f} \tag{643}$$

and, because of (640), the stability for the future implies the stability for the past (and conversely).

We have also symmetrically :

$$\overline{D}_{\varepsilon p} \supset D_{\varepsilon f} \tag{644}$$

and then :

$$\overline{D}_{\varepsilon p} = \overline{D}_{\varepsilon f} \tag{645}$$

These two sets are invariant in the Hamiltonian motion of interest.

Thus the past-future symmetry conjecture is true for the stability and for the exponential instability. The demonstration remains to be done for systems with either a power-instability or an Arnold diffusion. However it would be a pity if this beautiful symmetry property was not true.

Notice that the above demonstration is valid for all conservative systems, even non-Hamiltonian, and it seems that the past-future symmetry conjecture is true for all analytic and conservative systems.

The analytic conservative systems are not necessarily volume preserving like the Hamiltonian systems, but they preserve in the motion a suitable measure of the measurable sets of phase space. This forbid them to have an asymptotic stability and their periodic orbits have only the four possible stability or instability types already met for analytic Hamiltonian systems :

The exponential instability

The power-m instability

The Arnold diffusion

The non-asymptotic stability

For conservative systems the sum of Liapunov characteristic exponents is zero and then the exponential instability for the future implies the exponential instability for the past and conversely.

Notice that the preserved measure is usually independent of the time (and written $\mu(\vec{x})$). However all the above results remain true if the preserved measure μ is periodic in terms of t provided that it remains integrable and almost everywhere non-zero in a vicinity of the trajectory of interest.

10.7.9 On the cases with multiple Floquet multipliers or multiple eigenvalues.

Since the middle of Section 10.7.7.3 the analysis assumes that all ε_K are zero, which happens especially when all eigen--values λ_K or all Floquet multipliers μ_K are different.

When ε_K is not zero we obtain $\varepsilon_K = 1$; $\lambda_K = \lambda_{K+1}$; $\mu_K = \mu_{K+1}$. The corresponding analyses have been carried out in Ref.67,183 for Hamiltonian systems with only two degrees of freedom but a general analysis remains to be done.

We can notice the following particular point : If the system of interest is Hamiltonian and has a differentiable integral of motion that is not stationary at the periodic solution of interest, this integral of motion gives at least two Floquet multipliers equal to one and two corresponding eigenvalues equal to zero. This property has been demonstrated in Section 10.7.4.4. Fortunately the integral of motion itself allows to reduce the number of degrees of freedom of the system and the two Floquet multipliers equal to one disappear.

For instance let us assume that $\partial H/\partial q_1 \equiv 0$. The parameter p_1 is then an integral of motion and q_1 is ignorable. Δp_1 is constant and the corresponding eigenvector gives the Floquet multipliers μ_1 and μ_1^{-1} equal to one.

For the given value of p_1 the study of the reduced Hamiltonian system $H(p_2 \ldots p_r, q_2 \ldots q_r, t)$ avoid these two Floquet multipliers μ_1 and μ_1^{-1}.

If the Hamiltonian H is itself an integral of motion, i.e if H is autonomous, and if H is non stationary at the periodic solution of interest the problem can be reduced exactly to that already discussed between the equations (91) and (95) in Chapter 7.

10.7.10 Example. The all-order stability of Lagrangian motions.

The stability of motions near the Lagrangian equilibrium points has been the subject of many studies (Ref. 60-67) and their results are presented in Section 8.2.

Section 8.2.1 gives the results of the first-order studies. The motions in the vicinity of collinear Lagrangian points are always exponentially unstable and so are most motions near the triangular Lagrangian points, as shown by Fig. 13.

Section 8.2.2 gives the results of the higher-order studies in the plane circular restricted case and we will extend these analyses by some considerations of the three-dimensional circular three-body case (either restricted or general).

The Hamiltonian form of the restricted three-body problem has already been used in Sections 9.3 and 10.7.4.4. Let us use it again in the simple case when $e = 0$ (circular problem) and $G = 1$; $m_1 + m_2 = 1$; $\omega = 1$; $r_{12} = 1$.

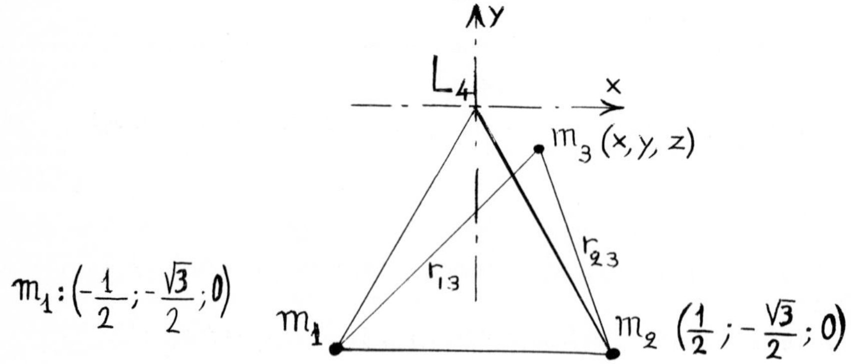

Fig. 71. The set of axes of the Lagrangian point L_4.

We will use the rotating set of axes of the Lagrangian point L_4 (Fig 71) and the parameters n, p, q will be the conjuguate parameters of the coordinates x, y, z of the third body.

The Hamiltonian H and the equations of motion of the problem are then the following.

$$\left.\begin{aligned}
H &= \frac{1}{2}\left[(n+y)^2 + (p-x)^2 + q^2 + z^2\right] - m_1\left(\frac{1}{r_{13}} + \frac{r_{13}^2}{2}\right) - \\
&\qquad - m_2\left(\frac{1}{r_{23}} + \frac{r_{23}^2}{2}\right) \\
&\text{with :} \\
r_{13}^2 &= 1 + x + y\sqrt{3} + x^2 + y^2 + z^2 \\
r_{23}^2 &= 1 - x + y\sqrt{3} + x^2 + y^2 + z^2 \\
dx/dt &= \partial H/\partial n = n + y \quad ; \quad dn/dt = - \partial H/\partial x \\
dy/dt &= \partial H/\partial p = p - x \quad ; \quad dp/dt = - \partial H/\partial y \\
dz/dt &= \partial H/\partial q = q \qquad ; \quad dq/dt = - \partial H/\partial z
\end{aligned}\right\} \tag{646}$$

The first simplification of Section 10.7.7.3 leads from H to H_a and is here only $H_a = H + 1.5$ with the same parameters n, p, q, x, y, z.

The development of H_a up to the fourth-order gives the following.

$$\begin{aligned}
\frac{1}{r_{13}} = (r_{13}^2)^{-1/2} &= 1 - (x+y\sqrt{3})/2 + (- x^2 + 6\sqrt{3}xy + 5y^2 - 4z^2)/8 + \\
&+ (7x^3 - 3\sqrt{3}x^2y - 33xy^2 - 3\sqrt{3}y^3 + 12xz^2 + 12\sqrt{3}yz^2)/16 + \\
&+ (- 37x^4 - 100\sqrt{3}x^3y + 246x^2y^2 + 180\sqrt{3}xy^3 + 3y^4)/128 + \\
&+ 3z^2(- x^2 - 10\sqrt{3}xy - 11y^2 + 6z^2)/16 + \{\text{fifth-order terms}\}
\end{aligned}$$

$$\begin{aligned}
\frac{1}{r_{23}} = (r_{23}^2)^{-1/2} &= 1 + (x-y\sqrt{3})/2 + (- x^2 - 6\sqrt{3}xy + 5y^2 - 4z^2)/8 + \\
&+ (-7x^3 - 3\sqrt{3}x^2y + 33xy^2 - 3\sqrt{3}y^3 - 12xz^2 + 12\sqrt{3}yz^2)/16 + \\
&+ (- 37x^4 + 100\sqrt{3}x^3y + 246x^2y^2 - 180\sqrt{3}xy^3 + 3y^4)/128 + \\
&+ 3z^2(- x^2 + 10\sqrt{3}xy - 11y^2 + 6z^2)/16 + \{\text{fifth-order terms}\}
\end{aligned}$$

Let us put :

$$R = 3\sqrt{3}(m_1 - m_2)/4 \tag{647}$$

Hence :

$$\left.\begin{aligned} H_a &= H + 1.5 = \tfrac{1}{2}(n^2+p^2+q^2+z^2) + ny - px + \tfrac{1}{8}(x^2-5y^2) - Rxy + \\ &+ \frac{Rx\sqrt{3}}{36}(-7x^2 + 33y^2 - 12z^2) + \frac{3y\sqrt{3}}{16}(x^2 + y^2 - 4z^2) + \\ &+ \frac{1}{128}(37x^4 - 246x^2y^2 - 3y^4 + 24x^2z^2 + 264y^2z^2 - 48z^4) + \\ &+ \frac{Rxy}{24}(25x^2 - 45y^2 + 60z^2) + \{\text{fifth-order terms}\} \end{aligned}\right\} \tag{648}$$

The next step is the first-order study with the second-order terms H_{a2} of H_a.

10.7.10.1 The first-order study

The second-order terms of H_a are the following.

$$H_{a2} = \frac{1}{2}(n^2 + p^2 + q^2 + z^2) + ny - px + \frac{1}{8}(x^2 - 5y^2) - Rxy \tag{649}$$

These terms give the following equation of motion

$$\frac{d}{dt}\begin{pmatrix} n \\ p \\ q \\ x \\ y \\ z \end{pmatrix} = N \begin{pmatrix} n \\ p \\ q \\ x \\ y \\ z \end{pmatrix} ; \text{ with : } N = \begin{pmatrix} 0 & 1 & 0 & -1/4 & R & 0 \\ -1 & 0 & 0 & R & 5/4 & 0 \\ 0 & 0 & 0 & 0 & 0 & -1 \\ 1 & 0 & 0 & 0 & 1 & 0 \\ 0 & 1 & 0 & -1 & 0 & 0 \\ 0 & 0 & 1 & 0 & 0 & 0 \end{pmatrix} \tag{650}$$

The eigenvalues λ_1 to λ_6 are given by the characteristic equation of the matrix N :

$$(\lambda^4 + \lambda^2 + \frac{27}{16} - R^2)(\lambda^2 + 1) = 0 \tag{651}$$

The discussion of this equation is classical :

If $R^2 < 23/16$, i.e. if the smallest primary is more than 3.852 % of the total mass there are two conjugate eigenvalues with positive real parts and the motion has an exponential instability.

If $R^2 \geq 23/16$ the problem is "critical" and the study of the higher orders is necessary. The six eigenvalues are purely imaginary, they are given by :

$$\left.\begin{aligned} \lambda_1, \lambda_2 &= \pm iK_1 \quad ; \quad K_1 = 0.5\left[2 + \sqrt{(16R^2 - 23)}\right]^{1/2} \\ \lambda_3, \lambda_4 &= \pm iK_2 \quad ; \quad K_2 = 0.5\left[2 - \sqrt{(16R^2 - 23)}\right]^{1/2} \\ \lambda_5, \lambda_6 &= \pm i \quad ; \quad i = \sqrt{(-1)} \end{aligned}\right\} \tag{652}$$

Hence, since $m_1 + m_2 = 1$ and $R^2 = 27(m_1 - m_2)^2/16 \in [23/16 ; 27/16]$:

$$0 \leq K_2 \leq \sqrt{2}/2 \leq K_1 \leq 1 \quad ; \quad K_1^2 + K_2^2 = 1 \tag{653}$$

10.7.10.2 <u>The second simplification</u>.

The second simplification of Section 10.7.7.3 is a real linear canonical transformation leading from n, p, q, x, y, z, H_a to n_b, p_b, q_b, x_b, y_b, z_b, H_b with, for the second-order terms :

$$H_{b2} = \frac{1}{2}\left[\omega_1(n_b^2 + x_b^2) + \omega_2(p_b^2 + y_b^2) + \omega_3(q_b^2 + z_b^2)\right] \tag{654}$$

The eigenvalues λ_1 to λ_6 are $\pm i\omega_1$, $\pm i\omega_2$, $\pm i\omega_3$ and thus :

$$\omega_1 = \pm K_1 \quad ; \omega_2 = \pm K_2 \quad ; \omega_3 = \pm 1 \tag{655}$$

The Hamiltonian H_a is autonomous, hence the canonical transfor--mation is independent of the time and so is H_b with :

$$\left.\begin{aligned} H_a(n,p,q,x,y,z) &\equiv H_b(n_b,p_b,q_b,x_b,y_b,z_b) \\ H_{a2}(n,p,q,x,y,z) &\equiv H_{b2}(n_b,p_b,q_b,x_b,y_b,z_b) \end{aligned}\right\} \tag{656}$$

The signatures of H_{a2} and H_{b2} must be the same, that is (+,+,+,+,-,-) hence among the three pulsations ω_1, ω_2, ω_3 two are positive and one is negative.

The obvious separation of the first-order problem into in-plane motion (n,p,x,y) and out-of-plane motion (q,z) shows that $\omega_3 = 1$ and hence $\omega_1\omega_2 = - K_1K_2$.

The linear canonical transformation $(n,p,q,x,y,z) \rightarrow (n_b,p_b,q_b,x_b,y_b,z_b)$ is obvious for the out-of-plane components :

$$q_b = q \quad ; \quad z_b = z \tag{657}$$

However it is not simple for the in-plane components, where we must use the symplectic matrices B and B^{-1} :

$$\begin{pmatrix} n \\ p \\ x \\ y \end{pmatrix} = B \begin{pmatrix} n_b \\ p_b \\ x_b \\ y_b \end{pmatrix} \quad ; \quad \begin{pmatrix} n_b \\ p_b \\ x_b \\ y_b \end{pmatrix} = B^{-1} \begin{pmatrix} n \\ p \\ x \\ y \end{pmatrix} \tag{658}$$

with :

$$\left.\begin{aligned} C_1 &= \left[4K_1(2K_1^2 - 1)(4K_1^2 + 3)\right]^{-1/2} \quad ; \\ C_2 &= \left[4K_2(1 - 2K_2^2)(3 + 4K_2^2)\right]^{-1/2} \end{aligned}\right\} \tag{659}$$

and :

$$B = \begin{bmatrix} (4K_1^2-3)C_1 & (3-4K_2^2)C_2 & 4RK_1C_1 & 4RK_2C_2 \\ -4RC_1 & 4RC_2 & (5K_1-4K_1^3)C_1 & (5K_2-4K_2^3)C_2 \\ -4RC_1 & 4RC_2 & 8K_1C_1 & 8K_2C_2 \\ (4K_1^2+3)C_1 & -(3+4K_2^2)C_2 & 0 & 0 \end{bmatrix} \tag{660}$$

$$B^{-1} = \begin{bmatrix} 8K_1C_1 & 0 & -4RK_1C_1 & (4K_1^3-5K_1)C_1 \\ 8K_2C_2 & 0 & -4RK_2C_2 & (4K_2^3-5K_2)C_2 \\ 4RC_1 & -(4K_1^2+3)C_1 & (4K_1^2-3)C_1 & -4RC_1 \\ -4RC_2 & (3+4K_2^2)C_2 & (3-4K_2^2)C_2 & 4RC_2 \end{bmatrix} \tag{661}$$

The research of the symplectic matrix B is simplified by the following : if $\vec{V}_1$, $\vec{V}_2$, $\vec{V}_3$, $\vec{V}_4$ are the four column vectors of B, the column vectors $\vec{V}_1 \pm i\vec{V}_3$ and $\vec{V}_2 \pm i\vec{V}_4$ are eigenvectors of the in-plane part of the matrix N of (650). The main difficulty is the determination of the coefficients C_1 and C_2.

From relation (656) it is now easy to verify that :

$$H_{b2} = \frac{1}{2}\left[K_1(n_b^2 + x_b^2) - K_2(p_b^2 + y_b^2) + (q_b^2 + z_b^2)\right] \tag{662}$$

Hence the three pulsations ω_1, ω_2, ω_3 are :

$$\omega_1 = K_1 \quad ; \quad \omega_2 = - K_2 \quad ; \quad \omega_3 = 1 \tag{663}$$

Notice that this canonical transformation (657)-(661) is possible only if C_1 and C_2 are defined, that is only if $K_2(1 - 2K_2^2)$ is not zero. This excludes the two limit cases $R^2 = 23/16$ and $R^2 = 27/16$, that is, if m_2 is the smallest primary , the cases $m_2 = 0$ and $m_2/(m_1+m_2) = (9-\sqrt{69})/18 = 0.03852...$

The limit case $m_2 = 0$ is simple. The mass m_3 then has an ordinary Keplerian motion about m_1. The distances r_{12} and r_{13} are stable, but the distance r_{23} has a linear instability that is characteristic of unstable systems with multiple eigenvalues.

The limit case $m_2/(m_1+m_2) = (9-\sqrt{69})/18$ is complicated and will not be considered. It is certainly a very interesting case.

10.7.10.3 The quasi-integrals I_N.

The quasi-integrals I_N of Section 10.7.7.4 are met in (597)-(601) and are expressed in (600). They are very simple and contain much informations.

We will look for rational relations between the four pulsations ω_0, ω_1, ω_2, ω_3 of the problem and we need first to choose $\omega_0 = 2\pi/T$. However the problem is autonomous and its period T is arbitrary.

We will of course choose ω_0 without any rational relation to ω_1, ω_2, ω_3 and we are thus led to the following three main cases.

First case : ω_1, ω_2 and ω_3 are not rationally related.

Let us recall that ω_1, ω_2, ω_3 are given by :

$$\left.\begin{array}{l} \omega_1 = K_1 = \left[1+\sqrt{(1-27m_1m_2)}\ /2\right]^{1/2} \\ \omega_2 = -K_2 = -\left[1-\sqrt{(1-27m_1m_2)}\ /2\right]^{1/2} \\ \omega_3 = 1 \end{array}\right\} \left.\begin{array}{l} m_1+m_2 = 1 \\ K_1^2+K_2^2 = 1 \\ 0 \leq K_2 \leq \sqrt{2}/2 \leq K_1 \leq 1 \end{array}\right\} \quad (664)$$

If the mass ration of the primaries is such that ω_1, ω_2, ω_3 have no rational relation there are three sequences I_N of quasi-integrals beginning at $(n_b^2 + x_b^2)$, $(p_b^2 + y_b^2)$, $(q_b^2 + z_b^2)$ and the motion has a "generalized Birkhoff differential rotation". The motion is "all-order stable" but we do not know if it has a non-asymptotic stability or an Arnold diffusion.

Second case : ω_1, ω_2, and ω_3 have one and only one rational relation.

This relation can be written with the suitable irreducible integers a_1, a_2, a_3 :

$$a_1\omega_1 + a_2\omega_2 + a_3\omega_3 = 0 \quad (665)$$

that is :

$$a_1K_1 + a_3 = a_2K_2 \quad (666)$$

There remain only two independent quasi-integrals of the I_N type :

$$\left.\begin{array}{l} a_1(p_b^2 + y_b^2) - a_2(n_b^2 + x_b^2) \\ a_1(q_b^2 + z_b^2) - a_3(n_b^2 + x_b^2) \\ a_2(q_b^2 + z_b^2) - a_3(p_b^2 + y_b^2) \end{array}\right\} \text{Only two independent quasi-integrals} \quad (667)$$

The rule of signs (620) of Section 10.7.7.4 leads to :
If :

$$\inf\{a_1,a_2,a_3\} < 0 < \sup(a_1,a_2,a_3) \quad (668)$$

i.e. here with (666) :

If :

$$a_1 a_3 < 0 \tag{669}$$

the motion has the "all-order stability".

Notice that the second quasi-integral of (667) does indeed then not allow an exponential escape **or** a power-m escape of $q_b^2 + z_b^2$ and $n_b^2 + x_b^2$ while the two other quasi-integrals have the same effect for $p_b^2 + y_b^2$.

If $a_1 a_3 \geq 0$ we can choose a_1, a_2 and a_3 positive or zero. The motion has not necessarily the all-order stability, but the analysis of Section 10.7.7.6 with $w = a_1 + a_2 + a_3 - 1$ shows that if the motion has a power-m instability we have necessarily :

$$m \geq a_1 + a_2 + a_3 - 1 \tag{670}$$

The symmetry (q,z)/(-q,-z) allows to improve this result to :

$$\{\text{If } a_3 \text{ is odd} : m \geq 2(a_1 + a_2 + a_3) - 1\} \tag{671}$$

Third case : ω_1, ω_2 and ω_3 have two rational relations.

Then K_1 and K_2 are both rational and only one quasi-integral I_N survives :

$$K_1(n_b^2 + x_b^2) - K_2(p_b^2 + y_b^2) + (q_b^2 + z_b^2) \tag{672}$$

This quasi-integral is $2H_{b2}$ and is insufficient for the "all-order stability".

The analysis of Section 10.7.7.6 leads to the following.

1) K_1 and K_2 are positive rational quantities satisfying $K_1 > K_2$ and $K_1^2 + K_2^2 = 1$. They are classically given by the following expressions with positive integers a and b :

$$\{K_1;K_2\} = \left\{\frac{|a^2 - 4b^2|}{a^2 + 4b^2} \; ; \; \frac{4ab}{a^2 + 4b^2}\right\} \qquad \text{a is an odd integer prime to b} \tag{673}$$

2) The relations $a_1\omega_1 + a_2\omega_2 + a_3\omega_3 = 0$ are equivalent to :

$$e_1(a^2 - 4b^2) + e_2(4ab) + e_3(a^2 + 4b^2) = 0 \qquad (674)$$

with $e_1 = \pm a_1$ or $\pm a_2$; $e_2 = - a_2$ or $+ a_1$; $e_3 = a_3$.

3) Because of the symmetry $(q,z)/(-q,-z)$ we need e_3 even and the simplest integer solutions of (674) with e_3 even are the following :

e_1	$2b$	$2a$	$2a + 2b$	$2a + 4b$
e_2	$-a$	$4b$	$4b - a$	$4b - 2a$
e_3	$2b$	$-2a$	$2b - 2a$	$4b - 2a$

(675)

4) The discussion of Section 10.7.7.6 in terms of the ratio b/a leads to the following.

If $b/a < (\sqrt{2}-1)/2$ we obtain :

$$K_1 = \omega_1 = \frac{a^2 - 4b^2}{a^2 + 4b^2} \quad ; \quad K_2 = - \omega_2 = \frac{4ab}{a^2 + 4b^2} \qquad (676)$$

and :

$$a_1 = e_1 \quad ; \quad a_2 = - e_2 \quad ; \quad a_3 = e_3 \qquad (677)$$

The first solution of (675) (that is $a_1 = e_1 = 2b$; $a_2 = - e_2 = = a$; $a_3 = e_3 = 2b$) has positive values of a_1, a_2, a_3 and limits the value of the parameter w of (635), (636) to :
$|a_1| + |a_2| + |a_3| - 1 = a + 4b - 1$

Hence if the solution has not the "all-order stability" it has a power-m instability with :

$$m \geqq a + 4b - 1 \qquad (678)$$

On the contrary if $b/a > (\sqrt{2} - 1)/2$ the first solution will always have $a_1a_3 < 0$ and will not limit the value of the parameter w which can go up to the limit given by the three other solutions ; we then obtain :

$$m \geqq \inf\left\{(4a+4b);(2a+2b+|4b-a|+|2b-2a|);(2a+4b+|8b-4a|)\right\} - 1 \qquad (679)$$

Hence the general discussion for a power-m instability give the following in terms of the ratio b/a :

$$
\begin{array}{lll}
0 < b/a < (\sqrt{2}-1)/2 & & m \geq a + 4b - 1 \\
(\sqrt{2}-1)/2 < b/a \leq 1/4 & \Longrightarrow & m \geq 5a - 4b - 1 \\
1/4 \leq b/a \leq 3/8 & \Longrightarrow & m \geq 3a + 4b - 1 \\
3/8 \leq b/a \leq 1/2 & \Longrightarrow & m \geq 6a - 4b - 1 \\
1/2 \leq b/a \leq 5/8 & \Longrightarrow & m \geq 12b - 2a - 1 \\
5/8 \leq b/a \leq 1 & \Longrightarrow & m \geq 3a + 4b - 1 \\
1 \leq b/a \leq 5/4 & \Longrightarrow & m \geq 8b - a - 1 \\
5/4 \leq b/a & \Longrightarrow & m \geq 4a + 4b - 1
\end{array}
\qquad (680)
$$

For the most usual cases this implies :

a	b	K_1	K_2	m
1	1	4/5	3/5	$m \geq 6$
3	1	12/13	5/13	$m \geq 12$
1	2	15/17	8/17	$m \geq 11$
3	2	24/25	7/25	$m \geq 16$
5	1	21/29	20/29	$m \geq 8$

(681)

The lower bounds given in (670), (671) and (680) generally lead to large values of m for a power-m instability. For instance only seven mass-ratios allow $m \leq 5$, the mass-ratios given by $K_2 = 1/2$ or 2/3 and those given by $K_2/K_1 = 1/2$, 1/3, 1/4, 1/5 or 2/3. Only a denumerable sequence of mass-ratios gives the possibility of a power-m instability and thus the critical cases generally lead to "all-order stability".

10.7.10.4 Extension to the circular Lagrangian motions of the general three-body problem.

We have seen in Section 8.2.1 that Eulerian motions are exponentially unstable and that the first-order analysis of general Lagrangian motions is identical to this of the restricted case. The first order stability depends only on the two following parameters : the eccentricity e of the orbits and the ratio N given in (125), that is :

$$N = (m_1^2+m_2^2+m_3^2-m_1m_2-m_1m_3-m_2m_3)^{1/2}/(m_1+m_2+m_3) \qquad (682)$$

The results are presented in Fig. 13 of Section 8.2.1.

In the first-order stable domains of Fig. 13 the problem is critical and higher-order effects must be analysed. This is very difficult for the elliptic case and we will consider only the circular case.

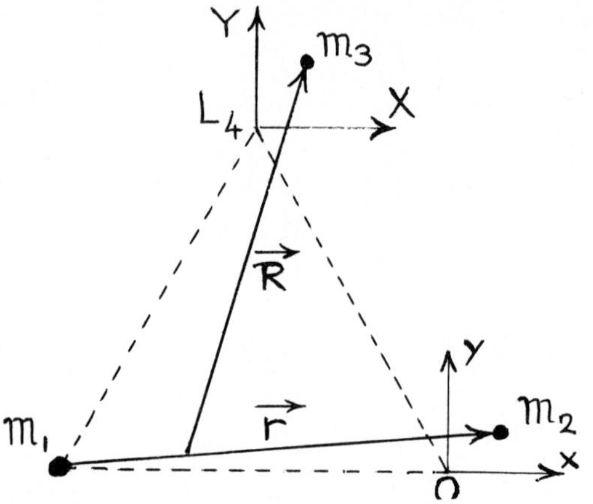

Fig. 72. Analysis of the circular Lagrangian motions of the general three-body problem.
The two rotating sets of axes Oxyz and L XYZ.

Let us use the rotating sets of axes of Fig. 72 similar to this of Fig. 71.

We will put :

$$\left.\begin{aligned} &G = \text{constant of Newtons's law} = 1 \\ &m_1 + m_2 + m_3 = \text{total mass} = 1 \\ &m_1/(m_1+m_2) = \alpha \\ &m_2/(m_1+m_2) = \beta = 1 - \alpha \\ &m_1 m_2/(m_1+m_2) = m_1\beta = m_2\alpha = m \\ &m_3(m_1+m_2)/(m_1+m_2+m_3) = \mathcal{M} \end{aligned}\right\} \quad (683)$$

The usual Jacobi vectors $\vec{r}$ and $\vec{R}$ of Fig. 72 have the following coordinates :

$$\vec{r} = (1+x;\ y;\ z) \quad (684)$$

$$\vec{R} = \left(\alpha-(1/2)+X;\ (\sqrt{3}/2)+Y;\ Z\right) \quad (685)$$

The parameters x,y,z,X,Y,Z are identically zero in the Lagrangian motion. In the other cases their Hamiltonian H is a generalization of the Hamiltonian of (646) :

$$H = H(x,y,z,X,Y,Z,n,p,q,N,P,Q) =$$
$$= \frac{1}{2m}\Big[(n+my)^2+(p-mx)^2+q^2+m^2z^2\Big] + \frac{1}{2\mathcal{M}}\Big[(N+\mathcal{M}Y)^2+(P-\mathcal{M}X)^2+Q^2+\mathcal{M}^2Z^2\Big] -$$
$$- m_1m_2\Big(\frac{1}{r} + \frac{r^2}{2}\Big) - m_1m_3\Big(\frac{1}{r_{13}} + \frac{r_{13}^2}{2}\Big) - m_2m_3\Big(\frac{1}{r_{23}} + \frac{r_{23}^2}{2}\Big) \qquad (686)$$

with :

$$\left.\begin{aligned} &r^2 = (1+x)^2 + y^2 + z^2 \\ &\vec{r}_{13} = \vec{R} + \beta\vec{r} \ ; \quad r_{13}^2 = \Big(\frac{1}{2} + X+\beta x\Big)^2+\Big(\frac{\sqrt{3}}{2} + Y+\beta y\Big)^2 + (Z+\beta z)^2 \\ &\vec{r}_{23} = \vec{R} - \alpha\vec{r} \ ; \quad r_{23}^2 = \Big(X- \frac{1}{2} -\alpha x\Big)^2+\Big(\frac{\sqrt{3}}{2} + Y-\alpha y\Big)^2+(Z-\alpha z)^2 \end{aligned}\right\} \qquad (687)$$

and with the Hamiltonian equations :

$$\left.\begin{aligned} &dx/dt = \partial H/\partial n = \frac{n}{m} + y \ ; \quad dn/dt = - \partial H/\partial x \\ &dy/dt = \partial H/\partial p = \frac{p}{m} - x \ ; \quad dp/dt = - \partial H/\partial y \\ &dz/dt = \partial H/\partial q = q/m \ ; \quad dq/dt = - \partial H/\partial z \\ &dX/dt = \partial H/\partial N = \frac{N}{\mathcal{M}} + Y \ ; \quad dN/dt = - \partial H/\partial X \\ &dY/dt = \partial H/\partial P = \frac{P}{\mathcal{M}} - X \ ; \quad dP/dt = - \partial H/\partial Y \\ &dZ/dt = \partial H/\partial Q = \frac{Q}{\mathcal{M}} \ ; \quad dQ/dt = - \partial H/\partial Z \end{aligned}\right\} \qquad (688)$$

This autonomous Hamiltonian system with six degrees of freedom has four integrals of motion, the Hamiltonian itself and the components of the angular momentum :

$$\left.\begin{aligned} &(x+1)(p+m) - yn + \Big(X+\alpha- \frac{1}{2}\Big)\Big(P+\mathcal{M}\big(\alpha- \frac{1}{2}\big)\Big) - \Big(Y+\frac{\sqrt{3}}{2}\Big)\Big(N-\mathcal{M}\frac{\sqrt{3}}{2}\Big) \\ &A \cos t + B \sin t \\ &- A \sin t + B \cos t \end{aligned}\right\} \qquad (689)$$

with :

$$A = zn - (x + 1)q + Z(N - \mathcal{M}\frac{\sqrt{3}}{2}) - (X + \alpha - \frac{1}{2})Q$$

$$B = yq - z(p + m) + (Y + \frac{\sqrt{3}}{2})Q - Z(P + \mathcal{M}(\alpha - \frac{1}{2})) \quad (690)$$

In the first-order analysis the characteristic equation of eigenvalues is the following :

$$\lambda^2(\lambda^2 + 1)^3\left(\lambda^4 + \lambda^2 + \frac{27}{4}(m_1m_2 + m_1m_3 + m_2m_3)\right) = 0 \quad (691)$$

This equation is similar to the characteristic equation of the restricted case given in (651) since $27/16 - R^2 = 27m_1m_2/4$. Hence, for $m_1 + m_2 + m_3 = 1$, the critical case corresponds to :

$$m_1m_2 + m_1m_3 + m_2m_3 \leq 1/27 \quad (692)$$

The multiple eigenvalues 0 and ± i are related to the existence of the four integrals of motion. They correspond to a kind of "orbital instability" of Lagrangian motions, let us consider for instance two neighbouring Lagrangian motions with slightly different periods : in the rotating set of axes one of them will escape slowly and linearly.

If we disregard these instabilities and look only for the stability of the three mutual distances and of the plane of the three bodies we can use reductions similar to those of Chapter 7. We then obtain autonomous Hamiltonian systems with three degrees of freedom for planar motions and four degrees of freedom for three-dimensional motions.

The characteristic equation then becomes :

$$(\lambda^2 + 1)^K\left[\lambda^4 + \lambda^2 + \frac{27}{4}(m_1m_2 + m_1m_3 + m_2m_3)\right] = 0 \quad (693)$$

with $\begin{cases} K = 1 \text{ in the plane case} \\ K = 2 \text{ in the three-dimensional case} \end{cases}$

The signature of the quadratic part H_2 of the Hamiltonian is (+,+,+,+,-,-) in the planar case and (+,+,+,+,+,+,-,-) in the three-dimensional case. Hence the generalization of the Hamiltonian H_{b2} of (662) is the following with only one negative pulsation (that is ω_2) :

$$H_{b2} = \frac{1}{2}\left[\omega_1(n_{b1}^2 + x_{b1}^2) + \omega_2(n_{b2}^2 + x_{b2}^2) + \omega_3(n_{b3}^2 + x_{b3}^2)\right] \quad (694)$$

in the planar case

$$H_{b2} = \frac{1}{2}\left[\omega_1(n_{b1}^2+x_{b1}^2) + \omega_2(n_{b2}^2+x_{b2}^2) + \omega_3(n_{b3}^2+x_{b3}^2) + \omega_4(q_b^2+z_b^2)\right] \quad (695)$$

in the three-dimensional case

with the following generalization of (663)-(664) :

$$\left.\begin{array}{ll} \omega_1 = K_1 = \{(1+\sqrt{K})/2\}^{1/2} & K=1-27(m_1m_2+m_1m_3+m_2m_3) \geq 0 \\ \omega_2 = -K_2 = -\{(1-\sqrt{K})/2\}^{1/2} & m_1+m_2+m_3 = 1 \ ; \ K_1^2 + K_2^2 = 1 \\ \omega_3 = \omega_4 = 1 & 0 < K_2 \leq \sqrt{2}/2 \leq K_1 < 1 \end{array}\right\} \quad (696)$$

Notice that $\omega_3 = \omega_4$ and hence that eigenvalues $\pm\ i\omega_3$ are multiple eigenvalues. Fortunately this does not forbid use of the analysis of Section 10.7.7.3 because here the coefficients ε_K of (516) and (542) are zero (indeed in the first-order analysis the out-of-plane motions related to ω_4 are independent of the in-plane motions).

The limiting case $K_1 = 1$; $K_2 = 0$ is simple and restricted (two masses are zero). The other limit case $K_1 = K_2 = \sqrt{2}/2$ has multiple eigenvalues with coefficients ε_K equal to one, it is a complex case that we will not consider.

The pulsations ω_3 and ω_4 are of course rationally related but fortunately they have the same sign and almost all conclusions of the previous Section 10.7.10.3 can be maintened :

A) If K_1, K_2 and 1 have no rational relation the system has the "all-order stability" and the quasi-integrals I_N begin with $(n_{b1}^2 + x_{b1}^2)$, $(n_{b2}^2 + x_{b2}^2)$ and $(n_{b3}^2 + x_{b3}^2 + q_b^2 + z_b^2)$, however the "generalized Birkhoff differential rotation" does not necessarily apply if the motion is not planar.

B) If K_1, K_2 and 1 have only one rational relation we can write it as in (665)-(666) with the irreducible integers a_1, a_2, a_3 :

$$a_1\omega_1 + a_2\omega_2 + a_3\omega_3 = 0 \ ; \ \text{that is :}\ a_1K_1 + a_3 = a_2K_2 \quad (697)$$

If $a_1a_3 < 0$ the system still has the "all-order stability", on the contrary if $a_1a_3 \geq 0$ the system can also have a "power-m instability" with :

$$m \geq |a_1| + |a_2| + |a_3| - 1 \tag{698}$$

But the relation (671) disappears.

C) Finally if K_1 and K_2 are both rational the analysis of the previous Section can be simplified (because an even e_3 is no longer necessary) and we can write :

$$\{K_1 \; ; K_2\} = \left\{\frac{c^2 - d^2}{c^2 + d^2} \; ; \; \frac{2cd}{c^2 + d^2}\right\} \quad \left\{\begin{array}{l} c \text{ and } d \text{ are integers} \\ 0 < d < c \\ c + d \text{ is odd} \\ c \text{ is prime to } d \end{array}\right. \tag{699}$$

where c and d are a and 2b of (673).

The relations $a_1\omega_1 + a_2\omega_2 + a_3\omega_3 = 0$ are equivalent to :

$$e_1(c^2 - d^2) + e_2(2cd) + e_3(c^2 + d^2) = 0 \tag{700}$$

with $e_1 = a_1$ or $- a_2$; $e_2 = - a_2$ or $+ a_1$; $e_3 = a_3$.

The simplest integer solutions of (700) are :

e_1	d	c	c + d
e_2	-c	d	d - c
e_3	d	-c	d - c

(701)

The discussion (680) reduces saying that the system has either an all-order stability or a power-m instability. In the latter case, the following lower bounds can be given with the analysis of Section 10.7.7.6 :

$$\left.\begin{array}{lll} 0 < d/c < (\sqrt{2} - 1) & \Longrightarrow & m \geq c + 2d - 1 \\ (\sqrt{2} - 1) < d/c \leq 1/2 & \Longrightarrow & m \geq 2c + d - 1 \\ 1/2 \leq d/c < 1 & \Longrightarrow & m \geq 3c - d - 1 \end{array}\right\} \tag{702}$$

In the simplest cases already considered in (681), these lead to :

c	d	K_1	K_2	a	b	m
2	1	4/5	3/5	1	1	$m \geqq 4$
3	2	12/13	5/13	3	1	$m \geqq 6$
4	1	15/17	8/17	1	2	$m \geqq 5$
4	3	24/25	7/25	3	2	$m \geqq 8$
5	2	21/29	20/29	5	1	$m \geqq 8$

(703)

These lower bounds are smaller than or equal to those given in (681). Nevertheless the general conclusions are the same :

If $27(m_1m_2+m_1m_3+m_2m_3) > (m_1+m_2+m_3)^2$, which applies in most cases, the system has not the first-order stability and the circular Lagrangian motions are exponentially unstable.

On the contrary if $27(m_1m_2+m_1m_3+m_2m_3) \leqq (m_1+m_2+m_3)^2$, the system is "critical".

These critical systems have the "all-order stability" for almost all mass-ratios, but we will see in the next Sections that several cases of resonance and "power-m instability" exist.

Notice that the critical cases imply a very strong asymmetry : let us call $\mathcal{R}$ the ratio of the largest mass to the total mass, it is a good indicator, because for :

$\mathcal{R}<(9+\sqrt{69})/18=0.9614791...$ => The system is not critical

$\mathcal{R}\geqq(3+\sqrt{32})/9=0.9618726....$ => The system is critical

These limits are very close to one and in a critical case the largest mass is much larger than the other two.

10.7.10.5 <u>The second-order study</u>.

The second-order study deals with the second and third-order terms of the Hamiltonian. Let us begin by the analysis of the restricted case.

Let us recall that our Hamiltonians were successively, with $R = 3\sqrt{3}(m_1-m_2)/4$:

$H = H_a - 1.5$, given in (646) (704)

$$H_a = H_{a2} + H_{a3} + \{\text{fourth-order}\} \tag{705}$$

$$H_{a2} = \frac{1}{2}(n^2+p^2+q^2+z^2) + ny - px + \frac{1}{8}(x^2-5y^2) - Rxy \tag{706}$$

$$H_{a3} = \frac{Rx\sqrt{3}}{36}(-7x^2+3y^2-12z^2) + \frac{3y\sqrt{3}}{16}(x^2 + y^2 - 4z^2) \tag{707}$$

$$H_b = H_{b2} + H_{b3} + \{\text{fourth-order}\} \tag{708}$$

$$H_{b2} = H_{a2} = \frac{1}{2}\left[\omega_1(n_b^2+x_b^2) + \omega_2(p_b^2+y_b^2) + \omega_3(q_b^2+z_b^2)\right] \tag{709}$$

where ω_1, ω_2, ω_3 are given in (664).

$$H_{b3} = H_{a3} = H_{b3}(n_b,p_b,q_b,x_b,y_b,z_b) \tag{710}$$

H_{b3} is given by (707), (710) and by the linear transformation (657)-(661) with the constants R, K_1, K_2, C_1, C_2 :

$$\begin{aligned} x &= -4RC_1n_b + 4RC_2p_b + 8K_1C_1x_b + 8K_2C_2y_b \\ y &= (4K_1^2 + 3)C_1n_b - (4K_2^2 + 3)C_2p_b \\ z &= z_b \end{aligned} \tag{711}$$

For the second-order study we need the simplification of the third-order terms given by the passage from H_b to H_3 (Section 10.7.7.3).

The parameters n_3,p_3,q_3,x_3,y_3,z_3 are equivalent to n_b,p_b,q_b,x_b,z_b (their differences are of second-order), they are given by the following generatrix function.

$$S = n_bx_3 + p_by_3 + q_bz_3 + tH_3 + \varepsilon(n_b,p_b,q_b,x_3,y_3,z_3) \tag{712}$$

where ε is a suitable homogeneous polynomial of degree three and the new parameters are given by the following relations :

$$\left.\begin{aligned}
x_b &= \partial S/\partial n_b = x_3 + \partial\varepsilon/\partial n_b \\
y_b &= \partial S/\partial p_b = y_3 + \partial\varepsilon/\partial p_b \\
z_b &= \partial S/\partial q_b = z_3 + \partial\varepsilon/\partial q_b \\
n_3 &= \partial S/\partial x_3 = n_b + \partial\varepsilon/\partial x_3 \\
p_3 &= \partial S/\partial y_3 = p_b + \partial\varepsilon/\partial y_3 \\
q_3 &= \partial S/\partial z_3 = q_b + \partial\varepsilon/\partial z_3
\end{aligned}\right\} \quad H_B = \partial S/\partial t = H_3 \tag{713}$$

The main part of H_b is of the second-order, hence the second-order terms of H_3 have the same expression :

$$H_{3.2} = \frac{1}{2}\left[\omega_1(n_3^2 + x_3^3) + \omega_2(p_3^2 + y_3^2) + \omega_3(q_3^2 + z_3^2)\right] \tag{714}$$

and, since $H_3 = H_b$, the third-order terms of H_3 are given by :

$$H_{3.3} = H_{b2} - H_{32} + H_{b3} + \{\text{fourth-order}\} \tag{715}$$

that is, since $x_b - x_3 = \partial\varepsilon/\partial n_b = \{\text{second-order}\}$, etc ... :

$$\left.\begin{aligned}
H_{3.3} &= H_{b3} + \omega_1\left(x_3 \frac{\partial\varepsilon}{\partial n_b} - n_b \frac{\partial\varepsilon}{\partial x_3}\right) + \omega_2\left(y_3 \frac{\partial\varepsilon}{\partial p_b} - p_b \frac{\partial\varepsilon}{\partial y_3}\right) + \\
&\quad + \omega_3\left(z_3 \frac{\partial\varepsilon}{\partial q_b} - q_b \frac{\partial\varepsilon}{\partial z_3}\right) + \{\text{fourth-order}\}
\end{aligned}\right\} \tag{716}$$

The expression (716) is not easy to use and a slight transformation is very useful.

Let us define the following complex variables :

$$\left.\begin{aligned}
n_b + ix_3 &= \mathfrak{z}_1 \\
p_b + iy_3 &= \mathfrak{z}_2 \\
q_b + iz_3 &= \mathfrak{z}_3
\end{aligned}\right\} \tag{717}$$

Hence :

$$\left.\begin{aligned} &n_b = \tfrac{1}{2}\,(\mathfrak{z}_1 + \bar{\mathfrak{z}}_1) \quad ; \quad x_3 = \tfrac{i}{2}\,(\bar{\mathfrak{z}}_1 - \mathfrak{z}_1) \\ &P_b = \tfrac{1}{2}\,(\mathfrak{z}_2 + \bar{\mathfrak{z}}_2) \quad ; \quad \text{etc} \ldots \end{aligned}\right\} \tag{718}$$

and, with (711) :

$$\left.\begin{aligned} x = {} & (-2RC_1-4iK_1C_1)\mathfrak{z}_1 + (-2RC_1+4iK_1C_1)\bar{\mathfrak{z}}_1 + (2RC_2-4iK_2C_2)\mathfrak{z}_2 + \\ & + (2RC_2+4iK_2C_2)\bar{\mathfrak{z}}_2 + \{\text{second-order terms}\} \end{aligned}\right\} \tag{719}$$

$$y = (2K_1^2+1.5)C_1(\mathfrak{z}_1+\bar{\mathfrak{z}}_1) - (2K_2^2+1.5)C_2(\mathfrak{z}_2+\bar{\mathfrak{z}}_2) \tag{720}$$

$$z = z_b = \frac{i}{2}\,(\bar{\mathfrak{z}}_3 - \mathfrak{z}_3) + \{\text{second-order terms}\} \tag{721}$$

On the other hand we can write ε in terms of $(\mathfrak{z}_1, \mathfrak{z}_2, \mathfrak{z}_3, \bar{\mathfrak{z}}_1, \bar{\mathfrak{z}}_2, \bar{\mathfrak{z}}_3)$, it is a homogeneous polynomial of degree three and, with the usual formal derivatives $\partial\varepsilon/\partial\mathfrak{z}_j$ and $\partial\varepsilon/\partial\bar{\mathfrak{z}}_j$:

$$\left.\begin{aligned} &\partial\varepsilon/\partial n_b = \partial\varepsilon/\partial\mathfrak{z}_1 + \partial\varepsilon/\partial\bar{\mathfrak{z}}_1 \\ &\partial\varepsilon/\partial x_3 = i(\partial\varepsilon/\partial\mathfrak{z}_1 - \partial\varepsilon/\partial\bar{\mathfrak{z}}_1) \\ &\partial\varepsilon/\partial p_b = \partial\varepsilon/\partial\mathfrak{z}_2 + \partial\varepsilon/\partial\bar{\mathfrak{z}}_2 \ ; \ \text{etc} \ldots \end{aligned}\right\} \tag{722}$$

Hence (716) can be written :

$$\left.\begin{aligned} H_{3.3} = H_{b.3} & + i\omega_1\Big(\bar{\mathfrak{z}}_1 \frac{\partial\varepsilon}{\partial\bar{\mathfrak{z}}_1} - \mathfrak{z}_1 \frac{\partial\varepsilon}{\partial\mathfrak{z}_1}\Big) + i\omega_2\Big(\bar{\mathfrak{z}}_2 \frac{\partial\varepsilon}{\partial\bar{\mathfrak{z}}_2} - \mathfrak{z}_2 \frac{\partial\varepsilon}{\partial\mathfrak{z}_2}\Big) + \\ & + i\omega_3\Big(\bar{\mathfrak{z}}_3 \frac{\partial\varepsilon}{\partial\bar{\mathfrak{z}}_3} - \mathfrak{z}_3 \frac{\partial\varepsilon}{\partial\mathfrak{z}_3}\Big) + \{\text{fourth-order terms}\} \end{aligned}\right\} \tag{723}$$

It is now easy to simplify $H_{b.3}$ into the form $H_{3.3}$:

A) From relations (707) and (719)-(721) we may write H_{b3} in terms of $\mathfrak{z}_1, \mathfrak{z}_2, \mathfrak{z}_3, \bar{\mathfrak{z}}_1, \bar{\mathfrak{z}}_2, \bar{\mathfrak{z}}_3$ and neglect the fourth-order. H_{b3} is then homogeneous polynomial of degree three :

$$\left.\begin{aligned} &H_{b3} = \Sigma\mathcal{J}(\vec{\alpha},\vec{\beta}) . \mathfrak{z}_1^{\alpha_1}\bar{\mathfrak{z}}_1^{\beta_1}\mathfrak{z}_2^{\alpha_2}\bar{\mathfrak{z}}_2^{\beta_2}\mathfrak{z}_3^{\alpha_3}\bar{\mathfrak{z}}_3^{\beta_3} \\ &\text{with} : \alpha_1 + \beta_1 + \alpha_2 + \beta_2 + \alpha_3 + \beta_3 = 3 \end{aligned}\right\} \tag{724}$$

B) With (723) the polynomial $H_{3.3}$ is then reduce to zero if we choose :

$$\varepsilon = \Sigma \frac{\mathcal{J}(\vec{\alpha},\vec{\beta}) z_1^{\alpha_1}\bar{z}_1^{\beta_1} z_2^{\alpha_2}\bar{z}_2^{\beta_2} z_3^{\alpha_3}\bar{z}_3^{\beta_3}}{i\left[\omega_1(\alpha_1-\beta_1)+\omega_2(\alpha_2-\beta_2)+\omega_3(\alpha_3-\beta_3)\right]} \tag{725}$$

In most cases the denominators are not zero and the choice of ε as given in (725) leads to $H_{3.3} = 0$. The second-order study then only shows that the system of interest **has no** power-two instability.

In the cases of resonance some denominators of (725) are zero and cannot be used ; the corresponding term of $H_{3.3}$ is then equivalent to this of $H_{b.3}$.

In the restricted three-body problem the differences $(\alpha_3-\beta_3)$ are always even and the only possibilities of a vanishing denominator for a term of degree 3 are the two following :

$$\left.\begin{array}{ll} & \text{either : } \alpha_1=1 \ ; \ \alpha_2=2 \ ; \ \alpha_3=0 \ ; \ \vec{\beta}=0 \\ \omega_1+2\omega_2=0, \text{ with} & \\ & \text{or : } \beta_1=1 \ ; \ \beta_2=2 \ ; \ \beta_3=0 \ ; \ \vec{\alpha}=0 \end{array}\right\} \tag{726}$$

The resonance condition $\omega_1+2\omega_2 = 0$ corresponds to $K_1 = 2K_2 = \sqrt{0.8}$ that is $m_1 = (45+\sqrt{1833})/90$ and $m_2 = (45-\sqrt{1833})/90 = 0.0242939.....$ (if we assume $m_1 \geqq m_2$).

The calculation of $H_{3.3}$ is simple and does not require ε. We need only the terms in $z_1 z_2^2$ and in $\bar{z}_1\bar{z}_2^2$ of (724) with $H_{a3}(=H_{b3})$ as given in (707) and with (719)-(721).

The computations must be done for the masses m_1 and m_2 of the resonance, that is for :

$$\left.\begin{array}{l} R = 3\sqrt{3}(m_1-m_2)/4 = \sqrt{611}/20 = 1.23592... \\ K_1 = \sqrt{0.8} = 0.89443... \\ K_2 = \sqrt{0.2} = K_1/2 = 0.44721... \\ C_1 = (14.88)^{-1/2}(0.8)^{-1/4} = 0.27411... \\ C_2 = (9.12)^{-1/2}(0.2)^{-1/4} = 0.49516... \end{array}\right\} \tag{727}$$

As in Section 10.7.7.3 we will put :

$$n_3 + ix_3 = z_{3.1} \quad ; \quad p_3 + iy_3 = z_{3.2} \tag{728}$$

Thus $z_{3.1}$ and $z_{3.2}$ are respectively equivalent to $\mathfrak{z}_1$ and $\mathfrak{z}_2$ and we obtain in the case of resonance :

$$H_{3.3} = Cz_{3.1}z_{3.2}^2 + \bar{C}\bar{z}_{3.1}\bar{z}_{3.2}^2 \tag{729}$$

with :

$$C = \sqrt[4]{5}\left[\frac{3371}{1368\sqrt{310}} - i\,\frac{17\sqrt{611}}{684\sqrt{62}}\right] = 0.20928.. - 0.11667i \tag{730}$$

The final simplification of Section 10.7.7.3 is the following :

$$z_{3.1} = Z_{3.1}\exp\{i\omega_1 t\}$$

$$z_{3.2} = Z_{3.2}\exp\{i\omega_2 t\} \tag{731}$$

$$z_{3.3} = Z_{3.3}\exp\{i\omega_3 t\}$$

This simplification leads to the Hamiltonian $\mathcal{H}_3$ without second-order terms :

$$\mathcal{H}_3 = CZ_{3.1}Z_{3.2}^2 + \bar{C}\bar{Z}_{3.1}\bar{Z}_{3.2}^2 + \{\text{fourth-order terms}\} \tag{732}$$

The "main problem" is then the problem of motions with the Hamiltonian h_3 :

$$h_3 = \text{main part of } \mathcal{H}_3 = CZ_{31}Z_{32}^2 + \bar{C}\bar{Z}_{31}\bar{Z}_{32}^2 \tag{733}$$

This main problem is simple :

$$\left.\begin{aligned} dZ_{3.1}/dt &= 2i\partial h_3/\partial\bar{Z}_{3.1} = 2i\bar{C}\,\bar{Z}_{32}^2 \\ dZ_{3.2}/dt &= 2i\partial h_3/\partial\bar{Z}_{3.2} = 4i\bar{C}\,\bar{Z}_{31}\bar{Z}_{32} \end{aligned}\right\} \tag{734}$$

This main problem has two integrals of motion : h_3 itself and the difference D where :

$$D = 2Z_{3.1}\bar{Z}_{3.1} - Z_{3.2}\bar{Z}_{3.2} \tag{735}$$

This integral D is of course a particular quasi-integral of I_N type and the motion has radial escapes (of power 2) for $D = h_3 = 0$.

The general instability of the motion is shown by the following :

$$\left.\begin{aligned} u &= Z_{3.1}Z_{3.2}^2/i\bar{C} \\ du/dt &= 2(Z_{3.2}\cdot\bar{Z}_{3.2})^2 + 8(Z_{3.1}\cdot\bar{Z}_{3.1}\cdot Z_{3.2}\cdot\bar{Z}_{3.2}) \geq 0 \end{aligned}\right\} \quad (736)$$

Here Im(u) is constant (it is : $-h_3/2C\bar{C}$) while $\mathcal{R}e(u)$ is never decreasing and generally goes to infinity (Fig. 73).

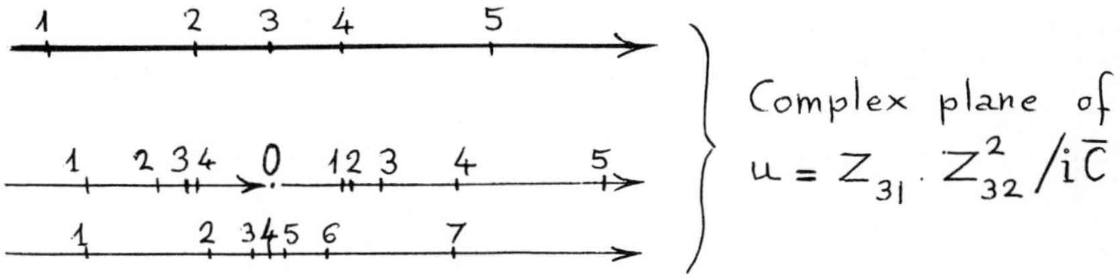

Fig. 73. System (733)-(736). Motion of $u = Z_{31}Z_{32}^2/iC$
$du/dt = 2Z_{32}^2\bar{Z}_{32}^2 + 8Z_{31}\bar{Z}_{31}Z_{32}\bar{Z}_{32} \geq 0$

The difference $D = 2Z_{31}\bar{Z}_{31} - Z_{32}\bar{Z}_{32}$ *is constant.*

All motions with $\mathcal{I}m(u) \neq 0$ *and/or* $D < 0$ *are diverging for both past and future.*

If $D \geq 0$ *and* $\mathcal{I}m(u) = 0$ *we obtain the following:*

A) $\mathcal{R}e(u) = 0$*, that is* $Z_{32} = 0$*,: equilibrium with* Z_{31} = *constant and* $Z_{32} = 0$.

B) $\mathcal{R}e(u) > 0$*: motion asymptotic to an equilibrium in the past and diverging in the future.*

C) $\mathcal{R}e(u) < 0$*: case opposite to the case B.*

Thus the second-order study gives only few results :

For all mass-ratios but one the Hamiltonian $H_{3.3}$ is zero and the system has no second-order escape, so we must study the higher-order effects.

For the resonance $\omega_1 + 2\omega_2 = 0$, that is $m_2 = (45-\sqrt{1833})/90 = 0.0242939...$ the system has a power-2 instability.

In the general three-body problem a new second-order resonance appears for the circular Lagrangian motions : this is the case when $2\omega_2 + \omega_3 = 0$, that appears even for plane motions.

These cases $\omega_1 + 2\omega_2 = 0$ and $2\omega_2 + \omega_3 = 0$ then correspond respectively to $m_1m_2+m_1m_3+m_2m_3 = 16/675$ and $1/36$ with $m_1+m_2+m_3 = 1$

These results reflect the usual features of most second-order analyses : such analyses generally lead to an identically zero $H_{3.3}$ and are then not able to give the first term of the quasi-integral $\mathcal{H}_N$. They thus cannot generally lead to decisive conclusions. However second-order resonances very often give a power-two instability and are one of the greatest danger for the stability of critical Hamiltonian systems.

Let us recall that there is a resonance when for a suitable set of integers a_K not-all-zero :

$$\sum_{K=0}^{r} a_K\omega_K = 0 \tag{737}$$

A second-order resonance is a resonance for which :

$$\sum_{K=1}^{r} |a_K| = 3 \tag{738}$$

Let us consider a critical Hamiltonian problem with one or several second-order resonances and assume that for all these resonances:

$$\{1 \leqq j \leqq K \leqq r\} \Longrightarrow a_j a_K \geqq 0 \tag{739}$$

Then it can be demonstrated that* :

Either $H_{3.3}$ is identically zero.

Or the critical Hamiltonian problem of interest has a power-two instability.

---

* Using relation (739) the terms of $\mathcal{H}_{3.3}$ have only then either Z_K's or $\overline{Z}_K$'s.

Let us call h the part of $\mathcal{H}_{3.3}$ that has only Z_K's :

$$\mathcal{H}_{3.3} = h + \overline{h}$$

The non-zero extrema of $\mathcal{Im}(h)$ for $\sum_{K=1}^{r} Z_K\overline{Z}_K = 1$ are then directions of radial escapes.

Furthermore in the "main problem": $dh/dt = \sum_{K=1}^{r}(\partial h/\partial Z_K)(dZ_K/dt) = 2i\sum_{K=1}^{r}\frac{\partial h}{\partial Z_K}\frac{\partial \overline{h}}{\partial \overline{Z}_K}$; obviously unstable with $\mathcal{Re}(h)$ constant and $d\,\mathcal{Im}(h)/dt \geqslant 18h\overline{h}/\sum_{K=1}^{r} Z_K\overline{Z}_K \geqslant 0.$

On the contrary resonances with integers a_K of opposite signs (for $K \geq 1$) are a factor of stability as in the following example :

$$\mathcal{H}_3 = AZ_1Z_2^2 + \bar{A}\bar{Z}_1\bar{Z}_2^2 + BZ_1\bar{Z}_2^2 + \bar{B}\bar{Z}_1Z_2^2 \tag{740}$$

The A and $\bar{A}$ terms imply :

$$a_0\omega_0 + \omega_1 + 2\omega_2 = 0 \tag{741}$$

and we will see they are destabilizing while the B and $\bar{B}$ terms are stabilizing and imply :

$$a'_0\omega_0 + \omega_1 - 2\omega_2 = 0$$

The equations of motion are :

$$\left.\begin{aligned} dZ_1/dt &= 2i\partial\mathcal{H}_3/\partial\bar{Z}_1 = 2i\left(\bar{A}\bar{Z}_2^2 + \bar{B}Z_2^2\right) \\ dZ_2/dt &= 2i\partial\mathcal{H}_3/\partial\bar{Z}_2 = 2i\left(2\bar{A}\bar{Z}_1\bar{Z}_2 + 2BZ_1\bar{Z}_2\right) \end{aligned}\right\} \tag{742}$$

This Hamiltonian system has two integrals of motion : the Hamiltonian $\mathcal{H}_3$ itself and the following sum Σ :

$$\left.\begin{aligned} &\Sigma = (B\bar{B} - A\bar{A})Z_2\bar{Z}_2 + 2W\bar{W} \\ &\text{with } W = AZ_1 + \bar{B}\bar{Z}_1 \end{aligned}\right\} \tag{743}$$

If $B\bar{B} - A\bar{A} > 0$, that is if $|B| > |A|$ the integral Σ forbids the escape of Z_2 and W, the motion is stable.

On the contrary if $|B| < |A|$ the motion has radial power-two escapes for $\Sigma = \mathcal{H}_3 = 0$ and is unstable.

This example emphasizes again the importance and the interest of the signs of the pulsations ω_K as defined in (565)-(572).

10.7.10.6 The third-order study.

In the non-resonant cases of the second-order study, i.e. when $H_{3.3}$ is identically zero, the system has no power-two insta--bility and the third-order study is necessary.

The analysis of the previous Section can be applied again and shows that it is not necessary to develop the canonical transformation from H_3 to H_4 : the terms of $H_{4.4}$ will be equivalent to the resonant terms of $H_{3.4}$ exactly as the terms of $H_{3.3}$ were equivalent to the resonant terms of $H_{b.3}$.

However it becomes necessary to develop $H_{3.4}$ and then, up to the fourth-order, the canonical transformation from H_b to H_3. This will be the subject of the thesis of my friend, Mr El Bakkali from Tetuan (Morocco).

We can already write the following.

A) There are many more resonant terms in the third-order study than in the second-order study and especially all terms of the following types :

$$\left.\begin{array}{ll} A z_{4j}^2 \bar{z}_{4j}^2 & \multirow{2}{*}{} \text{with } \{j,K\} = \{1,2,3\} \text{ in the restricted case} \\ B z_{4j} \bar{z}_{4j} z_{4K} \bar{z}_{4K} & \text{and } = \{1,2,3,4\} \text{ in the general case} \end{array}\right\} \quad (744)$$

Because of these terms the Hamiltonian $H_{4.4}$ is never identically zero and is the main part of the quasi-integral $\mathcal{H}_N$. It is also the Hamiltonian h_4 of the "main problem".

B) In the general three-body case the equality $\omega_3 = \omega_4$ and the z/-z symmetry give the following two resonant terms :

$$C\, z_{4.3}^2 \bar{z}_{4.4}^2 + \bar{C}\, \bar{z}_{4.3}^2 z_{4.4}^2 \tag{745}$$

C) Finally, for particular mass-ratios, there are some exceptional resonant terms :

C.1. The terms :

$$D\, z_{4.1} z_{4.2}^3 + \bar{D}\, \bar{z}_{4.1} \bar{z}_{4.2}^3 \tag{746}$$

when $\omega_1 + 3\omega_2 = 0$ (that is when $m_2/(m_1+m_2)=(15-\sqrt{213})/30=0.01352\ldots$ in the restricted case and $(m_1m_2+m_1m_3+m_2m_3)/(m_1+m_2+m_3)^2=1/75$ in the general case).

C.2. The terms :

$$E\, z_{4.2}^3 z_{4.3} + \bar{E}\, \bar{z}_{4.2}^3 \bar{z}_{4.3} \qquad (747$$

when $3\omega_2 + \omega_3 = 0$. (This case is possible only for the general three-body problem and corresponds to $(m_1m_2+m_1m_3+m_2m_3)/(m_1+m_2+m_3)^2 = 32/2187$).

C.3. The terms :

$$F\, z_{4.1}^2 z_{4.2} \bar{z}_{4.3} + \bar{F}\, \bar{z}_{4.1}^2 \bar{z}_{4.2} z_{4.3} \tag{748}$$

when $2\omega_1 + \omega_2 - \omega_3 = 0$.

Again this case is possible only for the general three-body problem. It corresponds to $\omega_1 = 0.8$; $\omega_2 = -0.6$ that is $(m_1m_2+m_1m_3+m_2m_3)/(m_1+m_2+m_3)^2 = 64/1875 = 0.034133$.

The analysis of the case $\omega_1 + 3\omega_2 = 0$ in the planar restricted three-body problem has already concluded to a power-three insta--bility (Ref. 65-67) and this of course can be extended to the three-dimensional problem.

On the contrary the resonance (748) doesn't give a power-three instability (as already notice in (703) with $K_1 = 4/5$; $K_2 = 3/5$) and higher-order terms must be analysed. It seems that the resonance (747) is in the same situation.

The general analysis of resonances such as (746) or (747) leads to the following Hamiltonian "main problem" with real constants A, B, C :

$$h_4 = A\, Z_1^2\bar{Z}_1^2 + B\, Z_1\bar{Z}_1 Z_2\bar{Z}_2 + C\, Z_2^2\bar{Z}_2^2 + D\, Z_1 Z_2^3 + \bar{D}\, \bar{Z}_1\bar{Z}_2^3 \tag{749}$$

The equations of motion are :

$$dZ_1/dt = 2i\,\partial h_4/\partial\bar{Z}_1 = 2i\left[2AZ_1^2\bar{Z}_1 + BZ_1Z_2\bar{Z}_2 + \bar{D}\bar{Z}_2^3\right] \tag{750}$$

$$dZ_2/dt = 2i\,\partial h_4/\partial\bar{Z}_2 = 2i\left[BZ_1\bar{Z}_1Z_2 + 2CZ_2^2\bar{Z}_2 + 3\bar{D}\,\bar{Z}_1\bar{Z}_2^2\right] \tag{751}$$

This problem has two integrals of motion : the Hamiltonian h_4 itself and the difference Δ where :

$$\Delta = 3Z_1\bar{Z}_1 - Z_2\bar{Z}_2 \tag{752}$$

The main problem is stable if $h_4 = 0$ and $\Delta = 0$ imply $Z_1 = Z_2 = 0$ that is if :

$$|A + 3B + 9C| > 6\sqrt{3}\,|D| \tag{753}$$

On the contrary if :

$$|A + 3B + 9C| < 6\sqrt{3}\ |D| \tag{754}$$

the main problem has power-three escapes in the manifold defined by $h_4 = 0$; $\Delta = 0$ (these escapes verify $Z_1\bar{Z}_1 = Z_2\bar{Z}_2/3 = [2K(t_f - t)]^{-1}$ with $K = \left(108D\bar{D} - (A+3B+9C)^2\right)^{1/2}$ and are asymptotic to $Z_1 = 0$; $Z_2 = 0$ but they are generally not radial). Hence the condition (754) is a sufficient condition for a power-three instability.

The general conclusion of the third-order study is that it gives much more information than the second-order study, particularly because it gives the main part of the quasi-integral $\mathcal{H}_N$ and leads to the "all-order stability" in almost all cases.

However let us recall that the "all-order stability" corresponds either to a true stability (non asymptotic) or to the extremely slow "Arnold diffusion". It would be very useful to investigate this theoretical point but notice that most researchers believe the Arnold diffusion to be the usual feature except for the planar restricted case.

10.8 THE SERIES OF SOME SIMPLE PERIODIC SOLUTIONS OF THE THREE-BODY PROBLEM.

Among the usual methods of analysis of mathematical problems, numerical methods benefit from the huge progress of computers and have an ever increasing importance, they become faster and more accurate by the days, and we will see some examples in Section 10.9. On the other hand the topological and qualitative methods have also recently made great progress and give much informations in spite of their inaccuracies, they will be the subject of Chapter 11.

However the traditional analytical methods retain their interest, they are much more accurate than the qualitative methods and give much broader views than the numerical methods, they allow assessment of the influence of each individual parameter (masses, initial conditions, etc...).

The method of perturbations and the method of series are the most popular. We have already met the Siegel series (Section 8.5 and Ref. 37, 48) for triple collisions and triple close approaches and the Brown series (Section 9.2.1 and Ref. 145) for the simplest periodic orbits of the Hill problem. We now consider some other simple cases.

10.8.1 The pseudo-circular orbits.

The pseudo-circular orbits belong to two families of plane periodic orbits, the families that goes to circular Keplerian motions as the Jacobi ratio r/R goes to zero (Fig. 74). The two families are the families or direct (or prograde) and retrograde pseudo-circular orbits.

These two families have three non-trivial parameters : the mean Jacobi ratio r/R and the two mass ratios.

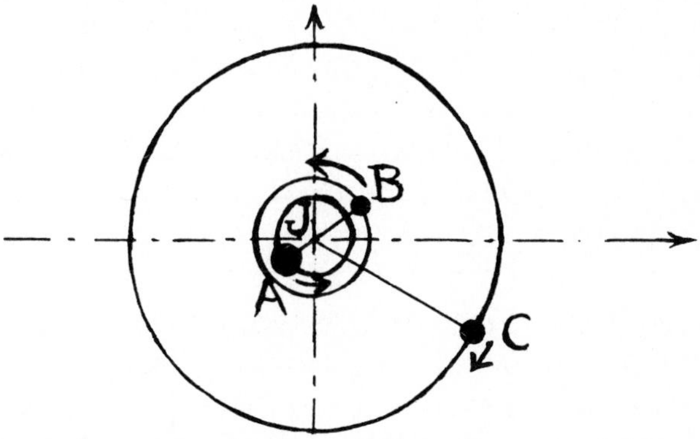

Fig. 74. A retrograde pseudo-circular orbit in the non-rotating axes of the center of mass **J** *of the close binary.*

If the ratio $\mathbf{r}^3(m_1 + m_2 + m_3)/\mathbf{R}^3(m_1 + m_2)$ *is small the variations of* $\mathbf{r}$ *and* $\mathbf{R}$ *are very small.*

The Keplerian approximation leads to the following.

$$r = a_i = \text{inner semi-major axis} \tag{755}$$

$$n_i = \text{inner mean angular motion} = \left(G(m_1+m_2)/a_i^3\right)^{1/2} \tag{756}$$

$$R = a_e = \text{exterior semi-major axis} \tag{757}$$

$$\left.\begin{aligned} &n_e = \text{exterior mean angular motion} = \pm\left(GM/a_e^3\right)^{1/2} ; \\ &M = m_1 + m_2 + m_3 \end{aligned}\right\} \tag{758}$$

We choose $n_e > 0$ for direct orbits and $n_e < 0$ for retrograde orbits.

The complex affixe of the Jacobi vectors $\vec{r}$ and $\vec{R}$ will be z and Z :

$$z \simeq a_i \exp\{in_i t\} \ ; \quad Z \simeq a_e \exp\{in_e t\} \tag{759}$$

The corresponding equations of motion are :

$$\frac{d^2z}{dt^2} = - G(m_1+m_2)\frac{z}{r^3} + Gm_3\left[\frac{Z - \alpha z}{r_{23}^3} - \frac{Z + \beta z}{r_{13}^3}\right] \tag{760}$$

$$\frac{d^2Z}{dt^2} = - GM\left[\alpha\frac{Z + \beta z}{r_{13}^3} + \beta\frac{Z - \alpha z}{r_{23}^3}\right] \tag{761}$$

with :

$$\alpha = m_1/(m_1+m_2) \quad ; \quad \beta = m_2/(m_1+m_2) = 1 - \alpha \tag{762}$$

$$\vec{r}_{13} = \vec{R} + \beta\vec{r} \quad ; \quad r_{13} = |\vec{r}_{13}| = |Z + \beta z| \tag{763}$$

$$\vec{r}_{23} = \vec{R} - \alpha\vec{r} \quad ; \quad r_{23} = |\vec{r}_{23}| = |Z - \alpha z| \tag{764}$$

The Fourier series of pseudo-circular orbits can be developed in terms of the angle nt, the angular velocity n being the relative angular velocity of inner to outer motion :

$$n = n_i - n_e \tag{765}$$

$$z = a_i \exp\{in_i t\} \cdot \sum_{K=-\infty}^{+\infty} b_K \exp\{Kint\} \tag{766}$$

$$Z = a_e \exp\{in_e t\} \cdot \sum_{K=-\infty}^{+\infty} B_K \exp\{Kint\} \tag{767}$$

The symmetries show that all b_K, B_K are real and if $m_1 = m_2$, then all b_K and B_K with odd K are zero.

The series (766) and (767) can be obtained by identification of the successive orders in the equations of motion (760)-(764). They give the following :

$$s = \text{main infinitely small} = \sup\left\{\frac{|m_1-m_2|a_i}{(m_1+m_2)a_e} \; ; \; \frac{a_i^2}{a_e^2} \; ; \; \frac{m_3\, a_i^3}{(m_1+m_2)a_e^3}\right\} \tag{768}$$

$$\left.\begin{aligned} b_0 &= 1 - \frac{m_3\, a_i^3}{6(m_1+m_2)a_e^3}\left[1 + O(s)\right] \\ b_2 &= \frac{m_3\, a_i^3}{16(m_1+m_2)a_e^3}\left[3 + 14\,\frac{n_e}{n} + O(s)\right] \\ b_{-2} &= \frac{m_3\, a_i^3}{48(m_1+m_2)a_e^3}\left[- 57 - 194\,\frac{n_e}{n} + O(s)\right] \end{aligned}\right\} \tag{769}$$

other $b_K = O\{s m_3 a_i^3/(m_1+m_2)a_e^3\}$

$$\left.\begin{aligned}
B_0 &= 1 + \frac{\alpha\beta a_i^2}{4a_e^2}\left[1 + O(s)\right] \\
B_2 &= \frac{3\alpha\beta a_i^2}{32a_e^2} \cdot \frac{n_e^2}{n^2}\left[1 - \frac{n_e}{n} + O(s)\right] \\
B_{-2} &= \frac{15\alpha\beta a_i^2}{32a_e^2} \cdot \frac{n_e^2}{n^2}\left[1 + \frac{n_e}{n} + O(s)\right] \\
\text{other } B_K &= O\{s\alpha\beta a_i^2 n_e^2/a_e^2 n^2\}
\end{aligned}\right\} \quad (770)$$

In the Hill case the expressions (769) are identical to the Brown series given in (159) in the Section 9.2.1 and for small n_e/n these expressions (769) and (770) are very accurate especially if $m_1 = m_2$.

It is surprising that if m_1 and m_2 are very unequal the series are less accurate and the "perturbations" or the "inequalities" are larger than when $m_1 = m_2$.

The pseudo-circular orbits of the three equal mass case have been computed in the references 133, 134, 184. They will be presented in Section 10.9.1.

10.8.2 A family of periodic orbits with the largest number of symmetries.

The families of symmetric periodic orbits were presented in section 10.6. All regular orbits of a family have the same space-time symmetries and if the three masses are unequal the number of space-time symmetries per period is always either 2 or 4.

If all three masses are equal the number of space-time symmetries per period can also be 6 or 12, however these symmetries are always of type III or IV (Fig. 75) with the three masses successively at the tip of the isoceles triangle.

At the circular triangular Lagrangian solution with three equal masses (Fig. 76), a family of symmetric-periodic orbits with twelve space-time symmetries per period arises.

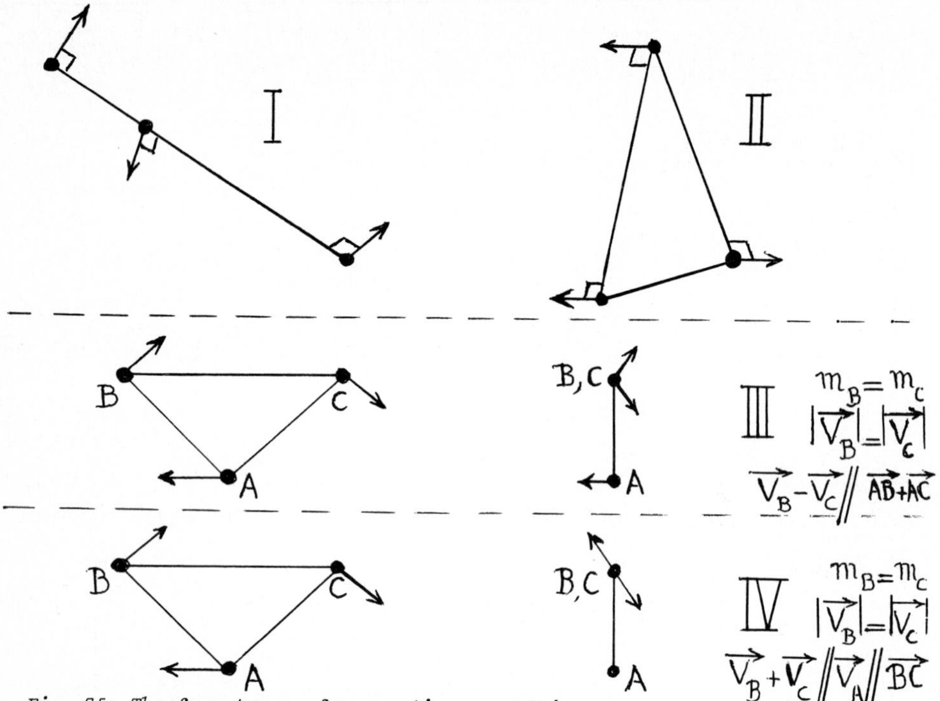

Fig. 75. The four types of space-time symmetries.

I. Collinear configuration with normal velocities.

II. Velocities normal to the plane of the three masses.

III and IV. Isosceles configuration with equal masses at the symmetrical corners and with symmetrical velocities.

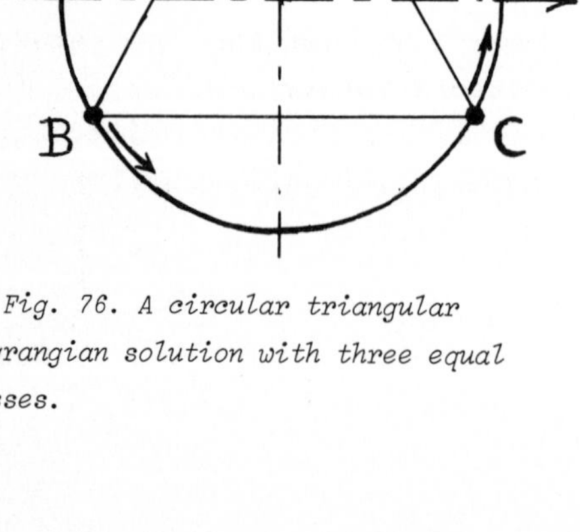

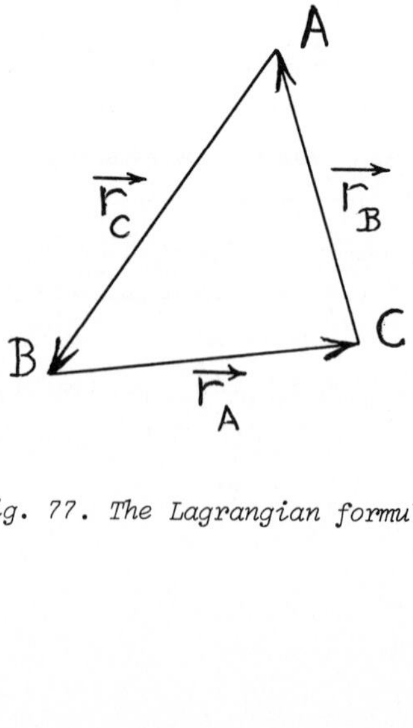

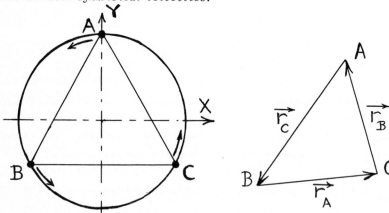

Fig. 76. A circular triangular Lagrangian solution with three equal masses.

Fig. 77. The Lagrangian formulation.

Let us use the Lagrangian formulation of the figure 77 and of Section 4.2 with a gravitational constant G equal to one and a total mass equal to one :

$$m_A = m_B = m_C = 1/3 \tag{771}$$

The equations of motion are then :

$$\frac{d^2\vec{r}_K}{dt^2} = \frac{1}{3}\left[\frac{\vec{r}_A}{r_A^3} + \frac{\vec{r}_B}{r_B^3} + \frac{\vec{r}_C}{r_C^3}\right] - \frac{\vec{r}_K}{r_K^3} \quad ; \quad K = \{A,B,C\} \tag{772}$$

A solution of the Lagrangian motion of Fig. 76 is the following :

$$\vec{r}_K = (X_K, Y_K, Z_K) \quad ; \quad K = \{A,B,C\} \tag{773}$$

$$\left.\begin{array}{l} X_A + iY_A = \exp\{it\} \\ \left.\begin{array}{l} X_B + iY_B = j\exp\{it\} \\ X_C + iY_C = j^2\exp\{it\} \end{array}\right\} j = \exp\{2i\pi/3\} = \frac{-1+i\sqrt{3}}{2} \\ Z_A = Z_B = Z_C = 0 \end{array}\right\} \tag{774}$$

The family of the symmetric periodic solutions of interest has one non-trivial parameter since the three masses are given. We will keep the period equal to 2 π, it is indeed a trivial parameter, and we will obtain the following Fourier series for the solutions.

$$\left.\begin{array}{l} X_A + iY_A = \sum_{K=-\infty}^{+\infty} a_{2K} \exp\left\{(2K + 1 + \frac{\alpha}{2\pi})it\right\} \\ Z_A = \sum_{K=0}^{\infty} c_{2K+1} \cos(2K+1)t \end{array}\right\} \tag{775}$$

Here α is the "angle of rotation" of the solution of interest and the space-time symmetries lead to the following :

$$\left.\begin{array}{l} X_B(t) + iY_B(t) = \left[X_A(t-\frac{2\pi}{3}) + iY_A(t-\frac{2\pi}{3})\right] . \exp\{(\frac{\alpha-2\pi}{3})i\} \\ Z_B(t) = Z_A(t - \frac{2\pi}{3}) \end{array}\right\} \tag{776}$$

$$\left.\begin{aligned} X_C(t) + iY_C(t) &= \left[X_A(t+\tfrac{2\pi}{3}) + iY_A(t+\tfrac{2\pi}{3})\right] . \exp\{(\tfrac{2\pi-\alpha}{3})i\} \\ Z_C(t) &= Z_A(t + \tfrac{2\pi}{3}) \end{aligned}\right\} \quad (776)$$

hence, with $j = \exp\{2i\pi/3\}$ that is $j^3 = 1$:

$$\left.\begin{aligned} X_B(t) + iY_B(t) &= \sum_{K=-\infty}^{+\infty} a_{2K}\, j^{(K+1)} \exp\{(2K+1+\tfrac{\alpha}{2\pi})it\} \\ Z_B &= \sum_{K=0}^{\infty} c_{2K+1} \cos\left[(2K+1)(t - \tfrac{2\pi}{3})\right] \\ X_C(t) + iY_C(t) &= \sum_{K=-\infty}^{+\infty} a_{2K}\, j^{(2K+2)} \exp\{(2K+1+\tfrac{\alpha}{2\pi})it\} \\ Z_C &= \sum_{K=0}^{\infty} c_{2K+1} \cos\,(2K+1)(t+ \tfrac{2\pi}{3}) \end{aligned}\right\} \quad (777)$$

The condition $\vec{r}_A + \vec{r}_B + \vec{r}_C = 0$ of Fig. 77 leads to :

$$\left.\begin{aligned} &\left.\begin{aligned} a_4 &= 0 \;;\; a_{10} = 0 \;;\; a_{16} = 0 \\ a_{-2} &= 0 \;;\; a_{-8} = 0 \;;\; a_{-14} = 0 \end{aligned}\right\} \text{ all } a_{6n+4} \text{ are zero} \\ &c_3 = 0 \;;\; c_9 = 0 \;;\; c_{15} = 0 \qquad \text{all } c_{6n+3} \text{ are zero} \end{aligned}\right\} \quad (778)$$

The remaining a_n and c_n are sufficient to satisfy the equations of motion (772) and the non-trivial parameter can be c_1. All a_n and c_n are real and of order $c_1^{|n|}$.

$$a_0 = 1 - \frac{27}{76} c_1^2 - \frac{4451}{109\ 744} c_1^4 + 0(c_1^6) \quad (779)$$

$$a_2 = - \frac{3}{76} c_1^2 - \frac{579}{109\ 744} c_1^4 + 0(c_1^6) \quad (780)$$

$$a_4 = 0 \quad (781)$$

$$a_6 = 0(c_1^6) \quad (782)$$

$$a_{2K} = 0(c_1^{2K}) \quad ; \quad K \geq 4 \quad (783)$$

$$a_{-2} = 0 \quad (784)$$

$$a_{-4} = (60/6859)c_1^4 + O(c_1^6) \tag{785}$$

$$a_{-6} = O(c_1^6) \tag{786}$$

$$a_{-8} = 0 \tag{787}$$

$$a_{-2K} = O(c_1^{2K}) \quad ; \quad K \geq 5 \tag{788}$$

$$c_1 = c_1 \tag{789}$$

$$c_3 = 0 \tag{790}$$

$$c_5 = (20/6859)c_1^5 + O(c_1^7) \tag{791}$$

$$c_7 = O(c_1^7) \tag{792}$$

$$c_9 = 0 \tag{793}$$

$$c_{2K+1} = O(c_1^{(2K+1)}) \quad ; \quad K \geq 5 \tag{794}$$

$$\alpha/2\pi = (3/19)c_1^2 + (2289/27436)c_1^4 + O(c_1^6) \tag{795}$$

The space-time symmetry at t = 0 gives the following :

t	X_A	Y_A	Z_A	X_B	Y_B	Z_B	X_C	Y_C	Z_C
-t	X_A	$-Y_A$	Z_A	X_C	$-Y_C$	Z_C	X_B	$-Y_B$	Z_B

(796)

This is a space-time symmetry of type III with the point A at the tip of the isoceles triangle.

Similarly the twelve space-time symmetries of the period are of the following types :

Instant of space-time symmetry	Type of space-time symmetry	Point at the tip of the isosceles triangle
0 ; π	III	A
$\pi/6$; $7\pi/6$	IV	B
$\pi/3$; $4\pi/3$	III	C
$\pi/2$; $3\pi/2$	IV	A
$2\pi/3$; $5\pi/3$	III	B
$5\pi/6$; $11\pi/6$	IV	C

(797)

The energy integral h and the angular momentum $\vec{c}$ are expressed by the relations (39) and (44) for the Lagrangian formulation (Section 5.1). They can easily be obtained in terms of c_1 :

$$\vec{c} = (0,0,c) \tag{798}$$

$$c = \frac{1}{3} - \frac{7}{38} c_1^2 + \frac{1153}{164616} c_1^4 + \mathbf{0}(c_1^6) \tag{799}$$

$$h = -\frac{1}{6} - \frac{1}{57} c_1^2 - \frac{91}{20577} c_1^4 + \mathbf{0}(c_1^6) \tag{800}$$

Let us also recall the following property related to the "eleventh local integral" :

$$c \, . \, d\alpha + 3T \, . \, dh + 2h \, . \, dT = 0 \tag{801}$$

This property was given in (327) in Section 10.5.2 and using (795), (799), (800), it can easily be verified up to order four along the family of periodic orbits of interest with $T = 2\pi$ and $dT = 0$.

For small c_1 the orbits of A, B, C are almost circular, with the same inclination nearly equal to c_1 and with ascending nodes at 120° from each other (Fig. 78).

These orbits rotate slowly about the **Z**-axis because of the small "angle of rotation" α.

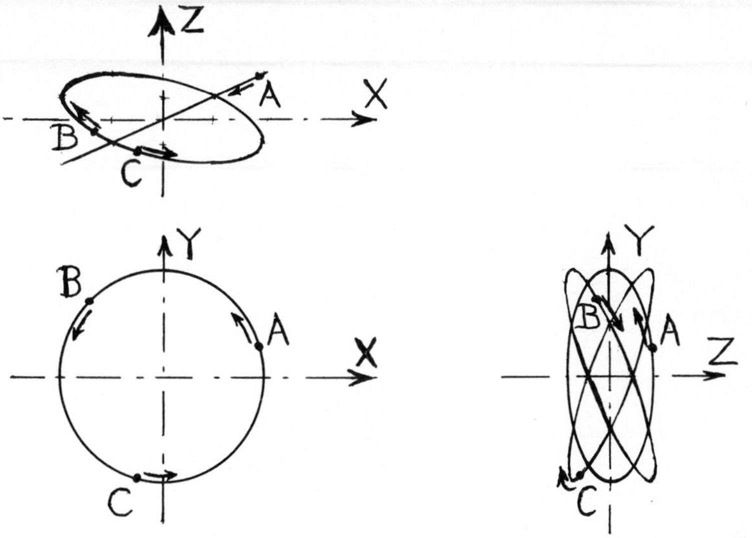

Fig. 78. Symmetric-periodic orbits with twelve space-time symmetries per period.

The three main projections of the orbits of A,B and C for small values of c_1 and t.

These almost circular orbits rotate very slowly about the Z-axis because of the small "angle of rotation".

For large c_1 the series (779)-(795) are of course less accurate and can no longer be used. However the periodic orbits of the family retain their twelve space-time symmetries per period and it would be interesting to look numerically for their evolution up to termination.

Notice that this termination is sometimes surprising as we will see in Section 10.9.1 for the family of retrograde pseudo-circular orbits that ends into rectilinear orbits !

10.8.3 The Halo orbits about the collinear Lagrangian points.

The Halo orbits are a family of simple periodic orbits of the circular restricted three-body problem. these orbits remain in the vicinity of a collinear Lagrangian point and are among the most useful orbits for many types of missions (Fig. 79). They have already been presented in Section 9.1 and in Fig. 24.

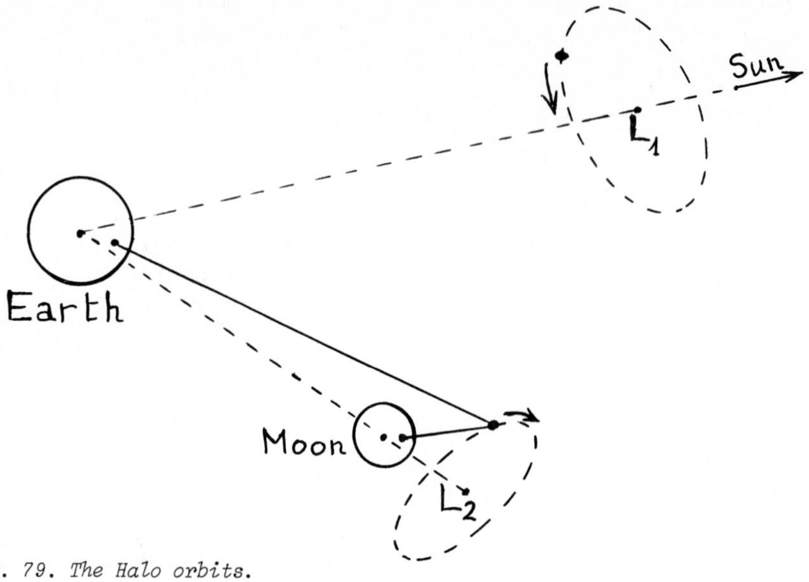

Fig. 79. The Halo orbits.

The Halo orbits about the Lagrangian point L_2 beyond the Moon can be used for continuous radio-communications with the lunar far side.

The Halo orbits about the point L_1 of the Sun-Earth system will be used for continuous observations of the Sun out of the Earth magnetosphere. The point L_1 itself will not be used : its radio-communications with Earth are perturbed by solar interferences.

The Halo orbits are three-dimensional periodic orbits and it is very often believed that their family begins at the Lagrangian point itself. However inifinitesimal motions in the vicinity of the Lagrangian point have slightly different periods in the in-plane and out-of-plane directions. Only sufficiently large orbits have the same two periods and are thus periodic.

Many people have studied the Halo orbits, either numerically or analytically (Ref. 68 - 74), and series giving these Halo orbits by a Lindstedt-Poincaré method can be obtained as follows.

Let us use the Hamiltonian system (439)-(441) of Section 10.7.4.4 and assume that :

A) The Lagrangian point L of interest is at the origin of the set of axes.

B) The axes and the unit of length are such that the nearest primary is at (-1,0,0) ; the other primary is along the **x**-axis at the point (s,0,0) (Fig. 80). The plane Lxy is the orbital plane of the primaries.

C) The units of mass and time are such that $G = 1$ and $\omega = 1$; the duration of the revolution of the primaries is then equal to 2π.

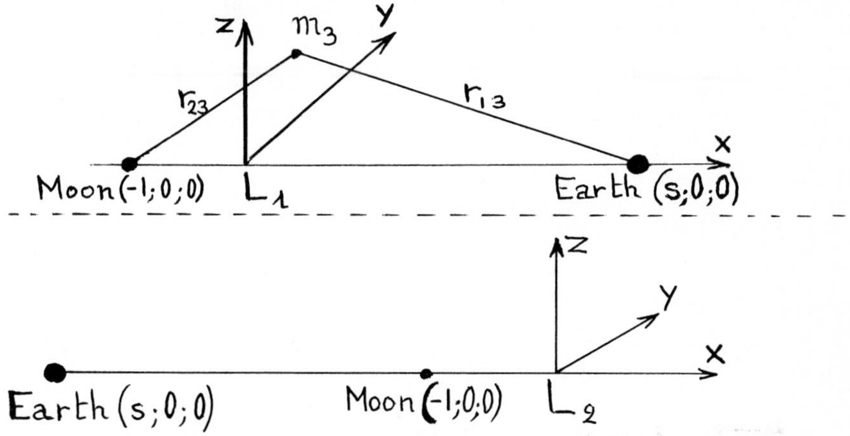

Fig. 80. The rotating set of axes of points L_1 and L_2 in the Earth-Moon System.

s is 5.625 404 in the first case and -6.958 314 in the second.

D) We will call m_2 the mass of the nearest primary (at (-1,0,0)) and m_1 the mass of the other primary (at (s,0,0)). In most interesting cases m_2 is smaller than m_1.

The Hamiltonian of the problem is then the following as easily deduced from (439) :

$$\left.\begin{aligned} H = \frac{1}{2}\left[(p_x-y)^2 + (p_y+x)^2 + p_z^2 + z^2\right] - \alpha\left[\frac{|s+1|^3}{r_{13}} + \frac{r_{13}^2}{2}\right] - \\ - \beta\left[\frac{|s+1|^3}{r_{23}} + \frac{r_{23}^2}{2}\right] \end{aligned}\right\} \qquad (802)$$

with :

$$\alpha = m_1/(m_1+m_2) \quad ; \quad \beta = m_2/(m_1+m_2) = 1 - \alpha \qquad (803)$$

$$r_{13}^2 = (x - s)^2 + y^2 + z^2 \quad ; \quad r_{23}^2 = (x + 1)^2 + y^2 + z^2 \qquad (804)$$

The Lagrangian equilibrium condition is :

$$m_1/m_2 = \alpha/\beta = \frac{s^2\{|s+1|^3 - 1\}}{3s^2 + 3s + 1} \qquad (805)$$

The developement of Hamiltonian H in the vicinity of the origin gives the following :

$$\left.\begin{aligned} H = H_0 &+ \tfrac{1}{2}(p_x^2+p_y^2+p_z^2) + y p_x - x p_y + K_2\left[-x^2 + \tfrac{1}{2}(y^2+z^2)\right] + \\ &+ K_3\left(x^3 - \tfrac{3}{2}x(y^2+z^2)\right) + K_4\left[-x^4+3x^2(y^2+z^2) - \tfrac{3}{8}(y^2+z^2)^2\right] + \ldots \end{aligned}\right\} \quad (806)$$

The constant H_0 is related to the value of H at the origin, it is useless in the following.

The factor of K_n is the homogeneous and harmonic polynomial of degree n only function of x and (y^2+z^2), whose x^n term is $-(-x)^n$.

The K_n are function of the mass ratio m_2/m_1 and are given by :

$$K_n = \frac{\alpha(-1)^n}{s^{(n+1)}}(s+1)^3 + \beta|s+1|^3 \quad (807)$$

Graphs of K_2, K_3, K_4 and s in terms of the mass ratios α and β are presented in Fig. 81 and 82.

The development (806) is convergent in the sphere $x^2+y^2+z^2 < 1$. The usual cases are as follows :

A) Earth-Moon system.

β = (mass of the Moon)/(total mass) = 0.01215057

$\alpha = 1 - \beta = 0.98784943$

A.1) Motions about L_1 :

unit of distance = mean distance L_1 - Moon = 58020 km

parameter s = 5.625404

$K_2 = 5.147594$; $K_3 = 3.246842$; $K_4 = 3.584729$

A.2) Motions about L_2 :

unit of distance = mean distance L_2 - Moon = 64515 km

parameter s = - 6.958314

$K_2 = 3.190426$; $K_3 = 2.659335$; $K_4 = 2.583011$

B) Sun-(Earth plus Moon) system.

β = (mass of (Earth plus Moon))/(total mass) = 1/328901.5

$\alpha = 1 - \beta$

B.1) Motions about L_1 :

unit of distance = mean distance L_1 - (Earth-Moon barycenter) = = 1 497 830 km

parameter s = 98.890341

$K_2 = 4.061074$; $K_3 = 3.020011$; $K_4 = 3.030538$

B2) - Motions about L_2

unit of distance = mean distance L_2-(Earth-Moon barycenter) = 1 507 894 km

parameter s = - 100.223 662

K_2 = 3.940 522 ; K_3 = 2.979 842 ; K_4 = 2.970 257

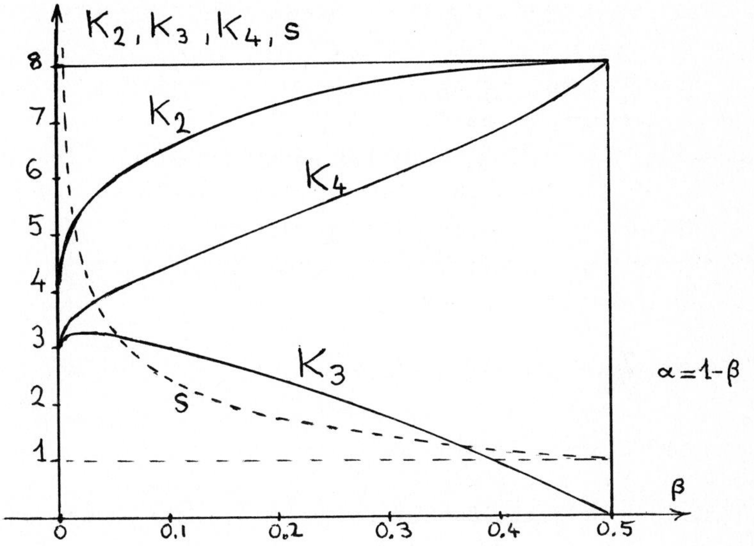

Fig. 81. Motions in the vicinity of L_1.
The parameters K_2, K_3, K_4, s in terms of the mass ratios α and β.

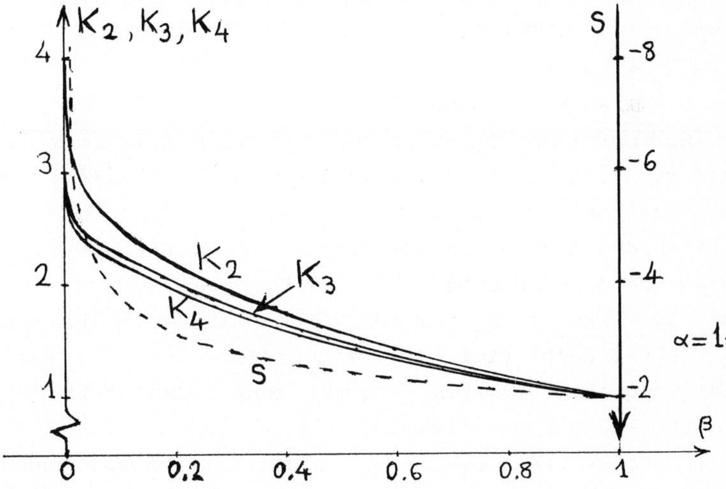

Fig. 82. Motions in the vicinity of L_2.
The parameters K_2, K_3, K_4, s in terms of the mass ratios α and β.

C) Hill case (equivalence between the points L_1 and L_2).
$\beta = 0$; $\alpha = 1$
parameter s : limit case when $s \to \pm \infty$; $\beta|s^3| \to 3$
$K_2 = 4$, the other $K_n = 3$

These five cases give very similar Hamiltonians in (806).

The first-order study of motions in the vicinity of L_1 or L_2 is classical and leads to the following with the six constants of integration g_1 to g_6 :

$$\begin{aligned} x = (1 - K_2 - \lambda^2)\{g_1\exp(\lambda t) + g_2\exp(-\lambda t)\} + \\ + (1 - K_2 + \mu^2)\{g_3\sin(\mu t) + g_4\cos(\mu t)\} \end{aligned} \quad (808)$$

$$\begin{aligned} y = 2\lambda\{g_1\exp(\lambda t) - g_2\exp(-\lambda t)\} + \\ + 2\mu\{g_3\cos(\mu t) - g_4\sin(\mu t)\} \end{aligned} \quad (809)$$

$$z = g_5\sin(\nu t) + g_6\cos(\nu t) \quad (810)$$

with :

$$\begin{aligned} \lambda &= \left\{[K_2 - 2 + \sqrt{(9K_2^2 - 8K_2)}]/2\right\}^{1/2} \\ \mu &= \left\{[2 - K_2 + \sqrt{(9K_2^2 - 8K_2)}]/2\right\}^{1/2} \\ \nu &= \sqrt{K_2} \end{aligned} \quad (811)$$

In all cases K_2 satisfies $1 < K_2 \leq 8$, λ is positive and the term $g_1\exp(\lambda t)$ represents an instability of the motion.

In-plane periodic motions are associated with g_3, g_4 and μ , while out-of-plane periodic motions are associated with g_5, g_6 and ν. The ratio μ/ν is always slightly larger than one $(1 < \mu/\nu \leq \sqrt{9/8})$ and the combination of in-plane and out-of-plane periodic motions gives usual quasi-periodic motions of the Lissajous type (Fig. 83).

In order to obtain the three-dimensional periodic motions of Halo orbits it is necessary to consider the upper order effects and to use sufficiently large orbits for which the in-plane and out-of-plane periods are equal.

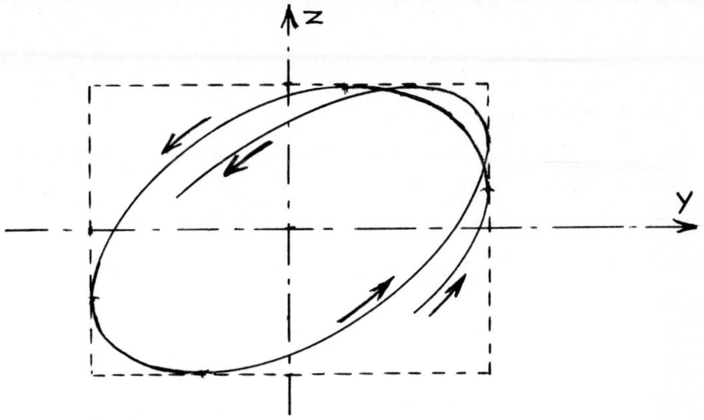

Fig. 83. A quasi-periodic solution of Lissajous type in the vicinity of the Lagrangian point L.

Usually the only Halo orbits that are considered are those with a symmetry about the x-z plane. Their Fourier development is the following :

$$\left.\begin{aligned} x &= a_0 + a_1\cos\theta + a_2\cos 2\theta + a_3\cos 3\theta + \ldots \\ y &= \phantom{a_0 +{}} b_1\sin\theta + b_2\sin 2\theta + b_3\sin 3\theta + \ldots \\ z &= c_0 + c_1\cos\theta + c_2\cos 2\theta + c_3\cos 3\theta + \ldots \end{aligned}\right\} \quad (812)$$

The angle θ is a linear function of the time t :

$$\theta = \theta_0 + kt \quad (813)$$

and then the period of the Halo orbit is the proportion $(1/k)$ of the period of the primaries.

If we choose a_1 as first-order quantity in (812) the parameters a_0 and c_0 are of the second order while the other a_n, b_n, c_n are of n-th order.

The usual Hamiltonian equations of motion are :

$$\left.\begin{aligned} &dx/dt = \partial H/\partial p_x \ ; \quad dy/dt = \partial H/\partial p_y \ ; \quad dz/dt = \partial H/\partial p_z \\ &dp_x/dt = -\,\partial H/\partial x \ ; \quad dp_y/dt = -\,\partial H/\partial y \ ; \quad dp_z/dt = -\partial H/\partial z \end{aligned}\right\} \quad (814)$$

and with (806) they give :

$$\left.\begin{aligned} d^2x/dt^2 &= 2y/dt + (1 + 2K_2)x + 1.5K_3(-2x^2 + y^2 + z^2) + \\ &\quad + K_4(4x^3 - 6xy^2 - 6xz^2) + \dots \\ d^2y/dt &= -2dx/dt + (1 - K_2)y + 3K_3xy + \\ &\quad + 1.5K_4(-4x^2y + y^3 + yz^2) + \dots \\ d^2z/dt^2 &= -K_2\mathbf{z} + 3K_3xz + 1.5K_4(-4x^2z + y^2z + z^3) + \dots \end{aligned}\right\} \quad (815)$$

Let us substitute in (815) the series (812) for x, y and z and let us neglect the fourth and higher order terms. We obtain :

$$\left.\begin{aligned} 0 &= \mathbf{k}^2(a_1\cos\theta + 4a_2\cos 2\theta + 9a_3\cos 3\theta) + 2\mathbf{k}(b_1\cos\theta + \\ &\quad + 2b_2\cos 2\theta + 3b_3\cos 3\theta) + (1 + 2K_2)(a_0 + a_1\cos\theta + \\ &\quad + a_2\cos 2\theta + a_3\cos 3\theta) + 1.5K_3(b_1^2\sin^2\theta + 2b_1b_2\sin\theta\sin 2\theta + \\ &\quad + c_1^2\cos^2\theta + 2c_0c_1\cos\theta + 2c_1c_2\cos\theta\cos 2\theta - 2a_1^2\cos^2\theta - \\ &\quad - 4a_1a_0\cos\theta - 4a_1a_2\cos\theta\cos 2\theta) + K_4(4a_1^3\cos^3\theta - \\ &\quad - 6a_1b_1^2\cos\theta\sin^2\theta - 6a_1c_1^2\cos^3\theta) \end{aligned}\right\} \quad (816)$$

and two other similar equations for d^2y/dt^2 and d^2z/dt^2.

The elimination of the angle θ then leads to the following eleven equations that are the third order equations of the Fourier series of the Halo orbits.

$$0 = (1 + 2K_2)a_0 + 0.75K_3(b_1^2 + c_1^2 - 2a_1^2) \quad (817)$$

$$\left.\begin{aligned} 0 &= \mathbf{k}^2a_1 + 2\mathbf{k}b_1 + (1 + 2K_2)a_1 + 1.5K_3(b_1b_2 + 2c_0c_1 + \\ &\quad + c_1c_2 - 4a_0a_1 - 2a_1a_2) + 1.5K_4(2a_1^3 - a_1b_1^2 - 3a_1c_1^2) \end{aligned}\right\} \quad (818)$$

$$0 = 4\mathbf{k}^2a_2 + 4\mathbf{k}b_2 + (1 + 2K_2)a_2 + 0.75K_3(c_1^2 - 2a_1^2 - b_1^2) \quad (819)$$

$$0 = 9k^2a_3 + 6kb_3 + (1 + 2K_2)a_3 + 1.5K_3(c_1c_2 - 2a_1a_2 - b_1b_2) + K_4(a_1^3 + 1.5a_1b_1^2 - 1.5a_1c_1^2) \tag{820}$$

$$0 = k^2b_1 + 2ka_1 + (1 - K_2)b_1 + 1.5K_3(2a_0b_1 - a_2b_1 + a_1b_2) + 0.375K_4(3b_1^3 + b_1c_1^2 - 4a_1^2b_1) \tag{821}$$

$$0 = 4k^2b_2 + 4ka_2 + (1 - K_2)b_2 + 1.5K_3a_1b_1 \tag{822}$$

$$0 = 9k^2b_3 + 6ka_3 + (1 - K_2)b_3 + 1.5K_3(a_1b_2 + a_2b_1) + 0.375K_4(b_1c_1^2 - 4a_1^2b_1 - b_1^3) \tag{823}$$

$$0 = - K_2c_0 + 1.5K_3a_1c_1 \tag{824}$$

$$0 = k^2c_1 - K_2c_1 + 1.5K_3(2a_0c_1 + 2a_1c_0 + a_1c_2 + a_2c_1) + 0.375K_4(b_1^2c_1 + 3c_1^3 - 12a_1^2c_1) \tag{825}$$

$$0 = 4k^2c_2 - K_2c_2 + 1.5K_3a_1c_1 \tag{826}$$

$$0 = 9k^2c_3 - K_2c_3 + 1.5K_3(a_1c_2 + a_2c_1) + 0.375K_4(c_1^3 - 4a_1^2c_1 - b_1^2c_1) \tag{827}$$

These eleven equations with twelve unknowns (k and a_0 to c_3) lead to a single parameters family of symmetrical Halo orbits.

This Lindstedt-Poincaré method was presented for Halo orbits in Ref. 70 with two minor simplifications (c_0 was zero and b_1/a_1 was fixed). However these simplifications were unfortunate because the eleven equations (817)-(827) can easily be solved exactly as now shown.

A) Determination of a_3, b_3, c_3 by (820), (823), (827) in terms of the other parameters.

$$a_3 = \{A(1 - K_2 + 9k^2) - 6Bk\}/C \tag{828}$$

$$b_3 = \{B(1 + 2K_2 + 9k^2) - 6Ak\}/C \tag{829}$$

$$c_3 = \{12K_3(a_1c_2+a_2c_1) + 3K_4(c_1^3-4a_1^2c_1-b_1^2c_1)\}/(8K_2-72k^2) \tag{830}$$

with :

$$A = 1.5K_3(2a_1a_2 + b_1b_2 - c_1c_2) + 0.5K_4(3a_1c_1^2 - 2a_1^3 - 3a_1b_1^2) \quad (831)$$

$$B = -1.5K_3(a_1b_2 + a_2b_1) + 0.375K_4(4a_1^2b_1 + b_1^3 - b_1c_1^2) \quad (832)$$

$$C = (1 - 9k^2)^2 + K_2(1 + 9k^2) - 2K_2^2 \quad (833)$$

B) Determination of a_0, c_0, a_2, b_2, c_2 by the equations (651), (653), (656), (658), (660) in terms of the parameters a_1, b_1, c_1 and the constants K_2, K_3, K_4, k.

$$a_0 = 3K_3(2a_1^2 - b_1^2 - c_1^2)/(4 + 8K_2) \quad (834)$$

$$c_0 = 3K_3a_1c_1/2K_2 \quad (835)$$

$$a_2 = \frac{3K_3}{4} \cdot \frac{(K_2 - 1 - 4k^2)(2a_1^2 + b_1^2 - c_1^2) - 8ka_1b_1}{2K_2^2 - K_2(1 + 4k^2) - (1 - 4k^2)^2} \quad (836)$$

$$b_2 = \frac{3K_3}{4} \cdot \frac{2k(2a_1^2 + b_1^2 - c_1^2) + (1 + 2K_2 + 4k^2)a_1b_1}{2K_2^2 - K_2(1 + 4k^2) - (1 - 4k^2)^2} \quad (837)$$

$$c_2 = 3K_3a_1c_1/(2K_2 - 8k^2) \quad (838)$$

C) The determination of c_1 in terms of k, a_1, b_1 (and the constants K_2, K_3, K_4) is given by the equation (825) in which a_0, c_0, a_2, c_2 have the values given by (834), (835), (836), (838). This leads to :

Either $c_1 = 0$ and hence $c_0 = c_2 = c_3 = 0$. (839)

These plane solutions are those of the Strömgren families a and c that are also the Matukuma family F (Fig. 84 and Ref. 104).

Or :

$$
\left.
\begin{aligned}
c_1^2 = {} & \frac{\frac{8}{9}(k^2 - K_2) + K_3^2\left[\frac{4a_1^2 - 2b_1^2}{1 + 2K_2} + \frac{4a_1^2}{K_2} + \frac{2a_1^2}{K_2 - 4k^2}\right]}{K_3^2\left[\frac{2}{1 + 2K_2} + \frac{1 - K_2 + 4k^2}{(1 - 4k^2)^2 + K_2(1 + 4k^2) - 2K_2^2}\right] - K_4} + \\
& + \frac{K_3^2\left[\frac{(1 - K_2 + 4k^2)(2a_1^2 + b_1^2) + 8ka_1b_1}{(1 - 4k^2)^2 + K_2(1 + 4k^2) - 2K_2^2}\right] + \frac{K_4}{3}(b_1^2 - 12a_1^2)}{K_3^2\left[\frac{2}{1 + 2K_2} + \frac{1 - K_2 + 4k^2}{(1 - 4k^2)^2 + K_2(1 + 4k^2) - 2K_2^2}\right] - K_4}.
\end{aligned}
\right\} \quad (840)
$$

D) It remains to obtain a_1 and b_1 in terms of the parameter k and the constants K_2, K_3, K_4 using the equations (818) and (821).

In these two equations (818) and (821) the six parameters a_0, a_2, b_1, c_0, c_1, c_2 can easily be expressed in terms of the parameters k, a_1, b_1 and the constants K_2, K_3, K_4 with (834)-(838) and with either (839) or (840).

In these conditions (818) becomes :

$$F_0a_1 + F_1b_1 + F_2a_1^3 + F_3a_1^2b_1 + F_4a_1b_1^2 + F_5b_1^3 = 0 \tag{841}$$

and (821) becomes :

$$G_0a_1 + G_1b_1 + G_2a_1^3 + G_3a_1^2b_1 + G_4a_1b_1^2 + G_5b_1^3 = 0 \tag{842}$$

The twelve functions F_0 to G_5 are rational fractions of k, K_2, K_3, K_4, for instance if c_1 is given by (840) then:

$$F_1 = G_0 = 2k + \frac{4kK_3^2(K_2 - k^2)}{K_3^2(K_2 - 1 - 4k^2) + \left[(1-4k^2)^2 + K_2(1+4k^2) - 2K_2^2\right]\cdot\left[K_4 - \frac{2K_3^2}{1+2K_2}\right]} \tag{843}$$

The complete expressions for F_0 to G_5 are given in the appendix 2.

The resolution of the equations (841), (842) for given k, K_2, K_3, K_4 is now easy.

A first solution is :

$$a_1 = b_1 = 0 \tag{844}$$

and hence : $a_3 = b_3 = c_0 = c_2 = 0$.

These solutions with four space-time symmetries per period are the eight-shaped solutions of Zagouras and Markellos (Fig. 84 and Ref. 72), they cross the x-axis twice per period when $\cos\theta = 0$.

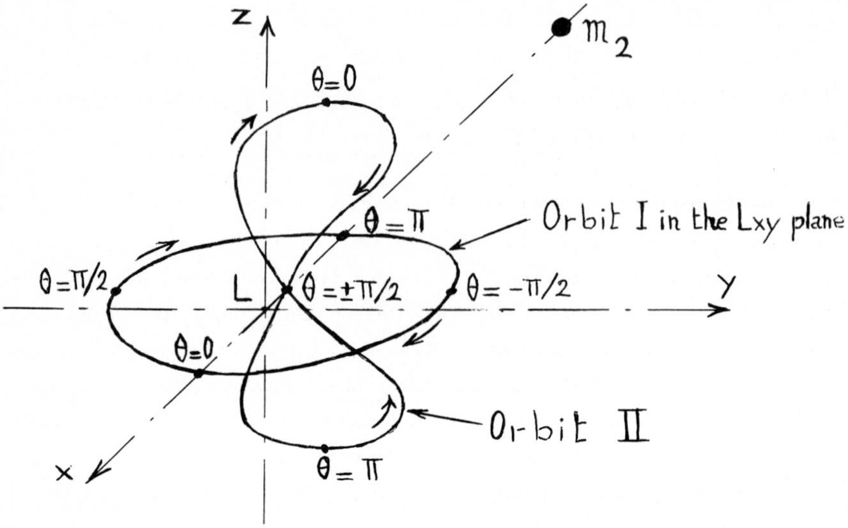

Fig. 84. Some periodic orbits in the vicinity of the Lagrangian point L.
Orbit I : plane periodic orbit of the Matukuma family F (i.e. Strömgren families a and c).
Orbit II : eight-shaped periodic orbit of Zagouras and Markellos.

For the other solutions let us put :

$$R = b_1/a_1 \tag{845}$$

R is a root of the following fourth-degree equation :

$$\left.\begin{aligned}(F_0 + F_1R)(G_2 + G_3R + G_4R^2 + G_5R^3) &= \\ = (G_0 + G_1R)(F_2 + F_3R + F_4R^2 + F_5R^3)&\end{aligned}\right\} \tag{846}$$

and this leads finally to :

$$a_1^2 = \frac{-(F_0 + F_1R)}{F_2 + F_3R + F_4R^2 + F_5R^3} \tag{847}$$

$$b_1 = Ra_1 \tag{848}$$

For usual Halo orbits R is the root of (846) in the vicinity of (- 3).

Thus the equations (817)-(827) of the third-order Lindstedt-Poincaré method have a very laborious but also straightforward solution (corresponding numerical program modules in fortran can be found in the Reference 73).

The simplest parameter of description of a family of Halo orbits is the angular velocity k of the angle θ. For given constants K_2, K_3, K_4 and a chosen k the fourth-degree equation (846) gives the ratio R and then the Fourier coefficients a_0 to c_3 of the development (812) are successively and directly given by (847), (848) for a_1, b_1, (840) for c_1, (834)-(838) for a_0, c_0, a_2, b_2, c_2 and finally (828)-(833) for a_3, b_3, c_3.

The most interesting parameter of description is not k which has only very small variations, it is c_1 that gives the width of the Halo orbit. The results are presented in Figs 85 to 88 for the four most interesting cases : the Hill case, the Halo orbits about the points L_1 or L_2 of the Sun-Earth system and the Halo orbits beyond the Moon that were presented in Fig. 79.

The two symmetries $(c_n/-c_n)$ and $(a_{2n+1}, b_{2n+1}, c_{2n+1}/-a_{2n+1}, -b_{2n+1}, -c_{2n+1})$ allow to consider only positive a_1 and c_1.

Figs. 89 and 90, as well as Fig. 24 of Section 9.1, give an idea of the families of Halo orbits ; their main features are as follows.

A) These orbits are near-ellipses with a major axis parallel to the y-axis.

B) The y-amplitude is always large and $|b_1|$ is always larger than 0.4 while b_1/a_1 remains in the vicinity of (- 3). On the contrary there is a broad choice of z-amplitude from zero to large positive or negative values.

C) The families of Halo orbits depend on the ratio m_2/m_1 of the masses of the primaries, but the influence of this ratio remains small and all families are qualitatively similar.

D) The smallest orbit of each family is a symmetrical plane periodic orbit that belongs also to the Matukuma-Strömgren families of Fig. 84 (bifurcation orbit).

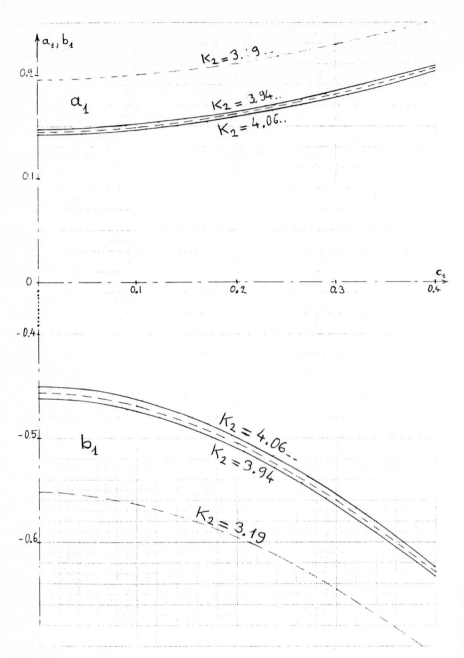

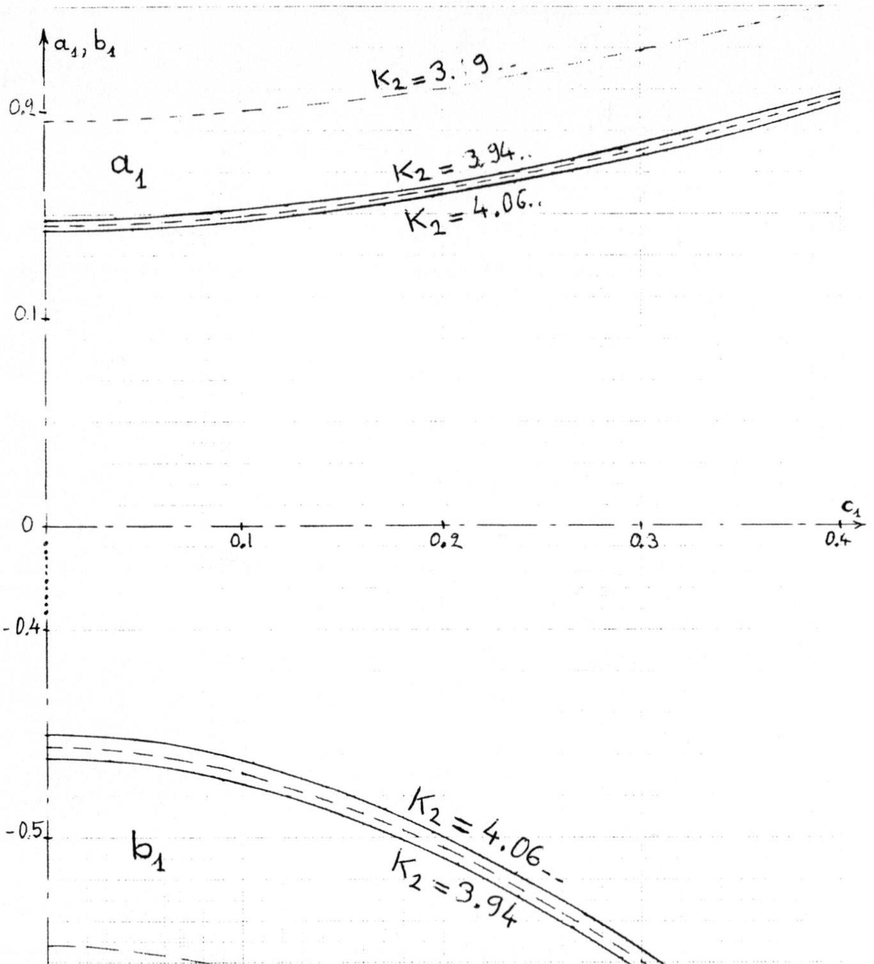

Fig. 85. The parameters a_1 and b_1 in terms of c_1 for the four following values of K_2.

K_2 = 3.190426 (point L_2 of the Earth-Moon system, beyond the Moon)
K_2 = 3.940522 (point L_2 of the Sun-(Earth plus Moon) system)
K_2 = 4 (points L_1 and L_2 of the Hill problem)
K_2 = 4.061074 (point L_1 of the Sun-(Earth plus Moon) system).

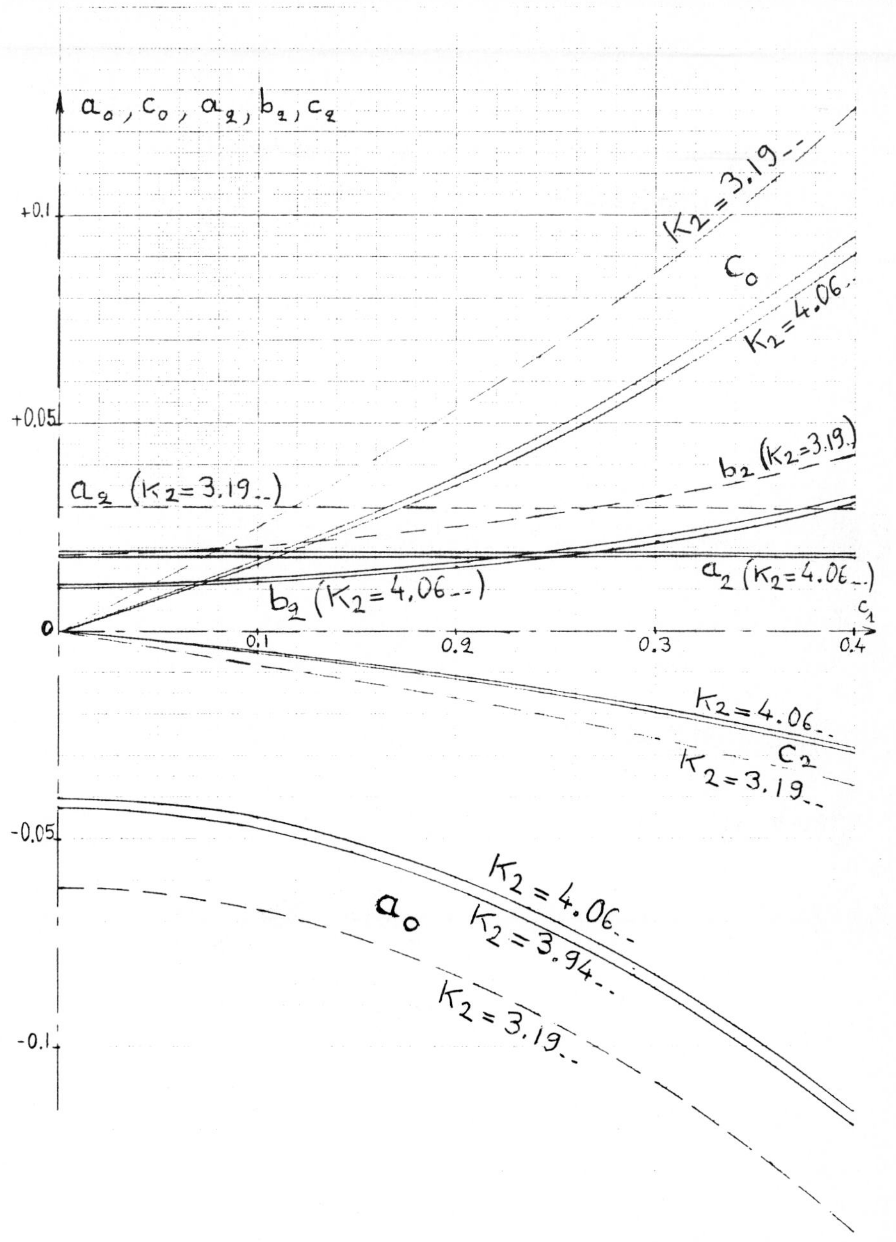

Fig. 86. The parameters a_0, c_0, a_2, b_2, c_2 in terms of c_1 for the three following values of K_2 :

K_2 = 3.190426 (point L_2 of the Earth-Moon system, beyond the Moon)

K_2 = 3.940522 (point L_2 of the Sun-(Earth plus Moon) system)

K_2 = 4.061074 (point L_1 of the Sun-(Earth plus Moon) system).

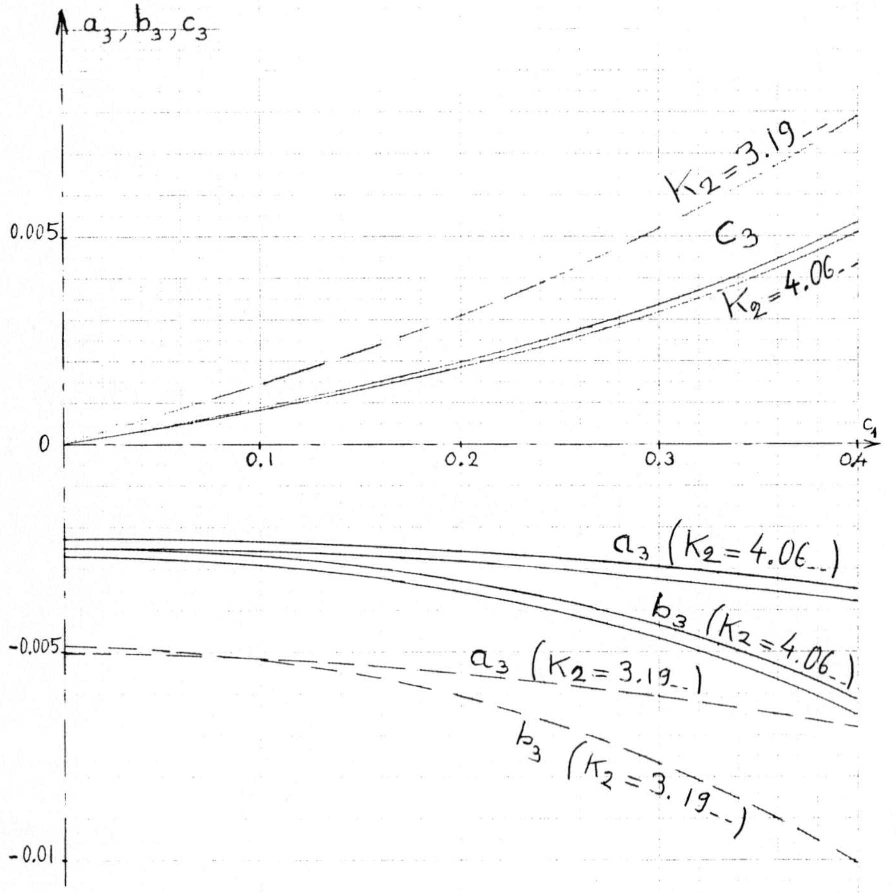

Fig. 87. The parameters a_3, b_3, c_3 in terms of c_1 for the three following values of K_2.

K_2 = 3.190426 (point L_2 of the Earth-Moon system, beyond the Moon)

K_2 = 3.940522 (point L_2 of the Sun-(Earth plus Moon) system)

K_2 = 4.061074 (point L_1 of the Sun-(Earth plus Moon) system).

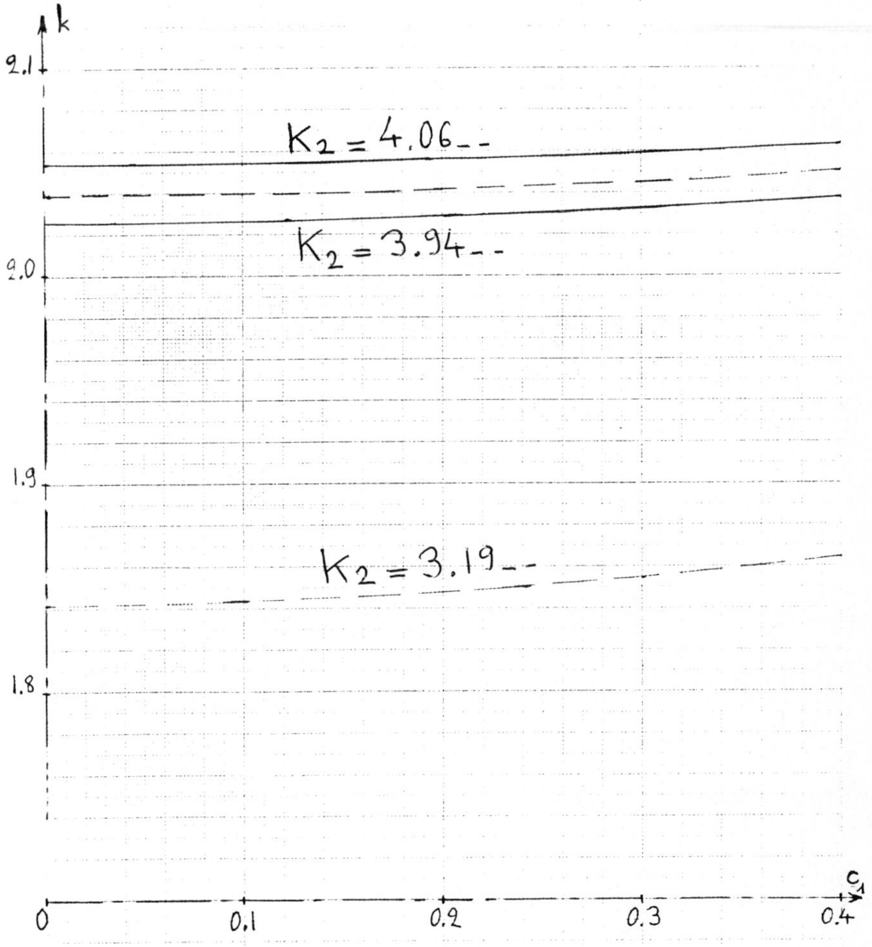

*Fig. 88. The parameter **k** in terms of c_1 for the four following values of K_2:*

K_2 = 3.190426 (point L_2 of the Earth-Moon system, beyond the Moon)
K_2 = 3.940522 (point L_2 of the Sun-(Earth plus Moon) system)
K_2 = 4 (points L_1 and L_2 of the Hill problem)
K_2 = 4.061074 (point L_1 of the Sun-(Earth plus Moon) system)

The period of the Halo orbit is the proportion (1/k) of the period of the primaries.

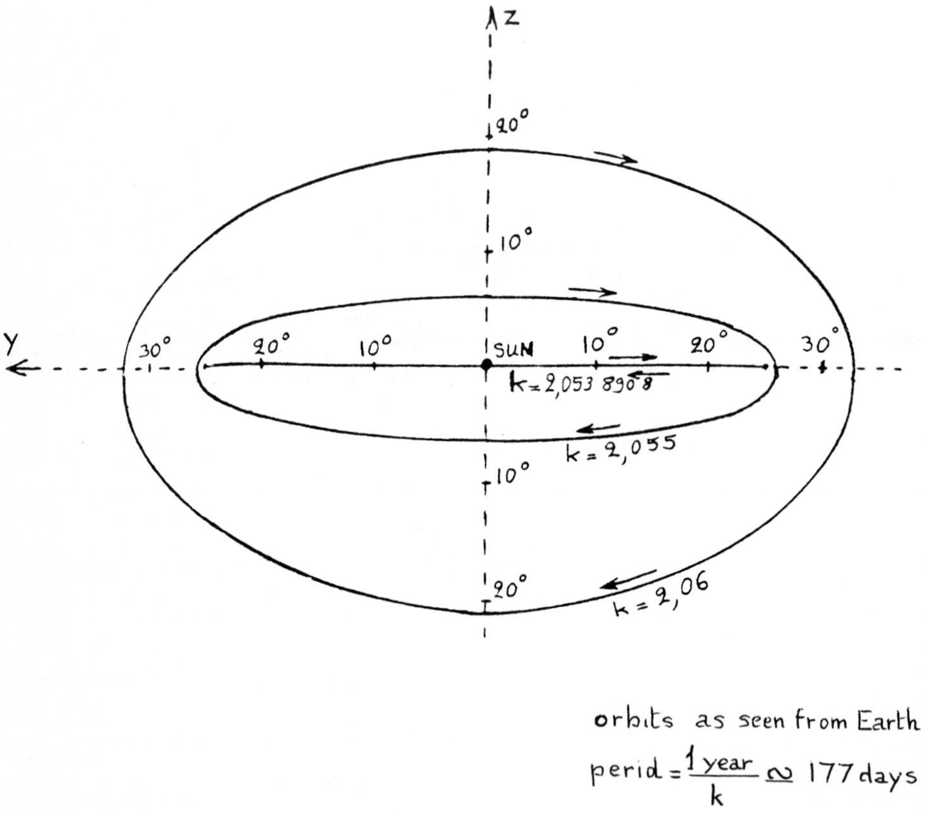

Fig. 89. Some Halo orbits in the vicinity of the Lagrangian point L_1 of the Sun-(Earth plus Moon) system.

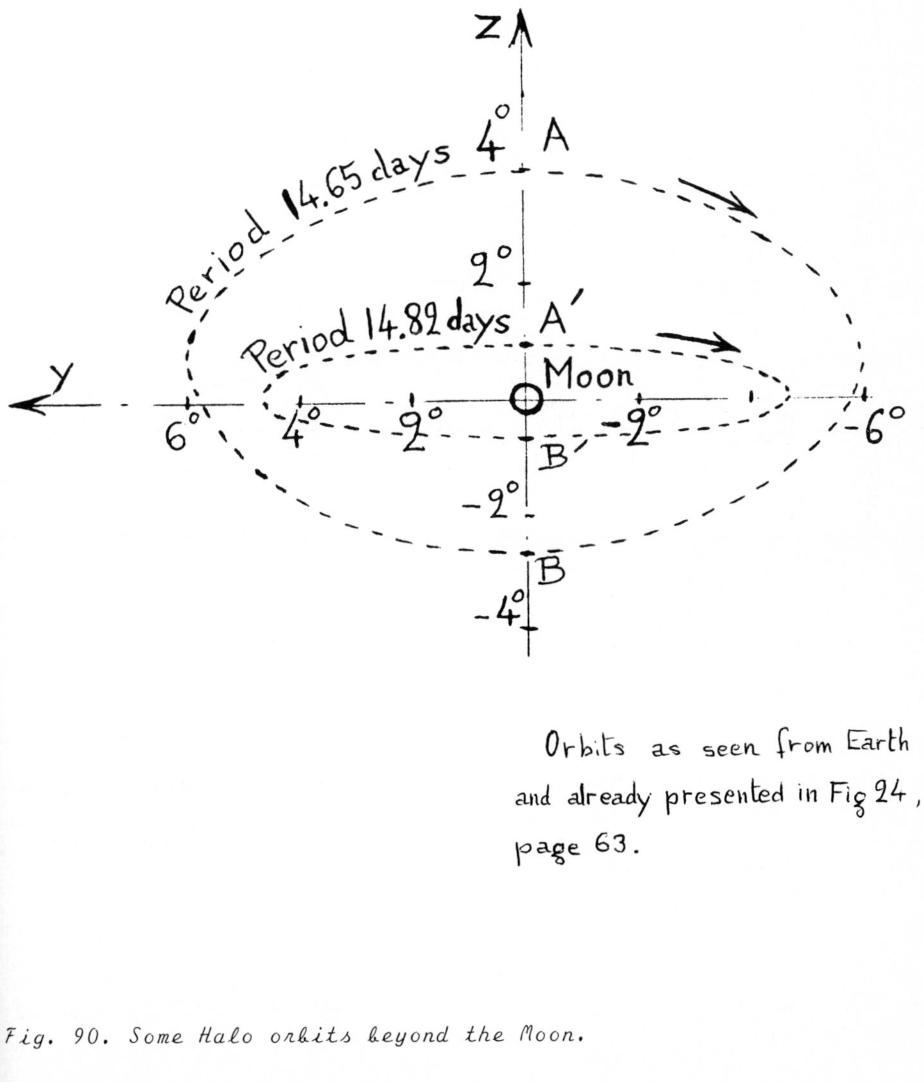

Fig. 90. Some Halo orbits beyond the Moon.

E) The numerical computations [Appendix 2 and Ref. 71-74, 104] show that the third-order method is excellent for the small orbits of Fig. 84 but its relative accuracy is only about 1 or 2% for the large Halo orbits. Hence for a given mission, after the choice of a suitable Halo orbit, an accurate numerical analysis will be necessary.

Let us add some practical considerations.

A) The Halo orbits undergo the influence of many sources of perturbations (solar or lunar effects, radiation pressure, etc...). Fortunately, the resulting perturbations remain small (Ref. 68,73)

B) The major source of perturbations is the ellipticity of the orbit of the primaries. It destroys the periodic character of Halo orbits, fortunately the resulting quasi-periodic orbits are simple and very near to the corresponding periodic orbits of the circular restricted three-body problem (Ref. 73).

C) All these periodic or quasi-periodic orbits are unstable and require a station-keeping procedure similar to those of the geostationary satellites (Ref 74).

Modern station-keeping procedures are accurate and requires very little propellant.

The three examples developed in this Section 10.8. are good illustrations of the interest of the analytical methods, which are usually less accurate than the numerical methods but give broader view of the set of solutions.

The symmetrical periodic orbits with the largest number of symmetries are given in terms of one non-trivial parameter (the parameter c_1), the Halo orbits are given in terms of two non-trivial parameters (m_2/m_1 and c_1) and the pseudo-circular orbits are given in terms of three non-trivial parameters (the two mass ratios and a_i/a_e). Numerically the same studies would be relatively easy for the one-parameter family, they would become more difficult for the two-parameter family and a clear view showing the influence of each parameter is numerically almost impossible for families with more than two non-trivial parameters.

Furthermore adjacent solutions as those of Fig. 84 appear during the analysis and their relationship with Halo orbits would of course be very difficult to guess numerically. It must also be noted that these adjacent solutions were known numerically only in the Hill case.

10.9 EXAMPLES OF NUMERICAL INTEGRATIONS

We have already met many numerical solutions of the three-body problem (Sections 8.5 ; 9.4 ; 10.4) and a general picture of the progress of numerical methods has been outlined in Sections 10.4 ; 10.5.

The numerical methods were initially a help to the analytical methods in providing a more accurate knowledge of solutions whose existence was already known.

The next step was the research by continuity, the best example being the construction of periodic orbits with their branchings, bifurcations, sub-families, etc...

Then came the "numerical experiments" i.e. the integration of a given arbitrary problem over very long intervals in order to see what happens. This allows us to verify the existence of given types of motion, to confirm or invalidate conjectures or even to find out new ideas, new types of motion, new conjectures.

To-day the most fruitful numerical method is the method of exploration that uses series of numerical experiments. This has led to fundamental results such as the discoveries of the importance of cahotic motions (see Section 9.4), strange attractors, fractal, gravitational scattering (Section 10.4.3), etc...

10.9.1 Researches by continuity. The retrograde pseudo-circular orbits of the three-body problem with three equal masses.

The pseudo-circular orbits have been studied numerically in many cases, for instance in the Hill case (Fig. 18 and 20, Section 8.5), in the restricted case (see the dotted orbit of Fig. 27, Sction 9.4) and also in the case of three equal masses (prograde motions in ref 184 , retrograde motions in ref. 133 , and both motions in ref. 134).

The most beautiful surprise appeared in the ref. 133 , where the initial orbit was a periodic and rectilinear Schubart orbit with two collisions per period (Fig. 91 and Ref. 135).

This Schubart orbit has first-order stability and belongs to a family of plane periodic orbits and Michel Hénon was very surprised to discover that the other end of the family was made of retrograde pseudo-circular orbits, the final limit corresponding to $a_i/a_e \to 0$ (Fig. 91, 92, 93).

Some informations summarizing the references 133, 134 and 184 are given in Table I.

Three accompanying points must be added.

A) There is a similar result for unequal masses and the families of retrograde pseudo-circular orbits end in (or begin by) periodic rectilinear Schubart orbits with two collisions per period, i.e. the orbits of Fig. 63 and 64 (Section 10.7.4.1).

Notice that the central mass of the Schubart motion becomes by continuity the outer mass of the pseudo-circular motion (Figs. 91-93).

B) Another surprise was the complexity of the family of prograde pseudo-circular orbits which contrasts with the simplicity of retrograde pseudo-circular orbits. In spite of many integrations the end of the family has not been found (Ref. 134, 184).

C) It is in this study that Michel Hénon has noticed the correlations between the maximums or minimums of the angular momentum c and the "critical orbits" of the family, i.e. the orbits at the limit of linear stability.

These maximums and minimums of c are obtained for constant energy and masses, for all orbits of the family, and the correspondence can be explained by the equation (317) of Section 10.5.2 i.e. by :

$$\vec{\delta}_0 = (I - A)^{-1} \cdot \vec{B} \cdot \delta c \tag{849}$$

The variation $\vec{\delta}_0$ of the Delaunay's elements along the family of periodic orbits is given by the expression (849) in terms of the variation δc of c and thus c cannot be an extremum for regular orbits i.e. when $(I - A)$ is invertible, that is when all eigenvalues of A are different from one.

Hence all extremums of c correspond to singular orbits, at which the branching of other families of periodic orbits arise. We have seen in Section 10.7.6 and in Fig. 70 that these singular orbits with Floquet multipliers equal to one are also very often "critical orbits" at the limit of linear stability, as they are in the Hénon family of Figs. 91-93.

Let us however recall that these two notions can be independent as shown by Fig. 70.

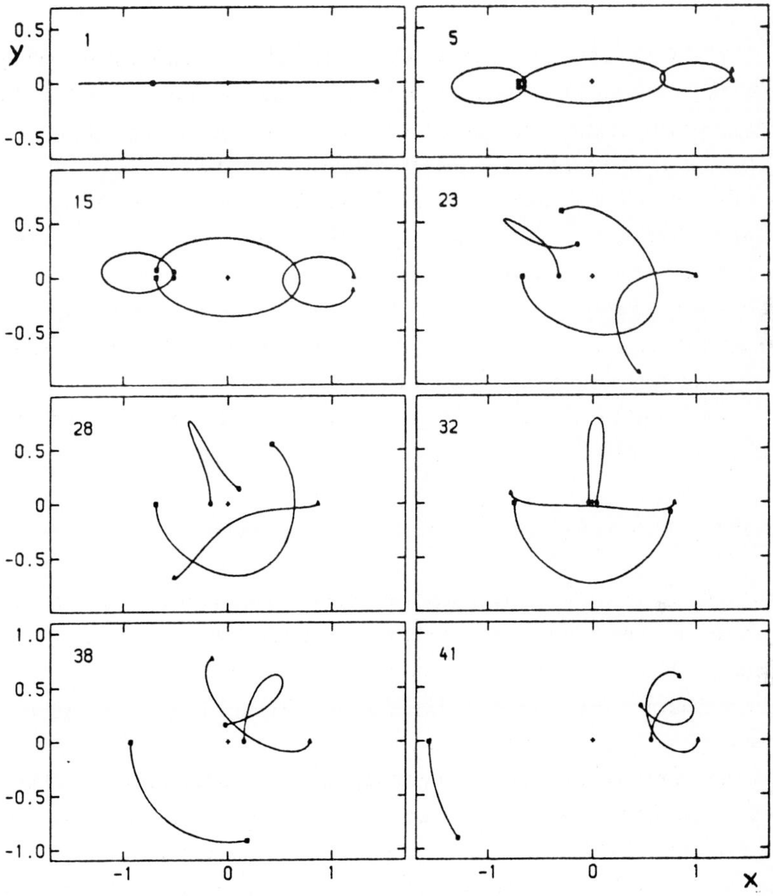

Fig. 91. Some orbits of the family of retrograde pseudo-circular orbits (three equal masses). Source : Hénon, M. [133].

The axes are the non-rotating axes of the center of mass and a full period is represented.

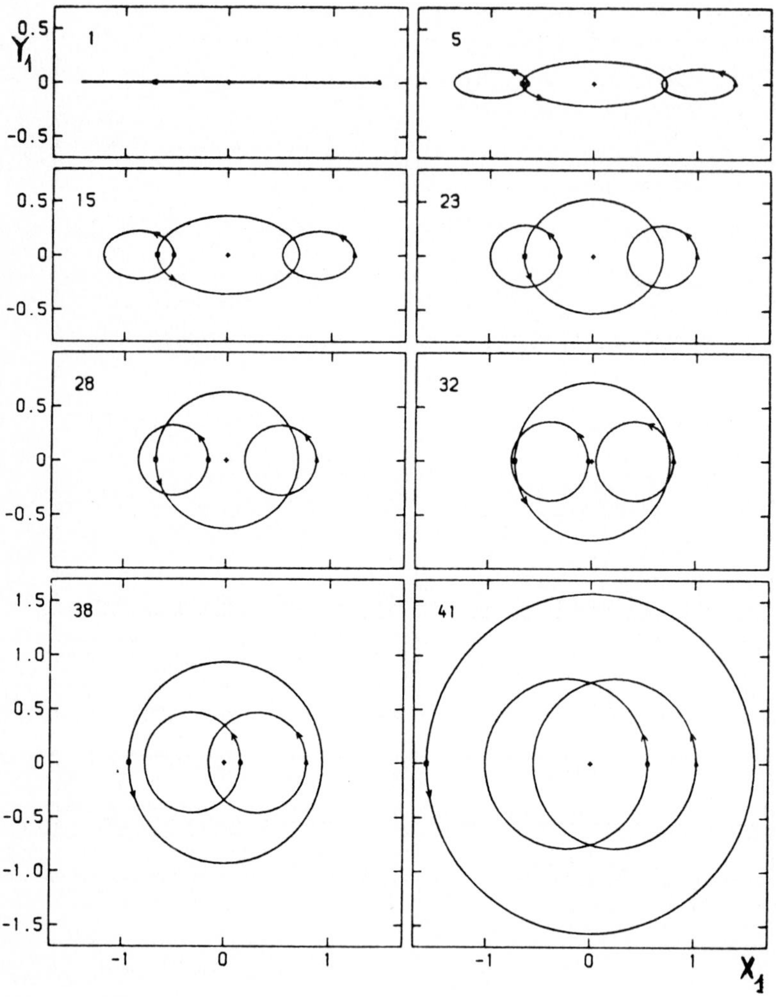

Fig. 92. The orbits of Fig. 91 in a uniformly rotating set of axes.
The rate of rotation corresponds to the "angle of rotation" and is zero for the rectilinear Schubart orbit. Source : Hénon, M. [133].

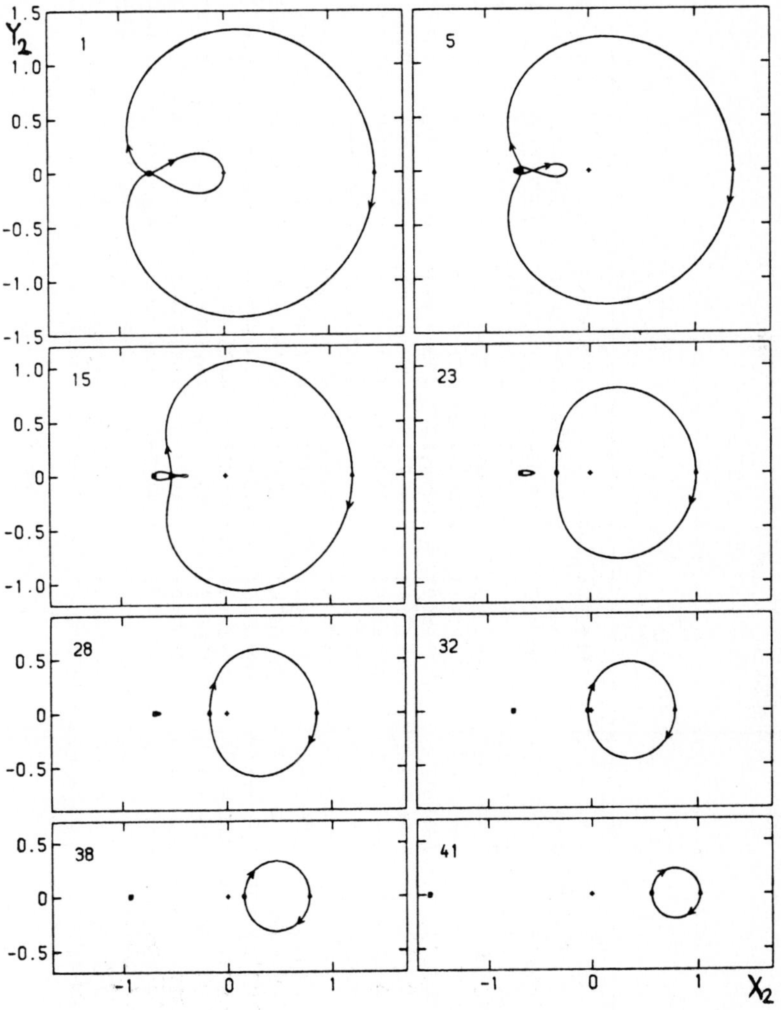

Fig. 93. The orbits of Figs. 91 and 92 in another uniformly rotating set of axes. Source : Hénon, M. [133].

The angle of rotation per period is 2π less than in Fig. 92.

Because of the symmetries two bodies have the same large orbit.

TABLE I. : Hénon's numerical results for the orbits of Figs. 91-93 and for some other orbits of the family of pseudo-circular retrograd orbits with three equal masses ($m_1 = m_2 = m_3 = 1/3$). The energy integral is always $h = -1/6$. Source : Hénon, M. [133].

Orbit number	C	x_1	v_1	x_2	v_2	T (period)	α (angle of rotation)	k_1	k_2
1	0.	−0.717 560 83	∞	−0.717 560 83	−∞	4.698 083 80	0.	−1.988 976	0.062 601
2	0.05	−0.707 716 75	6.328 009 73	−0.715 942 47	−6.373 788 14	4.700 788 51	0.036 312 27	−1.966 584	0.095 367
3	0.10	−0.675 216 05	2.986 507 53	−0.710 740 24	−3.078 989 56	4.707 639 64	0.067 383 12	−1.797 152	0.182 801
4	0.12	−0.653 074 33	2.385 677 68	−0.707 335 46	−2.497 176 15	4.709 672 73	0.073 651 42	−1.585 142	0.224 118
5	0.123 378 55	−0.648 558 04	2.299 666 66	−0.706 656 20	−2.414 382 52	4.709 739 12	0.073 834 10	−1.530 925	0.230 658
6	0.13	−0.638 850 75	2.140 088 21	−0.705 214 50	−2.261 094 50	4.709 413 09	0.072 984 16	−1.400 727	0.242 462
7	0.14	−0.621 388 22	1.915 512 98	−0.702 685 97	−2.045 904 87	4.706 991 47	0.067 049 02	−1.116 143	0.255 641
8	0.15	−0.598 417 94	1.698 383 01	−0.699 491 39	−1.837 733 41	4.699 486 05	0.049 891 95	−0.632 671	0.256 177
9	0.155	−0.582 882 12	1.584 329 26	−0.697 416 08	−1.727 697 04	4.691 102 21	0.031 595 61	−0.228 560	0.245 027
10	0.158 614 12	−0.567 855 11	1.491 823 13	−0.695 472 38	−1.637 555 21	4.680 100 18	0.008 233 45	0.225 044	0.225 044
11	0.159 484 57	−0.563 306 30	1.466 654 18	−0.694 895 89	−1.612 810 37	4.676 171 28	0.	0.374 696	0.217 192
12	0.16	−0.560 329 27	1.450 816 33	−0.694 521 40	−1.597 177 23	4.673 444 46	−0.005 689 85	0.475 734	0.211 599
13	0.162	−0.544 995 64	1.376 384 39	−0.692 627 12	−1.522 895 13	4.657 406 99	−0.038 868 46	1.034 184	0.177 183
14	0.163	−0.528 533 51	1.307 922 52	−0.690 652 38	−1.452 940 59	4.636 415 08	−0.081 899 78	1.700 334	0.130 125
15	0.163 092 67	−0.521 581 13	1.282 039 76	−0.689 834 94	−1.425 929 19	4.626 381 17	−0.102 411 16	2.	0.107 296
16	0.163	−0.514 500 99	1.257 323 92	−0.689 011 77	−1.399 770 38	4.615 466 37	−0.124 723 44	2.314 724	0.082 365
17	0.162	−0.496 638 59	1.201 531 74	−0.686 974 01	−1.339 085 36	4.584 950 97	−0.187 290 24	3.141 449	0.012 634
18	0.16	−0.478 383 06	1.152 869 53	−0.684 947 23	−1.283 725 66	4.549 762 30	−0.260 120 16	4.010 024	−0.067 174
19	0.155	−0.447 766 03	1.086 414 82	−0.681 692 75	−1.202 837 18	4.483 471 90	−0.400 347 25	5.417 933	−0.214 794
20	0.15	−0.422 945 39	1.043 465 06	−0.679 240 85	−1.145 947 30	4.425 000 22	−0.528 145 37	6.405 341	−0.342 288
21	0.14	−0.376 725 42	0.982 246 60	−0.675 430 62	−1.055 526 77	4.310 221 43	−0.792 162 67	7.579 348	−0.588 814

22	0.135	−0.352 673 26	0.957 526 21	−0.674 034 19	−1.014 729 78	4.249 185 92	−0.940 175 59	7.763 579	−0.720 133
23	0.13	−0.326 401 91	0.934 675 28	−0.673 144 89	−0.974 077 58	4.182 235 69	−1.108 685 36	7.615 196	−0.865 884
24	0.125	−0.295 878 95	0.912 470 77	−0.673 151 72	−0.931 195 63	4.104 351 41	−1.312 458 19	7.019 772	−1.038 365
25	0.12	−0.255 884 00	0.888 907 04	−0.675 262 14	−0.880 946 31	4.002 058 00	−1.591 201 75	5.692 509	−1.267 800
26	0.118	−0.233 130 60	0.877 758 83	−0.677 726 93	−0.854 940 81	3.943 533 06	−1.755 214 10	4.752 062	−1.397 642
27	0.116	−0.193 116 68	0.861 379 31	−0.684 668 05	−0.813 114 05	3.839 475 95	−2.052 095 54	2.969 901	−1.615 567
28	0.115 713 76	−0.170 963 74	0.853 831 43	−0.690 107 61	−0.791 865 62	3.780 993 49	−2.220 426 34	2.	−1.724 356
29	0.116	−0.148 550 84	0.847 156 59	−0.696 883 57	−0.771 605 58	3.721 042 48	−2.392 982 36	1.086 557	−1.820 454
30	0.118	−0.106 719 97	0.836 982 46	−0.713 236 16	−0.736 771 02	3.606 824 85	−2.718 812 58	−0.319 573	−1.947 506
31	0.12	−0.082 151 62	0.832 233 39	−0.725 241 76	−0.717 919 76	3.538 330 45	−2.910 752 03	−0.914 802	−1.984 752
32	0.125	−0.037 658 98	0.825 617 24	−0.751 774 38	−0.686 384 88	3.412 088 65	−3.254 684 18	−1.495 139	−1.998 805
33	0.13	−0.002 769 91	0.821 965 99	−0.777 026 87	−0.663 686 74	3.311 924 41	−3.516 728 52	−1.525 269	−1.970 533
34	0.135	0.027 607 80	0.819 704 31	−0.802 220 47	−0.645 168 52	3.224 734 42	−3.736 179 69	−1.376 257	−1.850 643
35	0.14	0.055 313 37	0.818 264 36	−0.827 764 02	−0.629 170 51	3.145 936 52	−3.927 277 77	−1.170 591	−1.648 703
36	0.145	0.081 260 11	0.817 362 75	−0.853 844 84	−0.614 882 45	3.073 338 98	−4.097 151 42	−0.943 993	−1.399 690
37	0.15	0.105 980 53	0.816 832 88	−0.880 562 60	−0.601 842 09	3.005 711 87	−4.250 024 60	−0.711 918	−1.130 193
38	0.16	0.153 007 80	0.816 488 79	−0.936 116 92	−0.578 440 90	2.882 552 58	−4.515 146 02	−0.262 848	−0.592 379
39	0.18	0.242 172 71	0.817 160 07	−1.056 305 60	−0.538 581 21	2.674 161 03	−4.925 096 98	0.484 762	0.303 561
40	0.20	0.329 490 34	0.818 243 51	−1.188 922 24	−0.504 610 61	2.505 242 79	−5.222 253 14	1.005 753	0.910 823
41	0.25	0.555 308 40	0.819 269 71	−1.574 218 98	−0.435 695 63	2.204 721 73	−5.673 359 08	1.644 458	1.624 717
42	0.3	0.803 874 70	0.817 314 45	−2.033 949 89	−0.382 454 11	2.016 097 91	−6.093 675 02	1.859 07	1.854 22
43	0.4	1.390 215 10	0.808 870 76	−3.168 086 74	−0.306 020 38	1.806 574 23	−6.108 685 63	1.9697	1.9692
44	0.5	2.105 927 47	0.799 195 32	−4.580 982 64	−0.254 392 79	1.700 090 25	−6.188 776 03	1.9911	1.9911
45	0.6	2.954 838 48	0.790 324 79	−6.268 440 65	−0.217 442 13	1.638 746 48	−6.226 339 96	1.9968	1.9968
46	0.7	3.938 098 29	0.782 642 84	−8.228 559 03	−0.189 773 16	1.600 148 59	−6.246 281 52	1.9986	1.9986
...									
	∞	∞	0.707 106 78	−∞	0.	1.480 960 98	−6.283 185 31	2.	2.

The initial coordinates and velocity components of m_1, m_2, m_3 are (x_1, y_1), (x_2, y_2), (x_3, y_3) and (u_1, v_1), (u_2, v_2), (u_3, v_3) with $x_3 = -x_1 - x_2$; $y_1 = y_2 = y_3 = 0$; $u_1 = u_2 = u_3 = 0$; $v_3 = -v_1 - v_2$. The Hénon's stability coefficient k_1 and k_2 give the plane first order stability. The orbit is first order stable iff k_1 and k_2 are real and between -2 and +2. The relation (327) of section 10.5.2 gives here along the family $c.d\alpha + 2hdT = 0$; also notice that the extremum of c corresponds to $k_1 = 2$.

10.9.2 A numerical experiment. The Pythagorean problem.

In 1913 Burrau (Ref 185) considered the Pythagorean problem, the initial conditions of which are given in Fig. 95, and he tried to integrate it. However his integration was too short for definitive conclusions.

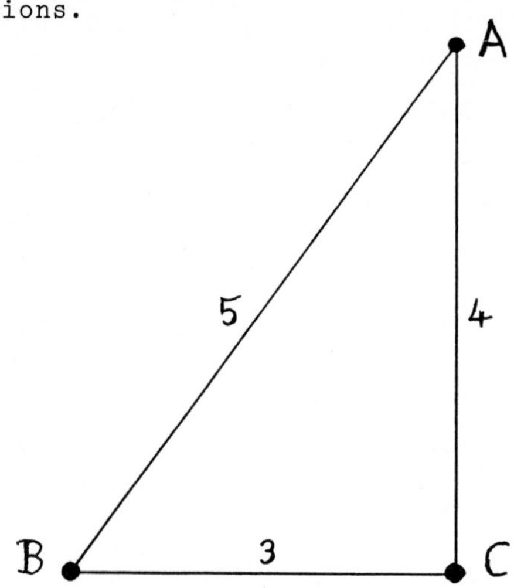

Fig. 95. The initial conditions of the Pythagorean problem. The three masses $m_A = 3$, $m_B = 4$ and $m_C = 5$ are intially motionless at the corners of a Pythagorean triangle 3, 4, 5.

After some theoretical studies (Ref.186 , 187) and several attempts (Ref.188-190) the problem was finally integrated in the Ref. 6 and leads to an hyperbolic-elliptic final motion, the smallest body escaping alone in the first quadrant while the bodies B and C escape together into the opposite direction (Fig. 102).

Figs. 96 to 102 present the motion in the non-rotating axes of the center of masses until the time t = 70 (with a constant G of the law of universal attraction equal to one). The orbit of A is the dotted line, the orbit of B the dashed line and the orbit of C the solid line.

The body C experiences many very close approaches with each of the other two bodies and hence the integration is extremely sensitive to small errors. This explains why the Pythagorean problem has required new integration methods and this is of course one of the interests of numerical experiments.

Further studies (Ref. 93) have shown that motions in the vicinity of the Pythagorean problem are indeed extremely sensitive to initial conditions.

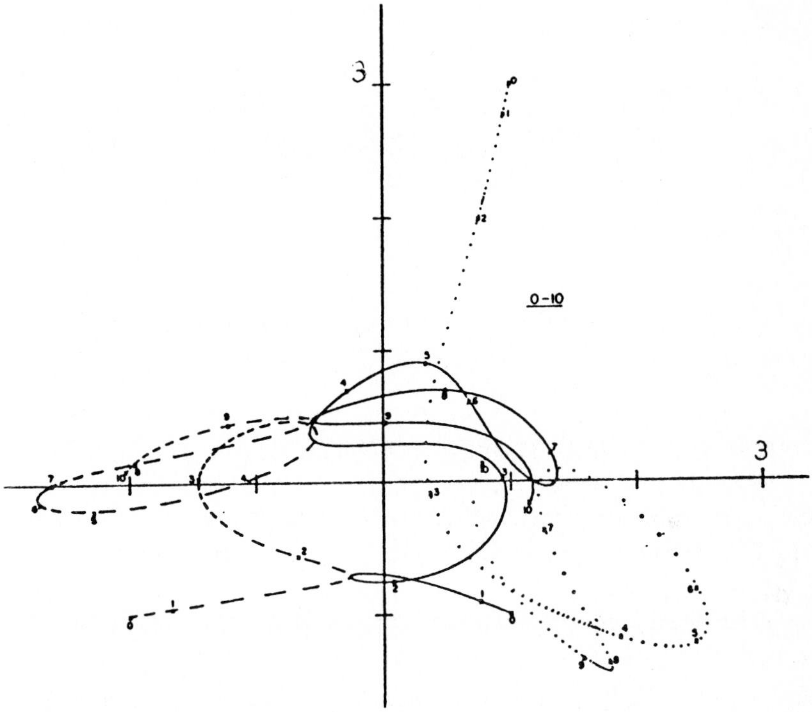

Fig. 96. The Pythagorean problem. Orbits for $0 \leq t \leq 10$.
Source : Szebehely, V. and Peters, C.F. [6].

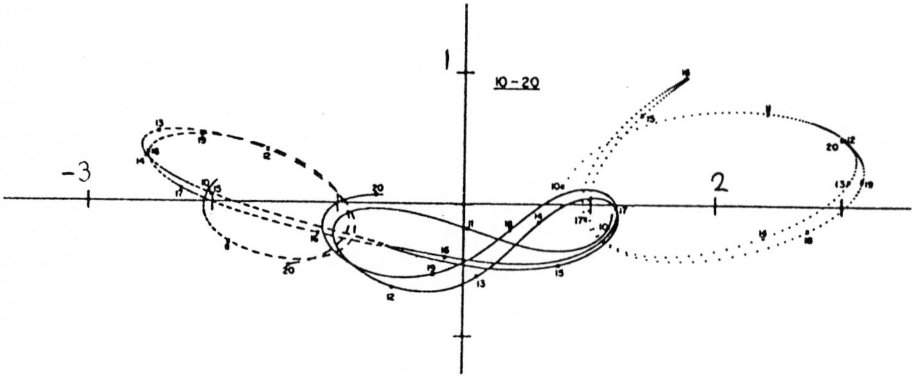

Fig. 97. The Pythagorean problem. Orbits for 10 ≤ t ≤ 20.
Source : Szebehely, V. and Peters, C.F. [6]

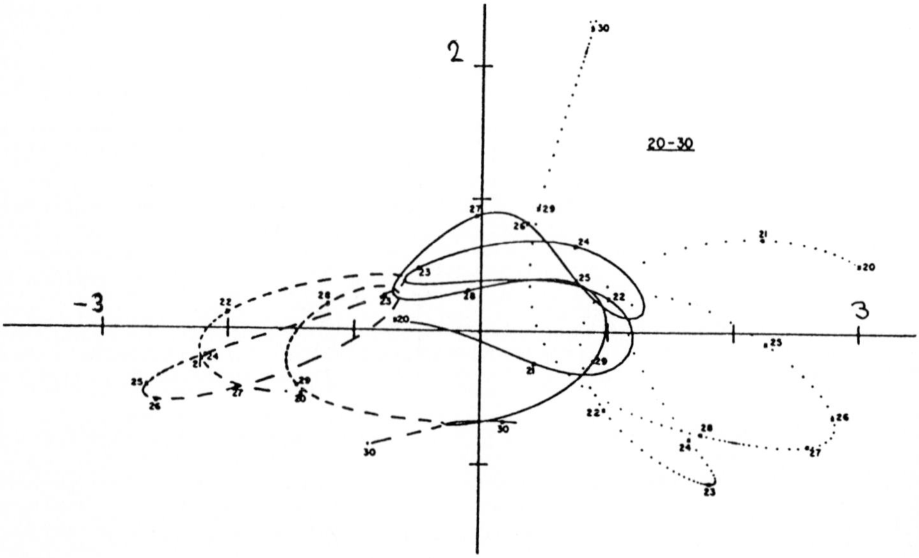

Fig. 98. The Pythagorean problem. Orbits for 20 ≤ t ≤ 30.
Source : Szebehely, V. and Peters, C.F. [6].

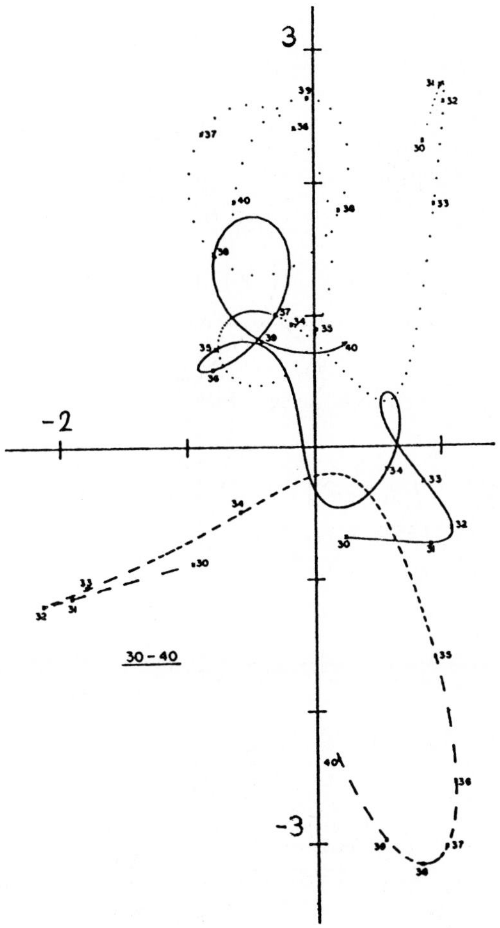

Fig. 99. The Pythagorean problem. Orbits for 30 ≦ t ≦ 40.
Source : Szebehely, V. and Peters, C.F. [6].

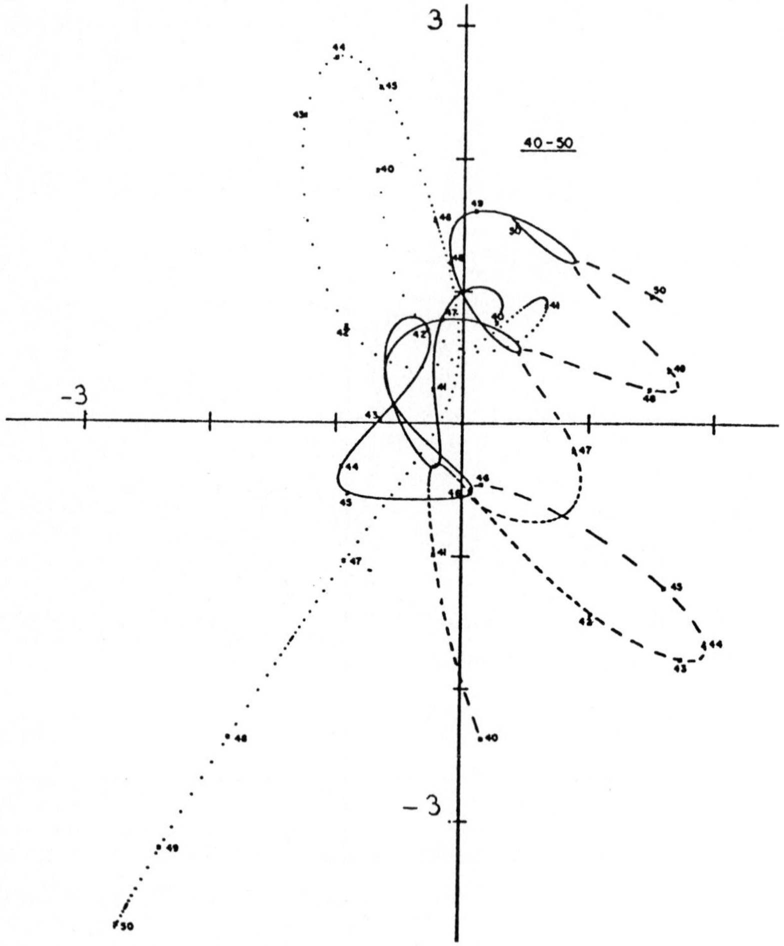

Fig. 100. The Pythagorean problem. Orbits for $40 \leq t \leq 50$.
Source : Szebehely, V. and Peters, C.F. [6]

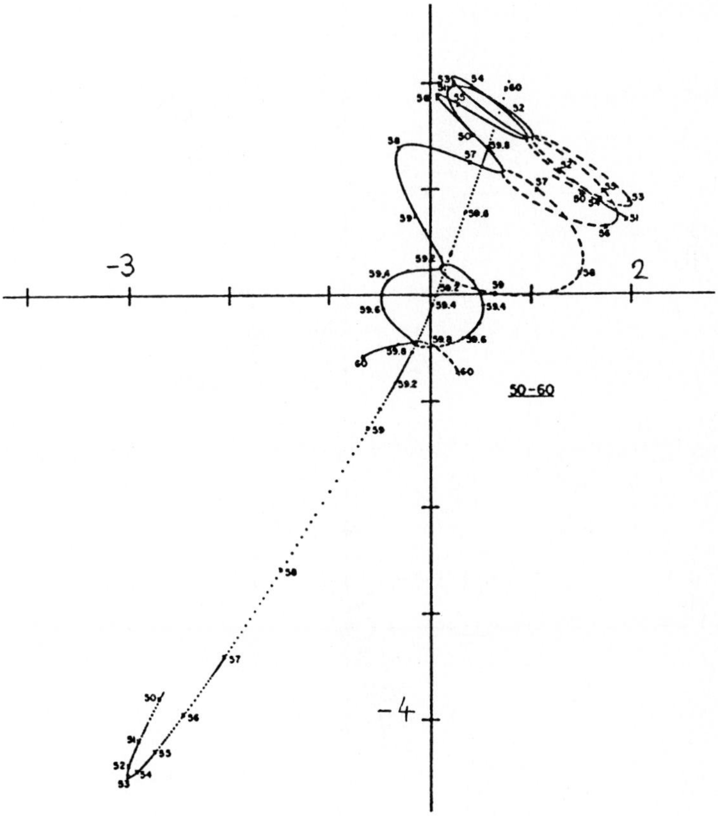

Fig. 101. The Pythagorean problem. Orbits for $50 \leqq t \leqq 60$.
Source : Szebehely, V. and Peters, C.F. [6]

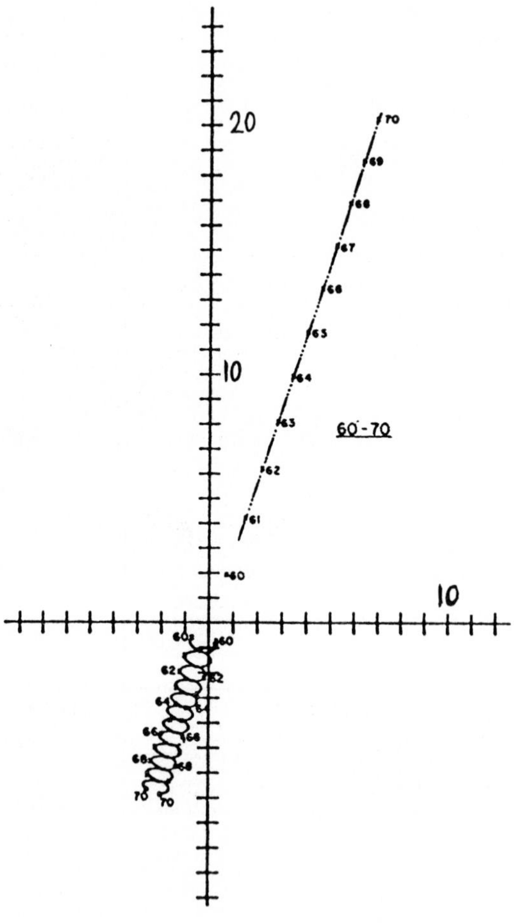

Fig. 102. The Pythagorean problem. The final escape of A. Orbits for 60 ≦ t ≦ 70. Source : Szebehely, V. and Peters, C.F. [6].

10.9.3 The method of numerical exploration. Encounters of satellites.

Two examples of the method of numerical exploration have already been presented, the study of the Copenhagen problem in Section 9.4 and the study of gravitational scattering in Section 10.4.3. There are many other examples of this method in the three-body problem especially in the study of the stability of triple systems (Ref.191)

Let us outline here a recent study on satellite encounters (Ref. 144 and 192).

Two very small satellites of a large planet such as Saturn move along very close circular coplanar orbits (Fig. 103). They move in the same direction of rotation and have slightly different periods, so that from time to time they have long close encounters. What will be the effect of their mutual attraction ?

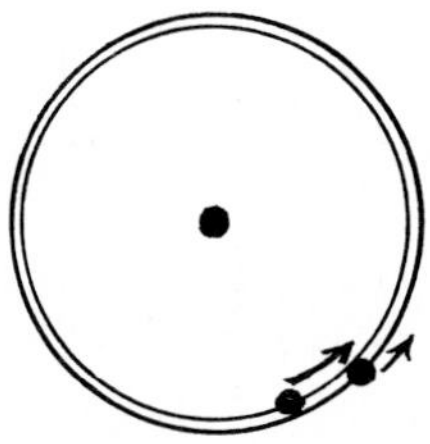

Fig. 103. Encounters of satellites on very close circular and coplanar orbits.

This is a plane problem and in a suitable uniformly rotating set of axes the radius vector $\vec{r} = (x,y)$ from one satellite to the other is governed by the Hill equations.

$$\left.\begin{aligned} d^2x/dt^2 &= -(x/r^3) + 2(dy/dt) + 3x \\ d^2y/dt^2 &= -(y/r^3) - 2(dx/dt) \end{aligned}\right\} \qquad (850)$$

with :

$$r = (x^2 + y^2)^{1/2} \qquad (851)$$

The Jacobi integral of motion is :

$$\Gamma = (2/r) + 3x^2 - (dx/dt)^2 - (dy/dt)^2 \tag{852}$$

x is the radial coordinate and y the circumferential coordinate.

Initially the motion is linear with a positive and very large y :

$$\left.\begin{aligned} x &= h \\ y &= y_0 - (3/2)ht \end{aligned}\right\} h > 0 \tag{853}$$

The point x,y always arrives from above in the first quadrant but, according to h, it experiences many possible orbits and the problem has a surprising variety with almost always a final escape either in the second quadrant or in the fourth (Fig. 104).

The complexities of the problem are related to the existence of asymptotical orbits such as that of Fig. 105 with $h = 1.718779940$, in the vicinity of which we find the escape orbits of Figs. 106 and 107.

The conclusion is that a close encounter is a complex phenomenon that can lead to chaotic motions and also that the method of numerical exploration can be very powerful.

All this, and also the analytical studies (Ref. 193 , 194 , 195), help us to understand the dynamics of Saturn rings and the phenomenon of shepherd satellites. However the natural limitations of the method of numerical exploration must be underlined (only plane motions, only initially circular orbits, only two satellites, etc...). In the study of gravitational scattering (Section 10.4.3 and Ref. 168) 1.7 million orbits have been computed for the only case of three equal masses. Similar studies with unequal masses will be very useful for understanding the mass effects in close encounters and in gravitational scattering.

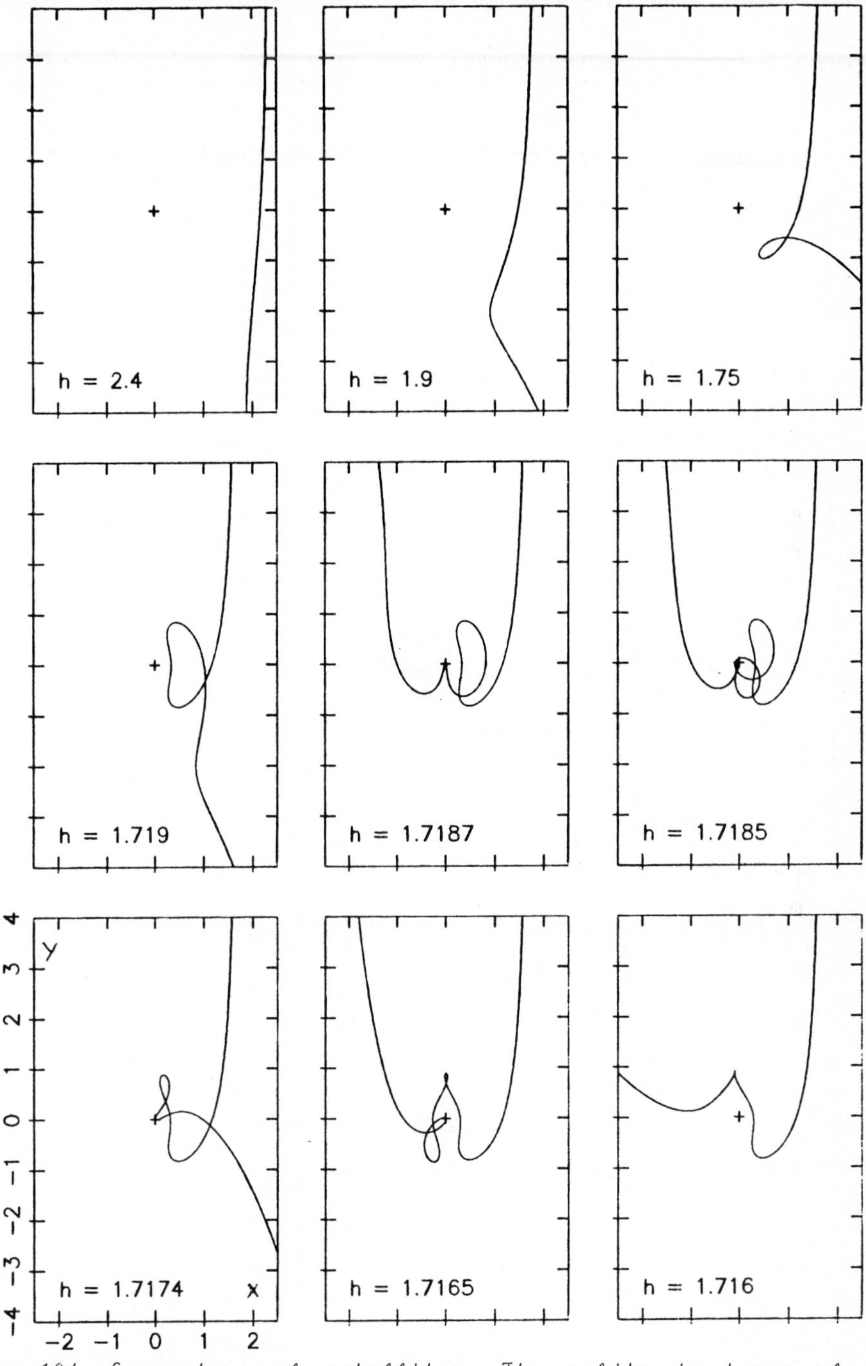

Fig. 104. Encounters of satellites. The orbits in terms of the initial radius difference h ; (initially : $x \simeq h$; $y \simeq y_0 - (3/2)ht$).
Source : Petit, J.M. and Hénon, M. [192].
Fig. 104a. Orbits for $1.716 \leq h \leq 2.4$

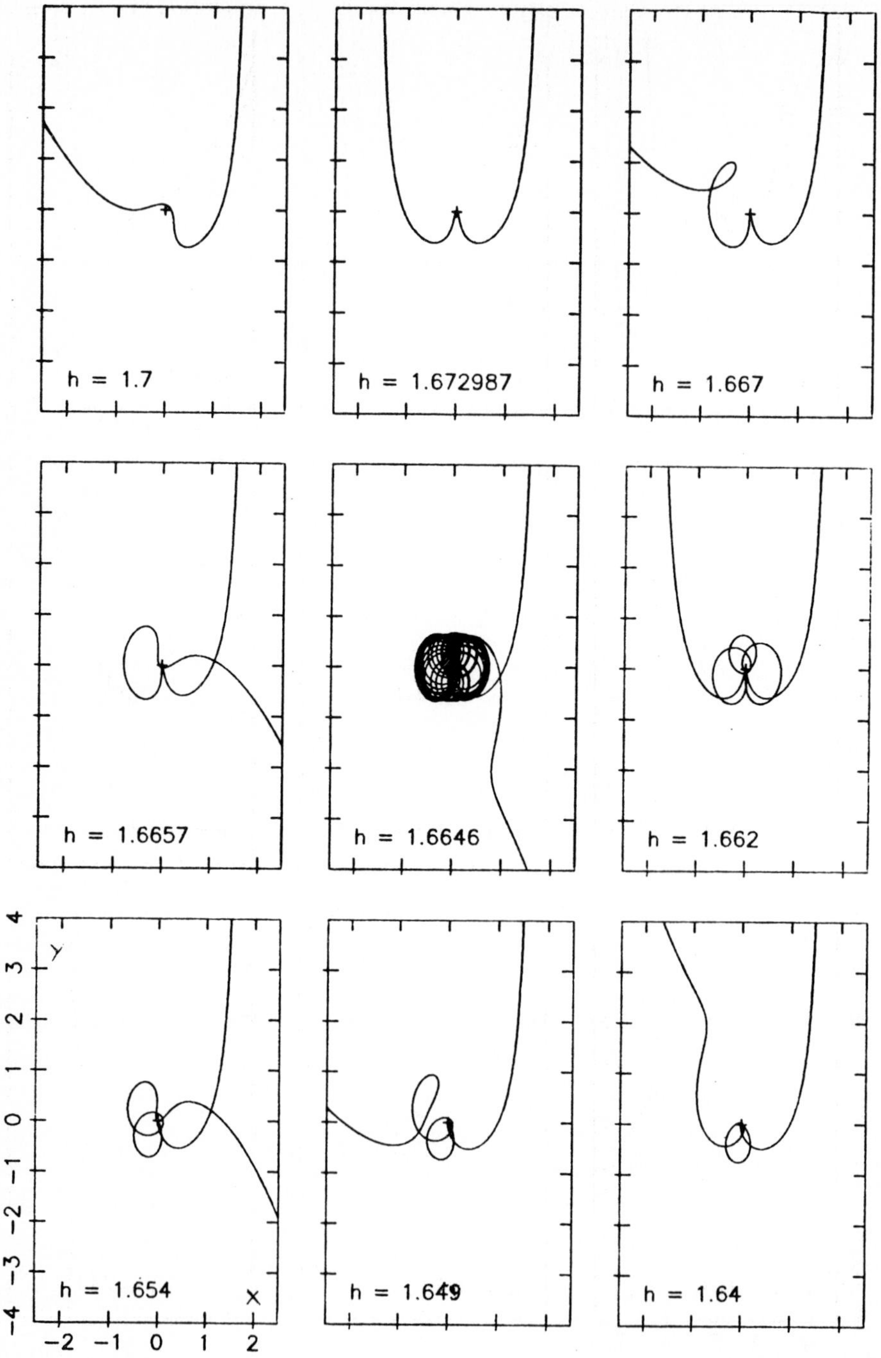

Fig. 104b. Orbits for 1.64 ≦ h ≦ 1.7

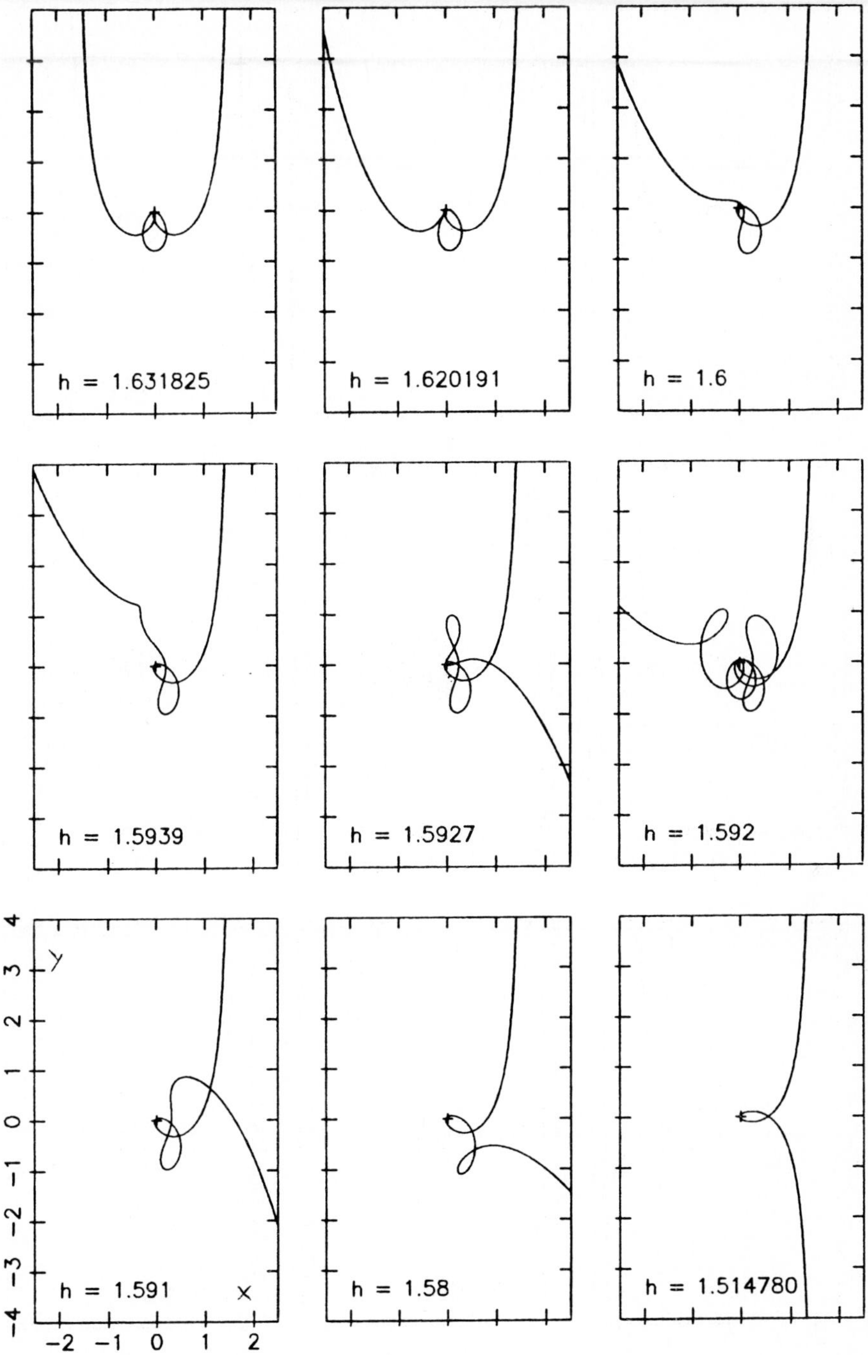

Fig. 104c. Orbits for 1.5 < h < 1.64

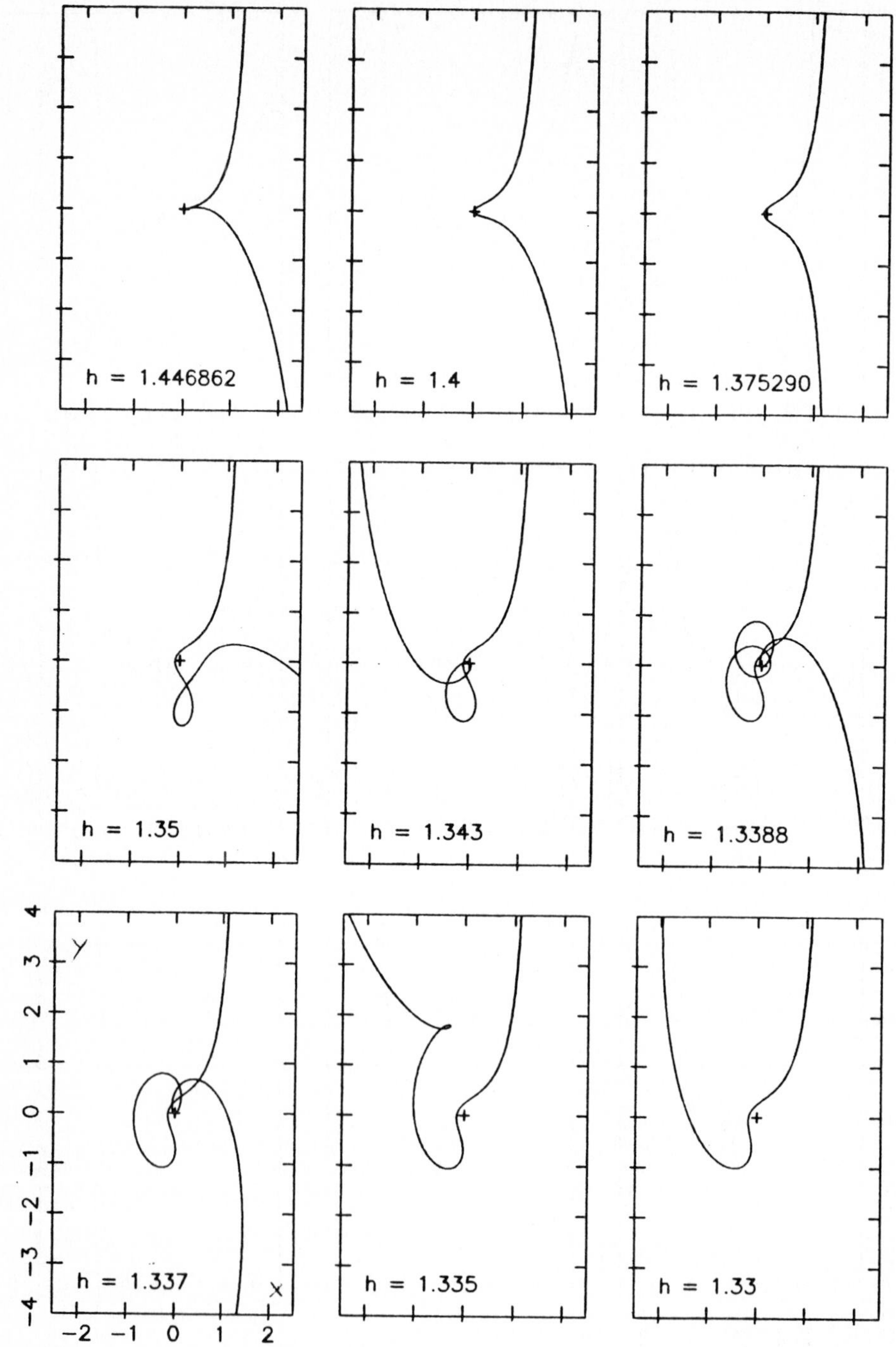

Fig. 104d. Orbits for $1.33 \leqq h < 1.45$

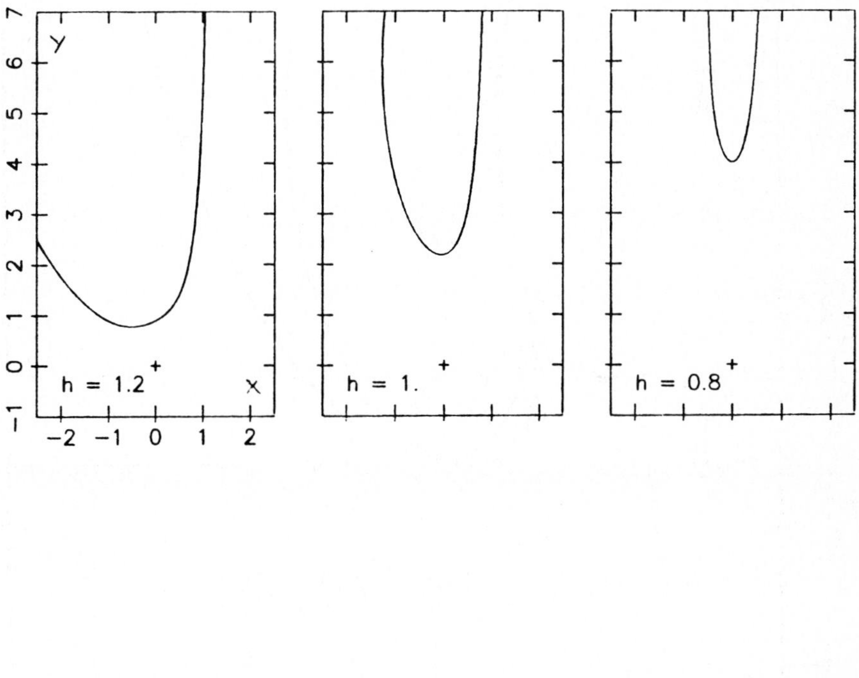

Fig. 104e. Orbits for $0.8 \leq h \leq 1.2$

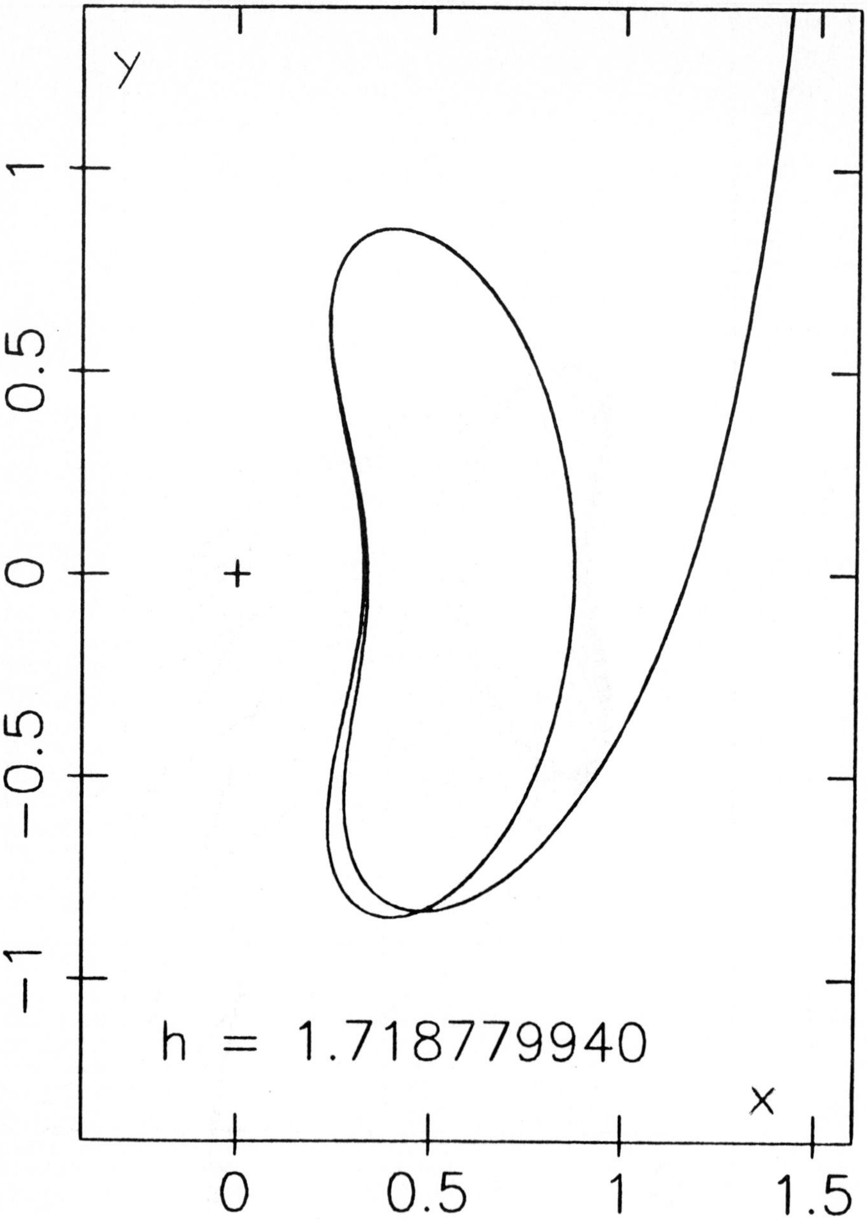

Fig. 105. The asymptotical orbit obtained for h = 1.718779940 and its unstable periodic limit.
Source : Petit, J.M. and Hénon, M. [192].

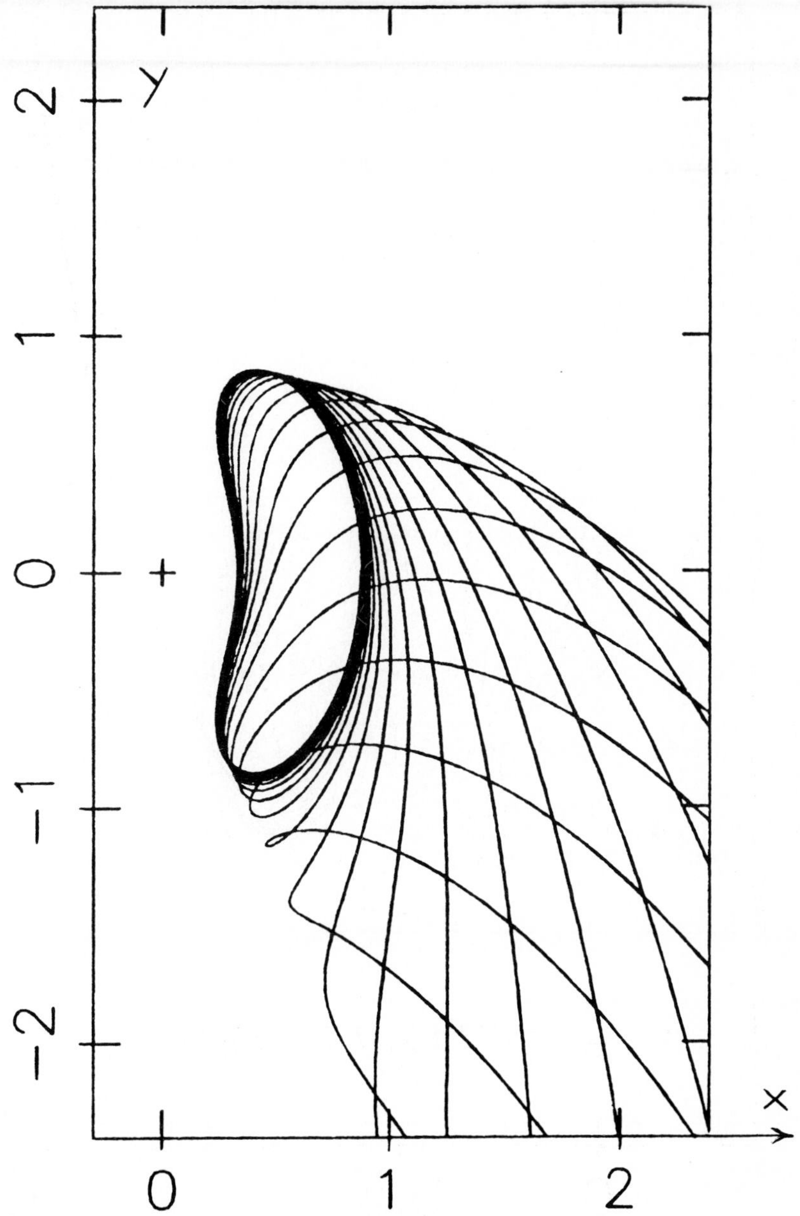

Fig. 106. Examples of escape from the vicinity of the limit periodic orbit of Fig. 105 when h is slightly larger than 1.718 779 940 Source : Petit, J.M. and Hénon, M. [192].

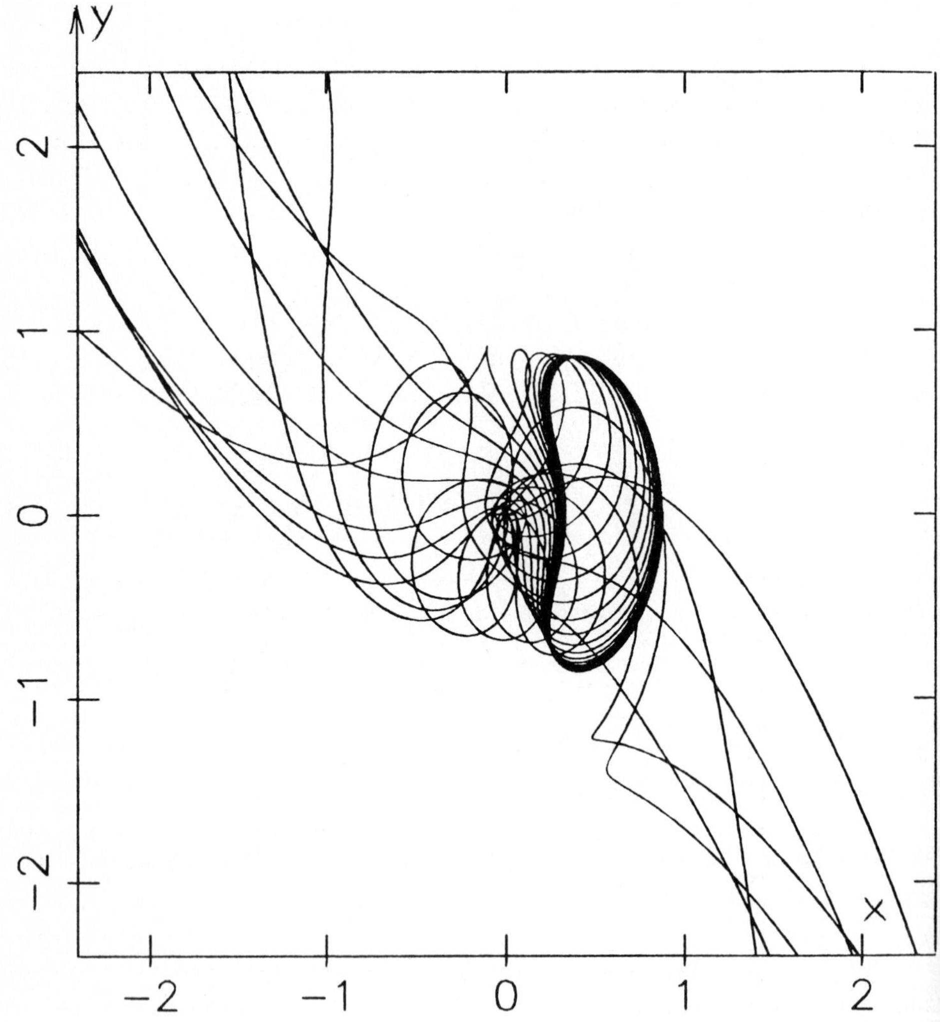

Fig. 107. Examples of escape from the vicinity of the limit periodic orbit of Fig. 105 when h is slightly smaller than 1.718779940. Source : Petit, J.M. and Hénon, M. [192].

Chapter 11

THE GENERAL THREE-BODY PROBLEM. QUALITATIVE ANALYSIS AND QUALITATIVE METHODS.

The usual quantitative methods, either analytical or numerical, generally give very accurate informations on the evolution of differential systems. However these informations are usually limited to the solution of interest and to a small vicinity, furthermore in most cases the accuracy decreases and disappear when the time increases.

The qualitative methods are the natural complement of quantitative methods ; they give partial but also rigorously demonstrated properties that are valid at least for very long periods of time and generally for all time. They give the general informations and the great classifications, they deal with questions of existence, uniqueness, integrals of motion, singularities, regularization, asymptotic properties, etc... In Celestial Mechanics, we can add periodic orbits and stability, symmetries and scale transformation, final evolutions and tests of escape, ergodic theorem, K.A.M. theorem and the structure of the set of solutions.

These methods have been developped by many famous scientists : Hill, Poincaré, Sundman, Birkhoff, Chazy, Levi-Civita, Merman, Khilmi, Sitnikov, Alexeev... The Kolmogorov-Arnold-Moser theorem can be considered as the result of a qualitative analysis.

The fantastic progress of computers have led to many improvements in the quantitative analysis of the n-body problem. These improvements have disclosed the extreme complexity of the set of solutions and have given new orientations and a new impetus to the qualitative analysis especially in the domain of final evolution and tests of escape.

Some of the famous old conjectures have recently been solved, but new ones have appeared.

11.1 THE PROTOTYPE OF QUALITATIVE METHODS

Let us consider the ordinary n-body problem with its usual notations :

$$U = G \sum_{1 \leq i<j \leq n} m_i m_j / r_{ij} = \text{"potential"} \tag{854}$$

$$h = 0.5 \sum_{j=1}^{n} m_j V_j^2 - U = \text{"energy integral"} \tag{855}$$

$$I = 0.5 \sum_{j=1}^{n} m_j r_j^2 = \text{"semi-moment of inertia"} \tag{856}$$

With $I'' = d^2I/dt^2$, the Lagrange-Jacobi identity is :

$$I'' = U + 2h \tag{857}$$

Hence, since $U > 0$, a double integration leads to :

$$I(t) \geq I_0 + I_0' t + ht^2 \tag{858}$$

Thus, for instance, the motion cannot remain forever bounded if the energy integral h is positive.

11.2 THE TRIVIAL TRANSFORMATIONS AND THE CORRESPONDING SYMMETRIES AMONG N-BODY ORBITS.

Since the time of Newton the unexpected contrast between the simplicity of the two-body problem and the complexity of the more-than-two-body problem has surprised and attracted many mathematicians.

Looking for simple elements the early researchers soon discovered most of the "trivial transformations" and the corresponding symmetries.

These simple transformations leading from one n-body solution to another are the following already noticed in Section 5.3.

Let us call $\vec{r}_j(t)$, with $j = \{1, 2, ..., n\}$, the radius-vectors of an arbitrary n-body solution. New solutions $\vec{r}_{jN}(t)$ can be obtain by the following transformations :

I) The space translation with two fixed vectors $\vec{A}$ and $\vec{B}$:

$$\vec{r}_{jN}(t) = \vec{r}_j(t) + \vec{A} + \vec{B}t \quad ; \quad j = \{1, 2, ..., n\} \tag{859}$$

II) The time translation with a fixed interval of time T :

$$\vec{r}_{jN}(t) = \vec{r}_j(t + T) \quad ; \quad j = \{1, 2, \ldots, n\} \tag{860}$$

III) The space symmetry about the Oxy plane :

$$\left.\begin{array}{l} \vec{r}_j = (x_j, y_j, z_j) \quad ; \quad \vec{r}_{jN} = (x_{jN}, y_{jN}, z_{jN}) \\ \forall t, \forall j : x_{jN}(t) = x_j(t) \; ; \; y_{jN}(t) = y_j(t) \; ; \; z_{jN}(t) = - z_j(t) \end{array}\right\} \tag{861}$$

IV) The space symmetry about the x-axis :

$$\forall t, \forall j : x_{jN}(t) = x_j t) \; ; \; y_{jN}(t) = - y_j(t) \; ; \; z_{jN}(t) = - z_j(t) \tag{862}$$

V) Similarly the space symmetries about any fixed plane, axis or point.

VI) The time symmetry about the instant T :

$$\vec{r}_{jN}(t) = \vec{r}_j(2T - t) \quad ; \quad j = \{1, 2, \ldots, n\} \tag{863}$$

VII) The space rotations, for instance the rotation of angle α about the z-axis :

$$\forall t, \forall j : \left\{\begin{array}{l} x_{jN}(t) = x_j(t)\cos\alpha - y_j(t)\sin\alpha \\ y_{jN}(t) = x_j(t)\sin\alpha + y_j(t)\cos\alpha \\ z_{jN}(t) = z_j(t) \end{array}\right\} \tag{864}$$

VIII) The scale transformations.

The transformation can be either on the lengths or on the masses or on the time or on all of them, with a suitable corresponding modification of the constant G of the law of Newton.

If we want to keep G and the masses we obtain the following most usual scale transformation, with a fixed real factor K :

$$\vec{r}_{jN}(t) = K^2 \, . \, \vec{r}_j(t/K^3) \quad ; \quad j = \{1, 2, \ldots, n\} \tag{865}$$

while the modification of masses m_j leads to :

$$\left.\begin{aligned} m_{jN} &= K^3\, m_j \\ \vec{r}_{jN}(t) &= K\, \vec{r}_j(t) \end{aligned}\right\} \quad j = \{1,\ 2,\ \ldots,\ n\} \tag{866}$$

IX) Finally the products of all these simple transformations also give new solutions of the n-body problem.

These simple transformations may seem trivial, but notice the two following points :

A) For a given value of the constant G of the law of Newton and from a given initial solution of the n-body problem these "trivial transformations" allow us to deduce four different twelve-parameter families of solutions of the n-body problem.

These four families are characterized by the inversion of time, or not and by the reversing of space or not.

B) These simple transformations are essential for an exhaustive study of the symmetries of the solutions of the n-body problem.

We can use the following definition :

If we consider a given solution of the n-body problem and if a "trivial transformation" allow us to fall back on the same solution this solution has a periodicity and/or a symmetry.

The simplest periodicity is the "absolute periodicity" given by the time translation (860) :

$$\vec{r}_j(t) \equiv \vec{r}_j(t + T) \quad ; \quad j = \{1,\ 2,\ \ldots,\ n\} \tag{867}$$

However we can also obtain a "relative periodicity" also called "periodicity in a suitable uniformly rotating set of axes" by the product of a space rotation (angle α) and a time translation (interval of time T).

In both cases the mutual distances have the period T and the previous Section 11.1 shows that these periodic solutions always have a negative energy integral h.

To these two classical cases of periodicity, studied in Sections 10.5 to 10.9.1, we can add the cases given by the supplementary use space translations. However these new cases only differ by the motion of the center of mass and are thus not really different.

For the same reason we will disregard the space translations in the study of the symmetries and we will only consider n-body motions with center of mass at the origin.

The scale transformations (865), (866) never allow us to fall back on the solutions of interest except for the very exceptional case of Euler or Lagrange radial motions of zero energy integral.

Hence for the study of the symmetries there remain only the time translation (860), the time symmetry (863) about an instant T, the space rotations and symmetries keeping the center of mass at the origin and the products of these trivial transformations.

The symmetries can be classified into the following three main categories.

11.2.1 The space-time symmetries

These symmetries correspond to the cases of reversion of the time and thus the use of the time symmetry (863) about an instant T :

At the instant T of space-time symmetry the positions of the n-bodies are symmetrical and their velocities are anti-symmetrical.

For instance :

A) At the instant T the positions of the n bodies are arbitrary and their velocities are zero (Fig. 108a).

This case has a pure time symmetry :

$$\vec{r}_j(t) \equiv \vec{r}_j(2T - t) \quad ; \quad j = \{1, 2, \ldots, n\} \tag{868}$$

B) At the instant T the positions of the n bodies are in the same plane and all velocities are normal to that plane (Fig. 108b).

This case is the Poincaré case of time-plane symmetry.

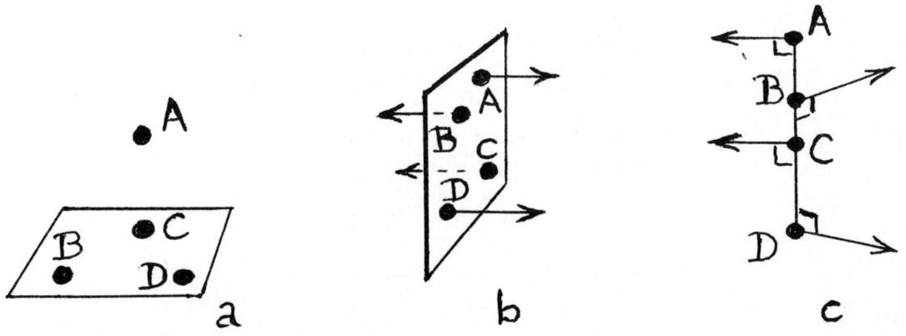

Fig. 108. Space-time symmetries. Positions and velocities at the instant of symmetry.
a) Pure time symmetry
b) Poincaré time-plane symmetry
c) Poincaré time-axis symmetry.

C) At the instant T the n bodies are along a given axis and their velocities are normal to that axis (Fig. 108c).

This case is the Poincaré case of time-axis symmetry.

In these three first cases all mutual distances are even functions of the time about the instant T :

$$r_{ij}(t) \equiv r_{ij}(2T - t) \quad ; \quad \{i,j\} = \{1, 2, \ldots, n\} \tag{869}$$

If two or several masses are equal they can be symmetrical to each other and this leads to many other cases of symmetry.

For instance in the case of time-plane symmetry the masses that are not at T in the plane of symmetry must be pairs of equal and symmetrical masses with antisymmetrical velocities (symmetry of type IV in Fig. 109).

Similar properties hold for the time-axis and time-center symmetries.

The symmetry of type III in Fig. 109 is a time-axis symmetry.

The numerical example presented in Section 10.4.1 and in Figs. 39-40 has a symmetry of both types III and IV while the numerical example presented in Section 10.4.2 and in Figs. 41-42 has a Poincaré time-axis symmetry.
time-axis symmetries and the pure time symmetries are also Poincaré time-plane symmetries ; there thus remain only four different types of space-time symmetries, the space-time symmetries of types I, II, III, IV presented in Fig. 109.

In the n-body problem more complex cases exist with space rotations or space anti-rotations* of angle $2\pi/K$ and with, at T, regular polygons or anti-polygons of K equal masses (Fig. 110).

---

* An anti-rotation is the product of a rotation by a symmetry about a plane or about a point.

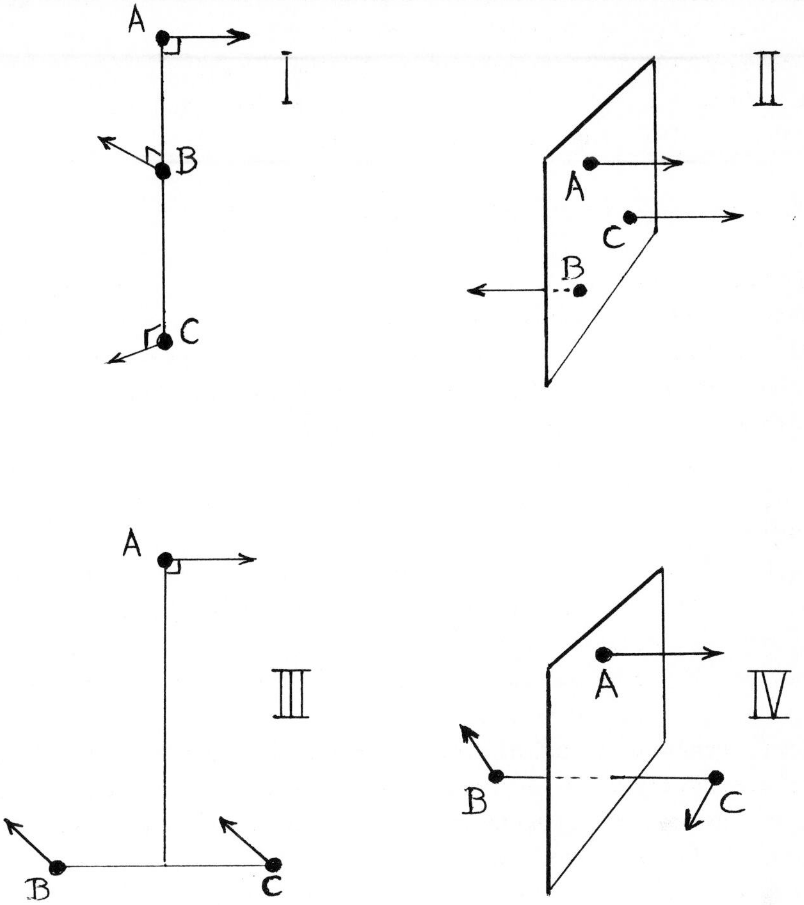

Fig. 109. The four space-time symmetries of the three-body problem at the instant T of space-time symmetry.
I. The Poincaré time-axis symmetry.
II. The Poincaré time-plane symmetry.
III. Time-axis symmetry. } Supplementary space-time symmetries with equal and symmetrical masses m_B and m_C (their velocities are antisymmetrical).
IV. Time-plane symmetry }

These space-time symmetries were already presented in Section 10.6.1 (Fig. 56) and in Section 10.8.2 (Fig. 75). The planar exchange orbit of Section 10.4.1 and the periodic orbits of Section 10.8.2 have space-time symmetries of types III and IV, while at $t = 0$ the orbit of Section 10.4.2 has a space-time symmetry of type I.

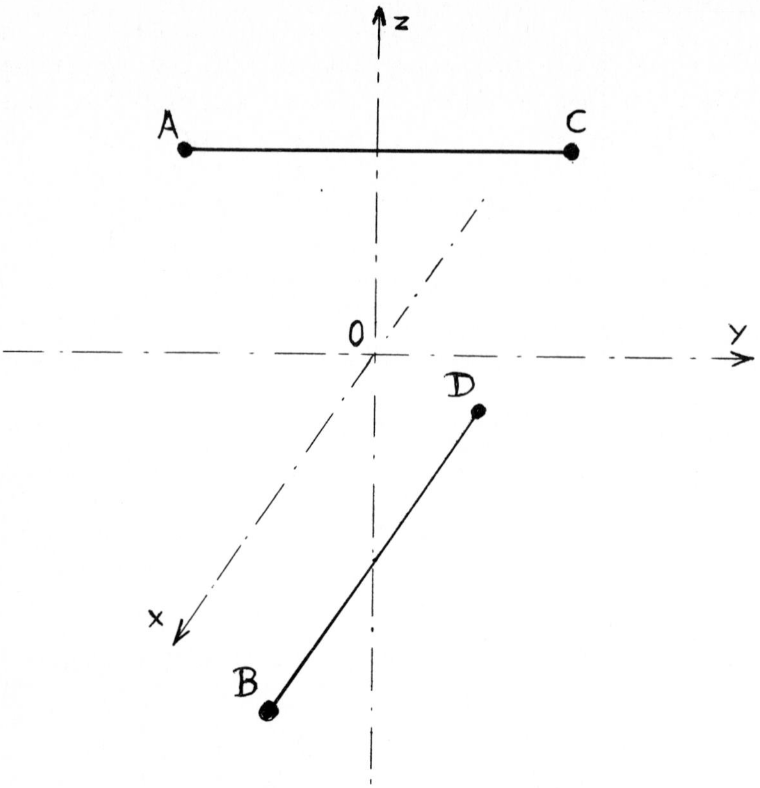

Fig. 110. A "time-anti-rotation symmetry" with four equal masses.
At the instant T :

$\vec{r}_A = (a, b, c)$; $\vec{V}_A(\alpha, \beta, \gamma)$

$\vec{r}_B = (-b, a, -c)$; $\vec{V}_B = (\beta, -\alpha, \gamma)$

$\vec{r}_C = (-a, -b, c)$; $\vec{V}_C = (-\alpha, -\beta, \gamma)$

$\vec{r}_D = (b, -a, -c)$; $\vec{V}_D = (-\beta, \alpha, \gamma)$

For all t the four vectors $\vec{r}_A(t)$, $\vec{r}_B(2T-t)$, $\vec{r}_C(t)$, $\vec{r}_D(2T-t)$ have this "anti-rotation symmetry".

If we choose $\gamma = 0$ (center of mass at rest) the functions $z_{ABCD}(t)$ are even functions of t about T.

11.2.2 The space symmetries

For "space symmetries" the trivial transformation allowing to fall back on the solution of interest has neither time-symmetry nor time-translation. The positions and velocities have a symmetry (or a rotation or an anti-rotation) at some instant t and hence at all times.

If all n masses are different each body must be self symmetrical and the results have little interest : we obtain either the plane motions (symmetry about a plane) or the collinear motions (symmetry about an axis).

If two or several masses are equal the possibilities are more interesting (Fig. 111).

The symmetry 111(a) about a plane leads to collisions of the two equal masses.

The symmetry 111(b) about an axis is called the Sitnikov symmetry because it was used by Sitnikov in his demonstration of the existence of oscillatory motions of the first kind (Section 11.7.8 and Ref. 196).

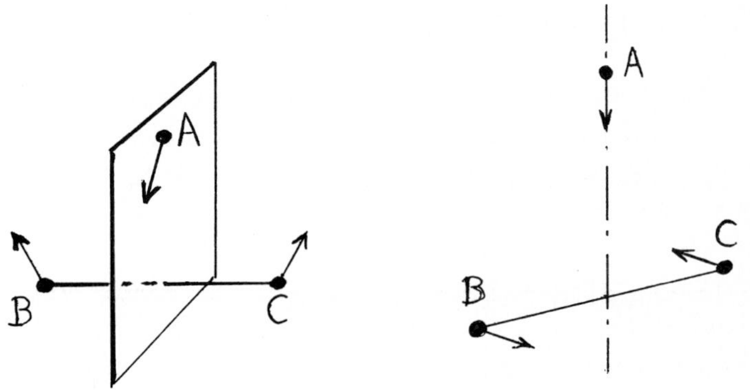

Fig. 111. Two space symmetries with equal masses m_B and m_C.
(a) Symmetry about a plane.
(b) Symmetry about an axis (Sitnikov symmetry).

Notice that the motion of Fig. 110 has a Sitnikov symmetry.

The more than-three body problem has many other possible "space symmetries" with either a rotation or an anti-rotation, but the three-body problem has only the above space symmetries (plane motions, collinear motions, Fig. 111) and the very particular space symmetry -in fact a 120° rotation- of the Lagrangian motions with three equal masses.

11.2.3 The remaining symmetries

We have already met the following trivial transformations leading back to the solution of interest :

A) The trivial transformations with a symmetry of time (space-time symmetries).

B) The trivial transformation with neither symmetry nor translation of the time (space symmetries).

C) The trivial transformations with a time-translation of interval T and a space-rotation of angle α.

If this trivial transformation allow us to fall back on the same bodies (and not on other bodies with the same mass) the solution of interest is a periodic solution of period T and "angle of rotation" α. This angle α can have any value, it is zero for the "absolute periodic orbits".

D) The remaining cases are those for which :

D_1) The time transformation is a non zero time-translation of interval T.

D_2) The space transformation is either a reversion (symmetry about a plane or a center or more generally anti-rotation of angle α) or, if it is a space rotation of angle α it leads to some exchange of equal masses.

Notice that, as in the case C, the knowledge of the solution over a time interval T provides full knowledge of the solution. Also notice that after several intervals of time T we reach a situation similar to the case C and then the solution of interest has a period KT and an angle of rotation $K\alpha$.

The positive integer K is at most 6 for three-body and four-body problems. It is 6 in the "family of periodic orbits with the largest number of symmetries" of Section 10.8.2, indeed a trivial transformation of type D with anti-rotation and circular exchange of bodies exists between the states at $t = 0$ and at $t = \pi/3$ (and also between the states at an arbitrary instant t and at $t + \pi/3$).

These "remaining symmetries" are complex and rarely useful. They are usually neglected and generally appear as the products of some ordinary symmetries or periodicities.

11.2.4 Multi-symmetries

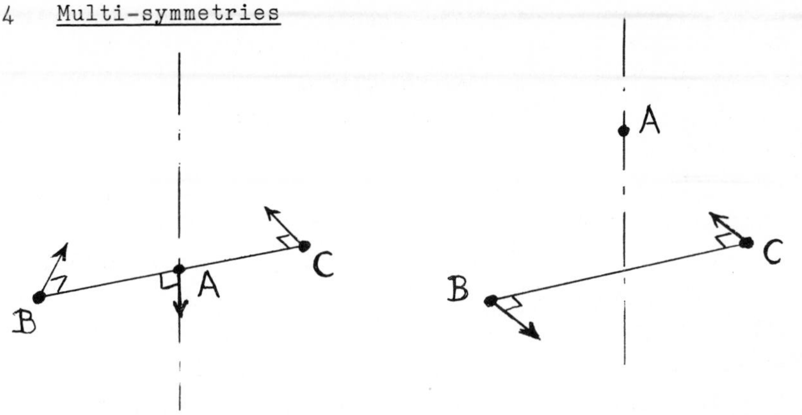

Fig. 112. Two cases of multi-symmetry.

The masses m_B and m_C are equal and the two figures have the Sitnikov symmetry of Fig.111.b, the first has also the space-time symmetry of type I and the second the space-time symmetry of type II (Fig.109)

It is of course very simple to obtain multi-symmetrical conditions as in Fig. 112. The corresponding orbits then have many properties that allow considerable simplification of the quantitative studies and lead to particular demonstrations (see Sections 11.4 and 11.7.8).

The most symmetrical three-body orbits are the circular Lagrangian motions with three equal masses. These motions belong to the family of periodic orbits of Section 10.8.2 with twelve space-time symmetries per period.

11.3 OTHER EARLY QUALITATIVE RESEARCHES

Beside the trivial transformation and the symmetries the early qualitative researches were oriented essentially towards the research of particular solutions and new integrals of motion.

11.3.1 The Eulerian and Lagrangian solutions. The central configurations.

Beside the extremely particular solutions of Section 8.4 the only known analytical solutions of the three-body problem are the Eulerian and Lagrangian solutions of Sections 8.1, 8.2, 8.3 and of Figs 9 to 12.

These solutions are characterized by a constant configuration, i.e. by constant ratios of mutual distances.

These configurations are the "central configurations" in which the gravitational acceleration of each body is proportional to its radius-vector (the acceleration are then "central" and directed toward the center of mass) :

$$j = \{1, 2, 3\} \Longrightarrow d^2\vec{r}_j/dt^2 = -K\vec{r}_j \qquad (870)$$

This condition (870) gives two well-known types of three-body central configurations (Fig. 113), the triangular type in which the three bodies are at the tip of an equilateral triangle and the collinear type.

All equilateral triangles satisfy the condition (870) but this condition imposes a unique "relative equilibrium condition" on the collinear central configurations.

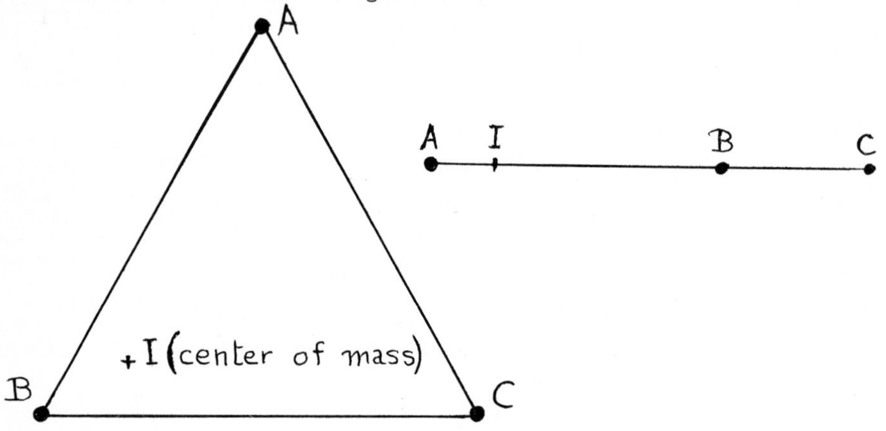

Fig. 113. The triangular and collinear central configurations.

This latter condition has many possible equivalent expressions, as presented in the equations (109) to (113) of Section 8.1. The simplest is the last one, i.e. the following :

If suitable origin and unit of length give the abscissas (+1) and (-1) to the two extreme masses, the collinear configuration is a central configuration if and only if the abscissa x of the medium mass and the abscissa x_I of the center of mass satisfy :

$$x_I = (x^5 - 2x^3 + 17x)/(x^4 - 10x^2 - 7) \qquad (871)$$

This relation (871) is presented graphically in Fig. 12 of Section 8.1.

Lagrange discovered that these two types of central configurations allow circular three-body motions about the center of mass, all three bodies revolving together, as a unique solid, in the same plane, in the same direction and with the same angular velocity.

The rate of rotation is the square root of the coefficient K of (870).

Euler extended these results to elliptic, parabolic and hyperbolic Keplerian motions. The configuration remains constant, but with a variable scale and a variable coefficient K, the three bodies have orbits in the same plane with the same eccentricity, the same attracting focus at the center of mass and the same time of pericenter passage.

Surprisingly it has become customary to speak of "Lagrangian motion" when the central configuration is an equilateral triangle and "Eulerian motion" when the central configuration is collinear (Figs. 114 and 115).

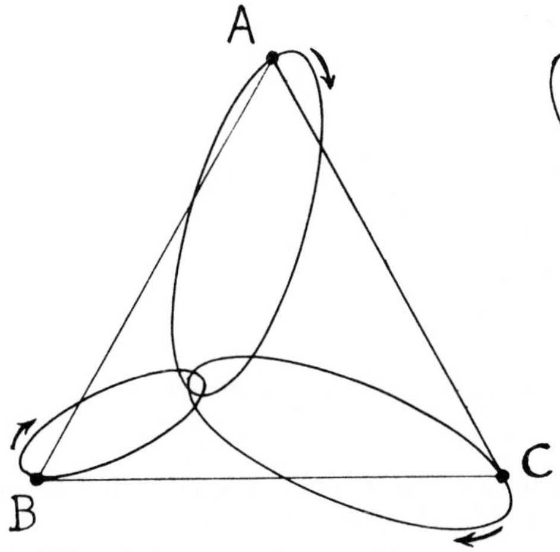

Fig. 114. A Lagrangian motion.

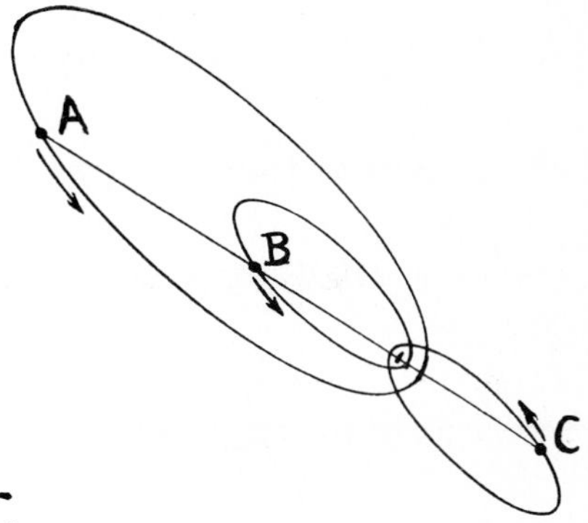

Fig. 115. A Eulerian motion.

Notice that for three given masses there exist three and only three collinear central configurations according to the mass that is between the two other masses. The equation (871) has indeed one and only one solution for given masses m_A, m_B, m_C in that order : the right-hand member of (871) is a decreasing function of x while the left-hand member, equal to $(m_C+xm_B-m_A)/m_A+m_B+m_C)$, is an increasing function of x.

The notion of central configuration also exists for the n-body problem :

$$j = \{1, 2, \ldots, n\} \quad \text{implies} \quad d^2\vec{r}_j/dt^2 = -K\vec{r}_j \tag{872}$$

These relations are equivalent to the following where G is the constant of the law of universal attraction and M the total mass equal to $m_1 + m_2 + m_3 + \ldots + m_n$.

$$F(\vec{r}_1, \vec{r}_2, \ldots, \vec{r}_n) = \sum_{1 \leq i < j \leq n} m_i m_j \left[\frac{2GM}{r_{ij}} + Kr_{ij}^2\right] \text{ : stationary} \tag{873}$$

Hence there is one and only one collinear central configuration of n given masses in a given succession : this configuration corresponds to the only minimum of F for the collinear succession of interest*.

For instance, if for ten equal masses we choose GM = K, the ten abscissas are the following :

$x_{10} = -x_1 = 1.33251896$

$x_9 = -x_2 = 0.97474812$

$x_8 = -x_3 = 0.67320742$

$x_7 = -x_4 = 0.39628903$

$x_6 = -x_5 = 0.13094092$

Figure 116 presents this central configuration and several four-body central configurations.

Notice that the Lagrangian and Eulerian motions of Figs. 114 and 115 can be extended to the central configurations of the n-body systems with the exception of the three-dimensional central configurations for which only radial Keplerian motions are possible.

For four given non-zero masses the classification of central configurations leads to the following.

A) There are 12 collinear central configurations, one for each possible succession of bodies. They generalize the Eulerian central configurations of three bodies.

---

* And thus for n given masses there are exactly n!/2 collinear central configurations.

B) The three-dimensional central configurations of four bodies are regular tetrahedron with all $r_{ij} = \Delta$ where the length Δ is given by :

$$\Delta = \{GM/K\}^{1/3}$$

C) The complexity arises for planar central configurations that can be divided into two types :

C.1) The convex quadrilaterals.

There are three and only three such central configurations : one for m_j opposed to m_1 with $j = \{2 ; 3 ; 4\}$.

These three central configurations correspond to the only minimum of the function F of (873) for the convex quadrilaterals of interest and satisfy always the following :

I) The four angles of the quadrilateral are between 60° and 150°.

II) The two diagonals r_{1j} and $r_{k\ell}$ (with $\{j,k,\ell\} = \{2,3,4\}$) are larger than the length Δ.

III) The four sides r_{1k}, r_{kj}, $r_{j\ell}$, $r_{\ell 1}$ are smaller than Δ.

C.2) The concave quadrilaterals.

One body is inside the triangle of the other three (Fig. 116). The three sides of this triangle are larger than the length Δ while the three other mutual distances are smaller than Δ.

For a given mass in the triangle there are generally two possible central configurations but this number can climb up to 16 if the four masses are sufficiently equal (Ref. 197, 198). Hence for four given unequal masses the total number of different configurations is a function of the mass ratios and varies from 24 to 80.

For four equal masses the symmetries reduce this number to seven (Fig. 116) : one collinear, one three-dimensional and five different planar central configurations.

The planar central configurations of four bodies satisfy always the following independently of the mass ratios :

$$(r_{12}^{-3} - \Delta^{-3})(r_{34}^{-3} - \Delta^{-3}) = (r_{13}^{-3} - \Delta^{-3})(r_{24}^{-3} - \Delta^{-3}) =$$

$$= (r_{14}^{-3} - \Delta^{-3})(r_{23}^{-3} - \Delta^{-3})$$

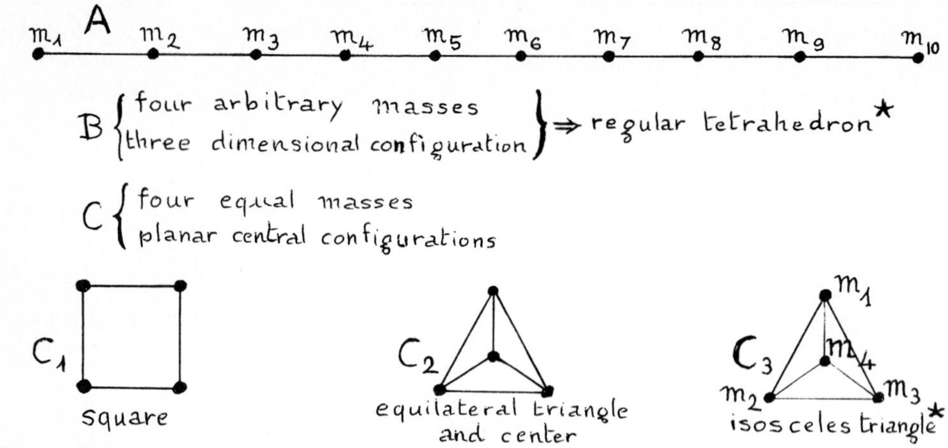

Fig. 116. A) Collinear central configuration of ten equal masses ; B) C) Some central configurations with four masses.

* *regular tetrahedron : all* $r_{ij} = \Delta$

* *isosceles triangle :*

$r_{12} = r_{13} = 1.28504675\ \Delta$

$r_{14} = 0.72291078\ \Delta$

$r_{23} = 1.23906931\ \Delta$

$r_{24} = r_{34} = 0.73903815\ \Delta$

with : $\Delta = \{GM/\mathbf{K}\}^{1/3}$

where :

$M = m_1 + m_2 + m_3 + m_4$

$d^2\vec{r}_j/dt^2 = -\mathbf{K}\vec{r}_j$

For four equal masses there are also two other planar central configurations (with scalene triangles).

The Lagrangian and Eulerian motions generalizing to four bodies those of Figs. 114-115 are always unstable except for suitable convex quadrilateral central configurations with very unequal masses : the largest mass must be more than 94% of the total mass.

The general study of n-body central configurations is very difficult and complex (Ref. 197,198).

11.3.2 The research of new integrals of motion

The classical integrals (center of mass, angular momentum, energy) are presented in Section 5.1, they can be considered as the result of qualitative researches.

The research of other integrals is also a qualitative research, however it has been unsuccessful in the ordinary meaning of this research (see Section 5.2).

The modern numerical computations have disclosed the extreme complexity of the set of three-body and n-body solutions and have dispelled all hopes of useful supplementary integral of motion.

11.4 PERIODIC ORBITS. THE METHOD OF POINCARÉ

Let us consider one of the Poincaré space-time symmetries of Fig. 108 (types I and II of Fig. 109).

If the symmetry occurs at the instant t = 0 the mutual distances r_{ij} are even functions of the time :

$$\forall i,j : r_{ij}(t) \equiv r_{ij}(-t) \tag{874}$$

Poincaré (Ref.199 p102) noticed that if another Poincaré space-time symmetry occurs at a later instant T all mutual distances have a period equal to 2T :

$$\forall i,j : \left\{ \begin{array}{l} r_{ij}(t) \equiv r_{ij}(-t) \\ r_{ij}(T+\theta) \equiv r_{ij}(T-\theta) \end{array} \right\} \left. \begin{array}{l} \text{hence, with} \\ T - \theta = -t \end{array} \right\} r_{ij}(t) \equiv r_{ij}(2T+t) \tag{875}$$

The solution is then a periodic solution of period 2T ; its angle of rotation can be obtained by a numerical integration over a half-period.

This method was used by Poincaré to demonstrate the existence of periodic solutions, and can be used for any number of bodies.

11.4.1 The three first species of Poincaré periodic orbits

Consider for instance a four-body problem with the Sun and three planets (Fig. 117).

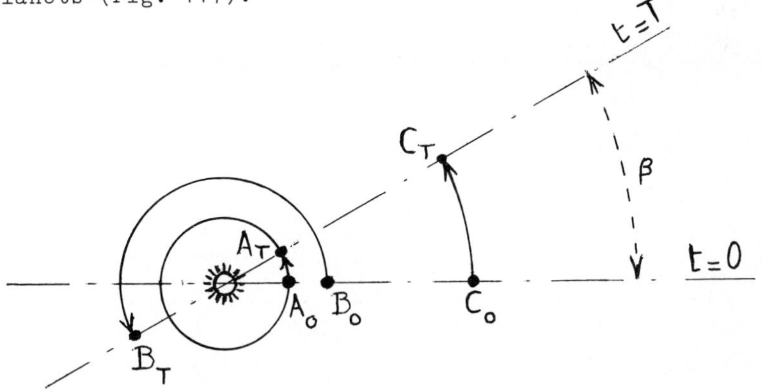

Fig. 117. The Poincaré method for the demonstration of the existence of periodic solutions : from a first space-time symmetry at t=0 to a second one at t=T.
The period is 2T and the angle of rotation is 2β.

Let us look for a planar periodic solution and let us start at $t = 0$ with a planar Poincaré symmetry condition : the Sun and the three planets are on the same straight line and their velocities are normal to that straight line.

Fig. 117 is presented in the heliocentric axes, axes in which the Poincaré space-time symmetry conditions remain the same.

Our purpose is to find the initial positions and velocities leading to a second Poincaré space-time symmetry at $t = T$ along the straight line determined by the angle β.

Let us first assume that the planets are infinitesimal and that $\sin\beta \neq 0$. All three orbits are then Keplerian orbits and obviously these orbits must be circular, their radius are determined by the final polar angle of each planet, that is β, $\beta+\pi$ and $\beta+2\pi$ in Fig. 117 (or in general $\beta+k\pi$).

If $\sin\beta = 0$ elliptic orbits are possible.

Let us now take account of the masses of the planets and of their mutual planetary perturbations.

If $\sin\beta \neq 0$ and if the planetary masses are sufficiently small the problem is "regular" and there are small successful modifications of the initial positions and velocities keeping the first Poincaré space-time symmetry and leading at $t = T$ to a second Poincaré symmetry.

The corresponding solutions are then periodic with the period $2T$ and the angle of rotation 2β. They are the "first species" of Poincaré periodic solutions with coplanar and almost circular orbits.

If $\sin\beta = 0$ the problem is singular because of the elliptic solutions and the demonstration does not work.

Another example is presented in Fig. 118 for a three-body and three-dimensional motion.

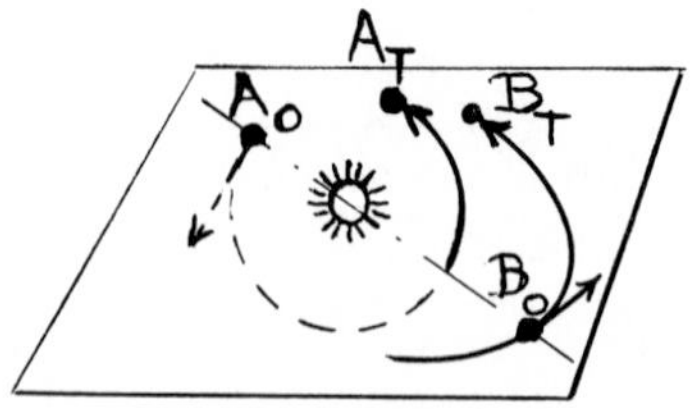

Fig. 118. The "third species" of Poincaré periodic solutions : the eccentricities are small and the inclinations are large.

The two planets A and B start at t = 0 at the points A_0 and B_0 on the line of nodes, their initial velocities are normal to this straight line and we have thus an initial Poincaré time-axis symmetry.

At the instant T the planets have reached the positions A_T and B_T with velocities normal to the Sun-A_T-B_T plane (Poincaré time-plane symmetry).

If the two planets are infinitesimal these two successive space-time symmetries are obtained with circular orbits and with an interval of time 0,T corresponding to odd numbers of quarters of revolution along each orbit.

If the two planets are sufficiently small the problem is again "regular" and there are small modifications of the initial positions and velocities which retain the two space-time symmetries.

The corresponding periodic solutions have the period 2T and an angle of rotation in the vicinity of π, they are the "third species" of Poincaré periodic solutions of the planetary case, with almost circular and non-coplanar orbits.

The "second species" is presented in Fig. 119, the planetary orbits are coplanar, co-axial and with arbitrary eccentricities.

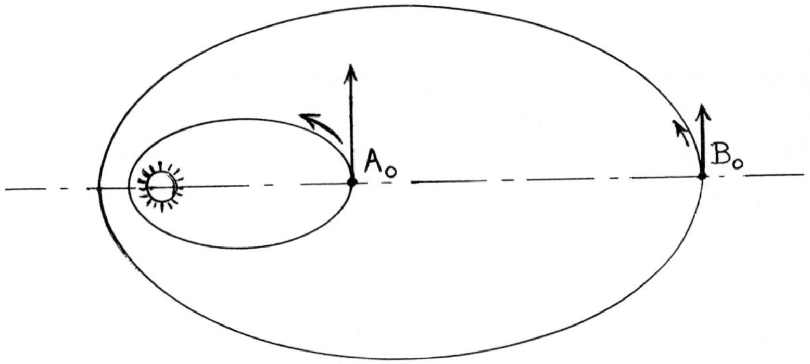

Fig. 119. The "second species" of Poincaré periodic solutions : the orbits are coplanar and co-axial, the eccentricities are large.

The initial conditions of Fig. 119 are obviously a time-axis Poincaré symmetry and these conditions will reappear in the future if the two periods of revolution are commensurable.

Notice however the effects of the small perturbations : the "advance of perihelion" can be different for the two planets and this would destroy the co-axiality.

Assume for instance that one of the two planets has a small but non-zero mass while the other planet has a mass equal to zero. The first planet then has no advance of perihelion (pure Keplerian motion) while the second planet has a small positive advance of perihelion (i.e. in the direction of orbital motion).

Hence for a given small sum $m_A + m_B$ of the two planetary masses, and if the orbital motions are in the same direction, there is one mass ratio m_A/m_B giving the same advance of perihelion and leading to periodic orbits. These periodic orbits have a very small angle of rotation.

Thus, for sufficiently small "planetary masses", the demonstrations of Poincaré lead to many families of periodic orbits of three different "species".

In the first species (Fig. 117) the orbits are coplanar and almost circular, the angle of rotation is different from zero and arbitrary.

In the second species (Fig. 119) the orbits are coplanar and with arbitrary eccentricities, the different families are characterized by the ratio of orbital periods* and the angle of rotation is near zero.

In the third species (Fig. 118) the orbits are almost circular with arbitrary inclinations, the ratio of nodal orbital periods** is equal to the ratio of two odd numbers and the angle of rotation is almost π.

Figure 117 presents the extension of the first species to the more-than-three-body problem, this can also be done for the two other species and in each case, for each family, the n-body problem has n non-trivial parameters :

The (n-1) mass ratios and the angle β for the families of the first species.

The (n-1) eccentricities and the mass ratio $(m_A + m_B + \ldots)/m_{SUN}$ for the families of the second species.

The (n-2) mutual inclinations and two suitable mass ratios for the families of the third species (one of them is $(m_A+m_B+\ldots)/m_{SUN}$).

---

* also called "anomalistic period" or period between two successive pericenter passages.
** also called "draconitic period".

11.4.2 The Poincaré conjecture

"Ce qui nous rend ces solutions périodiques si précieuses, c'est qu'elles sont, pour ainsi dire, la seule brèche par où nous puissions essayer de pénétrer dans une place jusqu'ici réputée inabordable" (The periodic orbits are extremely precious because they are our best and almost our only mean to penetrate this fortress previously considered unapproachable).

This famous proposition of Henri Poincaré (Ref. 199 p 82) can be extended to the n-body problem, indeed the periodic orbits have many advantages :

A) Their existence can be rigourously demonstrated.

B) The knowledge of one period gives the full knowledge of the solution.

C) They can be computed for all time with any given accuracy.

D) The families of periodic orbits are easy to follow step by step by the method of analytic continuation and they are connected to each other by the phenomenon of "branching" (see Section 10.5 and 8.5).

E) The unstable periodic orbits are limit of asymptotic orbits (see Section 10.7.7.2) while most of the first-order stable periodic orbits are, according to the Kolmogorov-Arnold-Moser theorem (Ref. 182), surrounded by families of tori of quasi-periodic orbits.

F) The most interesting feature of periodic orbits remains the still undemonstrated "Poincaré conjecture" :

The periodic solutions of the three-body problem, or even of the n-body problem, are certainly dense in the set of bounded solutions.

If true this conjecture would show that the periodic solutions are a very efficient mean of "exploration".

The very large number of families of periodic orbits already computed (Ref. 75 - 137), both in the three and more-than-three-body problem, seems to show that the Poincaré conjecture is true.

However notice that the periodic solutions are not dense among all solutions but only among the bounded solutions (and perhaps also among the "oscillatory solutions", see Section 11.7.7.6). For instance Section 11.1 shows that periodic solutions are impossible when the energy integral h is positive or zero.

11.5 UNSYMMETRICAL PERIODIC ORBITS. THE BROWN CONJECTURE

The method of Poincaré presented in Section 11.4 obviously leads to symmetrical periodic orbits.

Several methods exist for the demonstration of the existence of periodic solutions, for instance the method of successive collisions of infinitesimal planets (Ref.23,p.362; ref105), the method of power series or Fourier series, the application of the fixed point theorem, etc...

The method of "analytic continuation" either with or without modifications of masses (Sections 10.5.1 to 10.5.3) allow us to follow step by step families of periodic orbits and the phenomenon of "branching" connects different families (Orbit g1 of Figs 18 and 19 of Section 8.5 ; Fig. 51 in Section 10.5).

For decades most of the known periodic orbits were symmetrical. To-day the number of known families of periodic orbits is very large, and most of them are without symmetry (Ref. 75 - 137).

The two first known families of unsymmetrical periodic orbits were the "short period" and "long period" families of plane periodic orbits in the vicinity of the triangular equilibrium point L_4 in the circular restricted three-body problem (Fig. 120).

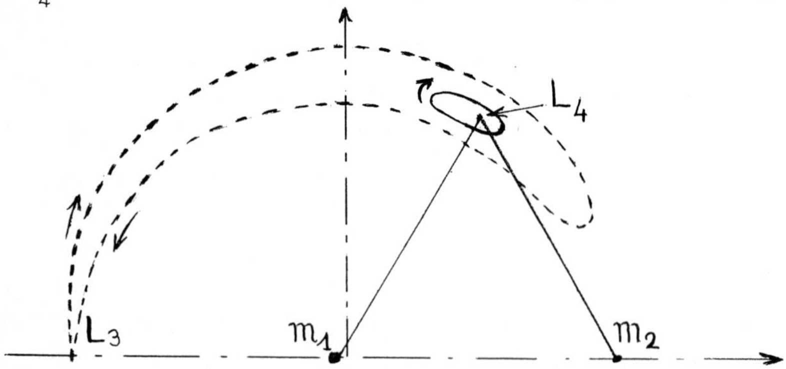

Fig. 120. Circular restricted three-body problem. A periodic orbit of the "long period" family and, in dotted line, the final orbit of the family according to the Brown conjecture.

These two families exist when the equilibrium point L_4 is linearly stable, i.e. when the mass ratio $m_2/(m_1+m_2)$ of the smallest primary to the total mass is smaller than 0.03852... (see Sections 8.2 or 10.7.10.1).

In 1911, Brown (Ref. 143) conjectured that the "long period" family has a final orbit doubly asymptotic to the unstable point L_3 (Fig. 120) ; such a doubly asymptotic orbit is called a "homoclinic orbit".

The numerical computations of J. Henrard (Ref. 139) have shown that, in the case of the Sun-Jupiter mass ratio $m_2/(m_1+m_2)$ = = 0.000953875, the Brown conjecture is not true (Fig. 121).

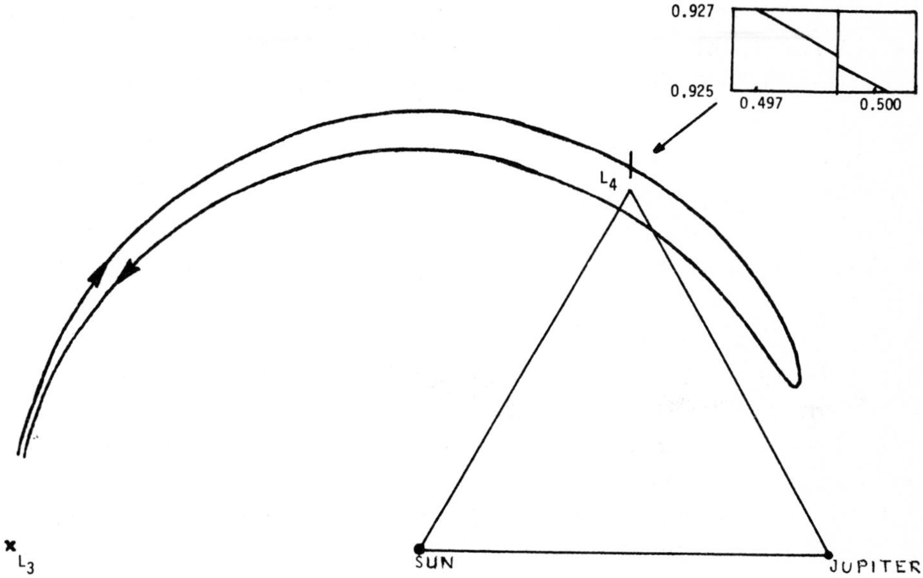

Fig. 121. Incoming and outgoing asymptotic orbits to L_3 when $m_2/(m_1+m_2)$ = 0.000953875. Source : Henrard, J. [139].
The outgoing orbit reaches the surface of Section $\{x=x(L_4)\ ;\ \dot{x} > 0\}$ at $y = 0.92592518\ ;\ \dot{y} = -\ 0.036877822$.
The incoming orbit starts at $y = 0.92572405\ ;\ \dot{y} = -\ 0.035097008$ along the same surface of section.
The two orbits are not very far apart but they are not the same.

The numerical computations also show that this long period family of periodic orbits begining at L_4 is much more complex than the short period family and than suggested by the Brown conjecture. Its "final orbit" is not unique and these final orbits seems to be homoclinic to short periodic orbits about L_3 (Ref. 139 ,141).

11.6 THE HILL STABILITY AND ITS GENERALIZATION

The circular restricted three-body problem has a well known integral, the Jacobi integral that is equivalent to the "energy integral in the rotating set of axes" (Fig. 122).

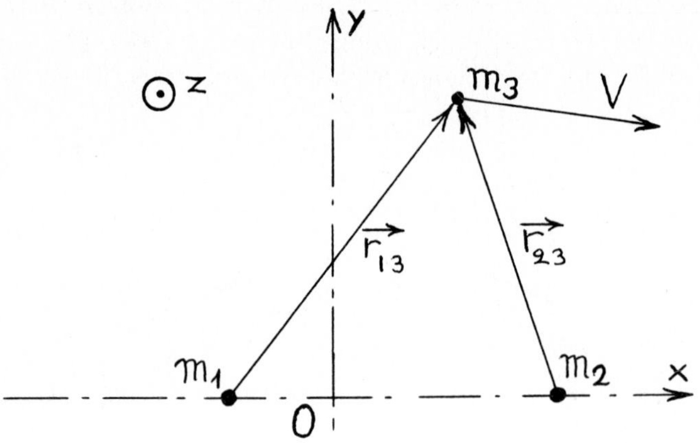

Fig. 122. The rotating set of axes.
Oxy is the plane of the circular orbits of m_1 and m_2 that are at fixed positions along the x-axis.

The distance r_{12} is $x_2 - x_1$, the mass m_3 is infinitesimal and the rate of rotation ω of m_1, m_2 and the axes is given by :

$$\omega^2 r_{12}^3 = G(m_1 + m_2) \tag{876}$$

We assume, as usual, that the unit of length, mass and time are chosen so that :

$$r_{12} = 1 \quad ; \quad m_1 + m_2 = 1 \quad ; \quad \omega = 1 \quad ; \quad G = 1 \tag{877}$$

If then x, y, z are the coordinates of m_3 and $\vec{V}$ its relative velocity (i.e. $\vec{V} = (dx/dt,\ dy/dt,\ dz/dt)$) the Jacobi integral Γ can be written :

$$\Gamma = m_1 \left(\frac{2}{r_{13}} + r_{13}^2\right) + m_2 \left(\frac{2}{r_{23}} + r_{23}^2\right) - z^2 - V^2 \tag{878}$$

The mechanical energy of m_3 (per unit of mass) in the rotating set of axes is $(-\Gamma/2)$ and we can write :

$$\left.\begin{aligned} &\Gamma = J - V^2 \leq J \\ &\text{with } J = \text{Jacobi function} = \\ &\qquad = m_1 \left(\frac{2}{r_{13}} + r_{13}^2\right) + m_2 \left(\frac{2}{r_{13}} + r_{23}^2\right) - z^2 \end{aligned}\right\} \tag{879}$$

The J = constant surfaces are the Hill surfaces and their intersection by the Oxy plane are the Hill curves (Fig. 123).

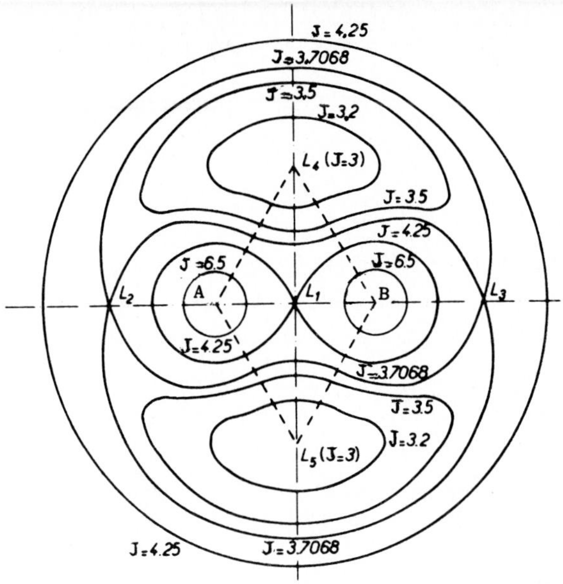

Fig. 123. The Hill curves of the circular rectricted three-body problem for equal primaries A and B.

Figure 23 of Section 9.1 presents the Hill curves in the case $m_1 = 10m_2$.

J is infinite at the two primaries and very far away ; its minimums in the Oxy plane are equal to 3 and occur at the triangular Lagrangian points L_4 and L_5 ; it has three saddle points at the collinear Lagrangian points L_1, L_2, L_3.

For a given Jacobi integral Γ the condition $J \geq \Gamma$ reduces the zone of possible motion to that limited by the $J = \Gamma$ Hill surface. If Γ is larger than $J(L_{1,2,3})$ the zone of possible motion of m_3 is divided into three disconnected parts : in the vicinity of m_1, in the vicinity of m_2 and very far away. A mobile m_3 initially near m_1 or near m_2 cannot escape, it has the "Hill stability".

The complete discussion can be found in Section 9.1. and in Ref. 1, p 159-207.

This classical notion of Hill stability has recently been extended to the general three-body problem (Ref.200-209 *). We will present it in the following simple way.

*We must also notice the numerous numerical computations of ROY A.E., VALSECCHI G.B. and CARUSI A, as in "The effect of Orbital eccentricities on the shape of the Hill-type Stability surfaces in the General Three-body Problem". Cel. Mech. 32, p.217 (1984).

11.6.1 The "generalized semi-major axis", the "generalized semi-latus rectum", the "mean quadratic distance", the "mean harmonic distance" and the Sundman function

Let us first consider some usual notation for three-body systems.

$$M = m_1 + m_2 + m_3 = \text{total mass} \tag{880}$$

$$\mu = GM = \text{gravitational constant of the system} \tag{881}$$

We will use the integrals of motion in the reference frame of the center of mass :

$$m_1\vec{r}_1 + m_2\vec{r}_2 + m_3\vec{r}_3 \equiv 0 \tag{882}$$

$$\vec{c} = \sum_{k=1}^{3} m_k\vec{r}_k \times \vec{V}_k = \text{angular momentum} \tag{883}$$

$$h = 0.5 \sum_{k=1}^{3} m_k V_k^2 - G\left(\frac{m_1 m_2}{r_{12}} + \frac{m_1 m_3}{r_{13}} + \frac{m_2 m_3}{r_{23}}\right) = \text{energy} \tag{884}$$

h is sometimes written :

$$h = 0.5 \sum_{k=1}^{3} m_k V_k^2 - U \tag{885}$$

with :

$$U = G\left(\frac{m_1 m_2}{r_{12}} + \frac{m_1 m_3}{r_{13}} + \frac{m_2 m_3}{r_{23}}\right) = \text{potential} \tag{886}$$

and finally as in (856) :

$$I = 0.5 \sum_{k=1}^{3} m_k r_k^2 = \text{semi-moment of inertia} \tag{887}$$

Let us now define the two following fixed lengths a and p by :

$$M^* = m_1m_2 + m_1m_3 + m_2m_3 \tag{888}$$

$$a = \text{"generalized semi-major axis"} = -\,GM^*/2h \tag{889}$$

$$p = \text{"generalized semi-latus rectum"} = Mc^2/GM^* \tag{890}$$

where a and p are indeed the semi-major axis and the semi-latus rectum of the relative orbit of the primaries when one of the three masses is infinitesimal (restricted case).

The variable lengths ρ and ν will be defined by :

$$M^* \rho^2 = m_1 m_2 r_{12}^2 + m_1 m_3 r_{13}^2 + m_2 m_3 r_{23}^2 \qquad (891)$$

ρ is the "mean quadratic distance", (ρ : Greek letter rho)

$$M^*/\nu = (m_1 m_2)/(r_{12}) + (m_1 m_3)/(r_{13}) + (m_2 m_3)/(r_{23}) \qquad (892)$$

ν is the "mean harmonic distance", (ν : Greek letter nu)

ρ and ν are related to the semi-moment of inertia I and the potential U :

$$M^* \rho^2 = 2MI \quad ; \quad GM^*/\nu = U \qquad (893)$$

The harmonic mean of positive quantities is always smaller than or equal to their quadratic mean, hence always :

$$0 \leq \inf(r_{12}, r_{13}, r_{23}) \leq \nu \leq \rho \leq \sup(r_{12}, r_{13}, r_{23}) \qquad (894)$$

The equality condition $\nu = \rho$ is obtained either for a two-body case ($\nu = \rho = r_{12}$) or for a restricted case ($\nu = \rho$ = distance of the primaries) or when all mutual distances r_{ij} are equal (equilateral triangle)**.

Finally we will also use :

$$\rho' = d\rho/dt \qquad (895)$$

$$j = (\rho/2a) + (p/2\rho) + (\rho\rho'^2/2\mu) = \text{Sundman function} \qquad (896)$$

11.6.2 The classical relations and the new notation

The notation of the previous Section does not lead to new equalities, but allows us to express the classical results by very simple and elegant relations.

For instance the inequalities

$$0 \leq \nu \leq \rho \qquad (897)$$

** Also notice that ν is of the order of the smallest mutual distance r_{ij} and ρ is of the order of the largest : $\{i \neq j\}$ implies: $m_i m_j \nu/M^* \leq r_{ij}$ and $r_{ij} \leq \rho\sqrt{M^* (m_i + m_j) / M m_i m_j}$.

corresponds to :

$$U > 0 \quad ; \quad IU^2 \geqq G^2M^{*3}/2M \tag{898}$$

For three given masses the minimum of the product IU^2 corresponds to the equilateral triangles and is indeed equal to the right-hand member of (898). However this complex expression is never used while we will see the interest of the inequality $\nu \leqq \rho$.

The Lagrange-Jacobi identity :

$$d^2I/dt^2 = U + 2h \tag{899}$$

is equivalent to :

$$d^2(\rho^2)/dt^2 = 2\mu(\frac{1}{\nu} - \frac{1}{a}) \tag{900}$$

In a two-body case and in restricted case ρ and ν are identical and then the relation (900) gives directly the evolution of the mutual distance in a two-body motion with the gravitational constant μ and the semi-major axis a.

In a general three-body case we have $\nu \leqq \rho$ and the two-body evolution appears as a limit.

Notice for instance that $\nu < a$ implies a concave up curve $\rho^2(t)$. Hence it is also the case if $\rho < a$ and the inequality $\rho < a$ can remain continuously satisfied for only limited periods of time, periods always less than $\pi\sqrt{a^3/\mu}$ (i.e. half the period of the two-body motion with the gravitational constant μ and the semi-major axis a).

Similarly the condition $\rho \geqq a$ remains satisfied for durations always larger than $\pi\sqrt{a^3/\mu}$, see Section 11.7.3.2.

The Sundman inequality :

$$c^2 + (dI/dt)^2 \leqq 4I(U + h) \tag{901}$$

is equivalent to :

$$\rho/\nu \geqq j \tag{902}$$

The function j is the Sundman function given in (896) and is proportional to the original Sundman function :

$$\left[c^2 + (dI/dt)^2 - 4hI\right] I^{-1/2} = jG\sqrt{(8M^{*3}/M)}.$$

The equalities (896) and (900) lead to :

$$\frac{dj}{dt} = \frac{\rho'}{\rho} \left(\frac{\rho}{\nu} - j\right) \qquad (903)$$

Hence the Sundman inequality (902) implies that ρ and j always vary in the same direction. This property gives much long-term information that can be extended to the n-body problem (Sections 11.7.1 to 11.7.3.4 and Ref. 210).

11.6.3 Hill-type stability in the general three-body problem

Let us consider three-body motions of three given masses m_1, m_2, m_3 with given integral of motion $\vec{c}$ and h in the axes of their center of mass.

Let us look for possible configurations of the three bodies and for their possible mutual distances r_{12}, r_{13}, r_{23} (Fig. 124).

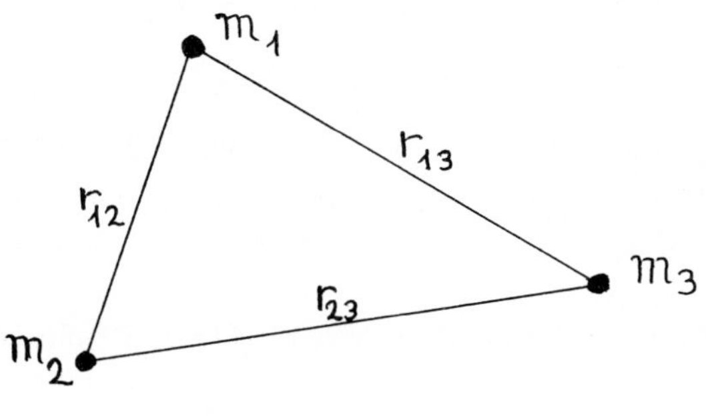

Fig. 124. A configuration and three mutual distances.

With the two fixed lengths a, p and the two variable lengths ρ, ν defined in (889)-(893) in Section 11.6.1 the answer is simple :

A configuration with the three mutual distances r_{12}, r_{13}, r_{23} is possible if and only if :

$$2/\nu \geqq (1/a) + (p/\rho^2) \qquad (904)$$

Hence a > 0 implies $0 \leqq \nu \leqq 2a$ always.

A) The condition is necessary.

Indeed the Sundman inequality (902) can be written :

$$\rho/\nu \geq (\rho/2a) + (p/2\rho) + (\rho\rho'^2/2\mu) \tag{905}$$

and if (904) is not satisfied (905) cannot be satisfied.

B) The condition is sufficient.

Indeed suitable orientation and velocities exist to obtain the integrals of motion : It is sufficient to put the three bodies in the "invariable plane" normal to $\vec{c}$ and to choose the following velocities :

$$\left.\begin{aligned} &\vec{V}_k = A\vec{c} \times \vec{r}_k + B\vec{r}_k \quad ; \quad k = \{1,\ 2,\ 3\} \\ &\text{with :} \\ &A = 1/2I = M/M^*\rho^2 \\ &B = \pm\sqrt{(\mu/\rho^2)\ \left[(2/\nu) - (1/a) - (p/\rho^2)\right]} \end{aligned}\right\} \tag{906}$$

In the limiting case when (904) is just satisfied the solution (906) gives $B = 0$ and is the only possible solution, but there are many other possible orientations and velocities when $2/\nu > 1/a + p/\rho^2$.*

Let us now disregard the scale of the triangle of the three bodies and let us look for the possible configurations. For a given configuration the two mean distances ρ and ν are proportional to the scale of the configuration but their ratio ρ/ν is constant and we will write (904) as :

$$\rho/\nu \geq (\rho/2a) + (p/2\rho) \tag{907}$$

If the energy integral h is positive or zero the quantity $1/a$, equal to $-2h/GM^*$, is negative or zero and (907) can always be satisfied for sufficiently large ρ.

---

* These solutions always satisfy the following :

$$\sum_{j=1}^{3} m_j z_j^2 \leq 2BC\ (U + h)\ /c^2$$

where z is the coordinate in the direction of $\vec{c}$ and A, B, C are the three usual main moments of inertia of the system (A=2I=B+C). All collisions (U infinite, CU=0) and all collinear alignments (C=0) occur in the invariable plane z=0.

If the energy integral h is negative the generalized semi-major axis a is positive and the minimum of the right-hand member of (907) is obtained for $\rho = \sqrt{ap}$ and is equal to $\sqrt{p/a}$.

In both cases we can write that a given configuration is possible (i.e. compatible with the integrals of motion) if and only if* :

$$\rho^2/\nu^2 \geqq p/a \tag{908}$$

Maps of ρ/ν = constant curves can be drawn in terms of the position of the mass m_1 relative to m_2 and m_3. These maps (Fig. 125) are generalizations of the maps of Hill's curves, indeed :

A) The ratio ρ/ν is minimum and equal to one at the triangular Lagrangian points L_4 and L_5.

B) The ratio ρ/ν is infinite at m_2, at m_3 and at infinity.

C) The ratio ρ/ν has three saddle points at the collinear Lagrangian points L_1, L_2, L_3.

D) These curves become the Hill curves of Fig. 123 as m_1 goes to zero.

Since the ratio ρ/ν is always larger than or equal to one the inequality (908) is useless when $p/a \leqq 1$, that is when $c^2h \geqq -G^2(M^*)^3/2M$, but it becomes very useful when $p/a > 1$. The configurations in the vicinity of equilateral triangles are then forbidden <u>for all time</u>. Furthermore if $\sqrt{p/a}$ is larger than $\sup_{j=\{1,2,3\}} (\rho/\nu)_{Lj}$ the possible configurations are divided into three disconnected parts, m_1 being either near m_2 or near m_3 or at a large distance : The motion remains <u>for all time</u> confined to one of these parts and there is forever a "small binary" that the "third isolated body" can neither approach nor disrupt.

This kind of stability will be called a "Hill-type stability". However notice that this Hill-type stability does not prevent an escape of the third isolated body, which is a difference with the usual Hill stability of the circular restricted three-body problem.

---

* In classical notation the relation (908) is $IU^2 + c^2h \geqq 0$.

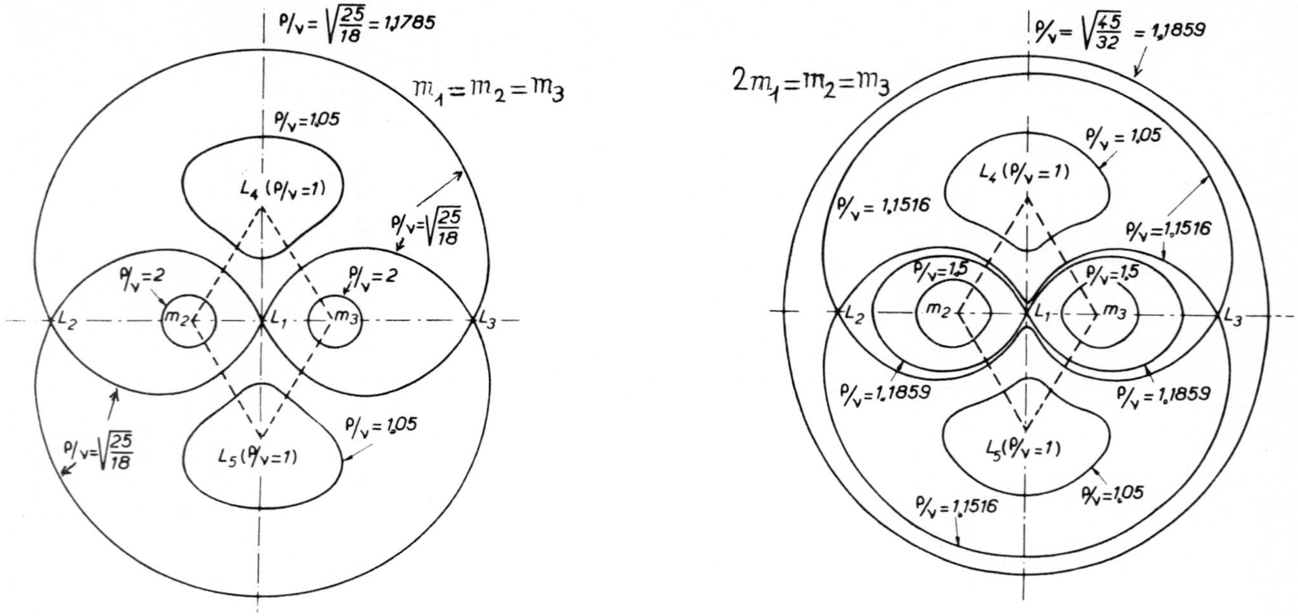

Fig. 125. Generalization of Hill curves in the cases $m_1=m_2=m_3$ and $2m_1=m_2=m_3$. Curves of constant ratio ρ/v in terms of the relative positions of m_1 with respect to m_2 and m_3. ρ/v is minimum and equal to one at the triangular Lagrangian points L_4 and L_5, it is infinite at m_2, at m_3 and at infinity, finally it has three saddle points at the collinear Lagrangian points L_1, L_2, L_3. The subscripts of the collinear Lagrangian points L_1, L_2, L_3 are the subscripts of the corresponding central masses.

If we give to the collinear Lagrangian points L_1, L_2, L_3 the subscript of the corresponding central mass (as in Fig. 125), the corresponding values of ρ/ν are in the opposite order of the masses, for instance :

$$\left.\begin{array}{l} m_1 \leqq m_2 \leqq m_3 \text{ implies } (\rho/\nu)_{L_1} \geqq (\rho/\nu)_{L_2} \geqq (\rho/\nu)_{L_3} \\ \text{and even :} \\ \sqrt{(343/243)}=1.1881\ldots \geqq (\rho/\nu)_{L_1} \geqq (\rho/\nu)_{L_2} \geqq (\rho/\nu)_{L_3} \geqq (\rho/\nu)_{L_4} = (\rho/\nu)_{L_5} = 1 \end{array}\right\} \quad (909)$$

The least upper bound of $(\rho/\nu)_{L_1}$ is obtained for $m_2 = m_3 = 1.6m_1$.

Notice that the Hill-type stability condition can be slitghtly extended : if $\sqrt{p/a}$ is between the largest and the second largest $(\rho/\nu)_{Lj}$ the zone of possible motion is no more divided into three disconnected parts but it is still divided into two parts and the Hill-type stability survives if the isolated mass is the smallest mass.

Furthermore the Hill-type stability survives even at the limit, i.e. when p/a is equal to its critical value, indeed then two neighbouring parts meet at one or several collinear Lagrangian points (see Fig. 125) but this or these Lagrangian points only corresponds then to circular Eulerian motions and remains barriers between the nearby parts.

Thus in all cases the Hill-type stability corresponds to $p/a \geqq$ "(p/a) critical" and the limit of the Hill-type stability corresponds to the collinear central configuration with the smallest mass of the binary between the two other masses. The critical value of the ratio p/a is equal to the ratio ρ^2/ν^2 for this central configuration, it is also equal to the p/a ratio of the circular Eulerian motion with this configuration (the circular Eulerian motions have $\nu \equiv a$ and $\rho \equiv \sqrt{ap}$).

The ratio p/a is proportional to the product c^2h with a negative coefficient :

$$p/a = -\, 2c^2hM/G^2(M^*)^3 \qquad (910)$$

Hence we can also write that the Hill-type stability occurs when the product c^2h is below than or equal to its negative critical value given by the corresponding circular Eulerian motion.

For a Hill-stable system it is of course very natural to use the classical Jacobi decomposition of Fig. 126 already considered in Section 4.3 with the corresponding "inner" and "outer" (or "exterior") osculating orbits.

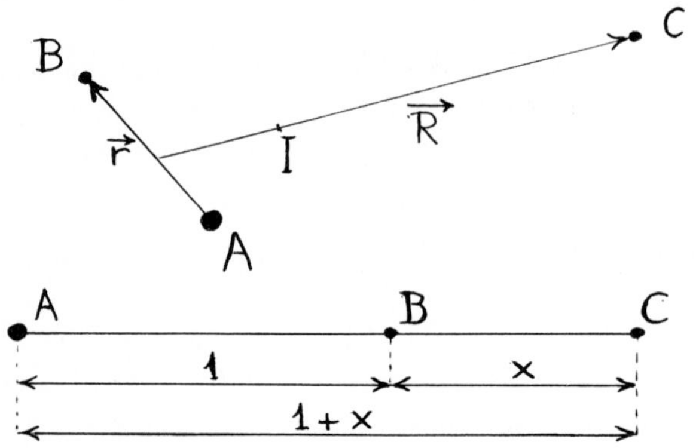

Fig. 126. The Jacobi decomposition of a Hill stable system (binary AB, isolated body C) and the corresponding collinear central configuration with $m_B \leq m_A$; B between A and C and $x = r_{BC}/r_{AB}$.

The "generalized semi-major axis" a and the "generalized semi-latus rectum" p are related to the (variable) elements of the inner and exterior orbits : the inner semi-major axis a_i of a Hill-stable system remains forever between 0 and a ; the exterior semi-latus rectum p_e has an almost symmetrical property :

$$\left.\begin{array}{l} m_c/M \leqq 0.75 \text{ implies : } p_e \text{ is forever larger than } p \\ m_c/M > 0.75 \text{ implies : } p_e \text{ is forever larger than } 0.98\, p \\ (m_c = \text{mass of the isolated body ; } M = \text{total mass}). \end{array}\right\} \qquad (911)$$

The critical values of the ratio p/a are only function of the mass ratios.

For three bodies A, B, C with B between A and C the collinear central configuration of Fig. 126 gives the following critical ratio $(p/a)_B$:

$$(p/a)_B = (\rho^2/\nu^2) \text{ minimum for B on the segment AC} =$$
$$= \inf_{x>0}\left\{\left[m_A m_B + m_A m_C(1+x)^2 + m_B m_C x^2\right]\cdot \left[m_A m_B + \frac{m_A m_C}{1+x} + \frac{m_B m_C}{x}\right]^2 \cdot \left[m_A m_B + m_A m_C + m_B m_C\right]^{-3}\right\} \tag{912}$$

x is here the ratio of mutual distances r_{BC}/r_{AB} (Fig. 126) and, since it corresponds to the minimum of ρ^2/ν^2, an obvious differentiation gives its value for the central configuration A, B, C :

$$(m_A + m_B)x^5 + (3m_A + 2m_B)x^4 + (3m_A + m_B)x^3 - (m_B + 3m_C)x^2 - (2m_B + 3m_C)x - (m_B + m_C) = 0 \tag{913}$$

This quintic equation always has one and only one positive root, it is the classical equation of central configurations.

Tables II and III give $(p/a)_B$ in terms of the mass ratios and are presented graphically in Fig. 127, while tables III and IV give the corresponding distances ratio x.

m_B/m_A ↓ \ m_C/m_A →	0	0.1	0.2	0.3	0.4	0.5	0.6	0.7	0.8	0.9	1
0	1	1	1	1	1	1	1	1	1	1	1
0.1	1	1,2316	1.2604	1,2574	1,2475	1,2363	1.2256	1,2157	1.2066	1,1984	1,1910
0.2	1	1,2327	1,3080	1,3332	1,3392	1.3369	1,3309	1,3234	1,3152	1,3069	1,2988
0.3	1	1,2126	1,3061	1.3496	1,3694	1,3771	1,3782	1,3756	1,3710	1,3653	1,3589
0.4	1	1,1920	1,2911	1.3444	1,3735	1.3889	1,3962	1,3985	1.3977	1,3950	1,3909
0.5	1	1.1741	1,2730	1,3313	1.3662	1.3870	1.3990	1,4053	1,4079	1.4079	1,4063
0.6	1	1,1589	1,2553	1,3155	1.3538	1.3783	1,3938	1,4033	1,4086	1,4110	1.4113
0.7	1	1,1461	1,2389	1,2994	1.3395	1.3664	1.3844	1.3962	1,4037	1,4080	1,4102
0.8	1	1,1351	1,2239	1,2838	1,3248	1,3532	1,3729	1,3865	1.3956	1,4015	1,4051
0.9	1	1,1256	1,2105	1,2692	1,3104	1,3397	1.3605	1.3754	1,3858	1,3930	1,3977
1	1	1,1174	1,1985	1,2557	1,2966	1,3263	1,3479	1.3637	1.3751	1,3832	1.3889

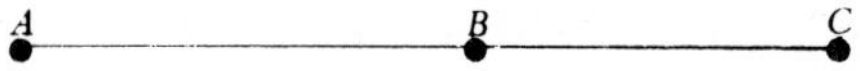

Table II. Ratio $(p/a)_B$ in terms of the mass ratios when m_A is the largest mass.
$(p/a)_B$ is equal to ρ^2/ν^2 for the collinear central configuration A, B, C.
This table gives the two critical (p/a) values for three given masses (see (914), (915)).

m_A/m_B \ m_C/m_B	0	0.1	0.2	0.3	0.4	0.5	0.6	0.7	0.8	0.9	1
0	1										
0.1	1	1.0891									
0.2	1	1.1129	1.1597								
0.3	1	1.1214	1.1823	1.2157							
0.4	1	1.1245	1.1938	1.2350	1.2604						
0.5	1	1.1250	1.1995	1.2462	1.2763	1.2960					
0.6	1	1.1243	1.2020	1.2525	1.2861	1.3089	1.3243				
0.7	1	1.1229	1.2025	1.2557	1.2920	1.3172	1.3347	1.3467			
0.8	1	1.1212	1.2018	1.2569	1.2952	1.3223	1.3414	1.3550	1.3644		
0.9	1	1.1193	1.2004	1.2567	1.2966	1.3251	1.3456	1.3604	1.3709	1.3782	
1	1	1.1174	1.1985	1.2557	1.2966	1.3263	1.3479	1.3637	1.3751	1.3832	1.3889

Table III. Ratio $(p/a)_B$ in terms of the mass ratios when m_B is the largest mass.
$(p/a)_B$ is equal to ρ^2/ν^2 for the collinear central configuration A, B, C.

m_B/m_A \ m_C/m_A	0	0.1	0.2	0.3	0.4	0.5	0.6	0.7	0.8	0.9	1
0	0	.394	.518	.611	.687	.752	.810	.863	.912	.958	1
0.1	.347	.477	.571	.648	.714	.772	.825	.873	.918	.960	1
0.2	.438	.534	.612	.678	.736	.789	.837	.882	.924	.963	1
0.3	.499	.577	.644	.703	.755	.804	.848	.890	.928	.965	1
0.4	.544	.612	.671	.724	.772	.816	.857	.896	.933	.967	1
0.5	.581	.641	.694	.742	.786	.827	.866	.902	.936	.969	1
0.6	.612	.665	.714	.758	.799	.837	.873	.907	.940	.970	1
0.7	.638	.687	.731	.772	.810	.846	.880	.912	.943	.972	1
0.8	.661	.705	.746	.784	.820	.854	.886	.916	.945	.973	1
0.9	.681	.721	.760	.795	.829	.861	.891	.920	.948	.974	1
1	.698	.736	.772	.805	.837	.867	.896	.923	.950	.975	1

Table IV. Values of the distances ratio $x = r_{BC}/r_{AB}$ in the central configuration A, B, C for the mass ratios of Table II.

m_A/m_B \ m_C/m_B	0	0.1	0.2	0.3	0.4	0.5	0.6	0.7	0.8	0.9	1
0	1										
0.1	.947	1									
0.2	.902	.953	1								
0.3	.865	.913	.958	1							
0.4	.832	.878	.921	.962	1						
0.5	.803	.847	.889	.928	.965	1					
0.6	.777	.820	.860	.898	.934	.968	1				
0.7	.755	.796	.835	.871	.906	.939	.970	1			
0.8	.734	.774	.812	.847	.881	.913	.943	.972	1		
0.9	.715	.754	.791	.825	.858	.889	.918	.947	.974	1	
1	.698	.736	.772	.805	.837	.867	.896	.923	.950	.975	1

Table V. Value of the distances ratio $x = r_{BC}/r_{AB}$ in the central configuration A, B, C for the mass ratios of Table III.

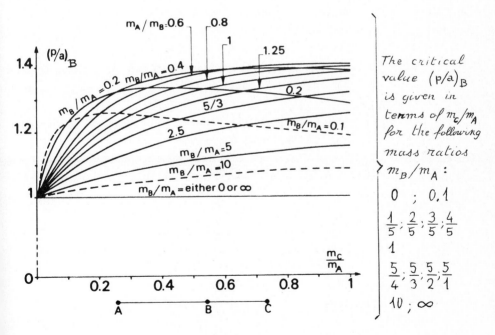

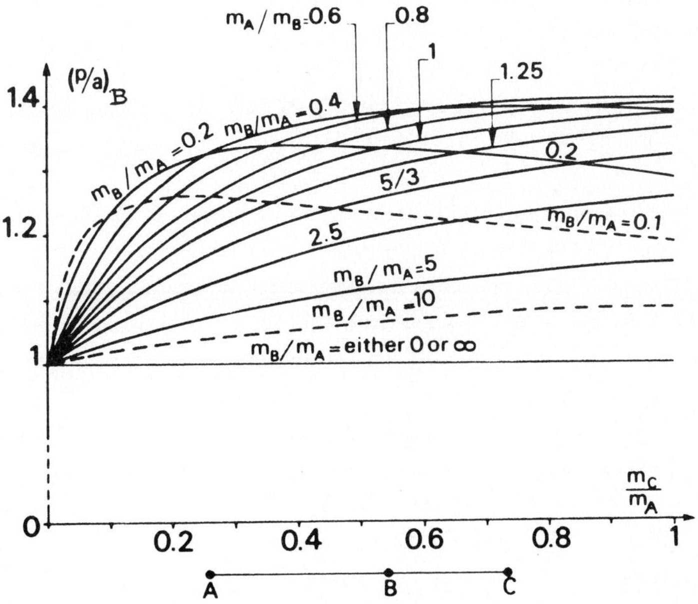

The critical value $(p/a)_B$ is given in terms of m_C/m_A for the following mass ratios m_B/m_A:
0 ; 0.1
$\frac{1}{5}$; $\frac{2}{5}$; $\frac{3}{5}$; $\frac{4}{5}$
1
$\frac{5}{4}$; $\frac{5}{3}$; $\frac{5}{2}$; $\frac{5}{1}$
10 ; ∞

Fig. 127. Ratio $(p/a)_B$ in terms of the mass ratios.
$(p/a)_B$ is equal to ρ^2/ν^2 for the collinear central configuration A, B, C.

In agreement with (909) we always find that

$$\left.\begin{array}{l} \{m_1 \leq m_2 \leq m_3\} \text{ implies :} \\ 343/243 = 1.4115\ldots \geq (p/a)_1 \geq (p/a)_2 \geq (p/a)_3 \geq (p/a)_4 = \\ \qquad = (p/a)_5 = 1 \end{array}\right\} \quad (914)$$

and the two critical values of p/a are then $(p/a)_1$ and $(p/a)_2$.

For instance :

$$\left.\begin{array}{l} \{m_1 = 2 \ ; \ m_2 = 5 \ ; \ m_3 = 10\} \text{ implies :} \\ (p/a)_1 = 1.3369 \ ; \ (p/a)_2 = 1.2730 \ ; \ (p/a)_3 = 1.1995 \end{array}\right\} \quad (915)$$

A simple example of a three-body system with Hill-type stability is provided by the solar system if we conider only the Sun, Jupiter and Saturn (99.99% of the total mass). The ratio p/a of the system is 1.0447, which is larger than the critical p/a ratio equal there to $(p/a)_{(Jupiter)}$ i.e. 1.0081.

The binary that can neither be approached nor disrupted by the third body, even over extremly long periods of time, is the Sun and Jupiter. Its semi-major axis will forever remain between 4.473 and 5.216 AU.

The Hill-type stability does not prevent an escape of Saturn.

Several triple stellar systems have been considered (Ref.211), most of them have Hill-type stability. On the contrary the Sun--Earth-Moon system and most Sun-planet-satellite systems are Hill-unstable, essentially because of the eccentricity of planetary orbits (Ref.212, p. 327).

This reference also presents a detailed study of the three limiting cases : the restricted case, the planetary case and the lunar case.

11.6.4 Scale effects

The condition $\rho^2/\nu^2 \geq p/a$ written in (908) is independant of the scale of the configuration of interest.

The stronger condition (904) can also be written as :

$$\left.\begin{aligned} &\rho/\nu \geqq (\rho/2a) + (p/2\rho) \\ &\text{or in terms of } \nu : \\ &\nu^2/\rho^2 \leqq \frac{\nu(2a - \nu)}{ap} \end{aligned}\right\} \quad (916)$$

These relations (916) are equivalent to $\rho^2/\nu^2 \geqq p/a$ when $\rho = \sqrt{ap}$ and/or when $\nu = a$, and are stronger in the other cases. Hence a given configuration with a given ratio ρ/ν can be reached only if ρ is sufficiently near to $\sqrt{ap}$ and/or only if ν is sufficiently near to a.

Note especially that the exchanges between bodies, which as seen in the previous section require $\rho/\nu \leqq \sqrt{(p/a)}$critical, can only happen for a limited range of values of ρ and ν.

11.6.5 Hill-type stability for systems with positive or zero energy integral

When the energy integral h is positive or zero the ratio p/a is negative or zero and the condition $\rho^2/\nu^2 \geqq p/a$ in (908) is always satisfied and thus completely useless.

Fortunately a second case of Hill-type stability can be obtained in the following way.

Let us consider first the Lagrange-Jacobi identity already given in (899)-(900) :

$$d^2(\rho^2)/dt^2 = 2\mu\left(\frac{1}{\nu} - \frac{1}{a}\right) \quad (917)$$

Since $0 \leqq \nu \leqq \rho$ and $-2\mu/a = 4Mh/M^* \geqq 0$, we can write :

$$d^2(\rho^2)/dt^2 \geqq 2\mu/\nu \geqq 2\mu/\rho > 0 \quad (918)$$

Hence the curve $\rho^2(t)$ is concave up and ρ has one and only one minimum : ρ_m at some time t_m. (Since ρ^2 is proportional to I this result was already obvious for $h > 0$, but not for $h = 0$, with the analysis of Section 11.1).

The limiting case $d^2(\rho^2)/dt^2 = 2\mu/\rho$ gives :

$$2(\rho - \rho_m)(\rho + 2\rho_m)^2 = 9\mu(t - t_m)^2 \quad (919)$$

Hence ρ goes up to infinity at least as $\{9\mu t^2/2\}^{1/3}$.

Let us now consider the relations (902) and (903) : the Sundman function j varies in the direction of ρ, hence it has one and only one minimum, j_m at the time t_m.

$$j_m = (\rho_m/2a) + (p/2\rho_m) \tag{920}$$

Let us give the subscript zero to the initial values at the time t_0 and let us assume that $\rho'_0 \geq 0$ i.e. that $t_0 \geq t_m$.
Hence at all later times :

$$\left.\begin{aligned} & t > t_0 \quad ; \quad \rho' > 0 \quad ; \quad \rho > \rho_0 \\ & \rho/\nu \geq j \geq j_0 = (\rho_0/2a) + (p/2\rho_0) + (\rho_0\rho_0'^2/2\mu) \end{aligned}\right\} \tag{921}$$

and at all former times, since $\rho_0 \geq \rho_m$ and $\rho_0/2a \leq \rho_m/2a$:

$$\left.\begin{aligned} & t < t_0 \\ & \rho/\nu \geq j \geq j_m = (\rho_m/2a) + (p/2\rho_m) \geq (\rho_0/2a) + (p/2\rho_0) \end{aligned}\right\} \tag{922}$$

Thus if $\rho'_0 \geq 0$ we obtain two lower bounds of the ratio ρ/ν :

$$\left.\begin{aligned} & \rho/\nu \geq j_0 \quad \text{if} \quad t \geq t_0 \\ & \rho/\nu \geq (\rho_0/2a) + (p/2\rho_0) \quad \text{if} \quad t < t_0 \end{aligned}\right\} \tag{923}$$

A symmetrical property is true if $\rho'_0 \leq 0$.

These lower bounds allow a discussion similar to that of Section 11.6.3 with its inequality $\rho/\nu \geq \sqrt{p/a}$ given in (908). If the lower bound is above its critical value the motion has a Hill-type stability and remains in one of the two or three disconnected zones of possible motion with a "binary" and an "isolated body".

The square of the critical values are given in the equation (912) and in the Tables II and III of Section 11.6.3 in terms of the mass ratios.

Notice that here the energy integral h is positive or zero and the mean quadratic distance goes up to infinity. Hence the isolated body escapes infinitely far from the binary, but this binary can be disrupted.

If h = 0 it is possible to demonstrate that the binaries of Hill-stable motions remain bounded.

11.7 FINAL EVOLUTIONS AND TESTS OF ESCAPE

The final evolutions and the tests of escape are generally considered to be the main domain of application of qualitative methods (note that these tests are also called criteria of escape).

It is of course very difficult to use quantitative methods, either analytical or numerical, to cover infinite periods of time and it is thus natural that qualitative methods find there their domain of excellence.

The final evolutions were classified by Chazy in 1922 (Ref. 213) as presented in Section 11.7.7. This classification has received some minor improvements and extensions and is summarized in Section 11.7.9.

The most surprising new element is the importance of "oscillatory evolutions of the second kind" that lead to infinitely close approaches, that is practically to collisions (Section 11.7.7.6).

All tests of escape obey the basic principle developed in Section 11.7.4. Section 11.7.5 presents the application of this basic principle to the construction of a simple and remarkable test of the n-body problem, with the three successive classical phase of the construction.

In recent years the progress of the tests of escape have been so large (Section 11.7.10) that they have almost reached their utmost possible limits. The result is a surprise : bounded motions are much rarer that was generally expected and the strict hierarchy of the Solar System is not only a result of its formation, but also a condition of its stability.

This Section 11.7 begins by looking at some classical results of the n-body problem expressed with the new notations of Sections 11.6.1 and 11.6.2. These classical results are improved in Section 11.7.3.

The new notations of these Sections have been found essential for the considerable recent progress of the tests of escape and for the analysis of final evolutions.

11.7.1 The new notations and the n-body problem

The extension of the notations of Section 11.6.1 to the n-body problem is obvious :

$$M = \sum_{K=1}^{n} m_K = \text{total mass} \tag{924}$$

$\mu = GM$ = gravitational constant of the system (925)

$\sum_{K=1}^{n} m_K \vec{r}_K \equiv 0 \quad ; \quad \sum_{K=1}^{n} m_K \vec{V}_K \equiv 0$, because we choose the axes of the center of mass (926)

$\vec{c} = \sum_{K=1}^{n} m_K \vec{r}_K \times \vec{V}_K$ = angular momentum (927)

$h = 0.5 \sum_{K=1}^{n} m_K V_K^2 - G \sum_{1 \leq j < K \leq n} m_j m_K / r_{jK}$ = energy (928)

$h = 0.5 \sum_{K=1}^{n} m_K V_K^2 - U$

with : $U = G \sum_{1 \leq j < K \leq n} m_j m_K / r_{jK}$ = potential (929)

$I = 0.5 \sum_{K=1}^{n} m_K r_K^2$ = semi-moment of inertia (930)

$M^* = \sum_{1 \leq j < K \leq n} m_j m_K$ (931)

$a = - GM^*/2h$ = "generalized semi-major axis" (932)

$p = Mc^2/G(M^*)^2$ = "generalized semi-latus rectum" (933)

$M^* \rho^2 = \sum_{1 \leq j < K \leq n} m_j m_K r_{jK}^2 = 2MI$

ρ is the "mean quadratic distance" (934)

$M^*/\nu = \sum_{1 \leq j < K \leq n} m_j m_K / r_{jK} = U/G$

ν is the "mean harmonic distance" (935)

$\inf_{1 \leq j < K \leq n} r_{jK} = r$ = "smallest mutual distance" (936)

$\sup_{1 \leq j < K \leq n} r_{jK} = R$ = "largest mutual distance" (937)

We have always :

$$0 \leq r \leq \nu \leq \rho \leq R \tag{938}$$

and even, independently of the n masses :

$$\rho^2 + (Rr(R+r))/\nu \leq R^2 + Rr + r^2 \tag{939}$$

The limit in (939) is obtained when all mutual distances r_{jK} are equal either to R or to r and then :

$$\left.\begin{array}{ll} \rho^2 = \alpha r^2 + \beta R^2 & \text{for some positive } \alpha,\beta \text{ depending} \\ 1/\nu = \alpha/r + \beta/R & \text{on the masses and such that } \alpha + \beta = 1 \end{array}\right\} \tag{940}$$

The elimination of α and β within the three linear equations of (940) gives the limit of (939).

If all masses are positive the mean quadratic distance ρ is of the order of the largest mutual distance R and the mean harmonic distance ν is of the order of the smallest mutual distance r. Indeed :

$$\{j \neq K\} \text{ implies : } m_j m_K \nu/M^* \leq r_{jK} \leq \rho\sqrt{M^*(m_j + m_K)/Mm_j m_K} \tag{941}$$

Hence, if m_1 and m_2 are the two smallest masses :

$$m_1 m_2 \nu/M^* \leq r \leq \nu \quad ; \quad \rho \leq R \leq \rho\sqrt{M^*(m_1 + m_2)/Mm_1 m_2} \tag{942}$$

These expressions become **worthless** if one of the n masses is infinitesimal.

Finally we find again :

$$\rho' = d\rho/dt \tag{943}$$

$$j = (\rho/2a) + (p/2\rho) + (\rho\rho'^2/2\mu) = \text{Sundman function} \tag{944}$$

The main relations remain as in Section 11.6.2 :

A) Lagrange-Jacobi identity :

$$d^2I/dt^2 = U + 2h \tag{945}$$

that is :

$$d^2(\rho^2)/dt^2 = 2\mu\big((1/\nu) - (1/a)\big) \tag{946}$$

B) Sundman inequality :

$$c^2 + (dI/dt)^2 \leq 4I(U + h) \tag{947}$$

that is :

$$\rho/\nu \geqq j \qquad (948)$$

C) Derivative of the Sundman function :

$$dj/dt = (\rho'/\rho)\big((\rho/\nu) - j\big) \qquad (949)$$

11.7.2 The classical result and the new notations

The oldest classical result on final evolutions is merely an obvious consequence of the expression of the energy integral :

"If the energy integral is negative the smallest mutual distance r, and the mean-harmonic distance ν, will remain forever bounded".

Indeed (929) can be written :

$$0 \leqq \sum_{K=1}^{n} m_K V_K^2 = 2U + 2h = GM^*((2/\nu) - (1/a)) \qquad (950)$$

If the energy integral h is negative the generalized semi-major axis a is positive and hence, since $0 \leqq r \leqq \nu$, at all times and for n arbitrary masses :

$$0 \leqq r \leqq \nu \leqq 2a \qquad (951)$$

The equality $\nu = 2a$ is obtained when all velocities are zero and we get $r = 2a$ if furthermore all mutual distances are equal (equilateral triangle or regular tetrahedron). The motion is then of radial Lagrangian type.

The generalized major axis 2a thus has a physical meaning : for the given energy it is the least upper bound of the smallest distance and of the mean harmonic distance.

The symmetrical proposition is related to Section 11.1 :

"If the energy integral h is positive or zero the largest mutual distance R, and the mean quadratic distance ρ, will go to infinity when the time t goes to plus or minus infinity".

This proposition assume of course that the motion avoids singularities and remains indefinitely defined, we will see that question in Section 11.7.6.

The discussion of the three-body case presented in Section 11.6.5 is also valid for the n-body case and thus, since

$0 \leqq \rho \leqq R$:

If the energy integral h is zero the distances ρ and R go up to infinity at least as $\{9\mu t^2/2\}^{1/3}$.

If the energy integral h is positive the second derivative $(\rho^2)''$ is always larger than the positive quantity $(-2\mu/a)$ and the distances ρ and R go up to infinity at least as $|t|.(-\mu/a)^{1/2}$.

In a two-body problem of positive energy (hyperbolic motion) the velocity $(-\mu/a)^{1/2}$ is the relative velocity at infinity. It becomes here a lower bound of the velocity $|\rho'|$ at infinity.

More accurate informations on the evolution of $\rho(t)$ and $R(t)$ will be given in the next Section 11.7.3.2 especially evolution in the vicinity of zero as already suggested in Section 11.6.2.

There remains one classical general proposition :

"If the energy integral h is negative, the smallest mutual distance r has a least upper bound (that is 2a) and it cannot remain forever above any fixed quantity larger than half this least upper bound".

This property is true even for the mean harmonic distance ν that is larger and has the same least upper bound 2a.

Indeed assume that exists a fixed length b such that :

$$t \geqq t_0 \quad \text{implies} \quad \nu(t) \geqq b > a > 0 \tag{952}$$

Hence :

$$d(\rho^2)/dt^2 = 2\mu\left((1/\nu) - (1/a)\right) \leqq 2\mu\left((1/b) - (1/a)\right) < 0 \tag{953}$$

and the mean quadratic distance ρ would reach the vicinity of zero and become smaller than ν, which is impossible.

If all n masses are positive (i.e. non zero) and if $h < 0$, it is even possible to demonstrate that :

$$\text{If } t \geqq t_0 \quad \text{implies} \quad \nu(t) > a \quad , \quad \text{then} \lim_{t\to\infty} \nu = a \tag{954}$$

Indeed we have :

A) The n velocities V_K remain bounded for all times t larger than t_0 and so is the velocity ν' equal to $d\nu/dt$:

$$\sum_{K=1}^{n} m_K V_K^2 = GM^* \left((2/\nu) - (1/a)\right) < GM^*/a \tag{955}$$

$$\left.\begin{array}{l} |\nu'| < (M^*/m_1 m_2)^{3/2} \sqrt{[G(m_1+m_2)]/a} \\ \text{If } m_1 \text{ and } m_2 \text{ are the two smallest masses} \end{array}\right\} \quad (956)$$

B) Let us put :

$$(1/a) - (1/\nu(t)) = f(t) > 0 \quad (957)$$

Hence

$$d^2(\rho^2)/dt^2 = 2\mu[(1/\nu) - (1/a)] = -2\mu f < 0 \quad (958)$$

The first derivative $d(\rho^2)/dt$ is decreasing and cannot become negative (it would lead to negative ρ^2), hence it has a positive or zero limit and :

$$0 \leq \int_{t_0}^{\infty} f(t)dt = (1/2\mu)\left((\rho^2)'_0 - (\rho^2)'_\infty\right) = F < \infty \quad (959)$$

On the other hand :

$$|df/dt| = |f'| = |\nu'/\nu^2| < Q = (M^*/m_1 m_2)^{3/2} \sqrt{G(m_1+m_2)/a^5} \quad (960)$$

If we assume $f(t_1) = f_1$, then, if $t_1 > t_0$

$$t > t_1 \quad \text{implies} \left\{\begin{array}{l} f(t) \geq f_1 - Q(t - t_1) \quad ; \quad f(t) > 0 \\ \int_{t_1}^{\infty} f(t)dt \geq f_1^2/2Q \end{array}\right\} \quad (961)$$

Hence :

$$[f(t_1)]^2 = f_1^2 \leq 2Q\int_{t_1}^{\infty} f(t)dt \quad (962)$$

Let us consider the bounded quantity $F(t_1) = \int_{t_1}^{\infty} f(t)dt$. Since $F(t_1)$ goes to zero when t_1 goes to infinity this then implies $f(t_1) \to 0$ and $\nu(t_1) \to a$, because of (962) and (957).

This demonstration can be extended to system with infinitesimal masses. However no n-body motion has been found with $a < \nu(t) \leq 2a$

for all $t \geq t_0$ and it seems that these motions do not exist.

This would imply that the mean harmonic distance ν would remain in the range $a < \nu \leq 2a$ for only bounded periods of time, while it can remains indefinitely in any small range $[b_1, b_2]$ with $0 \leq b_1 < b_2 \leq a$. However notice that between two extremums of ρ the equation (946) implies that the mean temporal value of $1/\nu$ is $1/a$.

If we consider a small range $[b_3, b_4]$ with $a \leq b_3 < b_4 \leq 2a$ it is always possible (see Section 11.7.3.2) to find n-body solutions with intervals of time of arbitrary length in which the conditions $b_3 < \nu(t) < b_4$ remain satisfied.

Let us recognize that these three general propositions are very weak. If $h < 0$ we do not know which mutual distance will remain bounded or even if there is such a mutual distance ; if $h \geq 0$ we do not know which mutual distance will go to infinity or even if there is such a mutual distance...

11.7.3 Improvements - (Three and n-body motions)

11.7.3.1 Limitations on the configuration, the scale, the orientation

We have seen in Section 11.6.3 and in inequality (904) that a given three-body configuration of given scale was possible (i.e compatible with the integrals of motion) if and only if :

$$2/\nu \geq (1/a) + (p/\rho^2) \tag{963}$$

This result can be extended to n-body systems and to the orientations :

A given n-body configuration of given scale and orientation (i.e. given positions of the n bodies about their center of mass...) is compatible with the integral of motion if and only if :

$$2/\nu \geq (1/a) + \frac{1}{GM^*}\left((c_A^2/A) + (c_B^2/B) + (c_C^2/C)\right) \tag{964}$$

where :

A, B, C are the three usual main moments of inertia :

$$2I \geq A \geq B \geq C \geq 0 \quad ; \quad 4I = A + B + C$$

$A = 2I$ for a planar configuration

(965)

$$\left.\begin{array}{l} C = 0 \;;\; B = A = 2I \text{ for a collinear configuration} \\ c_A, c_B, c_C \text{ are the components of the angular momentum} \\ \vec{c} \text{ on the three main axes of inertia} \end{array}\right\} \quad (965)$$

A collinear configuration without collision, implies $\nu > 0$, $C = 0$, $c_C = 0$ and the n bodies are necessarily in the invariable plane normal to $\vec{c}$. This property is true for all collinear configurations even those with collisions.

Since $GM^*((2/\nu) - (1/a)) = \Sigma m_K V_K^2$ it is easy to see that, in the limiting case when (964) is an equality, the velocities $\vec{V}_K$ have only one possible solution and are given by a suitable instantaneous rotation* :

$$\left.\begin{array}{l} K = \{1, 2, \ldots, n\} \quad \text{implies} \quad \vec{V}_K = \vec{\Omega} \times \vec{r}_K \\ \text{where :} \\ \Omega_A = c_A/A \;;\; \Omega_B = c_B/B \;;\; \Omega_C = c_C/C \end{array}\right\} \quad (966)$$

Since $c_A^2 + c_B^2 + c_C^2 = c^2$ and $c^2/2IGM^* = p/\rho^2$ the condition (964) leads easily to the condition (963) in the three-body case when the orientation is disregarded.

Unfortunately the condition (964)-(965) has much less interest in the more-than-three-body problem than in the three-body problem. It never leads then to a separation of the zone of possible motion into two or three disconnected parts and the Hill-type stability only exists in the three-body problem.

The Hill-type stability is one of the major differences between the problems of three and of more-than-three bodies.

---

* In the three-body case this rotation (966) always give the true rotation of the plane of the three bodies and hence the true velocity components normal to that plane, even if the inequality (964) is not at its limit.
If, improperly, we call "line of nodes" the intersection of the invariable plane and the plane of the three bodies this line always rotates about $\vec{c}$ in the positive direction with a rate of rotation larger than or equal to c/I_c where I_c is the moment of inertia about $\vec{c}$.

11.7.3.2 <u>On the evolution of the semi-moment of inertia I and the mean quadratic distance ρ.</u>

Let us write again the five main relations (938), (944), (946), (948), (949) with the three constants μ, a, p and the three variables ρ, ν, j :

$$\left.\begin{aligned} & 0 \leqq \nu \leqq \rho \\ & d^2(\rho^2)/dt^2 = 2\mu\big((1/\nu) - (1/a)\big) \\ & j = (\rho/2a) + (p/2\rho) + (\rho\rho'^2/2\mu) \quad ; \quad \text{where } \rho' = d\rho/dt \\ & \rho/\nu \geqq j \\ & dj/dt = (\rho'/\rho)\,\big((\rho/\nu) - j\big) \end{aligned}\right\} \qquad (967)$$

These relations contain much more informations on the evolution of the mean quadratic distance ρ and the related parameter I (I = semi-moment of inertia = $M^*\rho^2/2M$) than the classical results of Section 11.7.2.

Let us eliminate the mean harmonic distance ν from (967) :

$$0 \leqq \nu \leqq \rho \quad \text{and} \quad \nu \leqq \rho/j \qquad (968)^*$$

Hence :

$$d^2(\rho^2)/dt^2 \geqq \sup\Big\{\big((2\mu/\rho)-(2\mu/a)\big) \;;\; \big((\mu p/\rho^2)-(\mu/a) + \rho'^2\big)\Big\} \qquad (969)$$

and with the auxiliary relations :

$$\rho' = d\rho/dt \quad ; \quad j = (\rho/2a) + (p/2\rho) + (\rho\rho'^2/2\mu) \qquad (970)$$

we obtain always :

$$dj/d\rho \geqq 0 \quad ; \quad d\big(\rho(j - 1)\big)/d\rho \geqq 0 \qquad (971)$$

---

*(968) can sometimes be improved (Ref.203), e.g. for a galactic system : $\rho/\nu \geqq (1.2)^{1.5} = 1.314...$

The inequalities (971) are equivalent to (969).

The main question is the following :

[Are the initial and final conditions ρ_0, ρ'_0, t_0, and ρ_f, ρ'_f, t_f (with $t_f > t_0$) compatible with the relation (969) ?

The answer to this question is very simple (Ref. 210).

Let us consider the limit of (969) :

$$d^2(\rho^2)/dt^2 = \sup\left\{\left[(2\mu/\rho)-(2\mu/a)\right] \ ; \ \left[(\mu p/\rho^2)-(\mu/a) + \rho'^2\right]\right\} \tag{972}$$

That is :

$$d^2\rho/dt^2 = \sup\left\{\left(\frac{\mu}{\rho^2} - \frac{\mu}{a\rho} - \frac{\rho'^2}{\rho}\right) \ ; \ \left(\frac{\mu p}{2\rho^3} - \frac{\mu}{2a\rho} - \frac{\rho'^2}{2\rho}\right)\right\} \tag{973}$$

Then the initial conditions ρ_0, ρ'_0, t_0 and the final conditions ρ_f, ρ'_f, t_f with $t_f > t_0$ are compatible with the relation (969) <u>if and only if</u> exist two velocities ρ'_1 and ρ'_2 such that :

A)

$$\rho'_1 \geq \rho'_0 \ ; \ \rho'_2 \leq \rho'_f \tag{974}$$

B) The limit differential equation (972) (or (973)) leads from ρ_0, ρ'_1, t_0 to ρ_f, ρ'_2, t_f.

Hence in order to obtain informations about the future of the mean quadratic distance ρ after the initial conditions ρ_0, ρ'_0, t_0 we must integrate (972) or (973) with the initial conditions ρ_0, ρ'_1 t_0 satisfying $\rho'_1 \geq \rho'_0$.

Let us put :

$$j_1 = (\rho_0/2a) + (p/2\rho_0) + (\rho_0\rho_1'^2/2\mu) \ ; \ (j_1 = j_0 \text{ if } \rho'_1 = \rho'_0) \tag{975}$$

The integration of (972) and (973) leads to the two following families of limit solutions.

A) $j_1 \leq 1$

The quantity $\rho(j - 1)$ remains constant and equal to $\rho_0(j_1 - 1)$, the mean quadratic distance ρ varies as the mutual distance in a two-body motion in which :

$$\left.\begin{aligned} \mu_1 &= \text{gravitational constant} = \mu \\ a_1 &= \text{semi-major axis} = a \\ p_1 &= \text{semi-latus rectum} = p + 2\rho_0(1 - j_1) \geqq p \end{aligned}\right\} \quad (976)$$

B) $j_1 \geqq 1$

The Sundman function j remains constant and equal to j_1, the mean quadratic distance ρ varies as the mutual distance in a two-body motion in which :

$$\left.\begin{aligned} \mu_1 &= \text{gravitational constant} = \mu j_1 \\ a_1 &= \text{semi-major axis} = a j_1 \\ p_1 &= \text{semi-latus rectum} = p/j_1 \leqq p \end{aligned}\right\} \quad (977)$$

The Eulerian and Lagrangian motions of the three or n-body problems belong to this latter limit family. They have a constant configuration and thus a constant ratio ρ/ν that remain identical to the Sundman function j and is larger than or equal to one.*

The three-body Lagrangian motions of Fig. 114 (Section 11.3.1) have a ratio ρ/ν and a Sundman function j identical to one, they are at the border of the two limit families (976) and (977) with, as in the classical two-body problem :

---

* The interest of the ratio ρ/ν in emphasized by the following :
In a system of n non-infinitesimal masses the central configurations are the configurations with stationary values of the ratio ρ/ν (stationarity with respect to the positions of the n bodies)

These stationary values of ρ/ν are of course function of the n masses. They may have any value in the range {1 ; 1.18808} when n = 3, any value in the range {1 ; 1.33031} when n = 4, any value in $\{1 ; \phi(n)\}$, with $\phi(n) \sim \sqrt{0.576}$ Log n when n is large.

If a > 0 the period of Eulerian and Lagrangian motions is $2\pi \cdot (\rho/\nu) \cdot \sqrt{a^3/\mu}$ and thus evolutions as slow as desired may be found for sufficiently large ρ/ν ratios.

$$\begin{array}{ll} \rho \text{ and } \nu \text{ move} & \left\{ \begin{array}{l} \text{gravitational constant} = \mu \\ \\ \text{semi-major axis} = a \\ \\ \text{semi-latus rectum} = p \end{array} \right\} \\ \text{as the mutual distance} & \\ \text{in a two-body motion with :} & \end{array} \qquad (978)$$

Notice that in all these Eulerian and Lagrangian motions, the two-body motion of the mean harmonic distance ν has a semi-major axis equal to a.

These limit motions (976) and (977) allow to draw in the ρ, ρ', t space the future attainable domain corresponding to ρ_0, ρ'_0, t_0, (Figs. 128 - 129 and Ref. 210).

The limit orbit starting with $\rho'_1 = \rho'_0$ gives a lower bound of $\rho(t)$ at least until its next extremum of ρ. Hence if $h \geqq 0$ and $\rho'_0 \geqq 0$ it gives a lower bound of $\rho(t)$ for all future times since when $h \geqq 0$ the curve $\rho(t)$ has no maximum and one minimum (Fig. 128).

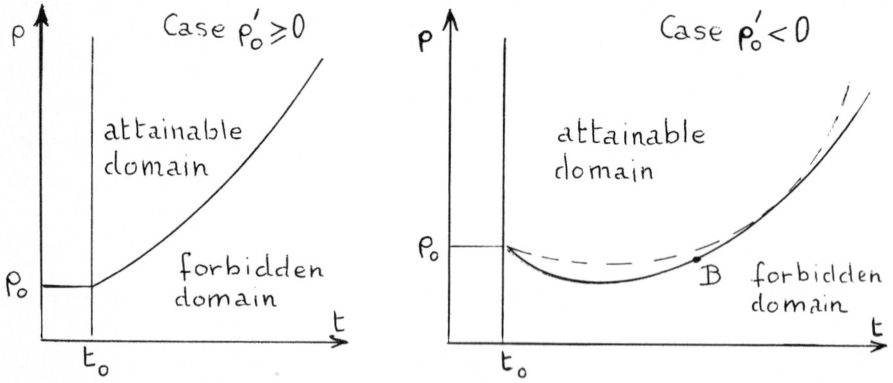

Fig. 128. Future attainable domain in the t, ρ plane when the energy integral h is positive or zero.
If $\rho'_0 \geqq 0$ the lower limit is given for all $t \geqq t_0$ by the two-body motion of (976) or (977) starting at ρ_0, ρ'_0, t_0.
This result remains true so long as :

$$\rho'_0 \geqq -\sqrt{-\mu/a} \quad \text{and} \quad j_0 \leqq 1$$

If not the lower limit is more complex. The above two-body motion remains the lower limit until its minimum and even until B a little later ; while for large t the relations (980), if h > 0, or (982), if h = 0, give excellent lower bounds.

If $a < 0$, that is if the energy integral h is positive, all two-body orbits (976)-(977) are hyperbolic, have a velocity at infinity equal to $\sqrt{-\mu/a}$ and verify for all t :

$$\rho(t) \geqq - \rho_0 + |t - t_0| \sqrt{- \mu/a} \tag{979}$$

This lower bound of $\rho(t)$ is valid for all solutions starting at (ρ_0, t_0) and not only for the limit solutions (976)-(977).

A much better bound, independent of ρ_0', is given by the following :

$$\left.\begin{aligned} &Z = \sup\left\{1 \; ; \; \big((p/2\rho_0) + (\rho_0/2a)\big) \; ; \; \big((p/2\rho) + (\rho/2a)\big)\right\} \\ &F \geqq 0 \quad ; \quad \cosh F = 1 - \big((\rho + \rho_0)/aZ\big) \\ &|t - t_0| \leqq Z(\sinh F - F) \sqrt{- a^3/\mu} \end{aligned}\right\} \tag{980}$$

This bound is excellent after the point B (Fig. 128). It gives even the exact limit if $j_1 < 1$ (family of orbits (976)).

For ρ' the integrals $\rho(j - 1)$ and j lead to the following :

$$\left.\begin{aligned} &\rho'^2 \geqq \left(1 - \frac{\rho_0}{\rho}\right)\left[- \frac{\mu}{a} + \frac{\mu}{\rho} \sup\left\{\left(2 - \frac{\rho_0}{a}\right) \; ; \; \frac{p}{\rho_0}\right\}\right] \\ &\rho \geqq \rho_0 \quad \text{implies} \quad \rho'(t - t_0) \geqq 0 \end{aligned}\right\} \tag{981}$$

Hence for all solutions the escape velocity of ρ is at least $\sqrt{- \mu/a}$. This lower limit is obtained if and only if all mutual distances go to infinity (complete scattering).

If the energy integral h is zero all two-body orbits (976)-(977) are parabolic.

The bounds (980)-(981) become :

$$(t - t_0)^2 \leqq \inf\{2p \; ; \; 4\rho_0 \; ; \; 4\rho\} \, . \, (\rho + \rho_0)^3/9\mu p \tag{982}$$

and :

$$\left.\begin{aligned} &\rho'^2 \geqq (\mu/\rho^2)(\rho - \rho_0) \, . \, \sup\{2 \; ; \; (p/\rho_0)\} \\ &\rho \geqq \rho_0 \quad \text{implies} \quad \rho'(t - t_0) \geqq 0 \end{aligned}\right\} \tag{983}$$

The mean quadratic distance ρ has then at least parabolic escapes (escapes as $|t|^{2/3}$).

We will see (Section 11.7.7.1) that if an escape of ρ is faster than $|t|^{2/3}$ it is at least as $|t|$. If the escape is parabolic it verifies $\rho \sim A|t|^{2/3}$ and $\nu \sim B|t|^{2/3}$ with $A \geq B$ and $A^2B = 4.5\mu$. The configuration of the n bodies then goes to a central configuration (n-parabolic evolution).

All these n-body motions with a positive or zero energy integral h have one and only one minimum ρ_m of $\rho(t)$ at some time t_m.

Given ρ_m and t_m better lower bounds can be found.

The simple relation (979) can be improved to :

$$\rho^2(t) \geq \rho_m^2 - (\mu/a)(t - t_m)^2 \tag{984}$$

To the complex bound (980) we can with advantage substitute the lower bound given by the two-body motion (976) or (977) starting at ρ_m, $\rho'_m = 0$, t_m.

For the case h = 0 this solution starting at ρ_m, $\rho'_m = 0$, t_m is an improvement of (982) and gives :

$$(t - t_m)^2 \leq \inf\left\{2p \; ; \; 4\rho_m\right\} . \; (\rho + 2\rho_m)^2(\rho - \rho_m)/9\mu p \tag{985}$$

If the generalized semi-major axis a is positive, that is if the energy integral h is negative, all two-body orbits of (976)-(977) are elliptic. The first have the period T given by the Keplerian relation :

$$T = 2\pi\sqrt{a^3/\mu} \tag{986}$$

The second orbits, those of (977), have the period T_1 :

$$T_1 = Tj_1 \geq T \tag{987}$$

Future attainable domains in the t,ρ plane are drawn in Fig. 129. Their lower limit begins by the curve $\rho(t)$ given by the two-body motion of (976) or (977) starting at ρ_0, ρ'_0, t_0 and proceeds, after the point B, by the envelope of other two-body motions of (976) or (977) starting at ρ_0, ρ_1', t_0 with $\rho_1' > \rho_0'$.

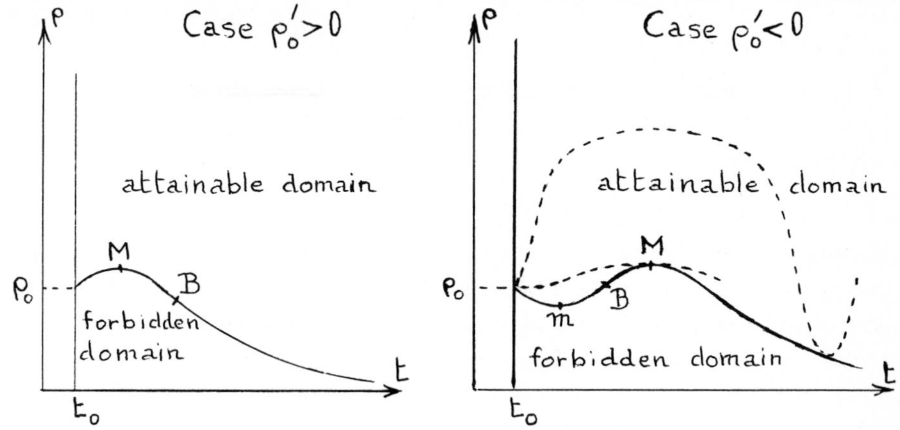

Fig. 129. Future attainable domain in the t,ρ plane when the energy integral h is negative.

The two-body motion of (976) or (977) starting at ρ_0, ρ_0', t_0 gives the lower limit until the point B between its next extremum and its second next.

After B the limit is given by the envelope of the two-body motions of (976) or (977) starting at ρ_0, ρ_1', t_0 with $\rho_1' > \rho_0'$.

If $\rho_0' \leq 0$ the maximum M corresponds to $\rho_1' = 0$ and is very simple :

$$\rho_M = \sup\left\{2a - \rho_0 \ ; \ \frac{ap}{\rho_0}\right\} \quad ; \quad t_M = t_0 + \frac{T}{2}\sup\left\{1; \ \frac{\rho_0}{2a} + \frac{p}{2\rho_0}\right\}$$

For large t the lower limit of ρ is equivalent to $Tp/2(t - t_0)$. The relation (988) gives an excellent lower bound of this lower limit.

All these limit orbits and also all non-limit orbits starting at ρ_0, t_0 always satisfy the following independently of ρ_0' :

$$\left.\begin{array}{l} \text{When } \rho < \sup\{2a - \rho_0 \ ; \ ap/\rho_0\} \text{ the point } t,\rho \text{ is attainable} \\ \text{only if, with :} \\ Z = \sup\left\{1 \ ; \ \left((\rho_0/2a) + (p/2\rho_0)\right) \ ; \ \left((\rho/2a) + (p/2\rho)\right)\right\} \ ; \\ \cos E = 1 - \left((\rho + \rho_0)/aZ\right) \ ; \quad 0 \leq E \leq \pi \end{array}\right\} \quad (988)$$

the time t satisfies :

$$\left\{\begin{array}{ll} \text{either} & |t - t_0| \leq TZ\left(\dfrac{E - \sin E}{2\pi}\right) \\ \text{or} & |t - t_0| \geq TZ\left(1 - \dfrac{E - \sin E}{2\pi}\right) \end{array}\right. \qquad (988)$$

This necessary condition is obviously an extension of the conditions (980) and (982) ; it is insufficient but gives an excellent lower bound of the attainable domain. It gives even the exact limit if $j_1 < 1$ (family of orbits (976)).

For large t the lower limit of ρ is equivalent to $Tp/2(t - t_0)$ and the n-body systems cannot go to a "complete collapse" (i.e. to $\rho = 0$) if the semi-latus rectum p, and then the angular momentum $\vec{c}$, are different from zero.*

A n-body system going down to a complete collapse satisfies :

$$\rho' < 0 \quad ; \quad p = 0 \quad ; \quad \vec{c} = 0 \quad ; \quad j \geq 1 \qquad (989)$$

The necessary condition (988) contains the following very simple information corresponding to the case Z = 1.

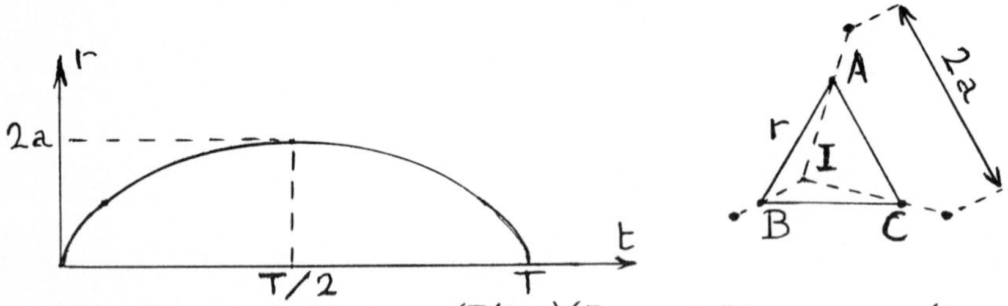

Fig. 130. The cycloid $t = (T/2\pi)(E - \sin E)$; $r = a(1 - \cos E)$ corresponding to the three-body radial Lagrangian motion (and to the two-body radial motion).

Let us consider (Fig. 130) the first arch of the cycloid defined by :

$$\left.\begin{array}{l} t = (T/2\pi)(E - \sin E) \\ r = a(1 - \cos E) \\ 0 \leq E \leq 2\pi \end{array}\right\} \qquad (990)$$

* It is even possible to demonstrate that then ρ has a positive lower bound (Section 11.8).

This arch corresponds to the motion r(t) of three-body radial Lagrangian motions and of two-body radial motions at the border between the two families (976) and (977) in the case when $p = 0$.

The condition (988) implies that for any n-body motion going from (t_0, ρ_0) to (t, ρ) it is impossible that the point $\{|t - t_0|\,;\,(\rho + \rho_0)\}$ lies inside the arch of cycloid of Fig. 130.

Hence for instance :

$$t - t_0 = \pm T/2 \quad \text{implies} \quad \rho_0 + \rho(t) \geq 2a \tag{991}$$

The three-body radial Lagrangian motions are thus a particular limit of the general n-body motions. This result can be extended, with (980) and (982), to systems of positive or zero energy.

The coefficient Z enlarges the arch of cycloid of Fig. 130 by a homothetic factor Z and then :

$$t - t_0 = \pm T\big((\rho_0/4a) + (p/4\rho_0)\big) \quad \text{implies} \quad \rho_0\rho(t) \geq ap \tag{992}$$

The condition (991) implies that the curve $\rho(t)$ cannot remain continuously below the generalized semi-major axis a for a period of time larger than T/2 (Fig. 131). Conversely $\rho(t)$ remains above a for intervals of time always larger than T/2.

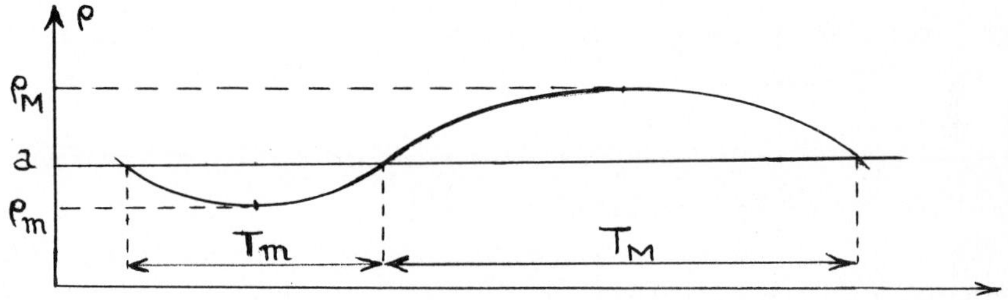

Fig. 131. The curve $\rho(t)$ for an arbitrary n-body motion.

When $\rho < a$ the second time-derivative $(\rho^2)''$ is positive and ρ has no maximum.

$$\rho < a \Longrightarrow T_m \leqq T\left(\frac{1}{2} - \frac{a - \rho_m}{\pi a}\right) < T/2$$

$$\rho > a \Longrightarrow T_M \geqq T\left(\frac{1}{2} + \frac{\rho_M - a}{\pi a}\right) > T/2$$

The limits are given by two-body motions of the family of (976), and when $\rho_M > 2a$ the greatest lower bound of T_M is even larger.

Let us end this Section by a surprising remark.

Many studies deal with the Sundman inequality $\rho/\nu \geqq j$ and reach conditions as (992). However notice that this condition (992) is useless when p = 0.

The conditions as (991) and the Fig. 131 remain useful even when p = 0. They correspond to the inequality $\nu \leqq \rho$ that has been used by only very few people.

The reason for this situation is certainly the complex expression of the inequality $\nu \leqq \rho$ in the classical notation :

$$\nu \leqq \rho \quad \text{is equivalent to } IU^2 \geqq G^2M^{*3}/2M \tag{993}$$

11.7.3.3 On the evolution of the potential U and the mean harmonic distance ν

The potential U and the mean harmonic distance ν are related by :

$$U = GM^*/\nu \tag{994}$$

The energy integral (929) leads to :

$$0 \leqq \sum_{K=1}^{n} m_K V_K^2 = 2U + 2h = GM^*\left(\frac{2}{\nu} - \frac{1}{a}\right) \tag{995}$$

This relation was used in 11.7.2, it gives $0 \leqq \nu \leqq 2a$ when the energy integral is negative, it is also gives an upper limit on the derivative ν' of ν :

$$\left.\begin{array}{l} |\nu'| \leqq (M^*/m_1m_2)^{3/2}\sqrt{G(m_1+m_2)\{(2/\nu)-(1/a)\}} \\ \text{If } m_1 \text{ and } m_2 \text{ are the two smallest masses} \end{array}\right\} \tag{996}$$

This upper limit corresponds to the case when r_{12} is the only bounded mutual distance. Unfortunately this is a very large upper bound, especially if the number of bodies is large, and usually ν has much more variation than the mean quadratic distance ρ. Its characteristic time of variation is much shorter than the period $T = 2\pi\sqrt{a^3/\mu}$ presented in Fig. 131 as the characteristic time of variation of ρ.

The only long-term informations on the evolution of ν and U are those related to the long-term informations of the evolution of ρ

presented in the previous Section and to the inequalities :

$$0 \leqq \nu \leqq \rho \ . \ \inf \left\{ 1 \ ; \ 1/j \right\} \tag{997}$$

Thus, starting at ρ_0, ρ_0', t_0, we obtain an upper bound $\nu_M(t)$ for any later $\nu(t)$; this upper bound is always given by orbits of the family (977) starting with ρ_1' positive or zero.

For these limit orbits the mean harmonic distance ν equal to ρ/j varies as the mutual distance in a two-body motion in which :

$$\left.\begin{aligned} \mu_\nu &= \text{gravitational constant} = \mu/j_1^2 \\ a_\nu &= \text{semi-major axis} = a \\ p_\nu &= \text{semi-latus rectum} = p/j_1^2 \end{aligned}\right\} \tag{998}$$

with, as given in (975) :

$$\left.\begin{aligned} & j_1 = (\rho_0/2a) + (p/2\rho_0) + (\rho_0\rho_1'^2/2\mu) \geqq 1 \ ; \ (j_1 = j_0 \text{ if } \rho_1' = \rho_0') \\ & \text{and} \\ & \rho_1' \geqq \sup(0, \rho_0') \end{aligned}\right\} \tag{999}$$

The case when $p = 0$ is simple : only one orbit of the family (977) is necessary, the orbit with the smallest ρ_1' satisfying the inequalities of (999). Hence j_1 has three possible values : either j_0 (if $\rho_0' \geqq 0$; $j_0 \geqq 1$) or $\rho_0/2a$ (if $\rho_0' \leqq 0$; $\rho_0/2a \geqq 1$) or 1 (remaining cases).

If $h \geqq 0$ this limit orbit will give an upper limit $\nu_M(t)$ going to infinity with t (as $t\sqrt{-\mu/a}/j_1$ if $h > 0$ and as $\{4.5\mu t^2/j_1^2\}^{1/3}$ if $h = 0$).

If $h < 0$ the limit orbit corresponds to an arch of cycloid (Fig. 132) similar to that of Fig. 130 but with a duration equal to Tj_1.

The attainable domain is limited to the arch until its top at B after which $\nu_M(t) = 2a$ and the gain of information is very small, it is represented by the zone ABC.

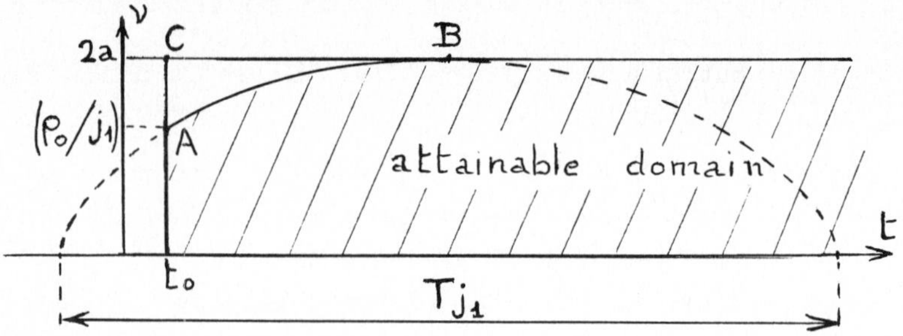

Fig. 132. Future attainable domain in the t,ν plane when p = 0 and h < 0.

When $p > 0$ the results are less simple and the gain of information remains generally very small (Fig. 133).

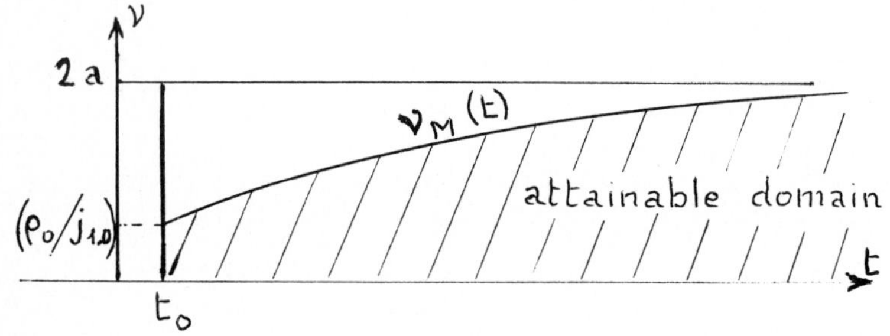

Fig. 133. Future attainable domain in the t,ν plane when p > 0 and h < 0.

For large t :

$$2a - \nu_M(t) \sim pT^2/8t^2$$

The upper limit $\nu_M(t)$ now uses many orbits of the family (977) even if it still begins with ρ_1' equal to the smallest possible value satisfying the inequalities of (999) and with the corresponding value $j_{1.0}$ of j_1.

The only interesting case is the case when $j_{1.0}$ is very large and $\rho_0/aj_{1.0}$ very small (for instance $\rho_0 \ll p$; $\rho_0^2 \ll ap$). The ratio ν/a will then remain small for a long interval of time, an interval of the order of $Tj_{1.0}$.

11.7.3.4 A psychological improvment, the use of r and R instead of ν and ρ

The smallest mutual distance r and the largest mutual distance R are less informative but more understandable than the mean harmonic distance ν and the mean quadratic distance ρ.

Their relations (931) - (942) have already been presented in Section 11.7.1 and we will consider here only those that are independent of the n masses, that is :

$$\left.\begin{aligned} &0 \leqq r \leqq \nu \leqq \rho \leqq R \\ &\rho^2 + \frac{Rr(R + r)}{\nu} \leqq R^2 + Rr + r^2 \end{aligned}\right\} \qquad (1000)$$

and also the relation (904), consequence of the Sundman inequality and valid even for more-than-three-body problems :

$$2/\nu \geqq (1/a) + (p/\rho^2) \qquad (1001)$$

In the ρ,ν plane the inequalities (1000) and (1001) lead to three possible types of picture according to the ratio p/a (Fig. 134).

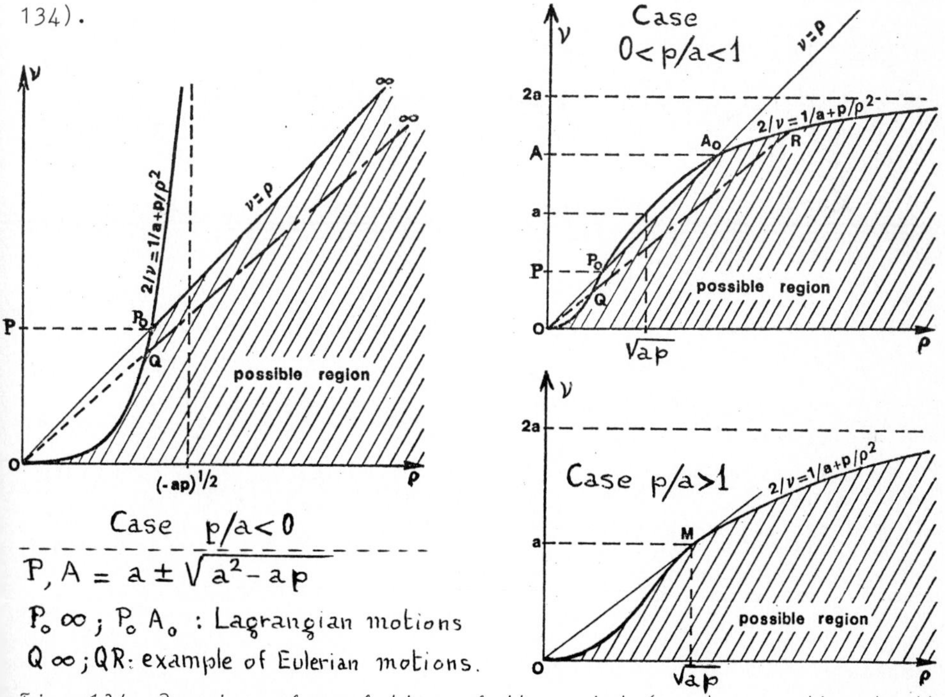

Fig. 134. Domains of evolution of the point (ρ,ν) according to the ratio p/a *(hatched zones).*

The elimination of ν and ρ among (1000), (1001) leads of course to a loss of information and leads to the following.

A) Elimination of ν.

$$2/r \geqq (1/a) + (p/\rho^2) \tag{1002}$$

$$\rho^2 + \frac{Rr(R + r)}{\rho} \leqq R^2 + Rr + r^2 \tag{1003}$$

$$\rho^2 + \frac{Rr(R + r)}{2}\left(\frac{1}{a} + \frac{p}{\rho^2}\right) \leqq R^2 + Rr + r^2 \tag{1004}$$

The inequality (1003) is always satisfied if :

$$0 \leqq r \leqq \rho \leqq R \tag{1005}$$

and there remain only (1002), (1004) and (1005).

B) Elimination of ρ.

That remaining question is the following : for given a, p, r, R does there exist ρ satisfying (1002), (1004), (1005) ?

Notice that (1004) and (1005) imply (1002), indeed :

$$\underbrace{\frac{Rr(R + r)}{2}\left(\frac{1}{a} + \frac{p}{\rho^2}\right) \leqq R^2 + Rr + r^2 - \rho^2}_{\text{from (1004)}} \leqq R^2 + Rr \tag{1006}$$

(the comparison $R^2 + Rr + r^2 - \rho^2 \leqq R^2 + Rr$ is from (1005))

The inequality between the left-hand and the right-hand member is equivalent to (1002).

Hence we are left with :

$$0 \leqq r \leqq \rho \leqq R \tag{1007}$$

$$\rho^2 + \frac{Rr(R + r)p}{2\rho^2} \leqq R^2 + Rr + r^2 - \frac{Rr(R + r)}{2a} \tag{1008}$$

For given R, r, p the left-hand member of (1008) is minimum when its two terms are equal, that is when :

$$2\rho^4 = Rrp(R + r) \tag{1009}$$

Hence :

If $2r^4 \geqq Rrp(R + r)$ the optimal ρ is $\rho = r$ and

the compatibility conditions are $0 \leqq r \leqq R \quad ; \quad \frac{p}{r^2} + \frac{1}{a} \leqq \frac{2}{r}$

If $2r^4 \leqq Rrp(R + r) \leqq 2R^4$ the optimal ρ is as given in

(1009) and the compatibility conditions are : (1010)

$0 \leqq r \leqq R \quad ; \quad \sqrt{2Rrp(R + r)} + \frac{Rr(R + r)}{2a} \leqq R^2 + Rr + r^2$

Finally if $2R^4 \leqq Rrp(R + r)$ the optimal ρ is $\rho = R$ and

the compatibility conditions are $0 \leqq r \leqq R \quad ; \quad \frac{p}{R^2} + \frac{1}{a} \leqq \frac{2}{R}$

The discussion :

$$2r^4 \gtrless Rrp(R + r) \gtrless 2R^4 \tag{1011}$$

leads to Fig. 135.

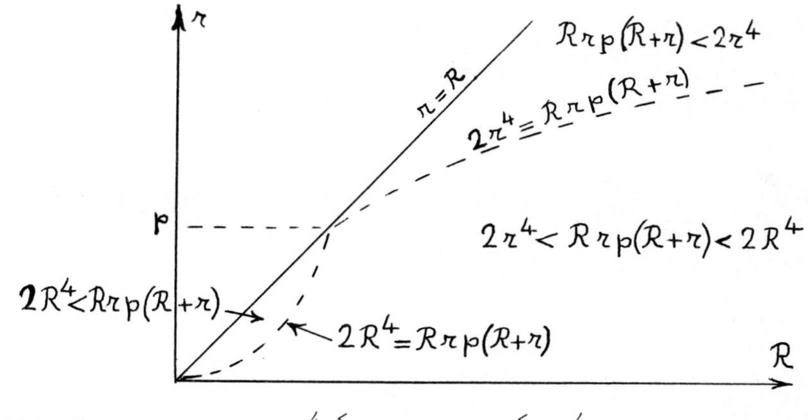

Fig. 135. The discussion $2r^4 \lesseqgtr Rrp(R + r) \lesseqgtr 2R^4$ only depends on the ratios R/p and r/p.

It is then easy to obtain, for given a and p, the domain of evolution of the point (R,r) for the three cases of Fig. 134 and the case p = a (Fig. 136) with the same points P_o and A_o of coordinates $\rho = \nu = R = r = a \pm \sqrt{a^2 - ap}$.

(P_o and A_o are of course for "pericenter" and "apocenter" since $a \pm \sqrt{a^2 - ap}$ are the pericenter and apocenter distance in a two-body motion with semi-major axis a and semi-latus rectum p. They are also the maximum and minimum mutual distances in a three-body Lagrangian motion).

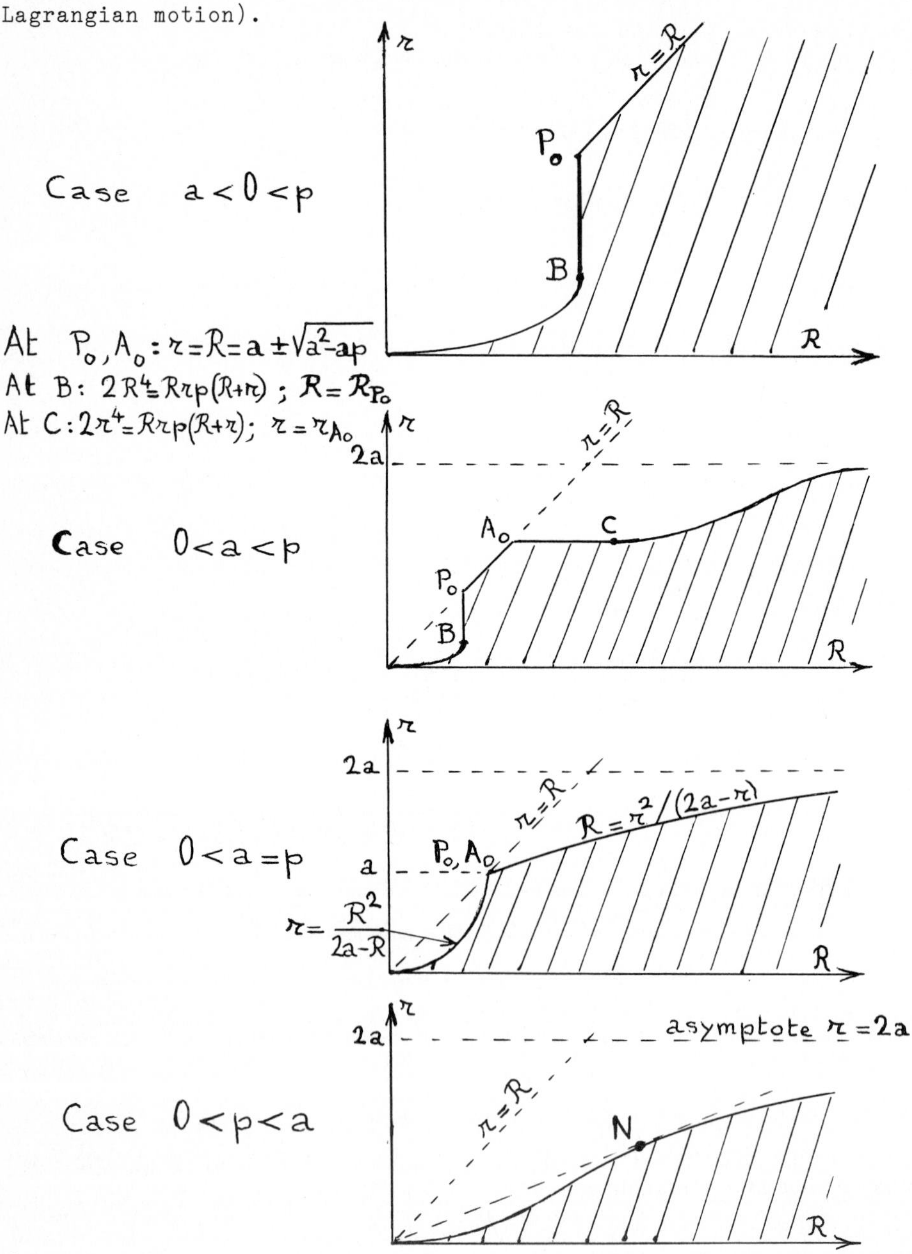

Fig. 136. Domains of evolution of the point (R, r) for several ratios p/a *(hatched zones). Some limits have segments such as BP_o, P_oA_o, A_oC.*

Let us recall that these domains of evolution are valid for any n-body problem. Their most interesting feature is the smallness of possible r/R ratios when p/a > 1 : the point N of Fig. 136 corresponds of course to the point M of Fig. 134. but its r/R ratio is much smaller than the ratio ν/ρ of M.

Consider for instance the system (1000) and a given value of the ratio r/R. What are the possible values of the ratio ν/ρ ?

The system (1000) gives

$$1 \leqq (\rho^2/\nu^2) \leqq \frac{R^2 + Rr + r^2}{\nu^2} - \frac{Rr(R + r)}{\nu^3} \tag{1012}$$

The maximum of the right-hand member is obtained when :

$$\nu = \big(3Rr(R + r)\big)\Big/\big(2(R^2 + Rr + r^2)\big) \tag{1013}$$

and the maximum of ρ^2 is then :

$$\rho^2 = (R^2 + Rr + r^2)/3 \tag{1014}$$

Thus for a given ratio $r/R = x$ the range of possible values of the ratio ν/ρ is given by :

$$1 \geqq (\nu^2/\rho^2) \geqq \frac{27R^2r^2(R + r)^2}{4(R^2 + Rr + r^2)^3} = \frac{27x^2(1 + x)^2}{4(1 + x + x^2)^3} \tag{1015}$$

Thus for instance :

$$x = 0.5 \quad \text{implies} \quad 1 \geqq (\nu/\rho) \geqq \sqrt{243/343} = 0.8417\ldots \tag{1016}$$

When $0 \leqq x \leqq 1$ the function $27x^2(1+x)^2/4(1+x+x^2)^3$ is monotonic and increasing, hence conversely :

$$\left.\begin{aligned} &\nu/\rho \leqq 0.8417\ldots \quad \text{implies} \quad r/R \leqq 0.5 \\ &\nu/\rho \leqq 0.9 \quad\quad\quad \text{implies} \quad r/R \leqq 0.5843 \end{aligned}\right\} \tag{1017}$$

When considering (1013) and (1014) the expression of r and R in terms of ν and ρ is simple :

Let us define the angle ϕ by :

$$\nu/\rho = \sin 3\phi \quad ; \quad 0° \leqq \phi \leqq 30° \tag{1018}$$

Then :

$$r = 2\rho\sin\phi \quad ; \quad R = 2\rho\sin(60°-\phi) \quad ; \quad x = r/R = \frac{\sin\phi}{\sin(60°-\phi)} \qquad (1019)$$

For a given ratio p/a the upper limits of ν/ρ and r/R correspond to the points M and N of Figs. 134 and 136 with $\rho = \sqrt{ap}$; $\nu = a$; $R = 2\rho\sin(60°-\phi)$; $r = 2\rho\sin\phi$.

Hence these upper limits correspond to the parametric expressions :

$$\left.\begin{array}{l} a/p = \sin^2 3\phi \quad ; \quad \nu/\rho = \sin 3\phi \quad ; \quad r/R = \sin\phi/\sin(60°-\phi) \\ \\ \text{with : } 0° \leq \phi \leq 30° \end{array}\right\} \qquad (1020)$$

These upper limits are presented graphically in Fig. 137.

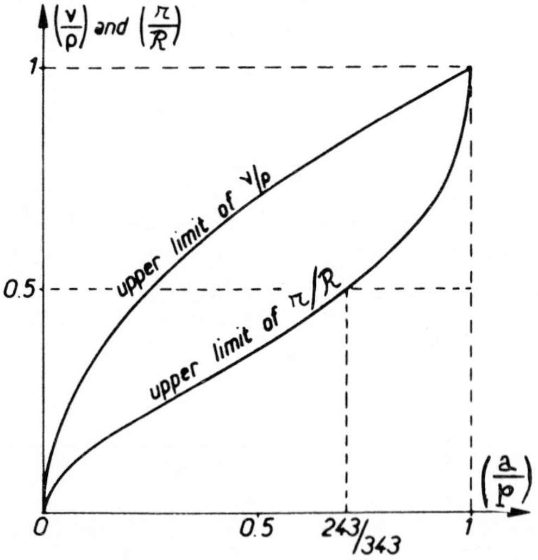

Fig. 137. The upper limits of ν/ρ and r/R in terms of a/p.

If there are only three bodies and if the upper limit of r/R is smaller than 0.5 (that is if p/a > 343/243) an exchange between the three bodies is impossible. The two masses initially nearest remain forever the two nearest (with $r < 2a$ forever) and there is "Hill-type stability".

Notice that with the more easily understood but less informative parameter r/R the Hill stability only appears for p/a > 343/243 which is later than with the parameter ν/ρ (see Section 11.6.3 and table II).

For a two-body system of eccentricity e the ratio p/a is $1 - e^2$ and is always less than or equal to one. However for all known natural more-than-two-body systems for which the sign of (p/a) - 1 is known this sign is always positive, as can be seen in the following list :

System	Value of the ratio p/a
Solar system	1.16
Jupiter and satellites	1.25
Saturn and satellites	1.043
Uranus and satellites	1.2
Multi stellar systems	from 1.15 to 7.2
Galactic system	about 1.2

The restrictions on possible configurations given by large values of p/a are certainly a factor of stability. However it is of course possible that these large values of the ratio p/a only reflect the conditions of formation of multi-systems.

We can make the following small improvement. If we consider the angular momentum as "vertical" and the invariable plane as "horizontal" we can give the subscript h to the "horizontal projections" as in Fig. 138.

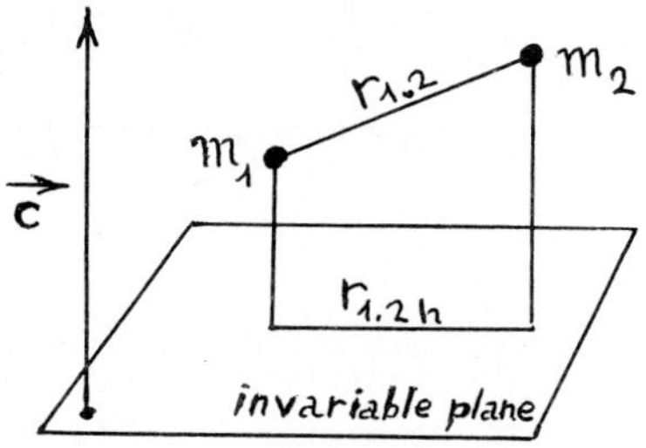

Fig. 138. The mutual distance r_{12} *and its "horizontal projection"* r_{12h} *on invariable plane.*

We can then define R_h and ρ_h :

$$\mathcal{R}_h = \sup_{1 \leq j < K \leq n} r_{jKh} \quad ; \quad \mathcal{R}_h \leq \mathcal{R} \tag{1021}$$

$$M^* \rho_h^2 = \sum_{1 \leq j < K \leq n} m_j m_K r_{jKh}^2 \quad ; \quad \rho_h \leq \rho \quad ; \quad \rho_h \leq \mathcal{R}_h \tag{1022}$$

The Sundman inequality $\rho/\nu \geq j$, that is :

$$2/\nu \geq (1/a) + (p/\rho^2) + (\rho'^2/\mu) \tag{1023}$$

can easily be improved into :

$$(2/\nu) \geq (1/a) + (p/\rho_h^2) + (\rho'^2/\mu) \tag{1024}$$

$$(2/\nu) \geq (1/a) + (p/\rho_h^2) + (\rho_h'^2/\mu) \tag{1025}$$

On the other hand the inequality $\rho^2 + \mathcal{R}r(\mathcal{R} + r)/\nu \leq \mathcal{R}^2 + \mathcal{R}r + r^2$ can easily be extended into :

$$\rho_h^2 + \frac{\mathcal{R}_h r(\mathcal{R}_h + r)}{\nu} \leq \mathcal{R}_h^2 + \mathcal{R}_h r + r^2 \tag{1026}$$

This inequality is obviously true if $\mathcal{R}_h \leq r$ and when $\mathcal{R}_h > r$ the "pessimization" of all r_{jK} and r_{jKh} leads immediately to $r \leq r_{jKh} = r_{jK} \leq \mathcal{R}_h$ that is to the case already encountered for ν, ρ, r, $\mathcal{R}$.

Hence the system (1000), (1001) becomes :

$$\left. \begin{array}{l} 0 \leq r \leq \nu \quad ; \quad 0 \leq \rho_h \leq \mathcal{R}_h \quad ; \quad (1026) \\ \text{and} \\ 2/\nu \geq (1/a) + (p/\rho_h^2) \end{array} \right\} \tag{1027}$$

The analysis of relations (1002) - (1010) can be done again with identical conclusions as long as $r \leq \mathcal{R}_h$. For $\mathcal{R}_h < r$ the only condition is $2/r \geq 1/a + p/\mathcal{R}_h^2$ (see the limit curves of Fig. 134).

The domains of Figs. 134 and 136 remain the same when $p/a \geq 1$ and this is an improvement since their abscissas are then the smaller lengths ρ_h and $\mathcal{R}_h$.

Figure 137 then gives also the upper limit of the ratios ν/ρ_h and $r/\mathcal{R}_h$ respectively equal to the upper limit of the smaller ratios ν/ρ and $r/\mathcal{R}$.

These improvements cannot be extended to the study of the evolution of ρ_h, like that of ρ in Section 11.7.3.2, because the Lagrange-Jacobi identity $(\rho^2)'' = 2\mu\left((1/\nu) - (1/a)\right)$ has no equivalent for ρ_h.

11.7.4 The principle of the tests of escape

A test of escape is : "A set of conditions on initial positions and velocities that, when satisfied, leads to an escape"*.

For instance :

A n-body system with a positive or zero energy integral h has a largest mutual distance going to infinity at least as $t^{2/3}$ (if its evolution remains defined, i.e. here if it avoids collisions).

For three-body systems of negative energy h an excellent example is given by the following Yoshida test (Ref. 225).

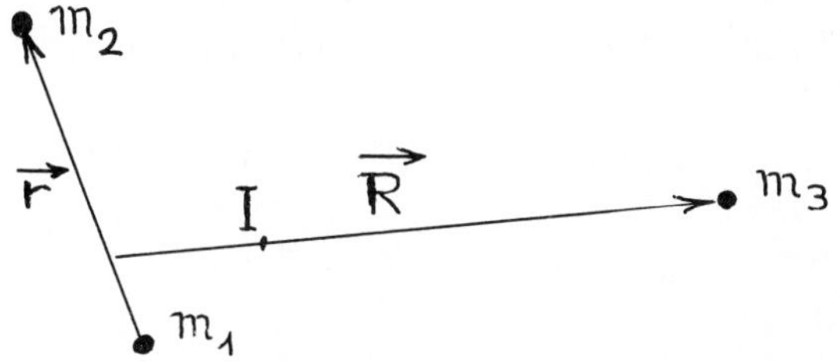

Fig. 139. The usual Jacobi decomposition with $m_1 \geq m_2$.

Let us use the usual Jacobi decompostion of Fig. 139 with the hypothesis $m_1 \geq m_2 > 0$; $m_3 > 0$.

Let us define the distance D by :

$$D = - Gm_1m_2/h \tag{1028}$$

The distance D is proportional to the "generalized semi-major axis" a of Section 11.6.1 :

$$D = 2am_1m_2/(m_1m_2 + m_1m_3 + m_2m_3) \tag{1029}$$

---

* We will see in Sections 11.7.7 to 11.7.10.1 the different meanings of the word "escape" and the corresponding classification of tests.

The Yoshida test is then :

The mass m_3 and the distance R will escape to infinity if at some arbitrary time the three following conditions are satisfied :

$$\left.\begin{aligned} & r < D \quad ; \quad R > D\alpha \\ & dR/dt \geq \left\{2GM \left[\frac{\beta}{R-D\alpha} + \frac{\alpha}{R+D\beta}\right]\right\}^{1/2} \\ & \text{with :} \\ & M = m_1 + m_2 + m_3 \; ; \; \alpha = m_1/(m_1 + m_2) \; ; \; \beta = m_2/(m_1 + m_2) = 1 - \alpha \end{aligned}\right\} \quad (1030)$$

The binary $m_1 m_2$ will then remain forever with a major axis smaller than D.

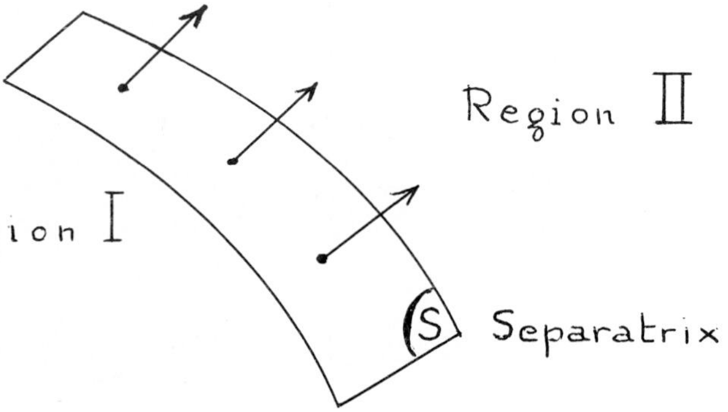

Fig. 140. The principle of the tests of escape. The phase space is divided into two regions and no motion is possible from region II to region I.

This Yoshida test is typical : as soon as the three conditions (1030) are satisfied they will remain forever satisfied. These conditions separate the phase space into two regions (Fig. 140) between which passages are possible only in one direction : from the region I (where the conditions (1030) are not satified) to the region II (where they are satisfied).

This first property is insufficient in itself. However the ergodic theorem (Section 11.7.10.2) shows that almost all orbits crossing the separatrix S are unbounded and even "open" (their representative point goes to infinity). It is then generally easy to decide which region, I or II, contains only escape orbits (escape in the past for the region I).

For instance in the Yoshida test the conditions (1030) imply forever $dR/dt > \sqrt{2GM/R}$ and R goes to infinity at least as $t^{2/3}$

For the test of systems of positive or zero energy h the "Region II" is defined by $\{dI/dt \geq A \; ; \; h \geq 0\}$ with A arbitrary and both I and dI/dt go to infinity.

Sometimes the division is not a division of the whole phase space but only of a part of it, a part defined by given values of the integrals of motion and/or by some symmetries or Hill-stability, but the principle remains the same.

Occasionally it happens, that a test does not satisfy the principle of Fig. 140. This is because the test is a simplification of a more efficient test that does satisfy the principle (see Fig. 141).

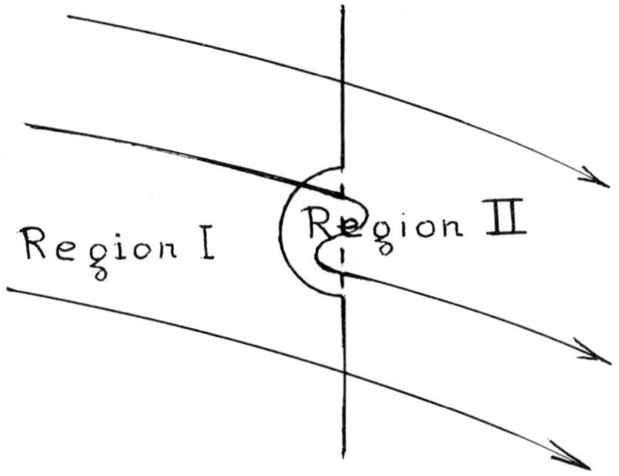

Fig. 141. The boundary of the two regions is complex and its simplification by use of the dotted line can lead to a simpler test that does not respect the principle of Fig. 140.

11.7.5 Example of the construction of a test of escape for the n-body problem

Let us consider first a ten-body problem with ten equal masses (Fig. 142) and let us try to construct a test of escape for this problem.

A test of escape gives only partial informations on the motion. We need it only because we cannot integrate the problem of interest. Thus we are led to the three successive phase of construction of the test : the simplification of the problem, the research of long-term valid results for the simplified problem and the improvement of the efficiency of the resulting test.

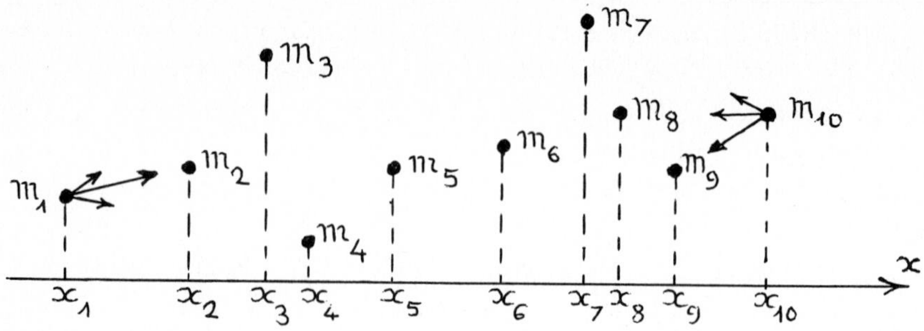

Fig. 142. A ten-body problem with ten equal masses. Research of tests of escape.

11.7.5.1 Simplification of the problem

The ten masses are equal, let us call m their common value and M the total mass ; $M = 10m$.

We will not consider the thirty coordinates of the ten bodies but, for instance, only the abscissas.

What are then the equations of motion ?

The initial equations of motion are :

$$d^2\vec{r}_j/dt^2 = Gm \sum_{K=1;K\neq j}^{10} \vec{r}_{jK}/r_{jK}^3 \quad ; \quad j = \{1, 2, \ldots, 10\} \tag{1031}$$

The projection on the x-axis gives :

$$d^2x_j/dt^2 = Gm \sum_{K=1;K\neq j}^{10} (x_K - x_j)/r_{jK}^3 \; ; \; j = \{1, 2, \ldots, 10\} \tag{1032}$$

The lengths r_{jK} are not given in terms of the ten abscissas, but we know that :

$$\forall\ j,K : r_{jK} \geq |x_K - x_j| \tag{1033}$$

Hence we can consider the following simplified system :

$$\left.\begin{array}{l} d^2x_j/dt^2 = Gm \displaystyle\sum_{K=1;K\neq j}^{10} (x_K - x_j)^{-2} \cdot \theta_{jK}\, \mathrm{sign}(x_K - x_j) \\ \text{with :} \\ \forall\ j,K : 0 \leq \theta_{jK} = \theta_{Kj} \leq 1 \end{array}\right\} \tag{1034}$$

We will label the ten abscissas x_1, x_2, ..., x_{10} in ascending order even if they do not always correspond to the same body :

$$\left.\begin{array}{l} x_1 \leq x_2 \leq x_3 \leq x_4 \leq x_5 \dots \leq x_{10} \\ \text{Hence for instance :} \\ x_1 \neq x_2 \text{ implies : } 0 \leq d^2x_1/dt^2 \leq Gm \sum_{K=2}^{10} (x_K - x_1)^{-2} \end{array}\right\} \quad (1035)$$

Thus the motion of x_1 is a succession of arcs satisfying (1035) and of discontinuities when $x_2 = x_1$; these discontinuities are negative discontinuities of dx_1/dt, the abscissa x_1 itself being continuous.

11.7.5.2 Research of long-term valid results

The upper bound of d^2x_1/dt^2 in (1035) can be very large but its lower bound is zero. This seems a simple and useful result ! Unfortunately the negative discontinuities of dx_1/dt when $x_2 = x_1$ prevent the long-term use of this property.

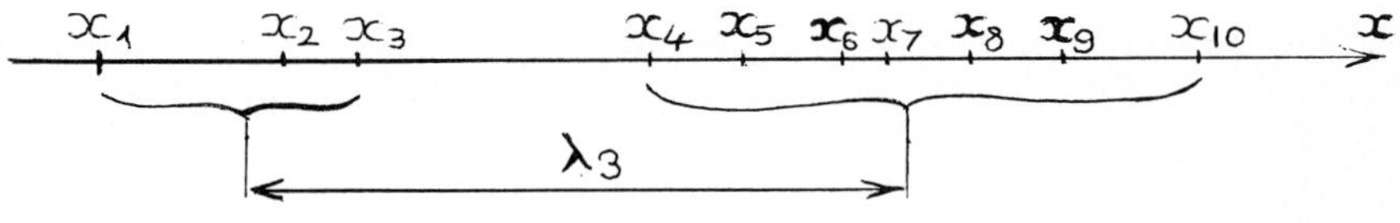

Fig. 143. The distance λ_3 between the center of mass of x_1, x_2, x_3 and the center of mass of the other abscissas.

Another example is given by the lengths λ_s, with s={1,2,...,9}.

A length λ_s is the distance between the abscissa of center of mass of x_1, x_2, ..., x_s and the center of mass of the (10 - s) other abscissas :

$$\lambda_s = \frac{1}{10 - s} \sum_{K=s+1}^{10} x_K - \frac{1}{s} \sum_{j=1}^{s} x_j \quad (1036)$$

Hence, if $x_s < x_{s+1}$:

$$- Gm\left[\frac{1}{10 - s} + \frac{1}{s}\right]\sum_{j=1}^{s}\sum_{K=s+1}^{10}(x_K - x_j)^2 \leq d^2\lambda_s/dt \leq 0 \tag{1037}$$

However we again find discontinuities of $d\lambda_s/dt$ when $x_s = x_{s+1}$ (this discontinuities are positive) and the question is how to avoid them.

Notice that <u>the largest λ_s</u> satisfies the relation :

$$x_{s+1} - x_s \geq \lambda_s/5 \tag{1038}$$

because $\lambda_s \geq \lambda_{s+1}$; $\lambda_s \geq \lambda_{s-1}$ and :

$$\left.\begin{aligned} &2s(10 - s)\lambda_s \equiv (s + 1)(10 - s - 1)\lambda_{s+1} + \\ &+ (s - 1)(10 - s + 1)\lambda_{s-1} + 10(x_{s+1} - x_s) \end{aligned}\right\} \tag{1039}$$

Hence if we define the length λ by :

$$\lambda = \sup \lambda_s \quad ; \quad s = \{1, 2, \ldots, 9\} \tag{1040}$$

we obtain when "the largest λ_s is alone" :

$$j \in \{1, 2, \ldots, s\} \quad ; \quad K \geqq (s + 1) \text{ imply } x_K - x_j \geq \lambda_s/5 \tag{1041}$$

and thus :

$$- Gm\left(\frac{1}{10 - s} + \frac{1}{s}\right)(10 - s)s.\ 25/\lambda^2 \leq d^2\lambda/dt^2 \leq 0 \tag{1042}$$

That is, with $M = 10m$ = total mass and $\mu = GM$ = gravitational constant of the system :

$$- 25\mu/\lambda^2 \leq d^2\lambda/dt^2 \leq 0 \tag{1043}$$

The lower bound can now be used at least "when the largest λ_s is alone". However we have exchanged one type of discontinuities (when $x_s = x_{s+1}$) for another that occurs when the two largest λ_s are equal.

Since $\lambda = \sup\{\lambda_1, \lambda_2, \ldots, \lambda_9\}$ the discontinuities are <u>positive discontinuities of $d\lambda/dt$</u> and the condition $d^2\lambda/dt^2 \leq 0$ of (1043) loses its interest.

Fortunately the condition $-25\mu/\lambda^2 \leqq d^2\lambda^2/dt^2$ remains useful and we finally reach a long term result. We will have a better idea of the evolution of λ if we notice the following property easy to demonstrate by similar considerations :

$$\left.\begin{array}{l} (\lambda - \lambda_s)/\lambda \leqq \varepsilon < 1/25 \text{ implies} \\ x_{s+1} - x_s \geqq \lambda\big(1 - \varepsilon(10s - s^2)\big)/5 \geqq \lambda(1 - 25\varepsilon)/5 \\ - 25\mu/\lambda^2(1 - 25\varepsilon)^2 \leqq d^2\lambda_s/dt^2 \leqq 0 \end{array}\right\} \quad (1044)$$

Thus, as long as the ten-body motion of interest remains defined the function $\lambda(t)$ has the following properties :
$\lambda(t)$ is continuous and non-negative.
$\lambda(t)$ has always a left-derivative $\lambda'_L(t)$ and a right-derivative $\lambda'_R(t)$ with :

$$\left.\begin{array}{l} \lambda'_L(t) = \lambda'_R(t) \text{ almost always} \\ \lambda'_L(t) \leqq \lambda'_R(t) \text{ always} \\ \lambda'_L(t) \text{ is left-semi continuous and lower-semi continuous} \\ \lambda'_R(t) \text{ is right-semi continuous and upper-semi continuous} \\ t_1 < t_2 \text{ implies } \lambda'_R(t_1) - 25\mu\int_{t_1}^{t_2} dt/\lambda^2 \leqq \lambda'_L(t_2) \leqq \lambda'_R(t_2) \end{array}\right\} \quad (1045)$$

Let us now consider the function F(t) :

$$F(t) = \lambda'_R(t)^{\,2} - 50\mu/\lambda \quad (1046)$$

and let us demonstrate that when $\lambda'_R(t) \geqq 0$ in $[t_0, t_f]$ the function F(t) is non-decreasing in this interval.

Indeed, if $t_0 \leqq t_1 < t_2 \leqq t_f$; we have :

$$\lambda'_R(t_2) \geqq \lambda'_R(t_1) - 25\mu\int_{t_1}^{t_2} dt/\lambda^2 \quad (1047)$$

Hence, since $\lambda'_R(t) \geqq 0$ in t_1, t_2, (consider first infinitesimal $t_2 - t_1$):

$$\lambda'_R(t_2)^{\,2} \geqq \lambda'_R(t_1)^{\,2} - 50\mu\int_{t_1}^{t_2} \lambda'_R(t)dt/\lambda^2 \quad (1048)$$

On the other hand :

$$- 50\mu/\lambda(t_2) = - 50\mu/\lambda(t_1) + 50\mu\int_{t_1}^{t_2} \lambda'_R(t)dt/\lambda^2 \tag{1049}$$

and, by addition of (1048) and (1049) :

$$F(t_2) \geqq F(t_1) \tag{1050}$$

Let us assume now that at $t = t_1$:

$$\lambda'_R(t_1) \geqq \sqrt{50\mu/\lambda(t_1)} \tag{1051}$$

The function $\lambda'_R(t_1)$ being right-semi continuous (and also lower-semi continuous) it will remain positive or zero in some interval $[t_1, t_2]$ with $t_2 > t_1$ and $F(t_2) \geqq F(t_1)$. However (1051) implies $F(t_1) \geqq 0$ and thus $F(t_2) \geqq 0$, which, with $\lambda'_R(t_2) \geqq 0$ implies :

$$\lambda'_R(t_2) \geqq \sqrt{50\mu/\lambda(t_2)} \tag{1052}$$

The conclusion is then : if (1051) is satisfied at some time t_1 the corresponding inequality <u>$\lambda'_R(t) \geqq \sqrt{50\mu/\lambda(t)}$ will remain forever satisfied and the function F(t) will remain forever non-decreasing and non-negative</u>.

We have thus divided the phase space into two parts : the region I where the inequality $\lambda'_R \geqq \sqrt{50\mu/\lambda}$ is not satisfied and the region II where this equality is satisfied. These two regions satisfy the principle of Fig. 140 and no motion from the region II to the region I is possible.

The region II is obviously a region of escape (if the motion remains indefinitely defined) and the length λ goes to infinity at least as $\{112,5\ \mu t^2\}^{1/3}$, and even as Kt if F(t) reaches the positive value K^2.

The condition :

$$d\lambda/dt \geqq \sqrt{50\mu/\lambda} \tag{1053}$$

is thus a test of escape of the ten-equal masses problem with $\mu = GM = 10$ Gm and λ defined in (1036), (1040) ; the non-decrease of F(t) give then a lower bound of future $d\lambda/dt$.*

* The opposite condition $d\lambda/dt \leqq -\sqrt{50\mu/\lambda}$ gives of course a "test of escape" for past motions.

11.7.5.3 Improvement of the efficiency of the test. Extension to the general n-body problem.

This third part has no theoretical interest but is essential for efficient tests. However the possibilities of improvement being infinite the discussion always end with a balance between simplicity and efficiency. The simplicity must be recognized as a part of the efficiency, it gives a better knowledge of the problem.

We will consider three types of improvement.

A) Improvements given by the research of the limits.

The equations (1041), (1042), (1043) give a lower bound of $d^2\lambda/dt^2$ but not the greatest lower bound. This greatest lower bound, for a given λ, is obtained when all ten bodies are along the x-axis and when all λ_s are equal, that is when all $x_{s+1} - x_s$ are equal to $\lambda/5$ (Fig. 144).

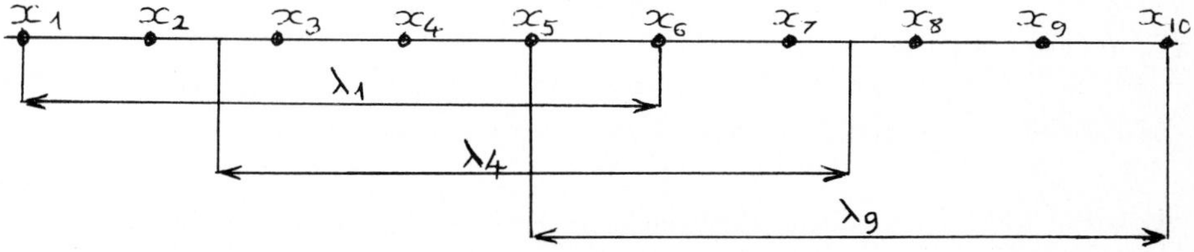

Fig. 144. The greatest lower bound of $d^2\lambda/dt^2$, for a given λ, is obtained when all ten bodies are along the x-axis and when all λ_s are equal, which implies the equality of all intervals $x_{s+1} - x_s$.
$\lambda = \lambda_s = 5(x_{s+1} - x_s)$; $s = \{1, 2, \ldots, 9\}$.

This collinear configuration gives the following second derivatives λ''_s :

$$\left.\begin{aligned}
\lambda''_1 = \lambda''_9 &= -\frac{25\mu}{9\lambda^2}\left(1 + \frac{1}{4} + \frac{1}{9} + \frac{1}{16} + \frac{1}{25} + \frac{1}{36} + \frac{1}{49} + \frac{1}{64} + \frac{1}{81}\right) = \\
&= -4.27713\mu/\lambda^2 \\
\lambda''_2 = \lambda''_8 &= -\frac{25\mu}{16\lambda^2}\left(1 + \frac{2}{4} + \frac{2}{9} + \frac{2}{16} + \frac{2}{25} + \frac{2}{36} + \frac{2}{49} + \frac{2}{64} + \frac{1}{81}\right) = \\
&= -3.22998\mu/\lambda^2 \\
\lambda''_3 = \lambda''_7 &= -\frac{25\mu}{21\lambda^2}\left(1 + \frac{2}{4} + \frac{3}{9} + \frac{3}{16} + \frac{3}{25} + \frac{3}{36} + \frac{3}{49} + \frac{2}{64} + \frac{1}{81}\right) = \\
&= -2.77260\mu/\lambda^2
\end{aligned}\right\} \quad (1054)$$

$$\left.\begin{aligned} \lambda_4'' = \lambda_6'' &= -\frac{25\mu}{24\lambda^2}\left(1+\frac{2}{4}+\frac{3}{9}+\frac{4}{16}+\frac{4}{25}+\frac{4}{36}+\frac{3}{49}+\frac{2}{64}+\frac{1}{81}\right) = \\ &= -2.56173\mu/\lambda^2 \\ \lambda_5'' &= -\frac{25\mu}{25\lambda^2}\left(1+\frac{2}{4}+\frac{3}{9}+\frac{4}{16}+\frac{5}{25}+\frac{4}{36}+\frac{3}{49}+\frac{2}{64}+\frac{1}{81}\right) = \\ &= -2.49926\mu/\lambda^2 \end{aligned}\right\} \quad (1054)$$

The smallest λ_s'' are λ_1'' and λ_9'', hence for a given λ the greatest lower bound of $d^2\lambda/dt^2$ is $-4.27713\mu/\lambda^2$ that is much better than the $-25\mu/\lambda^2$ given in (1043) and the test of escape (1053) becomes (with $8.55426 = 2 \times 4.27713$) :

$$d\lambda/dt \geq \sqrt{8.55426\mu/\lambda} \quad (1055)$$

B) Improvements given by a slight modification of the parameters of the test.

It is of course a pity that the second derivatives given in (1054) are different. There is an obvious loss there.

We can avoid it if instead of λ defined as $\sup(\lambda_s)$ we use the length L defined by the following with suitable coefficients K_s :

$$L = \sup_{s=\{1,2,\dots,9\}} K_s\lambda_s \quad (1056)$$

The greatest lower bound of d^2L/dt^2 for a given L is obtained when all bodies are along the x-axis and when all nine lengths $K_s\lambda_s$ are equal. The best coefficients K_s are those giving then the same value to the nine accelerations $K_s\lambda_s''$ and the configuration of the ten bodies is then the central configuration of Fig. 145 and already presented in Section 11.3.1 and in Fig. 116.

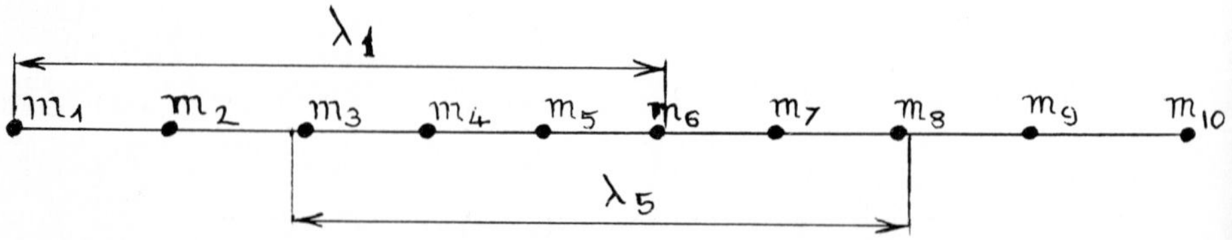

Fig. 145. The collinear central configuration for ten equal masses. The coefficients K_s of (1056), (1057) are equal to λ_5/λ_s for this configuration.

The coefficients K_s are easy to compute, being equal to the ratios λ_5/λ_s for the central configuration :

$$\left.\begin{aligned} K_1 &= K_9 = 0.94766 \\ K_2 &= K_8 = 0.97298 \\ K_3 &= K_7 = 0.98859 \\ K_4 &= K_6 = 0.99723 \\ K_5 &= 1 \end{aligned}\right\} \qquad (1057)$$

The above expression $\lambda'' \geqq -4.27713\mu/\lambda^2$ is improved to :

$$L'' \geqq -2.76216\mu/L^2 \qquad (1058)$$

and the test of escape (1055) becomes the more efficient test :

$$dL/dt \geqq \sqrt{5.52432\mu/L} \qquad (1059)$$

The velocity $\sqrt{5.52432\mu/L}$ can be called the escape velocity.

The improvement between (1055) and (1059) may seem small, the gain on the escape velocity being between 12.9 % and 19.6 %. However we will see that this gain is much larger when the number of bodies is very large.

C) Improvements given by the extension of the test.

Let us consider now an arbitrary n-body problem and its projection on the **x**-axis (Fig. 146).

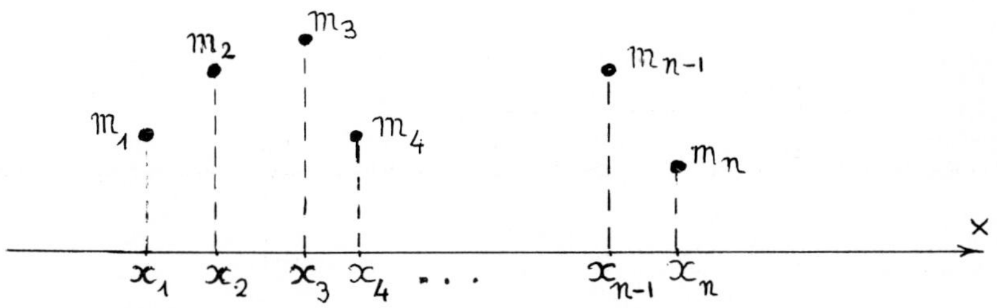

Fig. 146. A n-body motion and its projection on the x-axis.

We will again label the abscissas from left to right and will call $m_1, m_2, \ldots, m_n$ the masses corresponding to $x_1, x_2, \ldots, x_n$. For a given body these names are of course transitory.

The length λ_s will again be defined as the distance between the center of mass of the s left abscissas and the center of mass of the (n - s) right abscissas :

$$\left.\begin{aligned} &\lambda_s = \Big(\sum_{K=s+1}^{n} m_K x_K\Big)/(M - M_s) - \Big(\sum_{j=1}^{s} m_j x_j\Big)/M_s \\ &\text{where } M_s = \sum_{j=1}^{s} m_j \quad ; \quad M = \sum_{j=1}^{n} m_j = \text{total mass} \end{aligned}\right\} \quad (1060)$$

Hence :

$$\left.\begin{aligned} (m_s + m_{s+1})(M - M_s)M_s\lambda_s \equiv\ & m_{s+1}M_{s-1}(M - M_{s-1})\lambda_{s-1} \\ &+ m_s M_{s+1}(M - M_{s+1})\lambda_{s+1} + \\ &+ M m_s m_{s+1}(x_{s+1} - x_s) \end{aligned}\right\} \quad (1061)$$

The length λ will of course be again* :

$$\lambda = \sup \lambda_s \quad ; \quad s = \{1, 2, \ldots, n-1\} \quad (1062)$$

If λ is equal to λ_s the identity (1061) and the relations $\lambda_s \geqq \lambda_{s+1}$, $\lambda_s \geqq \lambda_{s-1}$ will give :

$$\left.\begin{aligned} &x_{s+1} - x_s \geqq \lambda_s(m_{s+1} + m_s)/M \geqq \lambda_s(m_A + m_B)/M \\ &\text{with } m_A \text{ and } m_B \text{ being the two smallest masses} \end{aligned}\right\} \quad (1063)$$

The inequality $d^2\lambda/dt^2 \geqq -25\mu/\lambda^2$ of (1043) then becomes :

* Notice that $x_n = \lambda_{n-1}(M-m_n)/M$ and $x_1 = -\lambda_1(M-m_1)/M$, hence, since $\lambda_1 \leqq \lambda$ and $\lambda_{n-1} \leqq \lambda$, we obtain $\lambda \leqq x_n - x_1 \leqq 2\lambda$

$$\left.\begin{array}{l} d^2\lambda/dt^2 \geq -(\mu/\lambda^2)\left[M/(m_A+m_B)\right]^2 \\ \text{with } \mu = GM \text{ gravitational constant of the system} \end{array}\right\} \quad (1064)$$

For a given λ the greatest lower bound of $d^2\lambda/dt^2$ is given by the collinear configuration of Fig. 147 similar to Fig. 144.

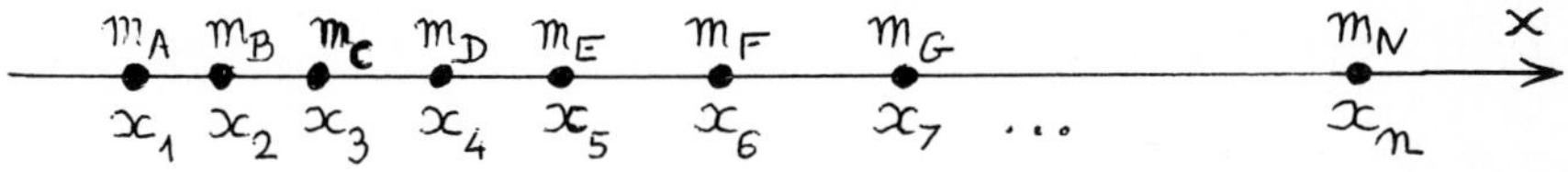

Fig. 147. The greatest lower bound of $d^2\lambda/dt^2$, for a given λ, is obtained when all n bodies are along the x-axis with the masses increasing from left to right and with all λ_s equal. (1061) then gives : $x_{s+1} - x_s = (\lambda/M).(m_s + m_{s+1})$.

The n-masses are is ascending order (i.e. $m_1=m_A$, $m_2=m_B$, etc...) and all λ_s are equal. Hence, with (1061), the distance between two successive bodies is proportional to the sum of their two masses.

The inequality (1064) is then improved to :

$$d^2\lambda/dt^2 \geq -\frac{\mu}{\lambda^2}\cdot\left[\frac{M^2}{M-m_A}\right]\cdot\left[\frac{m_B}{(m_A+m_B)^2}+\frac{m_C}{(m_A+2m_B+m_C)^2}+\frac{m_D}{(m_A+2m_B+2m_C+m_D)^2}+\dots+\frac{m_N}{(2M-m_A-m_N)^2}\right] \quad (1065)$$

with $m_A \leq m_B \leq m_C \leq \dots \leq m_N$

This greatest lower bound generalizes the greatest lower bound λ_1'' given in (1054). Its expression is complex but it is a good improvement of (1064) since it implies always* :

$$d^2\lambda/dt^2 \geq -(\pi^2/8)(\mu/\lambda^2)(M/(m_A+m_B)) \quad (1066)$$

The improvement given by the parameter L of (1056)-(1059) can easily be extended to the case of n equal masses with their corresponding collinear central configuration. The extension is not so easy for n unequal masses since they have n!/2 different collinear central configurations.

---

* This limit corresponds to the case when n is very large and when $m_A = 0$; $m_B = m_C = m_D = \dots = m_N$.

A good and simple approximation of the length L for a general n-body system can be obtained as follows :

$$\left.\begin{array}{l} L = \sup\left(\lambda_s \cos^2\alpha_s\right) \quad ; \quad s = \{1, 2, \ldots, n-1\} \\ \text{with : } \sin 3\alpha_s = (2M_s - M)/M \quad ; \quad |\alpha_s| \leq 30° \end{array}\right\} \qquad (1067)$$

Hence, since $x_n = \lambda_n(M - m_n)/M$ and $x_1 = -\lambda_1(M - m_1)/M$, we have:

$$0.75\lambda \leq L \leq \lambda \leq x_n - x_1 \leq 2\lambda \qquad (1068)$$

and the inequality (1058) becomes :

$$d^2L/dt^2 \geq -\frac{\mu}{L^2}\left[1 + \frac{81}{64}\,\mathrm{Log}\left(\frac{M}{m_A + m_B}\right)\right] \qquad (1069)$$

with the corresponding "escape velocity" :

$$dL/dt = \left\{\frac{2\mu}{L}\left[1 + \frac{81}{64}\,\mathrm{Log}\left(\frac{M}{m_A + m_B}\right)\right]\right\}^{1/2} \qquad (1070)$$

The expression (1069) is an interesting improvement of (1064), (1065), (1066) ; it can be used even if the number of bodies is very large. For instance in the galactic system the number n is about 10^{11} and :

$$\left.\begin{array}{l} L \text{ is about 40 000 light-years} \\ dL/dt \text{ is small} \end{array}\right\} \text{ for all directions of the galactic plane.}$$

Hence, with (1069), L remains larger than 0.96 times its present value for all the interval ± 3 millions of years, while a multiplicity of triple close approach followed by very fast escapes would lead to very large L.

The definition (1067) of the length L can certainly be slightly modified in order to give less importance to the smallest masses in (1069) and to reach a coefficient such as $(1 + \frac{81}{64}\,\mathrm{Log}\,\frac{n}{2})$ or even better. The coefficients of the λ_s can even be function of the initial conditions and not function only of the masses. All this illustrates our initial sentence : the possibilities of improvement of a test are infinite and a balance must be made between efficiency and simplicity.

11.7.6 Final evolutions : the singularities

The equations of motions of the n-body problem are singular at collision hence a n-body motion can have a singularity at $t = t_s$ if for at least one mutual distance r_{jk} : $\liminf r_{jk} = 0$ when $t \to t_s$.

We have already in Chapter 6 met the singularities for the three-body problem ; their occurrence can be studied easily, even in the n-body case, with the results of the previous Sections.

Let us demonstrate the following :

A) If there is a singularity at the instant t_σ, this singularity is either an "infinite expansion in a bounded interval of time" or a collision.

B) Three-body motions have no "infinite expansion in a bounded interval of time".

C) For collisions all non-infinitesimal bodies go to a definite final position when $t \to t_\sigma$.

D) Non-infinitesimal bodies entering into a collision approach a "central configuration" (This is not true for infinitesimal masses).

For the demonstrations we will made a broad use of the property (1064) or (1066), that is with some constant K :

$$d^2\lambda/dt^2 \geq - K/\lambda^2 \qquad (1071)$$

However the expression of K has the sum $(m_A + m_B)$ of the smallest masses at the denominator and we must first consider the case of n-body system with several infinitesimal masses, i.e. with $m_A + m_B = 0$.

If there are p infinitesimal masses with $p \geq 2$ the motion can be decomposed :

I) The (n-p) non infinitesimal masses have an ordinary (n-p)-body motion.

II) The motion of the p infinitesimal masses can be decomposed into p independant (n-p+1)-body motions, the infinitesimal masses interacting only at collision.

III) For each of these (n-p+1)-body motions an equation such as (1071) is valid with a bounded K (the same K for these p problems).

IV) For the general n-body problem the length λ is supremum of all p lengths λ of the p different (n-p+1)-body motions. The equation (1071) remains valid whatever be the model of collision of infinitesimal masses (the elastic collision model is the model generally used for collinear motions).

According to (1066) we can now choose K as :

$$K = \pi^2 \mu M/8(m_A + m_B) \tag{1072}$$

with the general rule : <u>one of the two masses m_A and m_B is the smallest non-infinitesimal mass and the other is the smallest among the (n-1) remaining masses</u>.

11.7.6.1 <u>The two types of singularity of the n-body problem</u>

Suppose we assume the existence of a singularity at $t = t_\sigma$; what will be the evolution of the evolution of the length λ relative to a given direction when t goes to t_σ ?

The function $\lambda(t)$ is continuous and we have already studied its evolution in Section 11.7.5.2 with the function F(t) :

$$\left.\begin{aligned} &F(t) = \lambda'^2 - 2K/\lambda \\ &\lambda' = \text{right-derivative of } \lambda \end{aligned}\right\} \tag{1073}$$

The function F(t) is non-decreasing in segments (t_1, t_2) where $\lambda' \geqq 0$, indeed, λ' being right-semi and upper-semi continuous :

$$\left.\begin{aligned} &dF/dt = 2\lambda'(\lambda'' + K/\lambda^2) \geqq 0 \\ &\Delta F = \Delta(\lambda'^2) \geqq 0, \text{ since } \Delta\lambda' \geqq 0 \end{aligned}\right\} \tag{1074}$$

We now consider the condition $\{F(t) \geqq 0 \;\; ; \;\; \lambda' > 0\}$, that is the inequality :

$$\lambda' \geqq \sqrt{2K/\lambda} \tag{1075}$$

If this inequality is satisfied at some time t_1 it will remain satisfied for all t verifying $t_1 \leqq t < t_\sigma$ and F(t) remains non--negative and non-decreasing after t_1.

We are thus led to two cases.

Either (1075) is satisfied at some time t_1 before t_σ. The derivative λ' is positive in (t_1, t_σ), λ is increasing and has a limit either finite or infinite.

Or alternatively (1075) is never satisfied before t_σ ; the continuous function $\lambda^{3/2} - t\sqrt{4.5K}$ is then decreasing (its derivative is $1.5\lambda'\sqrt{\lambda} - \sqrt{4.5K} < 0$). This continuous and decreasing function has a limit when t goes to t_σ and so has $\lambda(t)$.

In both cases $\lambda(t)$ has a limit when $t \to t_\sigma$ and this limit gives the two types of singularity :

If $\lambda \to +\infty$, for at least one direction of space, the singularity is "an infinite expansion in a bounded interval of time" and both F(t) and λ' also go to infinity.

If for all directions the length λ has a bounded limit we will see in Section 11.7.6.3 that all non-infinitesimal bodies go to a definite final position when $t \to t_\sigma$. Furthermore if these non-infinitesimal bodies remain at mutual distances bounded away from zero all n bodies go to a definite final position.* Hence the singularity has necessarily a collision : two or several bodies have the same final position.

11.7.6.2 Impossibility of the "infinite expansion in a bounded interval of time" for three-body motions.

We already know that the evolution of the continuous function $\lambda(t)$ is a succession of arcs on which :

$$-K/\lambda^2 \leqq d^2\lambda/dt^2 \leqq 0 \tag{1076}$$

and of positive discontinuities of $d\lambda/dt$ when the two largest λ_s are equal.

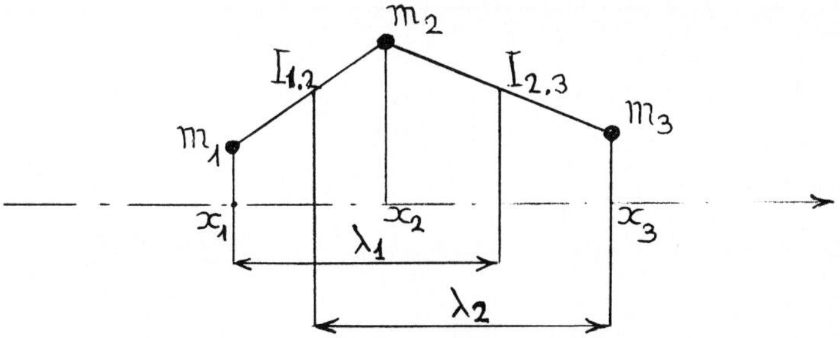

Fig. 148. The length λ equal to $\sup(\lambda_1, \lambda_2)$ in the three-body case.

---

* Provided that t_σ be not a time of accumulation of collisions of infinitesimal bodies.

in a three-body case (Fig. 148) we have then :

$$\left.\begin{aligned} &\lambda = \lambda_1 = \lambda_2 \\ &x_2 - x_1 = (m_1 + m_2)\lambda/M \\ &x_3 - x_2 = (m_2 + m_3)\lambda/M \\ &M = m_1 + m_2 + m_3 \end{aligned}\right\} \tag{1077}$$

and the energy integral gives :

$$\left.\begin{aligned} h \geq 0.5(m_1 x_1'^2 + m_2 x_2'^2 + m_3 x_3'^2) - G\Bigg[&\frac{m_1 m_2}{|x_2 - x_1|} + \frac{m_1 m_3}{|x_3 - x_1|} + \\ &+ \frac{m_2 m_3}{|x_3 - x_2|}\Bigg] \end{aligned}\right\} \tag{1078}$$

which implies for the left and right derivatives λ_L' and λ_R' :

$$\left.\begin{aligned} &x_1' = -\lambda_1'(m_2+m_3)/M \ ; \ x_3' = \lambda_2'(m_1+m_2)/M \ ; \ m_1x_1'+m_2x_2'+m_3x_3' = 0 \\ &m_1x_1'^2 + m_2x_2'^2 + m_3x_3'^2 = \frac{(m_1+m_2)(m_2+m_3)}{m_2M^2} \cdot \Big[m_1(m_2+m_3)\lambda_1'^2 + \\ &\qquad + m_3(m_1+m_2)\lambda_2'^2 - 2m_1m_3\lambda_1'\lambda_2'\Big] \\ &\lambda_L' = \inf\{\lambda_1' \ ; \ \lambda_2'\} \leq \left\{\frac{2M^2}{(m_1+m_2)(m_1+m_3)(m_2+m_3)}\left[h + \frac{\mu}{\lambda}\left(\frac{m_1m_2}{m_1+m_2} + \right.\right.\right. \\ &\qquad \left.\left.\left. + \frac{m_1m_3}{M+m_2} + \frac{m_2m_3}{m_2+m_3}\right)\right]\right\}^{1/2} \\ &\lambda_R' = \sup\{\lambda_1' \ ; \ \lambda_2'\} \leq \left\{\frac{2M}{m_1m_3+m_2 \cdot \inf\{m_1;m_3\}}\left[h + \frac{\mu}{\lambda}\left(\frac{m_1m_2}{m_1+m_2} + \right.\right.\right. \\ &\qquad \left.\left.\left. + \frac{m_1m_3}{M+m_2} + \frac{m_2m_3}{m_2+m_3}\right)\right]\right\}^{1/2} \end{aligned}\right\} \tag{1079}$$

Assume now that the motion is an "infinite expansion in a bounded interval of time". The length λ and its derivatives λ_L' and λ_R' will go to infinity when $t \to t_\sigma$ and after some time t_2, smaller than t_σ, the condition (1079) for λ_L' will no longer be satisfied, and the same for its equivalents by permutation of m_1, m_2, m_3.

Hence for $t_2 < t \leqq t_\sigma$ the condition (1076) remains satisfied. The derivatives λ'_L and λ'_R remain equal and non-increasing ; they cannot go to plus infinity which is contradictory to the assumption. Thus three-body motions with infinite expansion in a bounded interval of time are impossible.

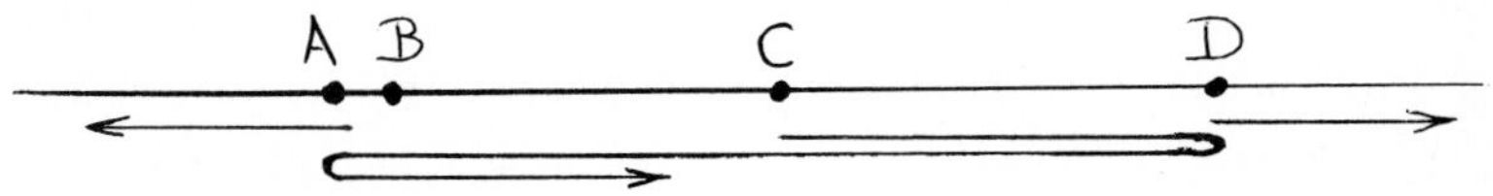

Fig. 149. An infinite expansion in a bounded interval of time.

A four-body example of an infinite expansion in a bounded interval of time has been given by Mather and Mac Gehee (Ref. 214). The close binary AB and the Body D are escaping in opposite directions (Fig. 149) while the light body C moves back and forth at an ever increasing rate between AB and D (with the energy extracted from successive collapses of AB on itself at suitable ABC close approach).

It is possible to demonstrate that this exceptional type of motion require extremely strong oscillations with at least two mutual distances r_{jK} satisfying :

$$\lim_{t\to t_\sigma} \sup\; r_{jK} = +\infty \quad ; \quad \lim_{t\to t_\sigma} \inf\; r_{jK} = 0 \tag{1080}$$

11.7.6.3 Analysis of a collision

Let us consider the second type of singularity : when t goes to the instant t_σ of the singularity the length λ goes to a bounded limit λ_σ.

Assume first that all n bodies are non infinitesimal and remember that :

$$\lambda = \sup \{ \lambda_1, \lambda_2, \ldots, \lambda_{n-1} \} \tag{1081}$$

Furthermore, with the two smallest masses m_A and m_B the generalization of (1044) leads to :

$$\left.\begin{aligned}&(\lambda - \lambda_s)/\lambda \leq 2m_A m_B/M^2 \text{ implies with (1061) :}\\ &x_{s+1} - x_s \geq (m_s + m_{s+1})\lambda/2M \geq (m_A + m_B)\lambda/2M\\ &\text{and : } - 4K/\lambda^2 \leq d^2\lambda_s/dt^2 \leq 0\end{aligned}\right\} \quad (1082)$$

Let us use the lengths λ and λ_s relative to the abscissas and let us demonstrate that all abscissas have a limit when $t \to t_\sigma$.

This property is obvious if $\lambda_\sigma = 0$: all abscissas x_j verify $|x_j| \leq \lambda$ and goes to zero.

If λ_σ is positive there exists a time t_3 such that :

$$1°)\ t_3 < t_\sigma \ ;\ 2°)\ t_3 \leq t \leq t_\sigma \text{ implies } \lambda(t) \geq \lambda_\sigma/2 \quad (1083)$$

In this interval of time $d^2\lambda/dt^2$ is bounded below by $- 4K/\lambda_\sigma^2$, hence $d\lambda/dt$ is bounded below by a constant B (for instance $B = \lambda'(t_3) - 4K(t_\sigma - t_3)/\lambda_\sigma^2$).

Assume that at $t = t_4$ in (t_3, t_σ) we have $\lambda = \lambda_s$, then, as long as (1082) can be used, we will have :

$$\lambda_s(t) \geq \lambda(t_4) + \lambda_s'(t_4)(t - t_4) - 8K(t - t_4)^2/\lambda_\sigma^2 \quad (1084)$$

$$\left.\begin{aligned}0 \leq \lambda(t) - \lambda_s(t) \leq \lambda(t) - \lambda(t_4) - \lambda_s'(t_4)(t - t_4) +\\ + 8K(t - t_4)^2/\lambda_\sigma^2\end{aligned}\right\} \quad (1085)$$

Since $\lambda_s'(t_4) \geq B$ the right-hand member of (1085) goes to zero when $t_4 \to t_\sigma$ with $t_4 \leq t \leq t_\sigma$. Hence for sufficiently large t_4 in the vicinity of t_σ the conditions (1082) can be used in entire interval (t_4, t_σ) and the corresponding λ_s has a definite final value when $t \to t_\sigma$.

Thus in the interval (t_4, t_σ) the n bodies are separated into two groups, the left group from m_1 to m_s and the right group from m_{s+1} to m_n. (Fig. 150).

The distance $x_{s+1} - x_s$ between the two groups remains at least $(m_A + m_B)\lambda_\sigma/4M$ and each group has a center of mass going to a definite abscissa $(- M_s\lambda_s(t_\sigma))/M$ and $(M - M_s).\lambda_s(t_\sigma)/M)$.

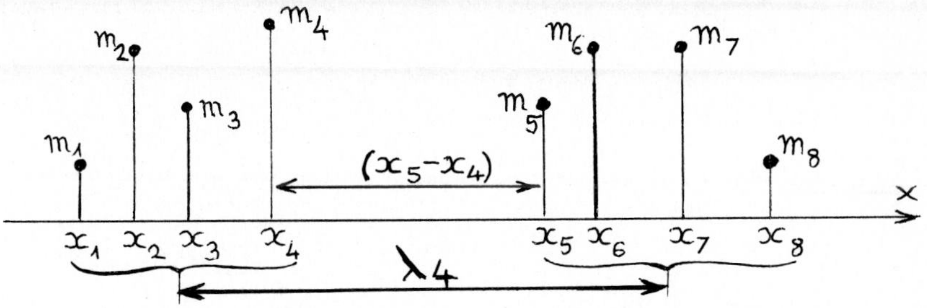

Fig. 150. The two groups m_1 to m_4 and m_5 to m_8 remain separated (when $t_4 \leq t \leq t_\sigma$) by a distance larger than $(m_A + m_B)\lambda_\sigma/4M$.

The next step is the study of one of these two groups and its division into two sub-groups with exactly the same properties by almost exactly the same analysis. The only difference is the influence of the other group and this influence remains limited, it is easy to take account of it and to develop the demonstration.

The analysis proceeds similarly with successive divisions of the groups and sub-groups until the individual bodies : each of them goes to a definite abscissa when the time t goes to t_σ.

This property is of course true in any direction and all bodies go to a definite final position when $t \to t_\sigma$.

This demonstration is not valid for the infinitesimal masses for which there are the following possibilities.

I) If the positive masses go to separate positions when $t \to t_\sigma$, each infinitesimal mass goes to a definite final position when $t \to t_\sigma$, at least if collisions of infinitesimal masses are avoided in some interval (t_5, t_σ) with $t_5 < t_\sigma$.

For each infinitesimal m_j the function

$$0.5W_j^2 - G \sum_{K\neq j} (m_K/r_{jK})$$

remains bounded in (t_5, t_σ) (with W_j = relative velocity of m_j with respect to the nearest positive mass).

II) It seems likely that the infinitesimal masses go to a definite final position even in the remaining cases (collision of positive masses at t_σ, accumulation of collision of infinitesimal masses before t_σ), however this has not yet been demonstrated.

III) In the three-body case this property is true and easy to demonstrate. For instance if two positive masses collide at t_σ :

Either the limit λ_σ is zero in the three mains directions of space and all three masses collide at the center of mass.

Or λ_σ is not zero in several directions : these λ_σ gives the final position of the third body.

Hence the only possible singularities of the three-body problem are the binary collision and the triple collision. In both cases all three bodies go to a definite position when $t \to t_\sigma$.

11.7.6.4 Collisions and central configurations

Let us consider a n-body system of n non-infinitesimal masses and let us assume that this system undergoes a "complete collapse" at the time $t = t_c$ (all n vectors $\vec{r}_K$ go to zero when $t \to t_c$).

We will use the notations and the main relations of Sections 11.6.1 and 11.6.2.

In a complete collapse all mutual distances go to zero and hence the mean quadratic distance ρ and the mean harmonic distance ν also go to zero.

With these notations the Lagrange-Jacobi identity is :

$$(\rho^2)'' = 2\mu((1/\nu) - (1/a)) \tag{1086}$$

where :

$$\left.\begin{aligned} &\mu = GM = G(m_1 + m_2 + \dots + m_n) = \text{gravitational constant} \\ &a = -\,GM^*/2h = \text{generalized semi-major axis} \\ &M^* = \sum_{i \le j < K \le n} m_j m_K \end{aligned}\right\} \tag{1087}$$

Since $0 \le \nu \le \rho$ the curve $\rho^2(t)$ is concave up when $\rho \le a$ and also when the energy integral h is positive or zero (i.e. $-1/a \ge 0$) Hence $\rho(t_c) = 0$ implies a continuous decrease of ρ from at least $|a|$ and we will call (t_1, t_c) the corresponding interval of time.

Let us now consider the Sundman function j :

$$j = (\rho/2a) + (p/2\rho) + (\rho\rho'^2/2\mu) \tag{1088}$$

with :

$$\left.\begin{aligned} &\rho' = d\rho/dt \\ &p = Mc^2/GM^{*2} = \text{"generalized semi-latus rectum"} \\ &\vec{c} = \text{angular momentum} \end{aligned}\right\} \quad (1089)$$

The Sundman inequality is :

$$\rho/\nu \geqq j \quad (1090)$$

and the Lagrange-jacobi identity (1086) implies :

$$dj/dt = (\rho'/\rho)\big((\rho/\nu) - j\big) \quad (1091)$$

In the interval (t_1, t_c) the derivative ρ' is negative or zero and so is dj/dt.

Let us consider two instants t_2 and t_3 in (t_1, t_c) :

$$t_1 < t_2 < t_3 < t_c \quad (1092)$$

Hence :

$$\begin{aligned} j_2 = j(t_2) &= (\rho_2/2a) + (p/2\rho_2) + (\rho_2\rho_2'^2/2\mu) \geqq j_3 = \\ &= (\rho_3/2a) + (p/2\rho_3) + (\rho_3\rho_3'^2/2\mu) \end{aligned} \quad (1093)$$

which implies :

$$j_2 \geqq (\rho_3/2a) + (p/2\rho_3) \quad (1094)$$

If $p > 0$ the mean quadratic distance ρ_3 cannot go to zero when $t_3 \to t_c$ and this implies the following first result :

<u>A n-body motion with a complete collapse has a generalized semi-latus rectum p and an angular momentum $\vec{c}$ equal to zero</u>.

Let us now consider the function $f(t)$ equal to $\rho(j - 1)$:

$$f(t) = \rho(j - 1) = (\rho^2/2a) + (\rho^2\rho'^2/2\mu) - \rho \quad (1095)$$

$$df/dt = \rho'(j - 1) + j'\rho = \rho'\{(\rho/\nu) - 1\} \tag{1096}$$

Since $0 \leq \nu \leq \rho$, the function $f(t)$ is non-increasing on (t_1,t_c) and since its limit when $t \to t_c$ is positive or zero we obtain :

$$t_1 < t < t_c \text{ implies } f(t) = \rho(j - 1) \geq 0 \tag{1097}$$

Hence $j \geq 1$ in (t_1,t_c) and this non-increasing function has a limit j_c when $t \to t_c$.

Thus :

When the motion goes to a complete collapse the Sundman function j goes down to a limit j_c larger than or equal to one.

The equation (1091) can be written :

$$dj/(d(\text{Log}\rho) = \{(\rho/\nu) - j\} \geq 0 \tag{1098}$$

On (t_1,t_c) the variation of j is bounded while the variation of $\text{Log}\rho$ is infinite, hence $\{(\rho/\nu) - j\}$ must generally be very near to zero. Let us demonstrate that $\{(\rho/\nu) - j\}$ goes to zero when $t \to t_c$ that is when $\text{Log}\rho \to -\infty$.

The derivative ρ' of ρ is given by the expression of j, that is here since $p = 0$:

$$(\rho/2a) + (\rho\rho'^2/2\mu) = j \tag{1099}$$

hence for small ρ, when t and j approach t_c and j_c :

$$\rho\rho'^2/2\mu \sim j_c \tag{1100}$$

which implies :

$$\left.\begin{aligned} &(\rho^{3/2})' = (3/2)\rho'\sqrt{\rho} \sim -\sqrt{4.5\mu j_c}\ ; \quad (\rho^{3/2}) \sim (t_c - t)\sqrt{4.5\mu j_c} \\ &\rho \sim \{4.5\mu j_c(t_c - t)^2\}^{1/3}\ ; \quad \rho' \sim -\{4\mu j_c/3(t_c - t)\}^{1/3} \end{aligned}\right\} \tag{1101}$$

On the other hand the derivative ν' of ν is bounded by the limit given in (996), which, for small ν and with a suitable constant K, is :

$$|d\nu/dt| = |\nu'| \leq K/\sqrt{\nu} \tag{1102}$$

The conditions (1098), (1100), (1102) imply the following for small ρ and ν when t approaches t_c (with another suitable constant K_1) :

$$\left.\begin{aligned} d\big((\rho/\nu) - j\big)/d(\mathrm{Log}\rho) &= (\rho/\rho')\,(\rho'/\nu - \rho\nu'/\nu^2) - \\ &- \big((\rho/\nu) - j\big) = (j - (\rho^2\nu')/(\rho'\nu^2)) \end{aligned}\right\} \quad (1103)$$

$$\left|d((\rho/\nu) - j)/d(\mathrm{Log}\rho)\right| \leq K_1(\rho/\nu)^{5/2} \quad (1104)$$

It is then easy to demonstrate that $(\rho/\nu) \to j_c$ when $t \to t_c$, indeed when $(\rho/\nu) \leq 2j_c$:

$$|d((\rho/\nu) - j)/d(\mathrm{Log}\rho)| \leq K_1(2j_c)^{5/2} = K_2 \quad (1105)$$

Let us put $\mathrm{Log}\rho = u$; $\mathrm{Log}\rho(t_2) = \mathrm{Log}\rho_2 = u_2$, etc ... and let us analyse in terms of u :

$$\int_{-\infty}^{u_2} \big((\rho/\nu) - j\big)du = j_2 - j_c \quad (1106)$$

If $j \leq 2j_c - \varepsilon$ and if $\big((\rho/\nu) - j\big) = \varepsilon$ for some $u = u_3$ we will have :

$$\left.\begin{aligned} &(\rho/\nu) - j = \varepsilon \quad \text{for} \quad u = u_3 \\ &(\rho/\nu) - j \geq \sup\{0 ; \varepsilon - K_2(u_3 - u)\} \;; \quad \text{for} \quad u \leq u_3 \end{aligned}\right\} \quad (1107)$$

$$\int_{-\infty}^{u_3} \big((\rho/\nu) - j\big)du \geq \varepsilon^2/2K_2 \quad (1108)$$

Since the integral of (1106) goes to zero when u_2 goes to minus infinity, u_3 characterized by $(\rho/\nu) - j = \varepsilon$ and bound by (1108) remains above some limit u_4. Since this is true for any ε the difference $(\rho/\nu) - j$ must go to zero when $u \to -\infty$ and $t \to t_c$.

Hence :

<u>When an n-body motion goes down to a complete collapse the ratio ρ/ν goes down to the limit j_c of the Sundman function and</u> :

$$\left.\begin{aligned} &\rho \sim \left\{4.5\mu j_c(t_c - t)^2\right\}^{1/3} \;;\; \rho' \sim -\left\{(4\mu j_c)/\big(3(t_c - t)\big)\right\}^{1/3} \;; \\ &\nu \sim \left\{(4.5\mu(t_c - t)^2)/j_c^2\right\}^{1/3} \end{aligned}\right\} \quad (1109)$$

Let us now consider the configuration of the n bodies.

We multiply all radius-vector $\vec{r}_K$ by $(t_c - t)^{-2/3}$:

$$\vec{s}_K = \vec{r}_K(t_c - t)^{-2/3} \quad ; \quad K = \{1, 2, \ldots, n\} \tag{1110}$$

The n vectors $\vec{s}_K$ remain bounded when $t \to t_c$ and the mutual distances s_{jK} remain bounded away from zero. They are excellent tools for the study of the configuration.

With the variable θ defined by :

$$\theta = \mathrm{Log}\big((1/(t_c - t)\big) \tag{1111}$$

the Newtonian equations of motion become :

$$\frac{d^2\vec{s}_K}{d\theta^2} = \frac{2}{9}\vec{s}_K + \frac{1}{3}\frac{d\vec{s}_K}{d\theta} + \sum_{j=1;j\neq K}^{n} Gm_j\,((\vec{s}_{Kj}/s_{Kj}^3)) \tag{1112}$$

The energy integral h becomes :

$$h = \exp(2\theta/3)\cdot\left\{\sum_{K=1}^{n} m_K\left[\frac{2}{9}s_K^2 - \frac{2}{3}\vec{s}_K\frac{d\vec{s}_K}{d\theta} + \frac{1}{2}\left(\frac{d\vec{s}_K}{d\theta}\right)^2\right] - \sum_{1\leq j<K\leq n} G\frac{m_j m_K}{s_{jK}}\right\} \tag{1113}$$

However when $t \to t_c$, that is when $\theta \to +\infty$, the equations (1109), (1110) give :

$$\left.\begin{aligned}
&\sum_{K=1}^{n} m_K s_K^2 = M^*\rho^2/M(t_c - t)^{4/3} \to M^*\{4.5\mu j_c\}^{2/3}/M \\
&\sum_{K=1}^{n} M_K\left(\vec{s}_K\frac{d\vec{s}_K}{d\theta}\right) = \frac{M^*}{M}\left[\frac{2}{3}\rho^2(t_c - t)^{4/3} + \rho\rho'(t_c - t)^{-1/3}\right] \to 0 \\
&\sum_{1\leq j<K\leq n} G\frac{m_j m_K}{s_{jK}} = \frac{GM^*}{\nu}(t_c - t)^{2/3} \to GM^*\left\{(j_c^2)/(4.5\mu)\right\}^{1/3}
\end{aligned}\right\} \tag{1114}$$

Hence :

$$\left.\begin{aligned}
\sum_{K=1}^{n} m_K(d\vec{s}_K/d\theta)^2 = {} & \frac{2GM^*}{\nu}(t_c - t)^{2/3} + (M^*/M)\left[\frac{4}{9}\rho^2(t_c - t)^{-4/3} + \right. \\
& \left. + \frac{4}{3}\rho\rho'(t_c - t)^{-1/3}\right] + 2h(t_c - t)^{2/3}
\end{aligned}\right\} \tag{1115}$$

which when $t \to t_c$ implies :

$$\sum_{K=1}^{n} m_K (d\vec{s}_K/d\theta)^2 \to 0 \tag{1116}$$

and also :

$$\int_{\theta_1}^{\infty} \sum_{K=1}^{n} m_K (d\vec{s}_K/d\theta)^2 < +\infty \tag{1117}$$

Indeed :

$$\int_{\theta_1}^{\infty} m_K (d\vec{s}_K/d\theta)^2 d\theta = -3h(t_c - t_1)^{2/3} - $$

$$- M^*/M \left[(\rho^2/3)(t_c - t_1)^{-4/3} + \right.$$

$$\left. + 2\rho\rho'(t_c - t_1)^{-1/3} + (4.5\mu j_c)^{2/3} \right]$$

When $\theta \to +\infty$ the n velocities $d\vec{s}_K/d\theta$ go to zero and in the equation (1112) it is necessary that $d^2\vec{s}_K/dt^2$ also go to zero, that is for all K = {1, 2, ..., n} :

$$\frac{2}{9}\vec{s}_K + \sum_{j=1;j\neq K} Gm_j \left((\vec{s}_{Kj}/s_{Kj})\right) \to 0 \tag{1118}$$

<u>This gives precisely an evolution towards a "central configuration" when the n bodies approach the complete collapse</u>. The Newtonian accelerations tend to become proportional to the radius-vectors.

If we consider collisions that are not complete collapses the demonstration must be slightly modified because of outer bodies and the angular momentum is arbitrary. However the main result remains the same : the non-infinitesimal bodies entering into collision approach a "central configuration".

Notice the following particular results :

A) We have divided the central configurations of the n-body problem into collinear central configurations of the n-body, planar central configuration and three-dimensional central configurations (Section 11.3.1) and it can be demonstrated that a complete collapse approaching a collinear central configuration is <u>necessarily</u> a collinear motion.

B) Similarly it seems that only planar motions can have a complete collapse approaching a planar central configuration.

C) In the three-body case the three $\vec{s}_K$ vectors have a limit when $t \to t_c$. This is very often but not always true in the general n-body case.

D) The demonstrations of this section are not valid for infinitesimal masses and indeed these masses can go to a collision by very different ways (Fig. 151). Such motions occur even in the three-body problem (Chapter 6).

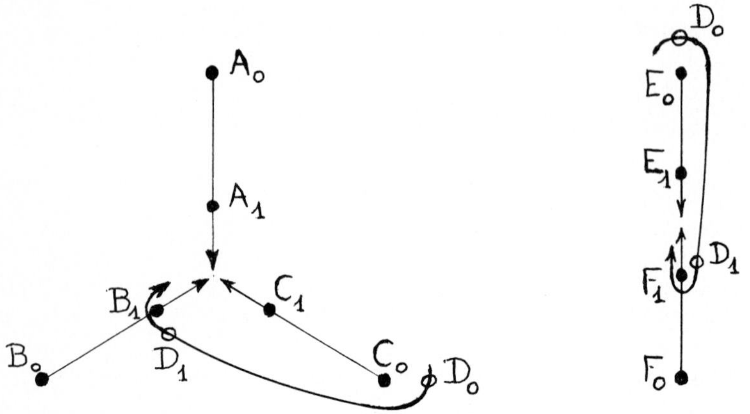

Fig. 151. The infinitesimal mass D can join the complete collapse by an infinite number of loops around three equal masses A, B, C or the two equal masses E and F.

11.7.6.5 On the regularization of singularities

Is it possible to extend naturally a solution after a singularity ? This question is the regularization, it has a theoretical interest and also a numerical one : Consider the regularized equations (88.1) to (88.3) at the end of Chapter 6 ; for close approaches their smoothness makes them much more easily and accurately integrable by numerical methods than the usual Newtonian equations.

The infinite expansion in a bounded interval of time is an "essential singularity" never regulaizable, it must be considered as the end of the corresponding solution.

The collision of three or more bodies can be regularized in extremely rare cases : only for problem of collinear motions and only for particular mass ratios (Ref.54,215). It seems obvious that collisions of the type of Fig. 151 are, a fortiori, not regularizable even in the three-body case.

It remains to examine binary collisions and we must consider two sub-cases.

A) The collision of two infinitesimal masses (Fig. 152).

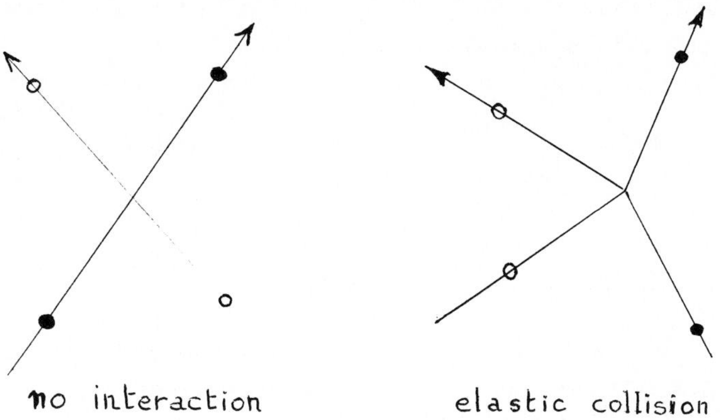

Fig. 152. Collision of two infinitesimal masses.

Two possibilities are generally considered for these collisions : either the two masses cross each other without perturbation or they have an "elastic collision".

The first case corresponds to the "Easton regularization" or regularization by continuity. The infinitesimal masses do not interact and an n-body problem with p non-infinitesimal masses can be decomposed into (n-p) problems of (p+1) masses : the p large masses and an infinitesimal one.

For elastic collisions the motion is not uniquely defined after the collision and supplementary hypotheses are needed except in the study of collinear motions where the uniqueness remains (Ref.54,215).

B) The collision of two masses of which at least one is not infinitesimal.

This final case is the true case of regularization in the n-body problem. These collisions already regularized by Sundman in 1912 are both Siegel ans Easton-regularizable (see Chapter 6). This regularization is so obvious and so natural (Fig. 5) that an n-body problem with only such collisions is sometimes considered as without singularity.

In the three-body problem we will henceforth consider that all binary collisions are regularized (except for two infinitesimal masses). Thus in three-body problems with at most one infinitesimal mass the only real singularity is the triple collision, that is here a "complete collapse".

Let us recall that a complete collapse require a zero angular momentum and, in the three-body case, a plane motion.

Hence a three-body motion with a non-zero angular momentum is uniquely defined for all values of the time from minus infinity to plus infinity.

11.7.7 Final evolutions. The Chazy classification of three-body motions

Almost all three-body motions are without the singularities of the previous Section and remain forever uniquely defined.

We will consider three-body problems with either three positive masses or two positive masses and one infinitesimal (restricted problem).

Let us label the three mases m_A, m_B, m_C in ascending order :

$$0 \leqq m_A \leqq m_B \leqq m_C < M = m_A + m_B + m_C \tag{1119}$$

The evolution of the lengths λ obtained for the different directions of space (Fig. 153) give immediately the main elements of the classification.

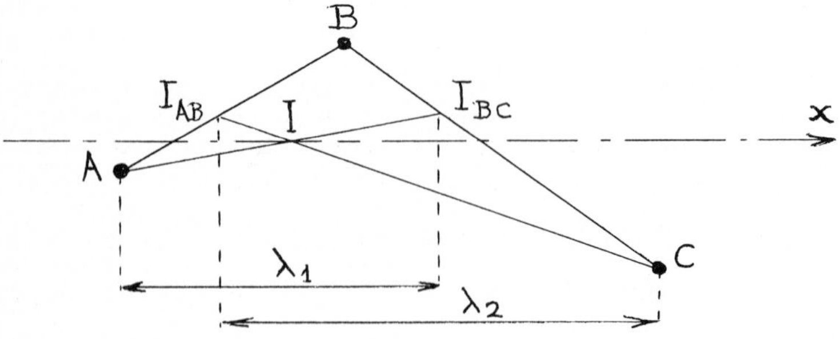

Fig. 153. The length λ in the direction of x.
$\lambda = \sup\{\lambda_1 ; \lambda_2\} = \sup_{J=A,B,C} \{M|x_J|/M - m_J)\}.$

Let us recall (see Section 11.7.5) that the function $\lambda(t)$ is continuous and is a succession of arcs satisfying :

$$\left.\begin{aligned}&- K/\lambda^2 \leq d^2\lambda/dt^2 \leq 0\\ &\text{with, from (1065) :}\\ &K = \frac{\mu M^2}{m_B + m_C}\left(\frac{m_B}{(m_A + m_B)^2} + \frac{m_C}{(M + m_B)^2}\right) \quad ; \quad \mu = GM\end{aligned}\right\} \qquad (1120)$$

and positive discontinuities of $d\lambda/dt$ from the left-derivative λ'_L to the right-derivative λ'_R.

These dicontinuities correspond to the equality $\lambda_1 = \lambda_2$ in Fig. 153 (with $\lambda'_L = \inf\{\lambda'_1 ; \lambda'_2\}$ and $\lambda'_R = \sup\{\lambda'_1 ; \lambda'_2\}$), they are possible only if λ'_L is smaller than or equal to the largest upper bound given by (1079), that is :

$$\left.\begin{aligned}&\lambda'_L \leq \{A + (B/\lambda)\}^{1/2}\\ &\text{with :}\\ &A = \frac{2M^2 h}{(M - m_A)(M - m_B)(M - m_C)} \quad ; \quad h = \text{energy integral}\\ &B = \frac{2M^2 \mu}{(M - m_A)(M - m_B)(M - m_C)}\left[\frac{m_A m_B}{m_A + m_B} + \frac{m_A m_C}{m_A + m_C} + \frac{m_B m_C}{M + m_A}\right]\end{aligned}\right\} \qquad (1121)$$

The evolution of λ is related to that of the function f(t) :

$$f(t) = \lambda_L'^2 - (2K/\lambda) \qquad (1122)$$

We have already seen that if $\lambda'_L > 0$ in $[t_1, t_2]$ the function f(t) is non-decreasing in this segment and if at some time t_1 there is $\{f(t_1) \geq 0 \; ; \; \lambda'_L(t_1) > 0\}$, that is :

$$\lambda'_L \geq \sqrt{2K/\lambda} \qquad (1123)$$

then λ has reached its "escape velocity" ; the function f(t) will remain forever non-negative and non-decreasing and the condition (1123) will remain forever satisfied. The length λ will go to infinity as $t^{2/3}$ if f(t) remains forever equal to zero and as t if f(t) is or become positive.

The final velocity λ'_∞ of λ is equal to $\sqrt{f_\infty}$, with $f_\infty = \lim_{t\to\infty} f(t) =$ final value of f.

Notice that f_∞ is always bounded ; indeed positive discontinuities of λ'_L are not possible when (1121) is not satisfied that is when :

$$\lambda'_L > \{A + (B/\lambda)\}^{1/2} \tag{1124}$$

which is equivalent to :

$$\lambda'_L > 0 \quad ; \quad f(t) > A + ((B - 2K)/\lambda) \tag{1125}$$

The difference B - 2K is negative or zero for all mass ratios (it is zero when $m_A = m_B = m_C$) and thus in the final phase of motion when $\{f(t) \geq 0 \; ; \; \lambda'_L > 0 \; ; \; \lambda$ sufficiently large$\}$:

Either f(t) remains forever less than or equal to A and is thus bounded.

Or f(t) reaches, for instance at $t = t_3$, a value larger than A. The function f will forever remain larger than A since it is then non-decreasing. Discontinuities of λ'_L are no longer possible and (1120) remains for ever satisfied which implies :

$$0 \leq \lambda'_\infty \leq \lambda'_L(t_3) \quad ; \quad f_\infty = \lambda'^2_\infty \leq \lambda'_L(t_3)^2 \tag{1126}$$

The function f(t) and f_∞ are again bounded.

We thus have essentially two main possibilities for the evolution of the length λ :

Either the strict inequality $\lambda'_L > \sqrt{2K/\lambda}$ is satisfied at some time t_1, it will remain forever satisfied after t_1 and λ has a "hyperbolic escape". The ratio λ/t and the derivative λ'_L go to the positive and bounded limit λ'_∞ .

Or, alternatively, the opposite inequality $\lambda'_L \leq \sqrt{2K/\lambda}$ is forever satisfied (which include the exceptional case when f(t) is identical to zero). The function $\lambda^{3/2} - t\sqrt{4.5K}$ is continuous and non-increasing and λ is at most of order $t^{2/3}$. The ratio λ/t again has a limit and this limit is zero.

Let us now consider the three-body system of interest and its lengths λ in all space directions.

11.7.7.1 Relations among the lengths λ. The limits of the vectors $\vec{r}_j/t$.

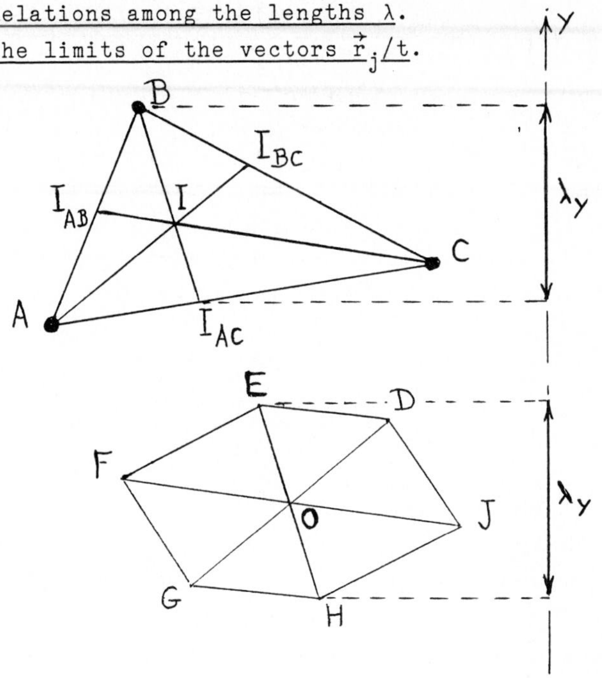

Fig. 154. Construction of the hexagon DEFGHJ with center of symmetry O.

$$\overrightarrow{DG} = \overrightarrow{I_{BC}A} = \vec{\lambda}_A = M\vec{r}_A/(M - m_A)$$

$$\overrightarrow{HE} = \overrightarrow{I_{CA}B} = \vec{\lambda}_B = M\vec{r}_B/(M - m_B)$$

$$\overrightarrow{FJ} = \overrightarrow{I_{AB}C} = \vec{\lambda}_C = M\vec{r}_C/(M - m_C)$$

The orthogonal projections of the hexagon give the lengths λ at the time of interest for all space directions.

Let us consider the Fig. 154 with the triangle ABC of the three bodies and their center of mass I.

The convex hexagon DEFGHJ has O as center of symmetry. Its three main diagonals are given by :

$$\left.\begin{aligned}
\overrightarrow{DG} &= \overrightarrow{I_{BC}A} = \overrightarrow{\lambda_A} = M\overrightarrow{r_A}/(M - m_A) \\
\overrightarrow{HE} &= \overrightarrow{I_{CA}B} = \overrightarrow{\lambda_B} = M\overrightarrow{r_B}/(M - m_B) \\
\overrightarrow{FJ} &= \overrightarrow{I_{AB}C} = \overrightarrow{\lambda_C} = M\overrightarrow{r_C}/(M - m_C)
\end{aligned}\right\} \qquad (1127)$$

By construction the orthogonal projection of the hexagon on an arbitrary direction as the x-axis gives the corresponding length λ at the time of interest.

Notice that conversely with (1127) the hexagon DEFGHJ easily gives the triangle ABC.

A similar construction is true for the ratios λ/t and for their limit when $t \to \infty$.

Can we conclude from this, if the motion remains uniquely defined for all future times, that all three vectors $\vec{r}_A/t$, $\vec{r}_B/t$ and $\vec{r}_C/t$ have a limit when $t \to \infty$?

This is obviously true if the final hexagon has a two--dimensional extension and also if it reduced to the point 0 (all $\vec{r}_j/t$ go to zero). However if the final hexagon has a one--dimensional extension (segment) it only gives only one of the three limits $\vec{r}_j/t$.

In this last case a sophistication as that giving the length L in (1056) with suitable factors K_s gives a second limit $\vec{r}_j/t$, this of the opposite side.

The third limit $\vec{r}_j/t$ is given by the integral of the center of mass ($\Sigma\, m_j\vec{r}_j = 0$) except when the third mass is infinitesimal. It thus remains to analyse the case of two primaries escaping hyperbo--lically into opposite directions with a third infinitesimal mass remaining in the vicinity of the segment joining the two primaries.

This final case is not too difficult. Solutions with an infinite number of loops such as those of Fig. 151 are impossible, and hence we reach the following general theorem :

<u>A three-body motion remaining uniquely defined for all future times has three vectors $\vec{r}_A/t$, $\vec{r}_B/t$, $\vec{r}_C/t$ going to bounded limits when $t \to \infty$</u>.

The extension of this theorem to n-body motions have been made only if all n masses are non-infinitesimal (Ref. 216) ; we then have the following.

Either for some directions the ratio λ/t goes to infinity with t. This is the "super hyperbolic expansion" very similar to the "infinite expansion in a bounded interval of time" of Fig. 149 and Sections 11.7.6.1 ; 11.7.6.2.

Or for all directions λ/t has a bounded limit when $t \to \infty$ and there exist n vectors $\vec{A}_j$ such that when t goes to infinity :

$$\vec{r}_j = \vec{A}_j t + O(t^{2/3}) \quad ; \quad j = \{1, 2, \ldots, n\} \tag{1128}$$

11.7.7.2 The hyperbolic final evolution

If the three vectors $\vec{r}_A/t$, $\vec{r}_B/t$, $\vec{r}_C/t$ have different limits when $t \to \infty$ the final evolution is called "hyperbolic final evolution" and all three mutual distances increase as t.

The three radius-vectors $\vec{r}_j$ can be developed in terms of t for large t :

$$\left.\begin{aligned} &\vec{r}_j = \vec{A}_j t + \vec{B}_j \mathrm{Log}t + \vec{C}_j + \vec{D}_j(\mathrm{Log}t/t) + \vec{E}_j/t + O(\mathrm{Log}t/t)^2 \\ &j = \{A,B,C\} \end{aligned}\right\} \tag{1129}$$

The Newtonian equations give the vectors $\vec{B}_j$, $\vec{D}_j$, $\vec{E}_j$, etc ... in terms of the six vectors $\vec{A}_j$, $\vec{C}_j$ themselves related by the integrals of motion :

$$\left.\begin{aligned} \vec{B}_A &= Gm_B(\vec{A}_A - \vec{A}_B).||\vec{A}_A - \vec{A}_B||^{-3} + \\ &\quad + Gm_C(\vec{A}_A - \vec{A}_C) \,.\, ||\vec{A}_A - \vec{A}_C||^{-3} \\ &\text{etc} \ldots \end{aligned}\right\} \tag{1130}$$

and :

$$\left.\begin{aligned} &\sum_{j=A,B,C} m_j\vec{r}_j = 0, \text{ that is : } \sum_j m_j\vec{A}_j = 0 \;;\; \sum_j m_j\vec{B}_j = 0, \text{etc} \ldots \\ &h = \sum_{j=A,B,C} 0.5 m_j A_j^2 \;;\; \vec{c} = \sum_j m_j\vec{C}_j \times \vec{A}_j \end{aligned}\right\} \tag{1131}$$

A main characteristic of the hyperbolic final evolution is that it requires a positive energy integral.

11.7.7.3 The hyperbolic-parabolic and the hyperbolic-elliptic final evolutions

These two types of final evolution occur if two of the vectors $\vec{r}_A/t$, $\vec{r}_B/t$, $\vec{r}_C/t$ have the same limit when $t \to \infty$ the limit of the third vector being different : there is a "small binary" and a remote escaping "third body". .

We will call m_3 the escaping third body and use the Jacobi dcomposition with m_1 and m_2 for the binary (Fig. 155).

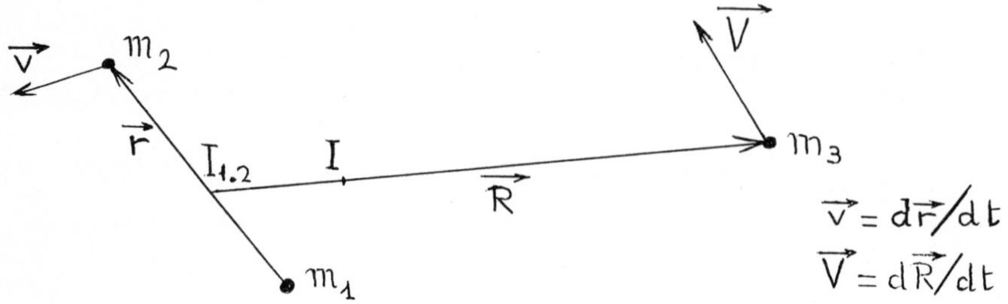

Fig. 155. The Jacobi decomposition of the three-body problem.

The motion of the binary is a two-body motion perturbed by a small and decreasing perturbation :

$$d^2\vec{r}/dt^2 = - G(m_1 + m_2)\vec{r}/r^3 + O(r/t^3) \tag{1132}$$

Since $(r/t) \to 0$ it is easy to show that $\vec{r}(t)$ approaches an elliptic or parabolic Keplerian motion with corresponding limit integrals of motion :

$$\left.\begin{aligned} &(v^2/2) - G\big((m_1 + m_2)/r\big) = h_r(t) \to h_{r\infty} \leq 0 \\ &\vec{r} \times \vec{v} = \vec{c}_r(t) \to \vec{c}_{r\infty} \end{aligned}\right\} \tag{1133}$$

Let us denote by $\vec{r}_a(t)$ the asymptotic Keplerian motion with integrals $h_{r\infty}$, $\vec{c}_{r\infty}$.

If $h_{r\infty} = 0$ the evolution is "hyperbolic-parabolic" and :

$$\left.\begin{aligned} &\vec{r}(t) - \vec{r}_a(t) = O(t^{-1/3}.\operatorname{Log} t) \\ &h_r(t) - h_{r\infty} = h_r(t) = O(t^{-5/3}) \\ &\vec{c}_r(t) - \vec{c}_{r\infty} = O(t^{-2/3}) \end{aligned}\right\} \tag{1134}$$

If $h_{r\infty} < 0$ the evolution is "hyperbolic-elliptic" and :

$$\left.\begin{aligned} \vec{r}(t) - \vec{r}_a(t) &= \begin{cases} O(t^{-2}) & ; \text{ if } \vec{c}_{r\infty} \neq 0 \\ O(t^{-4/3}) & ; \text{ if } \vec{c}_{r\infty} = 0 \end{cases} \\ h_r(t) - h_{r\infty} &= O(t^{-3}) \\ \vec{c}_r(t) - \vec{c}_{r\infty} &= O(t^{-2}) \end{aligned}\right\} \quad (1135)$$

On the other hand for the vector $\vec{R}(t)$:

$$d^2\vec{R}/dt^2 = -\mu\vec{R}/R^3 + O(r^2/t^4) \quad (1136)$$

The motion $\vec{R}(t)$ approaches an asymptotic Keplerian hyperbolic motion $\vec{R}_a(t)$:

$$\vec{R}(t) - \vec{R}_a(t) = \begin{cases} O(t^{-2/3}) & ; \text{ if } h_{r\infty} = 0 \\ O(t^{-2}) & ; \text{ if } h_{r\infty} \neq 0 \end{cases} \quad (1137)$$

$\vec{R}(t)$ can be developed :

$$\left.\begin{aligned} \vec{R}(t) &= \vec{A}t + \vec{B}\mathrm{Log}\,t + \vec{C} + O(t^{-2/3}) \\ d\vec{R}/dt &= \vec{A} + \vec{B}t^{-1} + O(t^{-5/3}) \\ d^2\vec{R}/dt^2 &= -\vec{B}t^{-2} + O(t^{-8/3}) \\ \vec{A} &\neq 0 \quad ; \quad \vec{B} = \mu\vec{A}/A^3 \end{aligned}\right\} \quad (1138)$$

The integrals of the three-body motion are :

$$\left.\begin{aligned} h &= \left(m_3(m_1 + m_2)/2M\right)A^2 + \left((m_1m_2)/(m_1 + m_2)\right)h_{r\infty} \\ \vec{c} &= \left(m_3(m_1 + m_2)/M\right)\vec{C} \times \vec{A} + \left((m_1m_2)/(m_1 + m_2)\right)\vec{c}_{r\infty} \end{aligned}\right\} \quad (1139)$$

Notice that, if $m_3 > 0$, the hyperbolic-parabolic evolution requires a positive energy integral while the hyperbolic-elliptic final evolution can have an energy integral of any sign.

11.7.7.4 The tri-parabolic final evolution

The remaining solutions have three vectors $\vec{r}_A/t$, $\vec{r}_B/t$ and $\vec{r}_C/t$ going to zero when $t \to \infty$, and even have $\lambda = O(t^{2/3})$ for all space directions ; thus all mutual distances r_{ij} remain of order $t^{2/3}$ or less.

The final evolution is of tri-parabolic type if and only if the three mutual distances are equivalent to $t^{2/3}$ (this is written $r_{ij} \approx t^{2/3}$ meaning that $r_{ij}.t^{-2/3}$ remains both bounded and bounded away from zero when $t \to \infty$).

The tri-parabolic type requires a zero-energy integral since $h < 0$ implies a bounded mean harmonic distance ν, and $h > 0$ implies a mean quadratic distance ρ increasing at least as the time t (Section 11.7.2).

The study of a tri-parabolic final evolution, and more generally of a n-parabolic final evolution of a n-body system is almost identical to the study of the complete collapse presented in section 11.7.6.4 and we find successively :

A) The mean quadratic distance ρ goes up to infinity as $t^{2/3}$.

B) Since $h = 0$ (instead of $\vec{c} = 0$) the Sundman function j is $(p/2\rho) + (\rho\rho'^2/2\mu)$.

C) When $t \to \infty$ the function j is non-decreasing and goes up to a limit j_∞.

D) j_∞ is bounded (because $\rho = O(t^{2/3})$).

E) The function $f = \rho(j - 1)$ is non-decreasing when ρ is non-decreasing, which is forever the case after the minimum of ρ. Hence f cannot go to minus infinity and this implies :

$$j_\infty \geqq 1 \qquad (1140)$$

F) $\rho\rho'^2/2\mu \to j_\infty$ implies when $t \to \infty$:

$$\rho \sim \{4.5\mu j_\infty t^2\}^{1/3} \quad ; \quad \rho' \sim \{4\mu j_\infty/3t\}^{1/3} \qquad (1141)$$

G) The following conditions with a suitable factor K :

$$\big((dj)/d(\mathrm{Log}\rho)\big) = \big((\rho/\nu) - j\big) \geqq 0 \text{ and } |d\nu/dt| \leqq K\sqrt{\nu} \qquad (1142)$$

imply :

$$\lim_{t\to\infty} (\rho/\nu) = j_\infty \qquad (1143)$$

and :

$$\nu \sim \{4.5\mu t^2/j_\infty^2\}^{1/3} \tag{1144}$$

Notice the similarity between (1141), (1144) and (1109).

H) Let us put :

$$\left.\begin{aligned} &\vec{s}_K = \vec{r}_K t^{-2/3} \quad ; \quad K = \{1, 2, \ldots, n\} \\ &\theta = \text{Logt} \end{aligned}\right\} \tag{1145}$$

The Newtonian equations of motion become :

$$\frac{d^2\vec{s}_K}{dt^2} = \frac{2}{9}\vec{s}_K - \frac{1}{3}\frac{d\vec{s}_K}{d\theta} + \sum_{j=1;j\neq K}^{n} Gm_j \frac{\vec{s}_{Kj}}{s_{Kj}^3} \quad ; \quad K = \{1, 2, \ldots, n\} \tag{1146}$$

and the energy integral h becomes :

$$\left.\begin{aligned} h = \exp\{-2\theta/3\}.\Bigg\{\sum_{K=1}^{n} m_K \Big[(2/9)s_K^2 + (2/3)\vec{s}_K \cdot d\vec{s}_K/d\theta + \\ + \frac{1}{2}(d\vec{s}_K/d\theta)^2\Big] - \sum_{1\leq j<K\leq n} G\big((m_j m_K)/s_{jk}\big)\Bigg\} \end{aligned}\right\} \tag{1147}$$

As in (1113) - (1117) and with h = 0 these expressions imply :

$$\left.\begin{aligned} &\sum_{K=1}^{n} m_K (d\vec{s}_K/d\theta)^2 \to 0 \\ &\text{and :} \\ &\int_{\theta_1}^{\infty} \sum_{K=1}^{n} m_K (d\vec{s}_K/d\theta)^2 d\theta < +\infty \end{aligned}\right\} \tag{1148}$$

For non-infinitesimal masses the conditions (1146), (1148) imply the approach of a "central configuration" : all s_{Kj} remain bounded and bounded away from zero and :

$$\forall\ K = \{1, 2, \ldots, n\} : (2/9)\vec{s}_K + \sum_{j=1;j\neq K}^{n} Gm_j(\vec{s}_{Kj}/s_{Kj}^3) \to 0 \tag{1149}$$

This property can also be demonstrated for infinitesimal masses and motions corresponding to those of Fig. 151, with an infinite number of loops around the main bodies, are impossible in the n-parabolic final evolution. This is because of the braking term - $(1/3)d\vec{s}_K/d\theta$ in (1146) instead of the opposite and accelerating term in the similar equation (1112).

In the tri-parabolic final evolution as well as in the triple collision the vectors $\vec{s}_K$ have a definite limit (Ref. 48).

$$\left.\begin{aligned} &\lim_{t\to\infty} \vec{s}_K = \vec{s}_{K\infty} \quad ; \quad K = 1, 2, 3 \\ &\vec{r}_K(t) = \vec{s}_{K\infty} t^{2/3} + O(t^{(2/3)-\varepsilon}) \quad ; \quad \varepsilon > 0 \end{aligned}\right\} \tag{1150}$$

This property is very often but not always true in the general n-body problem (it is true for all collinear central configurations).

The tri-parabolic final evolutions are classified in terms of the coresponding limit central configuration that give the final value of ρ/ν and j. We have then the Euler and the Lagrange types.

In the Lagrange type the final configuration is an equilateral triangle and $(\rho/\nu)_\infty = j_\infty = 1$. In the Euler type the final configurtion is a collinear central configuration and $(\rho/\nu)_\infty = j_\infty \in [1 ; 1.1881...]$ according to (909) and to tables II and III of Section 11.6.3.

11.7.7.5 The parabolic-elliptic final evolution

With the vectors $\vec{s}_K = \vec{r}_K t^{-2/3}$ of (1145) this evolution corresponds to the case when two bodies have the same final $\vec{s}_K$ and the third has another limit.

The Jacobi decomposition of Fig. 155 is again well adapted to the description of the motion : there is a "small binary" m_1, m_2 and an escaping "third-body" m_3.

The vector $\vec{r}$ of the small binary has a perturbed two-body motion with a decreasing perturbation :

$$d^2\vec{r}/dt^2 = - G(m_1 + m_2)\vec{r}/r^3 + O(r/t^2) \tag{1151}$$

The motion $\vec{r}(t)$ approaches an asymptotic elliptic Keplerian motion $\vec{r}_a(t)$ with integrals of motion $h_{r\infty}$ and $\vec{c}_{r\infty}$

$$\left.\begin{aligned} &(v^2/2) - G(m_1 + m_2)/r = h_r(t) \to h_{r\infty} < 0 \\ &\vec{r} \times \vec{v} = \vec{c}_r(t) \to \vec{c}_{r\infty} \end{aligned}\right\} \text{with } \vec{v} = d\vec{r}/dt \tag{1152}$$

and :

$$\left.\begin{array}{l}\vec{r}(t) - \vec{r}_a(t) = \begin{cases} 0(t^{-1}) & ; \text{ if } \vec{c}_{r\infty} \neq 0 \\ 0(t^{-2/3}) & ; \text{ if } \vec{c}_{r\infty} = 0 \end{cases} \\ h_r(t) - h_{r\infty} = 0(t^{-2}) \\ \vec{c}_r(t) - \vec{c}_{r\infty} = 0(t^{-1}) \end{array}\right\} \quad (1153)$$

The motion $\vec{R}(t)$ approaches an asymptotic parabolic Keplerian motion $\vec{R}_a(t)$:

$$\left.\begin{array}{l} d^2\vec{R}/dt^2 = -\mu\vec{R}/R^3 + 0(r^2/t^{8/3}) \\ \vec{R}(t) - \vec{R}_a(t) = 0(t^{-2/3} \,.\, \text{Log } t) \\ \vec{R}(t) = \vec{A}t^{2/3} + \vec{B}t^{1/3} + \vec{C} + 0(t^{-1/3}) \\ \text{with :} \\ \qquad ||\vec{A}|| = (9\mu/2)^{1/3} \\ \qquad \vec{A} \,.\, \vec{B} = 0 \\ \qquad \vec{C} = -3\vec{A} \,.\, B^2/4A^2 \end{array}\right\} \quad (1154)$$

The integrals of the three-body motion are :

$$\left.\begin{array}{l} h = m_1 m_2 h_{r\infty}/(m_1 + m_2) \\ \vec{c} = \big((m_1 m_2)/(m_1 + m_2)\big)\vec{c}_{r\infty} + \big(m_3(m_1 + m_2)/3M\big)\vec{B} \times \vec{A} \end{array}\right\} \quad (1155)$$

If $m_1 m_2 > 0$ the energy integral h is negative.

11.7.7.6 The bounded evolution, the two oscillatory evolutions and the collisions of stars

The remaining final evolutions are much less known than those above from the collisions to the parabolic-elliptic evolution. This is because these remaining evolutions have both an infinite duration and large perturbations of close bodies.

These evolutions have the following common characteristics :

A) The energy integral h is negative.

The corresponding generalized semi-major axis a is then positive and 2a is an upper bound for the smallest mutual distance r and the mean harmonic distance ν (see (951) and Sections 11.7.1 ; 11.7.2).

$$\text{At any time : } 0 \leqq r \leqq \nu \leqq 2a = -GM^*/h \tag{1156}$$

B) No hyperbolic or parabolic escapes are possible.

Hence in any direction the lengths λ of Figs. 153, 154 forever satisfy :

$$d\lambda/dt < \sqrt{2K/\lambda} \tag{1157}$$

the constant K being given in (1119) (1120).

All mutual distances are thus, at most, of order $t^{2/3}$ when the time t goes to infinity and the relations (1154) allow us even to write :

$$\mathcal{R} = \sup\{r_{jK}\} \quad ; \quad \lim_{t\to\infty} \sup\{\mathcal{R}t^{-2/3}\} \leqq \{9\mu/2\}^{1/3} \tag{1158}$$

Let us try to have a better idea of the possible evolutions of the largest mutual distance $\mathcal{R}$.

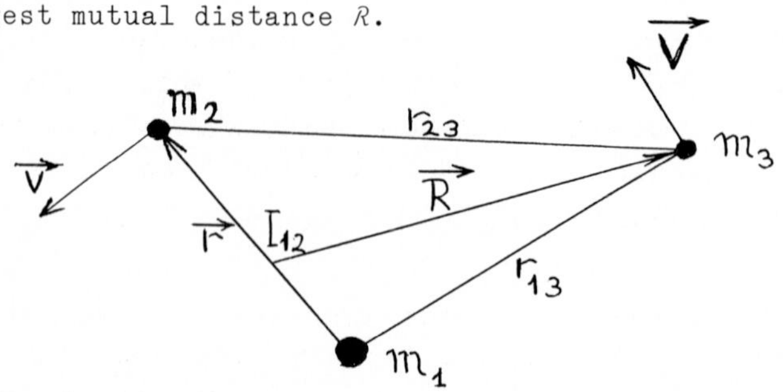

Fig. 156. The Jacobi decomposition of the three-body problem with $m_1 \geqq m_2$.

We will use again the Jacobi decomposition of the three-body problem (Fig. 156) with its "binary" m_1, m_2, where m_1 will be the largest mass, and its "third mass" m_3.

We have already met many times the Jacobi equations of motion :

$$\left.\begin{aligned}
&\mu = GM \quad ; \quad M = m_1 + m_2 + m_3 \\
&\alpha = m_1/(m_1 + m_2) \geq 1/2 \quad ; \quad \beta = m_2/(m_1 + m_2) = 1 - \alpha \\
&\vec{r}_{1.2} = \vec{r} \quad ; \quad \vec{r}_{1.3} = \vec{R} + \beta\vec{r} \quad ; \quad \vec{r}_{2.3} = \vec{R} - \alpha\vec{r} \\
&d^2\vec{R}/dt^2 = -\mu(\alpha\vec{r}_{1.3}r_{1.3}^{-3} + \beta\vec{r}_{2.3}r_{2.3}^{-3}) \\
&h = 0.5\left(\frac{m_1m_2}{m_1+m_2}v^2 + \frac{m_3(m_1+m_2)}{M}V^2\right) - G\left(\frac{m_1m_2}{r} + \frac{m_1m_3}{r_{1.3}} + \frac{m_2m_3}{r_{2.3}}\right) \\
&\vec{c} = \frac{m_1m_2}{m_1 + m_2}\vec{r} \times \vec{v} + \frac{m_3(m_1 + m_2)}{M}\vec{R} \times \vec{V}
\end{aligned}\right\} \quad (1159)$$

Assume that initially :

$$r \leq 2a \quad ; \quad R > 4a \quad (1160)$$

Hence r is obviously the smallest mutual distance $\mathscr{r}$ and, because of (1156), it will remain equal to $\mathscr{r}$ and smaller than or equal to 2a at least as long as R > 4a.

This property can be extended as follows. Assume that initially :

$$r \leq 2a \quad ; \quad R > 2.5a \quad (1161)$$

The length r is now not necessarily the smallest mutual distance $\mathscr{r}$ but it is easy to verify that (1161) implies $r \leq \nu$. This is implied even by the only condition $r \leq 0.8R$. Hence, since forever $\nu \leq 2a$, the length **r** will remain smaller than or equal to the mean harmonic distance ν and the generalized major axis 2a at least as long as R > 2.5a.

Thus we can study the "large ejections" of m_3 far away from the binary m_1, m_2 by analysis of the function R(t) under the assumption $r(t) \leq 2a$; $R(t) > 2.5a$. The inequality $r \leq 2a$ will remain satisfied all along the interval in which $R(t) > 2.5a$.

The equation (1159) of $d^2\vec{R}/dt^2$ gives :

$$d^2\vec{R}/dt^2 = \underbrace{-\mu\vec{R}/R^3}_{\text{Keplerian term}} + \underbrace{\text{remainder}}_{\text{perturbation term}} \tag{1162}$$

For given R and r the largest remainder is obtained when the configuration is collinear with m_2 between m_1 and m_3. Hence in all cases :

$$\left.\begin{aligned} & d^2\vec{R}/dt^2 = -\mu\vec{R}/R^3 + \vec{\varepsilon} \\ & \text{with :} \\ & \|\vec{\varepsilon}\| \leq \mu\left(\frac{\alpha}{(R+\beta r)^2} + \frac{\beta}{(R-\alpha r)^2} - \frac{1}{R^2}\right) = O\left(\mu\frac{r^2}{R^4}\right) \end{aligned}\right\} \tag{1163}$$

Since $r \leqq 2a$ the acceleration of $\vec{R}$ remains near the Keplerian acceleration, especially when R is large.

Let us now consider the acceleration d^2R/dt^2 of the length R.

$$\left.\begin{aligned} & d^2R/dt^2 = -\mu\frac{\vec{R}}{R}\left((\alpha\vec{r}_{1.3}/r_{1.3}^3) + (\beta\vec{r}_{2.3}/r_{2.3}^3)\right) + (V_N^2/R) \\ & \text{with : } \vec{V}_N = \text{component of } \vec{V} \text{ normal to } \vec{R} \\ & V_N^2 = V^2 - (dR/dt)^2 \end{aligned}\right\} \tag{1164}$$

For given R and r the greatest lower bound of d^2R/dt is obtained for the collinear configuration used for (1163) and when $V_N = 0$

$$d^2R/dt^2 \geqq -\mu\left(\frac{\alpha}{(R+\beta r)^2} + \frac{\beta}{(R-\alpha r)^2}\right) \geqq -\mu\left(\frac{\alpha}{(R+2\beta a)^2} + \frac{\beta}{(R-2\alpha a)^2}\right) \tag{1165}$$

This lower bound has its corresponding escape velocity

$$V_L = \left\{2\mu\left(\frac{\alpha}{R+2\beta a} + \frac{\beta}{R-2\alpha a}\right)\right\}^{1/2}$$

and its test of escape :

If at some time :

$$\left.\begin{array}{l} r \leqq 2a \quad ; \quad R > 2.5a \quad ; \quad dR/dt \geqq \left\{2\mu\left(\frac{\alpha}{R+2\beta a} + \frac{\beta}{R-2\alpha a}\right)\right\}^{1/2} \\ \text{the mass } m_3 \text{ will escape to infinity and the conditions} \\ \text{(1166) will remain forever satisfied.} \end{array}\right\} \quad (1166)^*$$

For an upper bound of d^2R/dt^2 we need an upper bound of V_N^2 and this can be given by the two integrals of motion h and $\vec{c}$ of (1159). When R is very large the upper bound of $|V_N|$ is of the order of R^{-1}.

A simple upper bound of d^2R/dt^2 is the following :

$$d^2R/dt^2 \leqq -(\mu/R^2) + Q(\mu/R^3) \quad (1167)$$

with the constant length Q in terms of the notations of Section 11.6.1. :

$$Q = \left[\frac{M^*}{m_3(m_1 + m_2)}\right]^2 \left[\sqrt{p} + \sqrt{\frac{m_1 m_2 M a}{(m_1 + m_2)M^*}}\right]^2 \quad (1168)$$

When R becomes larger than Q its acceleration is negative and either it remains forever increasing or it reaches a maximum R_M and fall back to values smaller than Q.

Let us demonstrate that the former hypothesis leads to parabolic or hyperbolic escapes.

Let us consider the two following functions :

$$\left.\begin{aligned} f(t) &= (dR/dt)^2 - 2\mu\left(\frac{\alpha}{R+2\beta a} + \frac{\beta}{R-2\alpha a}\right) = \\ &= (dR/dt)^2 - (2\mu/R) - \left[(8\mu\alpha\beta a^2)/R^3\right] - \ldots \end{aligned}\right\} \quad (1169)$$

and :

$$F(t) = (dR/dt)^2 - (2\mu/R) + (Q\mu/R^2) > f(t) \quad (1170)$$

Their derivatives are :

---

* The Yoshida test (1028) - (1030) is always more efficient than this test (1166).

$$\left.\begin{aligned} df/dt &= 2(dR/dt)\left[d^2R/dt^2 + \mu\left\{\frac{\alpha}{(R+2\beta a)^2} + \frac{\beta}{(R-2\alpha a)^2}\right\}\right] \\ dF/dt &= 2(dR/dt)\left[d^2R/dt^2 + \mu/R^2 - Q\mu/R^3\right] \end{aligned}\right\} \quad (1171)$$

Since $dR/dt > 0$, by hypothesis, the inequalities (1166) and (1167) implies that $f(t)$ is non-decreasing and $F(t)$ is non--increasing. They both have a limit when the time t goes to infinity and since for any time $f(t) < F(t)$ the limits f_∞ and F_∞ satisfy :

$$-\infty < f_\infty \leqq F_\infty < +\infty \quad (1172)$$

Hence at any time :

$$\left.\begin{aligned} (2\mu/R) - (Q\mu/R^2) + F_\infty \leqq (dR/dt)^2 \leqq f_\infty + 2\mu\left(\frac{\alpha}{R+2\beta a} + \frac{\beta}{R-2\alpha a}\right) = \\ = f_\infty + \frac{2\mu}{R} + \frac{8\mu\alpha\beta a^2}{R^3} + \ldots \end{aligned}\right\} \quad (1173)$$

If $F_\infty \geqq 0$ the integral of dR/dt diverges, $R(t)$ goes up to infinity and $f_\infty = F_\infty$. The escape is either hyperbolic ($F_\infty > 0$) or parabolic ($F_\infty = 0$).

If $F_\infty < 0$ the limit f_∞ is also negative. The sum $f_\infty + (2\mu/R) + \ldots$ must remain non negative and $R(t)$ cannot climb above $2\alpha a - 2\mu/f_\infty$. The acceleration d^2R/dt^2 remains negative and bounded away from zero and $R(t)$ necessarily reaches a maximum in a bounded interval of time and fall back afterwards.

The inequalities (1172), (1173) give some obvious tests either of hyperbolic escape or of "ejection without escape" (with bounds on the distance of ejection). These tests have received improvements but cannot establish parabolic escape in a given problem.

The same discussion applies to the ejection of masses m_1 and m_2 with their corresponding lengths Q_1 and Q_2. Hence three-body motions without parabolic or hyperbolic escape can have large ejections but always fall back to small sizes and, taking account of the possible difference of $2a$ between R and $\mathcal{R}$ during an ejection we can write :

$$\liminf_{t\to\infty} R(t) \leq \sup\{Q\ ;\ Q_1\ ;\ Q_2\ ;\ 2.5a\} + 2a \qquad (1174)$$

Note that if the three-body system of interest has the Hill-type stability (Section 11.6.3) only one of the three bodies can escape and then two of the three lengths Q, Q_1, Q_2 disappear from (1174).

This property of the largest mutual distance $R(t)$ is at the origin of the classical discussion of final evolutions without hyperbolic or parabolic escapes :

Either $R(t)$ remains forever bounded and the evolution is of "bounded type (e. g. the periodic orbits).

Or $R(t)$ is unbounded even if its lim inf is bounded. The system has an infinite number of larger and larger ejections but it has no escape, its always come back to small sizes and the evolution is of "oscillatory type". (Fig. 157).

This oscillatory type appears as a possible type in the Chazy classification of 1929 (Ref. 217). It was considered and analysed by Merman and Khilmi (Ref.218 - 220) and the first known example was described by Sitnikov (Section 11.7.8 and Ref. 221).

Another qualitative type of final evolution may be of interest. Assume that a solution usually considered as a bounded solution has the property that :

$$\liminf_{t\to\infty} r(t) = 0 \qquad (1175)$$

as for instance the numerical solution presented in Section 10.4.2. The system then has an infinite number of closer and closer approaches and its representative point in phase-space has an unbounded motion because the velocities are unbounded.

This type of evolution can be called "oscillatory type of the second kind" and, beside its existence, very little is known about it (Ref. 222), (Fig. 158).

The Kolmogorov-Arnold-Moser theorem (Ref.182) proves that the bounded type has a positive measure in phase space, it is thus one of the main types of final evolution like the hyperbolic type and the hyperbolic-elliptic type.

It seems that the oscillatory type of the first kind corresponds to a set of measure zero in phase space while the oscillatory type of the second kind would correspond to a set of positive measure.

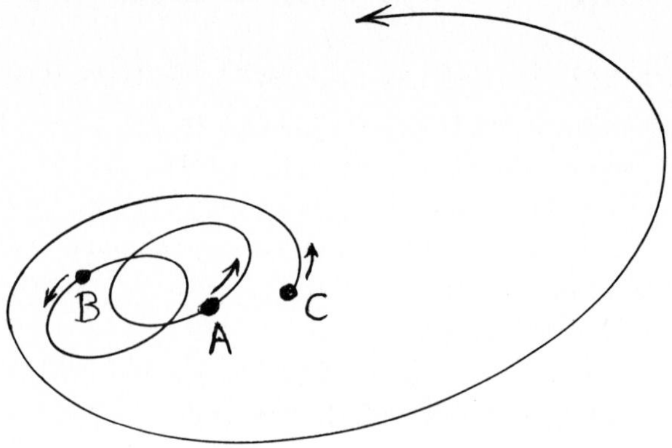

Fig. 157. An oscillatory motion of the first kind.
At least one of the three masses undergoes an infinite number of ejections as large as desired but it always come back to small distances.

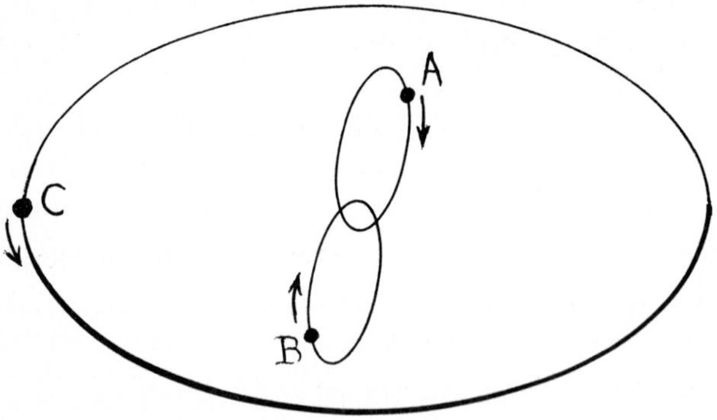

Fig. 158. An oscillatory motion of the second kind.
The "inner orbit" has an almost constant semi-major axis but also large perturbations of inclination and eccentricity (from 97.7° to 142° and from 0.2 to 0.9993 in the short example of Section 10.4.2). The system undergoes an infinite number of approaches as closed as desired, but orbits with actual collisions remain exceptional.

The natural bodies are of course not point-masses and "infinitely close approaches" lead to collisions.

The existence of oscillatory motions of the second kind increases considerably the possibilities of collisions in a system such as the galactic system : when two binary stars meet an exchange can lead to the formation of a new triple system and if this new system has an oscillatory motion of the second kind it will sooner or later lead to a collision.

The numerical example of Section 10.4.2 shows a close binary perturbed by a third body moving around. The semi-major axes of inner and outer orbits are almost constants as well as the eccentricity of the outer orbit, but the mutual inclination is large (97.7° initially) and this circumstance implies very large perturbations in both the mutual inclination itself and the eccentricity of the inner orbit (the general mechanism is presented in Sections 10.2.3 to 10.3).

When the eccentricty of the inner orbit approaches one the system undergoes very close approaches (Fig. 158). This phenomenon multiply by a very large factor the probability of collisions in the galactic system.

In the nineteenth century the novae and supernovae were considered to be the result of star collisions but this explanation was forsaked when the probability of direct collisions was found to be extremely small. These spectacular phenomena are now considered as physical instabilities during the evolution of a star and this is certainly true for the supernovae : they realease an energy of the order of 10^{42} Joules in light, 10^{44} Joules in kinetic energy of the explosion and 10^{46} Joules in neutrinos* which is about 10^{4} times the energy of a star collision. However the energy released by a nova is of the order of that of a star collision and since the probability of these collisions is much greater than the initial estimates it is possible that a significant fraction of these novae are the result of star collisions.

---

* These 10^{46} Joules for the neutrinos only concern Supernovae of the second type.

It would be interesting to research an evaluate of the probability of these indirect collisions*, to look for the physical effect of collisions of two stars and to compare these effects with those of the different kinds of novae. However I must recognize that most astrophysicists give litle credit to these ideas (Ref. 223), even if the probable presence of a great number of planetary systems increases very much the probability of collisions. The stability of the Solar System would hardly survive the passage of a star at two or three times the distance of Pluto.

Another effect of oscillatory orbits of the second-kind must be considered.

Let us consider a very close approach of the two body bodies of the inner binary : these bodies are neither point-masses nor rigid bodies and the tidal forces must have a very large effect during the close encounters.

These tidal forces keep the angular momentum of inner orbit constant but lead to a decrease of its energy and our three-body problem is modified.

Notice that oscillatory motions of the second kind appear when the modulus of the angular momentum of the system is sufficiently near to the modulus of the angular momentum of the outer orbit (Section 10.2.4 and Ref. 222). This condition survives to the action of tidal forces and future close encounters must be expected.

Thus the major effect of the existence of oscillatory motions of the second kind is perhaps not a very large increase of the number of star collisions but rather a very large increase in the number of very close binary stars, these systems that are unstable when they are too close and that lead to so many astrophysical phenomena...

---

* If we consider a triple system of given inner and outer periods T_i and T_e and given masses m_1, m_2, m_3. If furthermore we assume an isotropic choice of orientations, the first order study (Section 10.2 and Ref. 222) leads generally to a probability of "oscillatory orbit of the second kind" larger than $9m_3T_i/8MT_e$.

11.7.8 Sitnikov motions and oscillatory evolutions of the first kind.

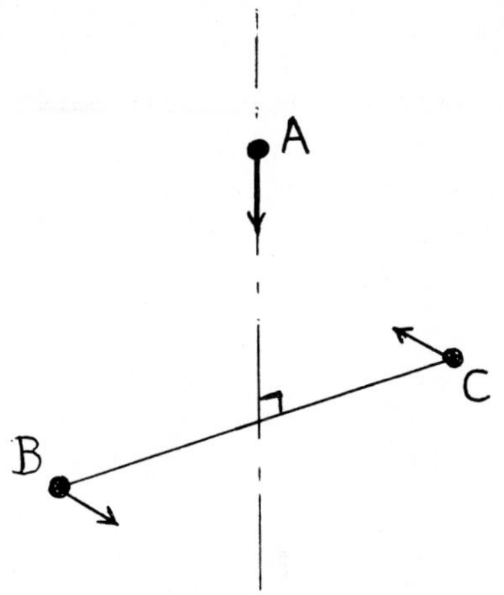

Fig. 159. The Sitnikov symmetry.
The body A moves along the axis of symmetry while B and C have equal masses and symmetrical motions.

We have already met the Sitnikov symmetry in Section 11.2.2. It is a symmetry about an axis (Fig. 159) with bodies B and C of the same mass and of symmetrical motions.

Let us choose the axis of symmetry as z-axis, the expression of the symmetry is then :

$$\left.\begin{aligned} & m_B = m_C \\ & x_A \equiv y_A \equiv 0 \\ & x_B \equiv - x_C \\ & y_B \equiv - y_C \\ & z_B \equiv z_C \end{aligned}\right\} \qquad (1176)$$

The angular momentum of the system is :

$$\vec{c} = \{0 \ ; \ 0 \ ; \ 2m_B(x_B y_B' - y_B x_B')\} \qquad (1177)$$

The equations of motion can be reduced as follows.
Let us put :

$$\left.\begin{aligned} X &= \{x_B^2 + y_B^2\}^{1/2} = r_{BC}/2 \geq 0 \\ Z &= z_A - z_B \\ R &= \{X^2 + Z^2\}^{1/2} \quad \text{this R is } r_{AB} \text{ and } r_{AC}\text{, it is not the usual R of the Jacobi decomposition} \end{aligned}\right\} \quad (1178)$$

We obtain :

$$\left.\begin{aligned} d^2X/dt^2 &= X'' = (H^2/R^3) - (Gm_B/4X^2) - (Gm_A X/R^3) \\ d^2Z/dt^2 &= Z'' = -(\mu Z/R^3) \quad ; \ \mu = G(m_A + m_B + m_C) = GM \end{aligned}\right\} \quad (1179)$$

with :

$$H = x_B y_B' - y_B x_B' = \mathbf{c}_z/2m_B = \text{constant} \quad (1180)$$

and with the energy integral h :

$$\left.\begin{aligned} h/m_B = K = \text{constant} &= X'^2 + (m_A/M)Z'^2 + (H^2/X^2) - \\ &- (Gm_B/2X) - (2Gm_A/R) \end{aligned}\right\} \quad (1181)$$

This equation gives the zone of possible motion in terms of the integrals H and K :

$$\left.\begin{aligned} &X \geq 0 \\ &K \geq (H^2/X^2) - (Gm_B/2X) - (2Gm_A/R) \end{aligned}\right\} \quad (1182)$$

The constant K is necessarily larger than or equal to $-G^2(4m_A + m_B)^2/16H^2$. This lower bound give $X \equiv 4H^2/G(4m_A + m_B)$; $Z \equiv 0$ and the motion is a circular Eulerian motion.

If K is the following range :

$$- G^2(4m_A + m_B)^2/16H^2 \leq K < - G^2m_B^2/16H^2 \qquad (1183)$$

the zone of possible motion remain bounded, indeed (1182)-(1183) imply :

$$R \leq - 2Gm_A/\left(K + (G^2m_B^2/16H^2)\right) < + \infty \qquad (1184)$$

This test of bounded motions is exceptional in the three-body problem. That rarity will be explained in Section 11.7.10.2.

The example of oscillatory evolution of the first kind given by Sitnikov (Ref. 221) uses an infinitesimal mass m_A and a simple two-body elliptic motion of masses m_B and m_C (Fig. 160).

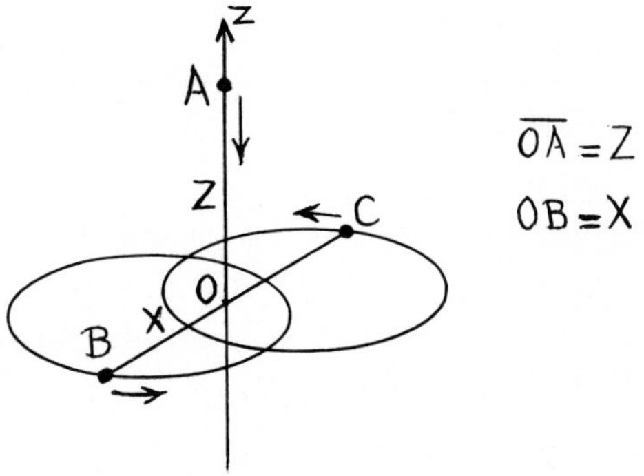

Fig. 160. The Sitnikov example of oscillatory motion of the first kind.

The distance OB = OC = X then has an ordinary periodic two-body evolution with a pericenter distance P and an apocenter distance A.

The real question concerns the evolution of the z-abscissa z_A of point A that is of the parameter Z of (1178)-(1181).

Let us assume that at t = 0 :

$$\left.\begin{array}{l} \text{B and C are at pericenter } (X_0 = P) \\ Z_0 = 0 \quad ; \quad Z'_0 \geq 0 \end{array}\right\} \qquad (1185)$$

and let us study the future evolutions for the different values of Z'_0.

Notice that the initial conditions (1185) have not only the Sitnikov symmetry but also the Poincaré time-axis symmetry of Section 11.2.1. This is an example of multi-symmetries of Section 11.2.4 and the function Z(t) is an odd function of t.

Since $Z'' = -\mu Z/R^3$ the first phase of motion is an increase of Z either up to infinity (escape) or to a maximum Z_1.

Figure 161 presents the graph of $Z_1(Z'_0)$ with the "escape velocity V_L.

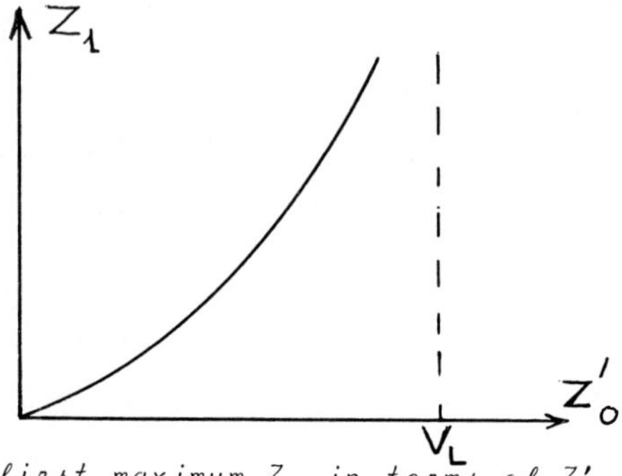

Fig. 161. The first maximum Z_1 in terms of Z'_0.

Note that during this first phase of motion the function $f_P = Z'^2 - 2\mu(Z^2 + P^2)^{-1/2}$ is increasing while the function $f_A = Z'^2 - 2\mu(Z^2 + A^2)^{-1/2}$ is decreasing, hence the escape velocity V_L satisfies :

$$\sqrt{2\mu/A} < V_L < \sqrt{2\mu/P} \tag{1186}$$

If Z(t) reaches a maximum Z_1 it falls back to zero, becomes negative and either go down to minus infinity or reaches a minimum $-Z_2$.

If the first return at Z = 0 occurs at the time of a pericenter or an apocenter passage of B and C it corresponds to a second Poincaré time-axis symmetry, the motion is periodic and $Z_2 = Z_1$.

If the first return occurs when B and C are falling towards 0 the body A loses some energy in the vicinity of 0 and $Z_2 < Z_1$.

Finally if the first return at Z = 0 occurs when X' > 0 the body A gains some energy in the vicinity of 0 and $Z_2 > Z_1$. In this final case if Z'_0 is sufficiently near to V_L the gain of energy will be sufficient to lead to an escape and the curve $Z_2(Z'_0)$ has an infinite number of vertical asymptotes (Fig. 162).

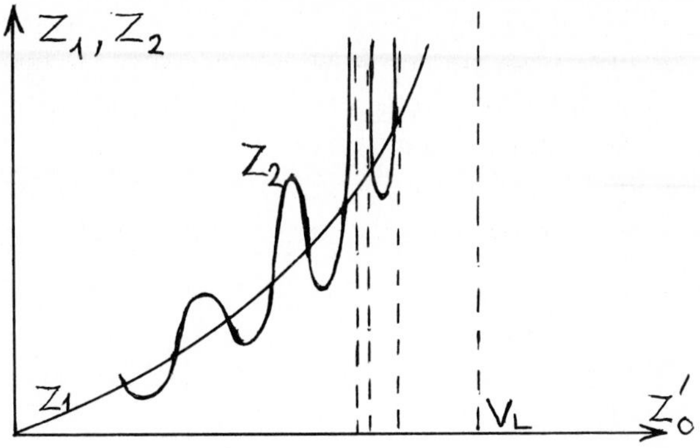

Fig. 162. The first maximum Z_1 and the first minimum - Z_2 in terms of Z'_0.

The same discussion holds for Z_3 the third extremum of Z(t) and it is thus possible to demonstrate the existence of many initial Z'_0 such that :

$$\left.\begin{array}{l} Z_1 < Z_2 < Z_3 < \dots < Z_n < Z_{n+1} < \dots \\ \text{with : } \{n \to \infty \Longrightarrow \mathbf{Z}_n \to \infty\} \end{array}\right\} \quad (1187)$$

that is precisely the existence of oscillatory orbits of the first kind with lim sup R = ∞ and lim inf R < ∞.

When the successive Z_n are large the variations of energy of body A in the intermediate passages near O are small and the passages at O occur either almost at pericenter time or almost at apocenter time of bodies B and C. Let us denote by x_1, x_2, x_3 the successive numbers of half-revolutions of B and C between two successive passages of A at O and by n_1, n_2, n_3, ... the successive nearby integers.

It is possible to demonstrate that one and only one "Sitnikow orbit" corresponds to a given succession of integers ... n_{-2}, n_{-1}, n_0, n_1, n_2, n_3, ... provided that :

I) One of these passages is specified : either near pericenter time or near apocenter time.

II) All these integers n_j are larger than an integer N function of the eccentricity of the orbits of B and C.

III) Orbits deduced from one another by rotations, symmetries and/or time-translation are considered as the same Sitnikov orbit.

Hence it is possible to build examples of orbits bounded in the past and oscillatory in the future (n_j bounded for negative j, unbounded for positive j). Or, by continuity, orbits with a parabolic or even a hyperbolic escape in the past and bounded in the future (case called "complete capture"), etc ...

If the succession of n_j is periodic the Sitnikov orbit is periodic and if that periodic succession has no time-symmetry, as 50, 51, 52, 50, 51, 52, ... the periodic Sitnikov orbit has no Poincaré symmetry.

All these results can be extended to Sitnikov orbits with non-infinitesimal mass m_A (Ref.224, p 43).

11.7.9 General table of final evolutions

As usual we will consider that the binary collisions are regularized and the only remaining singularity of the three-body problem is the triple collision.

The Chazy classification of 1922 (Ref. 213) has been slightly improved and was presented in Sections 11.7.6 and 11.7.7. It is summarized in the following Tables.

A) General three-body problem (all three masses positive) (see Table VI, in the following page).

B) Hyperbolic restricted three-body problem.

The possible final evolutions are the hyperbolic expansions of the Table VI.

C) Parabolic restricted three-body problem.

The energy integral h is zero but the possible types of final evolution are the parabolic expansions and the hyperbolic-parabolic type (with then A = Q).

D) Elliptic or circular restricted three-body problem.

Notice that the demonstration of lim inf $\mathcal{R} < +\infty$ for the evolutions without hyperbolic or parabolic escapes (Section 11.7.7.6) is not valid here and thus beside the types of the Table VI that allow $h < 0$ we must not exclude the possibility of solutions with $\mathcal{R}(t)$ going to infinity more slowly than $t^{2/3}$ (sub-parabolic escapes).

E) Restricted three-body problem with collinear motions of the primaries.

TABLE VI

Final evolutions of the general three-body problem

Class	Type	Measure in phase-space	Conditions on integrals of motion	Evolutions of $r = \inf(r_{jk})$ and $R = \sup(r_{jk})$.
Singular evolution	Triple collision at $t = t_c$	zero	$\vec{c} = 0$	$r \sim A(t_c - t)^{2/3}$, $R \sim B(t_c - t)^{2/3}$; $0<A \leq Q \leq B<2Q$, $Q = \{4.5\mu\}^{1/3}$
Hyperbolic expansions	Hyperbolic	positive	$h > 0$	$r \sim Ct$; $R \sim Dt$; $0 < C \leq D$
	Hyperbolic-parabolic	zero	$h > 0$	$r \sim At^{2/3}$; $R \sim Dt$; $0 < A < Q$
	Hyperbolic-elliptic	positive	$h \gtrless 0$	r bounded ; $R \sim Dt$
Parabolic expansions	Tri-parabolic	zero	$h = 0$	$r \sim At^{2/3}$; $R \sim Bt^{2/3}$; $0<A \leq Q \leq B<2Q$
	Parabolic-elliptic	zero	$h < 0$	r bounded ; $R \sim Qt^{2/3}$
Sub-parabolic expansions	Bounded	positive	$h < 0$	$0<E \leq r \leq R \leq F<\infty$
	Oscillatory type I	probably zero	$h < 0$	r bounded ; $\limsup R = \infty$, $\liminf R < \infty$
	Oscillatory type II	probably positive	$h < 0$	$\limsup r > 0$, $\liminf r = 0$; R bounded

A, B, C, D, E, F are positive parameters ; $Q = \{4.5\mu\}^{1/3}$; $\mu = G(m_1 + m_2 + m_3)$; A and B are equal to Q iff the limit configuration is an equilateral triangle of Lagrange; if not : $2A \leq B$; $Q<B<Q+A$.

To the possibilities of the above cases B, C or D, according to the sign of h, we must add the possibilities of triple collisions.

Notice that these triple collisions are of two types : the usual triple collisions with the three masses approaching a central configuration when going down to the triple collision (Fig. 163) and the special triple collisions already presented in Fig. 151 with the infinitesimal mass making faster and faster an infinite number of smaller and smaller loops around the two primaries falling towards the center of mass.

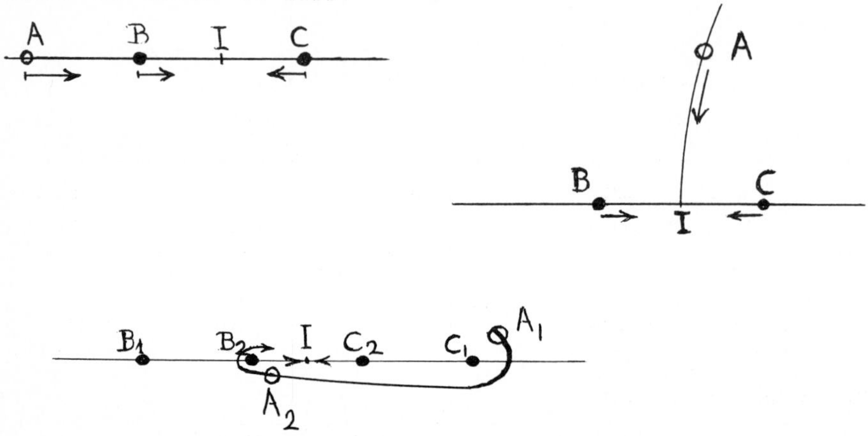

Fig. 163. The triple collisions.
1. The three bodies have a collinear motion and approach a Eulerian central configuration.
2. The three bodies approach a Lagrangian central configuration (equilateral triangle).
3. The infinitesimal mass A joins the complete collapse by an infinite number of smaller and smaller loops around the primaries.

The restricted three-body problem with collinear motions of the primaries has an obvious simple integral of motion : the component along the axis of motion of the primaries of the angular momentum of the infinitesimal mass.

For triple collisions this component must be zero (plane motion).

The measure in phase space of a type of evolution gives an idea of its importance, but the types of measure zero are not equivalent, they can have different "numbers of co-dimensions".

"Co" means here "complement" as in cosine and in a R^n space a manifold with co-dimensionality d has the dimensionality (n-d), hence the number of co-dimensions is the number of independent conditions that define the manifold of interest. This number is the same when we consider a general three-body problem either with three arbitrary masses or with three given masses.

The main features of the general three-body problem have the following number of co-dimensions (Ref. 48).

TABLE VII

Co-dimensionality of various manifolds of the general three-body problem.

Manifold leading to the following	Collinear problem	Plane problem	Three-dimensional problem
Binary collision	0	1	2
Triple collision (Euler type)	1	4(a)	7(b)
Triple collision (Lagrange type)		3(c)	5(d)
Hyperbolic-parabolic escape	1	1	1
Parabolic-elliptic escape	1	1	1
Tri-parabolic escape (Euler type)	2(e)	2(e)	2(e)
Tri-parabolic escape (Lagrange type)		3(e)	3(e)

with :

(a) : among these 4 conditions 3 are for the collinearity

(b) : among these 7 conditions 6 are for the collinearity

(c) : among these 3 conditions 1 is for a zero angular momentum

(d) : among these 5 conditions 3 are for a zero angular momentum

(e) : among these 2 or 3 conditions 1 is for a zero energy.

The co-dimensionality of these different manifolds remain the same in the restricted three-body problem except in the two following cases :

A) Parabolic restricted three-body problem.

All parabolic escapes lose one co- dimension (co-dimensionality zero for the hyperbolic-parabolic and parbolic-elliptic escapes, one for the tri-parabolic escape of Euler type and two for the tri-parabolic escape of Lagrange type).

B) Restricted three-body problem with collinear motion of the primaries.

The co-dimensionality of triple collision manifolds becomes then :

TABLE VIII

Co-dimensionality of the triple collision manifolds when the primaries have a collinear motion.

	Collinear problem	Plane problem	Three-dimensional problem
Triple collision (Euler's type)	1	3(a)	5(b)
Triple collision (Lagrange's type)		2	3(c)
Triple collision (special type of Fig. 163)	0	1	2(d)

with :

(a) : among these 3 conditions 2 are for the collinearity

(b) : among these 5 conditions 4 are for the collinearity

(c) : among these 3 conditions 1 is for a plane motion

(d) : among these 2 conditions 1 is for a plane motion.

It is a surprise to notice that the "special type" is more general than the two other triple collision types.

11.7.10 Progress in the tests of escape

A considerable number of new and efficient tests of escape have been established in recent years (Ref. 203,218-220,225-236).

These new tests are of course a consequence of the progress of computers that have provided new means of exploration of the three-body problem and disclosed its extreme complexity. The next step of the knowledge is obviously a rearrangement and a classification, in terms of final evolutions that require various and efficient tests.

An idea of the recent progress is given in Sections 11.7.10.4 and 11.7.10.5, see especially Fig. 194.

10.7.10.1 Classification of tests

There are of course various criteria for the classification of tests.

A) Domain of application.

A test can be applicable to the general three-body problem, to a particular restricted three-body problem or to some special cases like the test of bounded motions (1183) - (1184) applicable only to Sitnikov motions.

B) Efficiency of the test.

The test T_1 is more efficient than the test T_2 if and only if it gives more informations, hence when the conditions of the test T_2 are satisfied those of the test T_1 are also satisfied.

The efficiency is very often paid for by a lack of simplicity and, generally, a balance between these two condradictory qualities must be made.

C) Type of final evolution.

Some tests lead to a partial classification, for instance the test $\{h > 0, \vec{c} \neq 0\}$ leads to one of the three hyperbolic expansions. Other tests, more restricted, lead to a definite final evolution, usually a hyperbolic-elliptic final evolution with a definite escaping mass.

Notice that the ten-body tests (1053) or (1055) or (1059) lead to an escape and give an indication on the direction of escape but do not give the escaping mass or masses.

There are a few tests of hyperbolic final evolution, for instance when the hexagon of Fig. 154 (Section 11.7.7.1) extends into all directions of a plane. But tests of bounded motions are exceptional : the Hill stability of the circular restricted three-body problem and the test (1183) - (1184) of Sitnikov motions. This property will be explained in the next Section 11.7.10.2.

Notice that the tests of "ejection without escape" (Ref. 230, page 388 and also the test related to the above equations (1172), (1173)) cannot be considered as tests of bounded motions. They are only transitory tests on a short phase of the motion.

The Hill-type stability (Section 11.6) and the space-symmetries (Section 11.2.2) allow us to eliminate many final evolutions especially when $h \geqq 0$.

A very interesting case is that when the energy integral h is zero.

The possible final evolutions are then the triple collisions, the hyperbolic-elliptic escapes and the tri-parabolic escapes. Hence in the figure of generalized Hill curves (opposite Fig. 164 and Fig. 125 of Section 11.6.3) the mobile point m_1 has eight possible limit points : the five Lagrangian points L_1 to L_5 (triple collisions and tri-parabolic escapes) and the points m_2, m_3 and at infinity (hyperbolic-elliptic escapes).

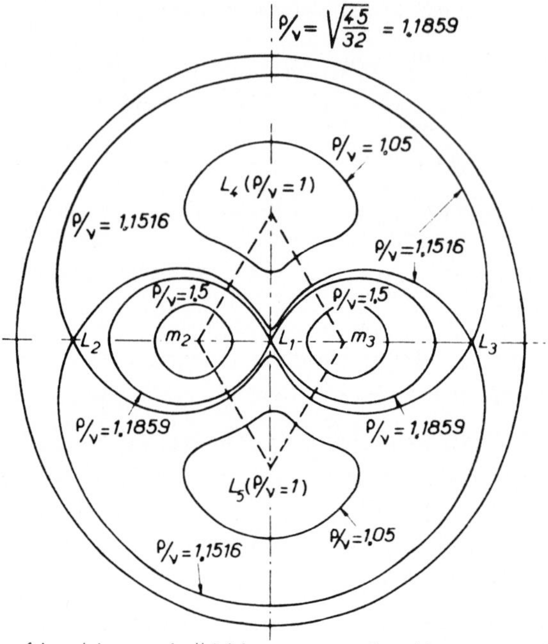

Fig. 164. Generalization of Hill curves in the case $2m_1 = m_2 = m_3$. Curves of constant ratio ρ/ν in terms of the position of m_1 with respect to m_2 and m_3.
ρ/ν is minimum and equal to one at the triangular Lagrangian points L_4 and L_5, it is infinite at m_2, at m_3 and at infinity, it has three saddle points at the collinear Lagrangian points L_1, L_2, L_3.

The discussion of Section 11.6.5 on Hill-type stability for systems of positive or zero energy and those of Sections 11.7.6.4 and 11.7.7.4 on triple collisions and tri-parabolic escapes leads to the following.

A) The mean quadratic distance ρ has one and only one minimum : ρ_m at some time $t = t_m$.

B) The Sundman function j is then $p/2\rho + \rho\rho'^2/2\mu$, it varies in the direction of ρ, it has then one and only one minimum : $j_m = p/2\rho_m$, at $t = t_m$.

C) A triple collision occurs at t_m when $\rho_m = 0$; the minimum j_m is then equal to the ratio ρ/ν at the limit Lagrangian point and the angular momentum $\vec{c}$ must be zero.

Hence before the time t_m of collision we must have one of the two following cases :

C.1) A collinear motion with $p = 0$, $\rho' < 0$, $j \geq j_m = \rho/\nu(L_{1,2 \text{ or } 3})$ if the triple collision is of Euler type (the subscript of L is that of the central mass since the succession of masses is invariable in collinear motions).

C.2) A plane but non collinear motion with $p = 0$, $\rho' < 0$, $j \geq 1$ if the triple collision is of Lagrangian type.

D) If the final evolution is not a triple collision, for instance if $\vec{c} \neq 0$ and/or $\rho' \geq 0$ and/or $j < 1$, the increasing values of the Sundman function j after t_m and the Hill-type stability related to $\rho/\nu \geq j$ give informations on this final evolution. The function j goes to infinity when the final evolution is of hyperbolic-elliptic type and it goes to $(\rho/\nu)_{L_K}$ of the limit Lagrangian point L_K when the final evolution is tri-parabolic. Hence as soon as j is above all $(\rho/\nu)_{L_K}$, the evolution is certainly of hyperbolic-elliptic type and the escaping body is known.

E) If furthermore the motion has one of the two space symmetries of Section 11.2.2 and Fig. 111 (the symmetry about a plane or the Sitnikov symmetry about an axis), two masses are equal, for instance m_2 and m_3 as in Fig. 164, and the triangle m_1, m_2, m_3 remains forever isosceles ($r_{12} \equiv r_{13}$). The mobile point m_1 moves on the straight line L_4, L_1, L_5 of Fig. 164 and the possible final evolutions are even more restricted.

These symmetries give an obvious "test of tri-parabolic escape (of Euler type)" : It is sufficient that at some arbitrary time the mobile point m_1 be in Fig. 164 between L_4 and L_5 with :

$$\left.\begin{array}{lll} j \geq 1 & \text{if} & \rho' \geq 0 \\ p \geq 2\rho & \text{if} & \rho' < 0 \end{array}\right\} \qquad (1188)$$

The demonstration for $\rho' < 0$ is similar to that of Section 11.6.5.*

* It is even possible to extend the condition of (1188) for $\rho'<0$ to : $\rho^2\rho'^2 \geq \mu c_1(2\rho - p)$ with $c_1 = (\rho/\nu)_{L_1}$, the derivative of j being then bounded by : $(\rho'/\rho)(c_1 - j) \leq dj/dt \leq 0$.

In the symmetrical motion of interest the only possible limit point of m_1 is then L_1.

This test of tri-parabolic escape is of course exceptional, it has been used for the construction of an example of motion from a "triple explosion" to a tri-parabolic escape (Ref. 48).

11.7.10.2 The ergodic theorem. The difficulty of a test of bounded motions

The first version of the ergodic theorem goes back to Poincaré (Ref. 23). Since then it has received several improvements (Ref.24 – 27) and has been extended to the comparison of past and future motions and to phase spaces of infinite measure (Ref.28, 29).

The ergodic theorem has several versions. It can be applied to autonomous conservative systems of differential equations as the usual three-body or n-body systems.

The word "conservative" means here that there exists a suitable "integral invariant" as the "volume in phase space" of Section 5.4. To this integral invariant corresponds the "measure in phase-space" of measurable subsets of that space (their "volume" or "hypervolume" for the n-body problem and for autonomous Hamiltonian problems). This measure is conserved by the conservative motion of interest, like the volume of water in a water flow.

Let us consider the intersections of the orbits of the system of interest with a given measurable subset E_0 of phase space.

The ergodic theorem specifies that, <u>if the measure of E_0 is bounded</u>, the orbits can be classified into the following four categories :

I) The "inner orbits".

They are defined for all times and belong entirely to the subset E_0.

II) The "intersecting orbits".

They are also defined for all times. They always return to E_0 when $t \to +\infty$ and also when $t \to -\infty$.

Let us call $p(t_1, t_2)$ the proportion of time, in the interval (t_1, t_2), during which the orbit $\vec{X}(t)$ of interest belongs to E_0.

The ergodic theorem specifies that for intersecting orbits :

$$\left.\begin{aligned}&\limsup_{t\to+\infty} p(t_0,t) = \liminf_{t\to+\infty} p(t_0,t) = \\ &\quad = \limsup_{t\to-\infty} p(t,t_0) = \liminf_{t\to-\infty} p(t,t_0) < 1\end{aligned}\right\} \quad (1189)$$

This common limit is the general proportion $p\left[\vec{X}(t),E_0\right]$. It can be zero if the measure of the total phase space is infinite but even in this case the measure of "positive times in E_0" is infinite and also the measure of "negative times in E_0".

III) The "outer orbits".

They intersect the subset E_0 during only a bounded period of time : there exist t_1 and t_2 such that before t_1 and after t_2 the orbit is always out of E_0.

Orbits entirely out of E_0 are of course also outer orbits.

These outer orbits are not necessarily defined for all times.

IV) The "abnormal orbits".

Orbits that are not in the above categories I, II, III are "abnormal" and the ergodic theorem specifies that these abnormal orbits fill a subset of measure zero in phase space.

Let us consider now a denumerable sequence of measurable subsets F_n of finite measure and of measurable subsets I_n of infinite measure.

We will assume that among the sequence of F_n exists a sub-sequence G_n such that :

$$\left.\begin{aligned}&n < m \Longrightarrow G_n \subset G_m \\ &\text{The subsets } F_n \text{ belong to sufficiently large } G_m \\ &\text{The whole phase-space} = \bigcup_{n=1}^{\infty} G_n\end{aligned}\right\} \quad (1190)$$

For instance if the phase-space is some usual R^N space we can assume that the subset F_n are the open spheres of rational radii with center of rational coordinates. They can also be the open, closed or semi-open "rational parallelepiped", etc... The subset G_n being the open spheres of integer radii and centers at origin, or some corresponding hypercubes.

The ergodic theorem then allow to classify the orbits into the following 5 main categories in terms of their intersection with the subsets F_n, G_n and I_n.

A) The "bounded recurrent orbits".

They are inner orbits of sufficiently large G_n and have definite proportions $p\left[\vec{X}(t),F_n\right]$ or $p\left[\vec{X}(t),I_n\right]$ for all F_n and I_n. Furthermore if $p = 1$ the orbit is entirely in the subset of interest and it is entirely out if $p = 0$.

B) The "oscillatory orbits chiefly at bounded distance".

These have the properties of bounded recurrent orbits except that for large G_n the proportion $p\left[\vec{X}(t),G_n\right]$ goes to 1 when $n \to \infty$, but is never equal to 1. Hence $p = 1$ only for subsets I_n.

C) The "oscillatory orbits chiefly at infinity".

If these orbits exist they fill in phase-space a measurable set of infinite measure.

These orbits have the qualitative properties of the above orbits of types A and B : they are entirely in or entirely out or of intersecting type with respect to any subset F_n or I_n (in the last case the measure of "positive times in the subset" is infinite as well as the measure "positive times out of the subset", "negative times in the subset" and "negative times out of the subset"). However the proportion rules are no longer satisfied : $p\left[\vec{X}(t),F_n\right]$ is always zero and it is generally not even defined for subsets I_n.

D) The "open orbits".

These orbits are of outer type with respect to all subsets F_n and of arbitrary types with respect to subsets I_n. They are not necessarily defined for all times.

E) The "abnormal orbits".

These orbits fill a set of measure zero in phase-space and correspond to all the remaining orbits.

This expression of the ergodic theorem for autonomous conservative differential systems can be extended with slight modifications to periodic conservative differential systems, such as the elliptic restricted three-body problem, and to conservative mappings (Ref. 29).

The vocabulary of bounded, recurrent, oscillatory, open, abnormal orbits come of course from usual problems as the three--body problem in the axes of the center of mass :

The open orbits arrive from infinity (in phase-space) and return back to infinity by an escape or a triple collision.

The orbits bounded in the past and open in the future are "abnormal" and infinitely rare.

The oscillatory orbits of type II with an infinite number of approaches as close as desired (and thus with unbounded velocities) are "oscillatory orbits chiefly at bounded distance", while the oscillatory orbits of type I are not yet classified and seem abnormal. And so on...

Notice the strong stability of the orbits of the above three first types A, B, and C : they always come back into any open neighbourhood of any past or future situation and they do it with a well defined frequency that is the same for the past and for the future*.

This property is the Poisson stability and the corresponding orbits, either bounded or oscillatory, are the recurrent orbits.

An orbit asymptotic to an unstable periodic orbit is neither open nor recurrent and is thus an abnormal orbit.

The ergodic theorem implies, for bounded recurrent orbits, the existence of limit temporal mean values equal for the past and for the future, for all continuous functions of the state (such as the Earth-Mars distance in the problem of the Solar System). Hence the small dissipative forces such as the tidal forces are then the real factors of evolution.

The ergodic theorem will be very useful for comparison of original and final evolutions (Section 11.9.5). It also explains the difficulty of tests for bounded or oscillatory motions.

---

* This likeness between the two types of oscillatory orbits can be emphasized by the following.

Let us consider some sufficiently large subset G_m and let us compare the proportions $p(t_1,t_2)$ for a subset F_n and for the subset G_m.

If the trajectory $\vec{X}(t)$ of interest is "oscillatory, chiefly at infinity" these two proportions go to zero when $t_2 \to \infty$, but their ratio has a limit :

$$\lim_{t_2 \to \infty} \left\{ \left(p(t_1,t_2),F_n\right) / \left(p(t_1,t_2),G_m\right) \right\}$$

which : 1°) exists

2°) is equal to the limit obtained when $t_2 \to -\infty$

3°) is non-zero when $\vec{X}(t)$ intersect F_n.

Unfortunately these results cannot be extended to subsets I_n of infinite measure.

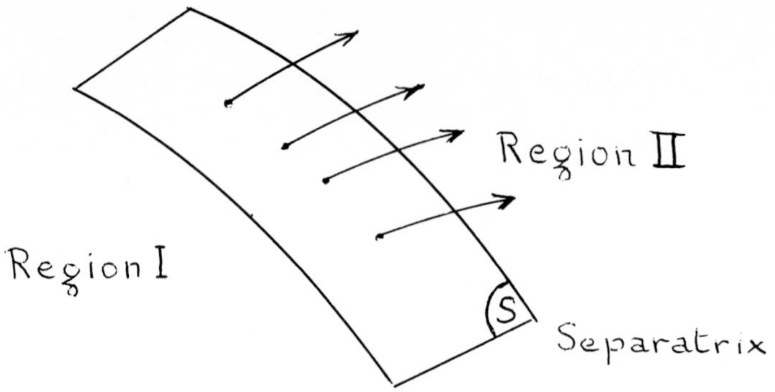

Fig. 165. The principle of the tests of escape.
The phase-space is divided into two regions and no motion is possible from region II to region I.

All demonstrations of known tests are based on the principle displayed in Section 11.7.4 and reproduced here in Fig. 165. The phase-space, or some invariant part of it, is divided by the test into two regions and no motion is possible from the region II to the region I : the conditions of the test remain forever satisfied after t_1 if they are satisfied at t_1.

In these conditions it is obvious that an orbit passing from the region I to the region II cannot come back and cannot be recurrent, this orbit is either abnormal or, more generally, open.

If we want that the region II of the test contains only bounded orbits it is necessary that no motion be possible from the region I to the region II and the separatrix S is then an invariant of motion.

These difficult conditions are obtained in some exceptional tests (Hill stability in the circular restricted three-body problem ; tests (1183) (1184) for Sitnikov motions...) but they are obviously much rarer than those of usual tests.

11.7.10.3 A test of escape valid even for very small mutual distances

This very efficient test of the general three-body problem has been set up with the help of Junzo Yoshida and Sun-Yi-Sui (Ref. 235). Some of its results are presented in Section 11.7.10.4. It uses essentially the three following elements :

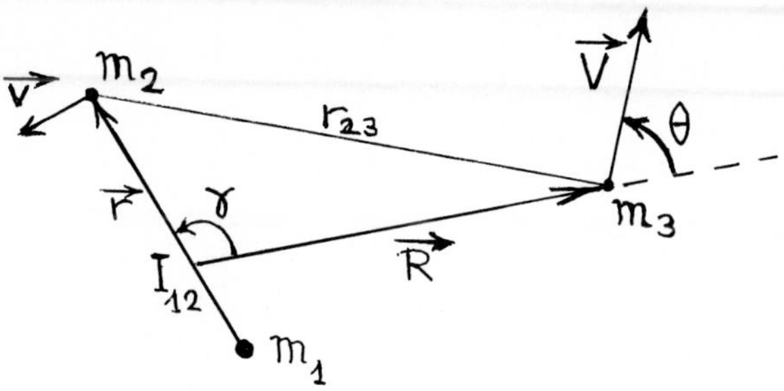

Fig. 166. The Jacobi decomposition.
$\gamma = (\vec{r},\vec{R})$; $\vec{v} = d\vec{r}/dt$; $\vec{V} = d\vec{R}/dt$; $\theta = (\vec{R},\vec{V})$.

I) The usual Jacobi decomposition (Fig. 166) with m_1 as the larger mass of the binary m_1, m_2 and with m_3 as the "third mass".

II) The new notations of Section 11.6.1, the constants M^*, μ, a, p and the variables ρ, ρ', ν, j, with the limits on the evolution of the mean quadratic distance ρ and its derivative ρ' given in Section 11.7.3.2.

III) The third element is the property of many three-body systems in which the Jacobi ratio r/R remain small for long intervals of time : systems with Hill-type stability (Section 11.6), hyperbolic-elliptic escapes, large ejections, close approaches and large values of the Sundman function j etc...

A)- <u>Limits on $d^2\vec{R}/dt^2$ when $r \leq KR$</u>

The acceleration $d^2\vec{R}/dt^2$ is given by the Jacobi expression

$$d^2\vec{R}/dt^2 = -\mu\left[\alpha\vec{r}_{13}\, r_{13}^{-3} + \beta\vec{r}_{23}\, r_{23}^{-3}\right] \qquad (1191)$$

with :

$$\left.\begin{array}{l} \mu = GM = G(m_1 + m_2 + m_3) = \text{gravitational constant of} \\ \qquad \text{the three-body system of interest.} \\ \alpha = m_1/(m_1 + m_2) \;\; ; \;\; 0.5 \leq \alpha < 1 \end{array}\right\} \qquad (1192)$$

$$\left.\begin{array}{l}\beta = m_2/(m_1 + m_2) = 1 - \alpha \\ \vec{r}_{13} = \vec{R} + \beta\vec{r} \quad ; \quad \vec{r}_{23} = \vec{R} - \alpha\vec{r}\end{array}\right\} \qquad (1192)$$

If $\vec{R}$ is given and $\vec{r}$ only bounded by $r \leq KR$ with $K < \alpha^{-1}$ the acceleration $d^2\vec{R}/dt^2$ is not known but its possible values remain confined. Let us look for all possible $d^2\vec{R}/dt^2$ for given $\vec{R}$, K and masses.

There are some simple cases.

A.1.- $\vec{r} = 0$.

The equation (1191) gives :

$$d^2\vec{R}/dt^2 = - \mu\vec{R}/R^3 \qquad (1193)$$

A.2.- Collinear case. $\vec{r} = \lambda\vec{R} \quad ; \quad |\lambda| \leq K$

$$d^2\vec{R}/dt^2 = - \mu \frac{\vec{R}}{R^3}\left[\frac{\alpha}{(1 + \lambda\beta)^2} + \frac{\beta}{(1 - \lambda\alpha)^2}\right] \qquad (1194)$$

The modulus of the acceleration is now larger than in (1193), it is maximum when $\lambda = K$ (Fig. 167) and we will then have :

$$d^2\vec{R}/dt^2 = - \mu(\vec{R}/R^3)(1 + \varepsilon) \qquad (1195)$$

with :

$$\left.\begin{array}{l}\varepsilon = \dfrac{\alpha}{(1 + K\beta)^2} + \dfrac{\beta}{(1 - K\alpha)^2} - 1 = \\ \qquad = 3K^2\alpha\beta + 4K^3\alpha\beta(\alpha - \beta) + O(K^4)\end{array}\right\} \qquad (1196)$$

ε is generally small ; $K \leq 0.5$ implies $\varepsilon \leq K^2$.

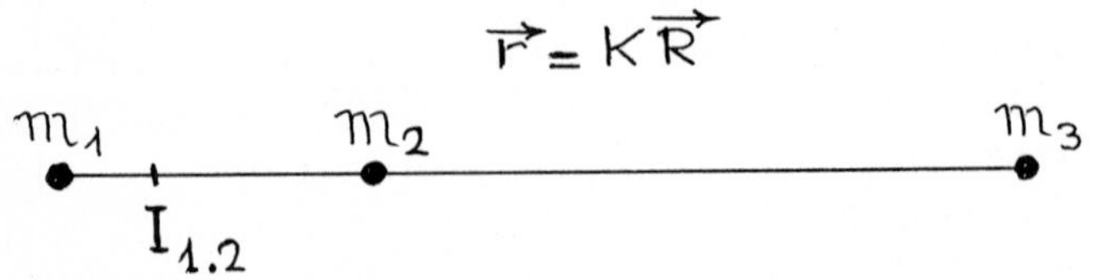

Fig. 167. Study of $d^2\vec{R}/dt^2$ when $r \leq KR$.
*The largest acceleration $|d^2\vec{R}/dt^2|$ for given K and **R** occurs when $\vec{r} = K\vec{R}$.*

A.3.- Isosceles case $r_{13} = r_{23}$:

$$d^2\vec{R}/dt^2 = -\mu\vec{R}/r_{13}^3 \tag{1197}$$

The modulus of the acceleration is smaller than in (1193) but the ratio of moduli remains larger than or equal to $(1 + K^2\alpha\beta)^{-3/2}$ that is larger than $1 - (\varepsilon/2)$.

In the remaining cases the acceleration is no longer radial, but inclined towards the nearest mass m_1 or m_2.

Let us call X the radial component of $d^2\vec{R}/dt^2$ and Y its circumferential component (Fig. 168), they can be expressed in terms of $\lambda = r/R$ and the angle $\gamma = (\vec{r},\vec{R})$

$$\left.\begin{aligned}
X &= -\mu R\left[\alpha r_{13}^{-3}(1 + \lambda\beta\cos\gamma) + \beta r_{23}^{-3}(1 - \lambda\alpha\cos\gamma)\right] \\
Y &= \mu R\lambda\alpha\beta\sin\gamma\left[r_{23}^{-3} - r_{13}^{-3}\right] \\
r_{13}^2 &= R^2 + 2\beta Rr\cos\gamma + \beta^2 r^2 \\
r_{23}^2 &= R^2 - 2\alpha Rr\cos\gamma + \alpha^2 r^2
\end{aligned}\right\} \tag{1198}$$

Thus when λ is small :

$$\left.\begin{aligned}
X &= -(\mu/R^2).\Big[1 + 1.5\lambda^2\alpha\beta(3\cos^2\gamma - 1) + \\
&\qquad + 2\lambda^3\alpha\beta(\alpha - \beta)(5\cos^3\gamma - 3\cos\gamma) + \dots\Big] \\
Y &= \mu/R^2\left[3\lambda^2\alpha\beta\sin\gamma\cos\gamma + 1.5\lambda^3\alpha\beta(\alpha - \beta)\sin\gamma(5\cos^2\gamma - 1) + \dots\right]
\end{aligned}\right\} \tag{1199}$$

When K and then λ are small the equations (1199) give an approximate idea of the region we are looking for : a prolate ellipsoid of revolution with a radial major axis.

The true region of $d^2\vec{R}/dt^2$ is complex its boundary corresponds to $\lambda = K$ and $r_{13} \geq r_{23}$ and we will only use the simple covering region of Fig. 168.

This covering region is a prolate ellipsoid with the major axis AB along the $\vec{R}$ direction from the abscissa $X_A = -\mu(1 + \varepsilon)R^{-2}$ to the abscissa $X_B = -\mu(1 - \varepsilon/2)R^{-2}$ and with two minor axes like CD whose lengths are two third of that of AB.

It is easy to verify that the accelerations given by (1193), (1194), (1195), (1197) belong to the prolate ellipsoid of Fig. 168 and it is not too difficult to verify that this ellipsoid also contains all the other possible accelerations $d^2\vec{R}/dt^2$ for given $\vec{R}$, K and masses.

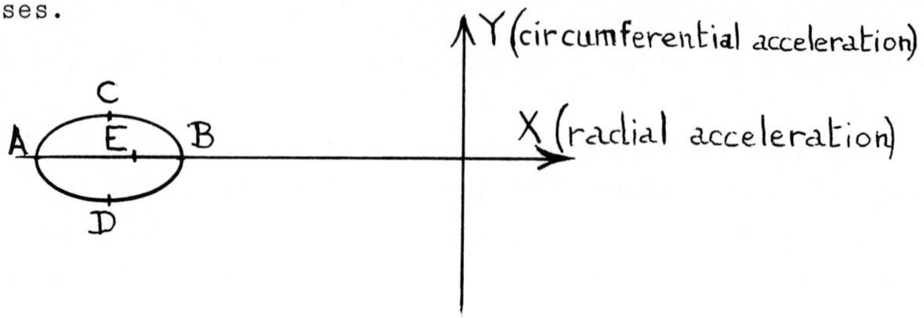

Fig. 168. The prolate ellipsoid containing all possible $d^2\vec{R}/dt^2$ for the given $\vec{R}$, K and masses.

$X_A = -\mu(1+\varepsilon)R^{-2}$; $X_B = -\mu(1-\varepsilon/2)R^{-2}$;

$X_C = X_D = -\mu(1+\varepsilon/4)R^{-2}$; $X_E = -\mu R^{-2}$; $y_C = -y_D = \mu\varepsilon R^{-2}/2$

B)- <u>Escape within the condition $r \leqq KR$</u>

Let us assume that by some exterior mean (test, numerical estimation, Hill-type stability, large value of the Sundman function j, etc ...) we have the certainty of :

$$\left.\begin{array}{l} r \leqq KR \text{ for all future times} \\ \\ K \text{ is a positive constant smaller than } \alpha^{-1} \end{array}\right\} \qquad (1200)$$

The limits on $d^2\vec{R}/dt^2$ given by Fig. 168 provide a rough picture of the motion and a basis for a test of escape of the mass m_3.

The dimensionless parameters ϕ and ψ proportional to the radial velocity $V\cos\theta$ and the circumferential velocity $V\sin\theta$ (Fig. 169) allow to simplify the analysis :

$$\phi = V\cos\theta \cdot \sqrt{R/\mu} \quad ; \quad \psi = V\sin\theta \cdot \sqrt{R/\mu} \geqq 0 \qquad (1201)$$

The evolution of $\vec{R}$ and $\vec{V}$ gives the following differential system :

$$\left.\begin{aligned} dR/dt &= \phi \cdot \sqrt{\mu/R} \\ d\phi/dt &= (\sqrt{\mu}/R^{3/2}).\left[(XR^2/\mu) + (\phi^2/2) + \psi^2\right] \\ d\psi/dt &= (\sqrt{\mu}/R^{3/2}).\left[(YR^2/\mu) - (\phi\psi/2)\right] \end{aligned}\right\} \qquad (1202)$$

The parameters X and Y are the accelerations of Fig. 168, they must of course correspond to a point of the ellipsoid.

The two brackets only contain the dimensionless parameters ϕ, ψ, XR^2/μ, YR^2/μ and the ratio $d\psi/d\phi$ is independant of R.

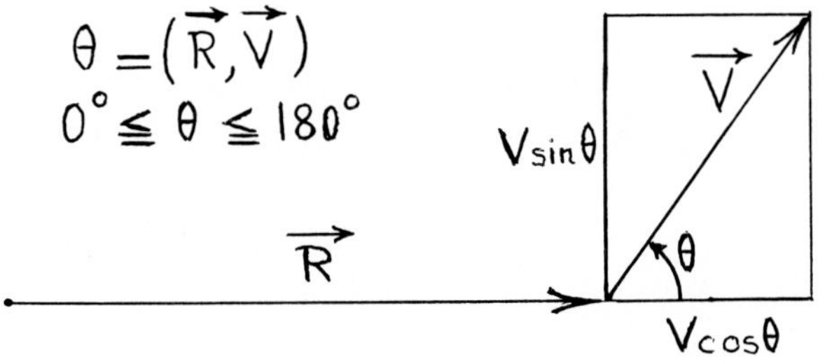

Fig. 169. The radial velocity $dR/dt = V\cos\theta$, the circumferential velocity $V\sin\theta$ and the dimensionless parameters $\phi = V\cos\theta\sqrt{R/\mu}$; $\psi = V\sin\theta\sqrt{R/\mu} \geq 0$.

The application of the principle of the tests of escape (Section 11.7.4) leads to an analysis in the ϕ, ψ half-plane (Fig. 170) and to limit curves which are functions only of the parameter ε of (1196) and Fig. 168.

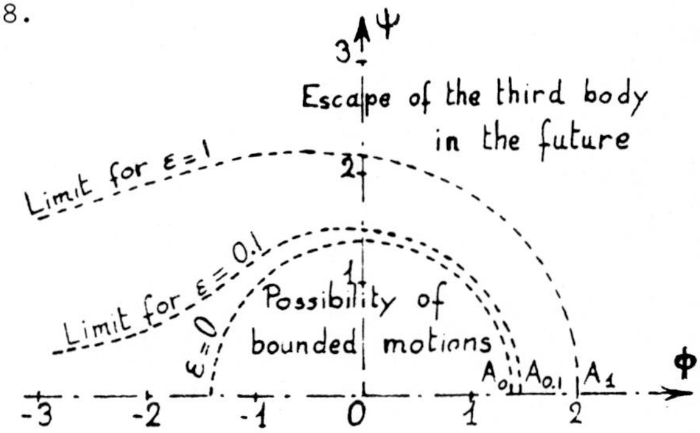

Fig. 170. Limit curves or escape velocity curves corresponding to the integration of (1203).

According to the principle of the tests of escape the limit curves are such that no motion is possible from outside to inside and this leads to the following.

A) For a given ε the limit curve starts at A_ε where $\phi = \sqrt{2 + 2\varepsilon}$ and $\psi = 0$.

B) The differential equation of the limit curve corresponds to the smallest possible value of $d\psi/d\phi$ as given by (1202) and the X and Y of Fig. 168 :

$$\left.\begin{array}{l}(2\phi^2 + 4\psi^2 - 4 - \varepsilon)(d\psi/d\phi) + 2\phi\psi + \\ \quad + \varepsilon\left[4 + 9(d\psi/d\phi)^2\right]^{1/2} = 0\end{array}\right\} \quad (1203)$$

We will call "escape velocity" V_{E+} corresponding to given $\mu, \vec{R}, \vec{V}, \varepsilon$ the limit velocity that can be deduced from Fig. 170 with the same value of angle $\theta = (\vec{R},\vec{V})$ of Figs. 166 and 169.

Hence :

$$V_{E+} = V_{E+}(\mu,R,\varepsilon,\theta) = (\sqrt{\mu/R})f(\varepsilon,\theta) \quad ; \quad \text{with } f^2 = \phi^2 + \psi^2 \quad (1204)$$

Thus, for any μ, R, K, ε, θ :

$$\left.\begin{array}{l}\{V \geq V_{E+} \text{ and the condition (1200)}\} \text{ implies } m_3 \text{ and R} \\ \text{escape to infinity when } t \to \infty\end{array}\right\} \quad (1205)$$

We will see that this escape is with forever $d^2(R^2)/dt^2 > 2\mu/R$; the distance R has no maximum and at most one minimum in the future.

Symmetrically, when $t \to -\infty$, we will define the escape velocity V_{E-}.

$$V_{E-}(\mu,R,\varepsilon,\theta) = (\sqrt{\mu/R}) \cdot f(\varepsilon,\pi-\theta) = V_{E+}(\mu,R,\varepsilon,\pi-\theta) \quad (1206)$$

The function $f(\varepsilon,\theta)$ must be deduced from (1203) ; it is defined for $\varepsilon \geq 0$; $0 \leq \theta < \pi$, with the following properties :

$$\left.\begin{array}{l} f \geq \sqrt{2 + 2\varepsilon} \quad ; \quad \partial f/\partial\varepsilon \geq 0 \quad ; \quad \partial f/\partial\theta \geq 0 \\ \{\varepsilon \text{ and/or } \theta = 0\} \quad \text{implies} \quad f = \sqrt{2 + 2\varepsilon} \\ \{\theta \to \pi \text{ and } \varepsilon > 0\} \quad \text{implies } f \sim \left[\frac{\varepsilon}{2(\pi-\theta)} \cdot \text{Log}\, \frac{1}{(\pi-\theta)}\right]^{1/2} \end{array}\right\} \quad (1207)$$

Numerical integration of (1203) is easy and shows that the "circumferential escape velocities", obtained for $\theta = \pi/2$, are not much larger than the "radial escape velocities" $\sqrt{(\mu/R)(2 + 2\varepsilon)}$ obtained for $\theta = 0$:

TABLE IX

ε	Ratio of circumferential to radial escape velocities.
0	1
0.1	1.0183
1	1.0818
∞	1.1389

The analytical integration of (1203) is more difficult and only approximations are known.

The first order approximation in ε gives an upper bound :

$$f(\varepsilon,\theta) \leq f_1(\varepsilon,\theta) = \sqrt{2 + 2\varepsilon\, g(\theta)} \quad (1208)$$

with :

$$\left.\begin{array}{l} g(\theta) = \frac{1}{4} + \frac{1}{2\sin^2\theta} \cdot \int_{\cos\theta}^{1} dc\ (4 + 5c^2) = \\ \quad = \frac{4 - \cos^2\theta - \cos\theta\sqrt{4 + 5\cos^2\theta}}{4\sin^2\theta} - \\ \quad - \frac{\sqrt{5}}{5\sin^2\theta}\, \text{Log}\left[\frac{\cos\theta + \sqrt{0.8 + \cos^2\theta}}{1 + \sqrt{1.8}}\right] \end{array}\right\} \quad (1209)$$

$g(\theta)$ increases from 1 to ∞ when θ increases from 0 to π, its graph is presented in Fig. 171 with $g(\pi/2) = 1.43041$.

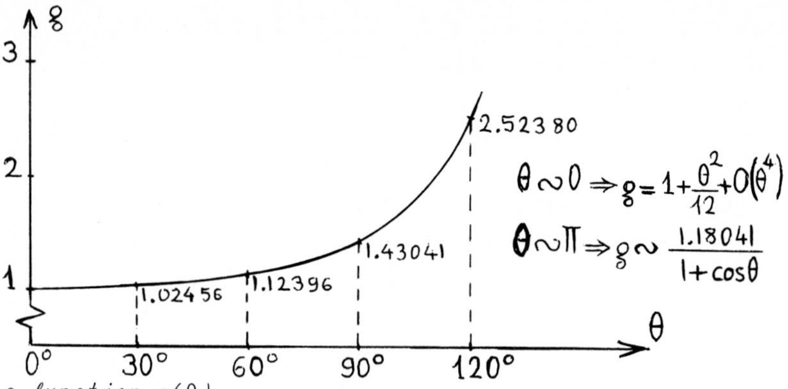

Fig. 171. The function g(θ).

Another upper bound is given by :

$$f(\varepsilon,\theta) \le f_2(\varepsilon,\theta) = \inf_{\theta_0 \in (\theta - \frac{\pi}{2}\,;\frac{\pi}{2})} \left[\frac{\sqrt{2 + 2\varepsilon g(\theta_0)}}{\cos(\theta - \theta_0)}\right] \tag{1210}$$

This upper bound is much better for acute angles θ and even up to the maximum of ψ.

As a lower bound we can use :

$$f \ge f_3(\varepsilon,\theta) = \{4 + (8\varepsilon + 4\varepsilon^2)g(\theta)\}^{1/4} \tag{1210.1}$$

These bounds are excellent for small θ, even if ε is large, but they are weak for large θ. In the case when $\theta = \pi/2$ they give the following bounds for the table IX :

TABLE X

ε	Lower bound given by f_3	ratio of circumferential to radial escape velocities.	upper bounds given by f_2	given by f_1
0	1	1	1	1
0.1	1.0182	1.0183	1.0185	1.0194
1	1.0724	1.0819	1.0855	1.1024
∞	1.0936	1.1389	1.1492	1.1960

It is certainly possible to improve and to simplify these approximate results.

C)- <u>Adjacent results</u>

Many adjacent results can be deduced from the previous Section as long as the condition $r \leqq KR$ remains satisfied.

Let us consider a given three-body motion and let us give the subscript zero to the initial conditions.

$V_0 \geqq V_{0E+}(\mu, R_0, \varepsilon, \theta_0)$ and $t \geqq t_0$ imply :

I) $V^2 - V^2_{E+}$ is a non-negative and non decreasing function of t.

II) The upper bounds f_1 and $f_2(\varepsilon, \theta)$ give :

$\{V^2 - \mu f_1^2/R\}$ and $\{V^2 - \mu f_2^2/R\}$ are two non-decreasing functions of t as soon as they are non negative.

III) :

$$\left.\begin{aligned} d^2(R^2)/dt^2 &= 2V^2 + 2RX \geqq 2V^2 - (2\mu/R)(1+\varepsilon) \geqq \\ &\geqq 2V^2 - 2V^2_{E+} + (2\mu/R)(1+\varepsilon) \geqq (2\mu/R)(1+\varepsilon) > 0 \end{aligned}\right\} \quad (1211)$$

The length R has at most one minimum in the interval of interest.

IV) The above properties I) and III) imply, with $R' = dR/dt$:

$$\left.\begin{aligned} &\{R^2R'^2 - 2\mu R(1+\varepsilon) - R^2(V_0^2 - V^2_{0E+})\} \\ &\text{has a derivative of the sign of } R' \end{aligned}\right\} \quad (1212)$$

Furthermore if $R'_0 \geqq 0$:

$$\left.\begin{aligned} &\{V^2 + V^2_{E+} - \mu(4+\varepsilon)/R\} \\ &\left\{R^2R'^2 - \mu R(2-\varepsilon) - R^2\left[V_0^2 + V^2_{0E+} - \mu(4+\varepsilon)/R_0\right]\right\} \end{aligned}\right] \text{are two non-increasing functions of t} \quad (1213)$$

V) If the condition $r \leqq KR$ remains forever satisfied the third mass m_3 escapes to infinity with a final velocity R'_∞ that can be estimated with (1212), (1213) :

$$\left.\begin{aligned} &\{V_0 \geqq V_{0E+}\} \quad \text{implies} \quad R'_\infty \geqq \sqrt{V_0^2 - V^2_{0E+}} \\ &\{V_0 \geqq V_{0E+} \text{ and } R'_0 \geqq 0\} \text{ implies } R'_\infty \leqq \sqrt{V_0^2 + V^2_{0E+} - (\mu/R_0)(4+\varepsilon)} \end{aligned}\right\} \quad (1214)$$

If ε and θ_0 are small the estimation can be very good and, if the energy integral h is positive, this estimation is very useful for the choice between hyperbolic and hyperbolic-elliptic escape since the final energy, $h_{1.2}$, of the binary is given by:

$$h_{1.2} = h - (m_3(m_1 + m_2)/2M) R'^2_\infty \qquad (1215)$$

All these results are of course reversible and can be used with V_{E-} for decreasing times.

D)- <u>The condition r ≦ KR</u>

Many possibilities can be used to ensure that the ratio r/R will remain forever small, the simplest being the Hill-type stability, either this of systems of negative energy (Section 11.6.3) or that of systems of positive energy (Section 11.6.5), but this can be enlarged:

Let us consider the collinear central configuration m_1, m_2, m_3 (Fig. 172), i.e. the central configuration with the smaller mass of the binary between the two other masses. This configuration has a ratio r/R that we will call λ_3 and a ratio ρ/ν that appears as $(\rho/\nu)L_2$ on generalized Hill curves (Fig. 125, Fig. 164).

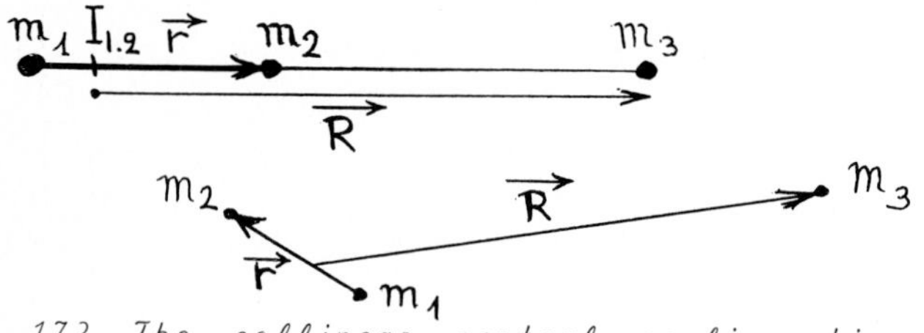

Fig. 172. The collinear central configuration m_1, m_2, m_3 with $r/R = \lambda_3$ and $\rho/\nu = (\rho/\nu)_{L_2}$.
In another configuration the mass m_3 is "isolated" iff : $r/R \leqq \lambda_3$ and $j \geqq (\rho/\nu)_{L_2}$.

According to (909) the ratio $(\rho/\nu)_{L_2}$ is always in the range 1 ; 1.1881..., its square is given in terms of the mass ratios in (912) and in Table II of Section 11.6.3.

The third mass m_3 will be defined as "isolated" if and only if at the time of interest :

$$r/R \leqq \lambda_3 \quad \text{and} \quad j \geqq (\rho/\nu)_{L_2} \qquad (1216)$$

Let us recall that j is the Sundman function given in (896) in Section 11.6.1, that is :

$$j = (\rho/2a) + (p/2\rho) + (\rho\rho'^2/2\mu) \tag{1217}$$

This Sundman function has two major interests : it is always less than or equal to the ratio ρ/ν (Sundman inequality (902) of Section 11.6.2) and its variations are slow and in the direction of those of the mean quadratic distance ρ. This situation gives large intervals of time in which j and ρ/ν are large (see the evolution of j and ρ in Section 11.7.3.2) and then r/R is small as can be seen in the figures of generalized Hill curves (Figs 125, 164) and as we will see more accurately in the next Section. These figures 125, 164 show that as soon as : $j \geqq \sup_{K=1,2,3} (\rho/\nu)_{L_K}$ one of the three bodies is isolated.

E)- The ratios r/R and ρ/ν

Let us consider three given positive masses m_1, m_2, m_3, with $m_1 \geqq m_2$, and a given value of the ratio r/R. What are the possible values of the ratio ρ/ν ? We will assume that $r/R < \alpha^{-1}$.

The mean quadratic distance ρ is a function of r, R and the three masses :

$$\left. \begin{aligned} M^*\rho^2 &= m_1m_2r_{12}^2 + m_1m_3r_{13}^2 + m_2m_3r_{23}^2 = \\ &= \frac{Mm_1m_2}{m_1 + m_2} r^2 + m_3(m_1 + m_2)R^2 \end{aligned} \right\} \tag{1218}$$

with, as in Section 11.6 :

$$M^* = m_1m_2 + m_1m_3 + m_2m_3 \quad ; \; M = m_1 + m_2 + m_3 \tag{1219}$$

The mean harmonic distance ν is not a function of r,R and the masses :

$$M^*/\nu = (m_1m_2/r_{12}) + (m_1m_3/r_{13}) + (m_2m_3/r_{23}) \tag{1220}$$

with :

$$\left.\begin{array}{l} r_{12} = r \quad ; \quad \alpha r_{13}^2 + \beta r_{23}^2 = \alpha\beta r^2 + R^2 \\ (\alpha = m_1/(m_1+m_2) \geq 0.5 \quad ; \quad \beta = m_2/(m_1+m_2) = 1 - \alpha) \end{array}\right\} \qquad (1221)$$

The largest ν is obtained in the isosceles case $r_{13} = r_{23}$ and the smallest ν in the collinear case with m_2 between m_1 and m_3.

Hence :

$$f(\lambda) \leq \rho/\nu \leq F(\lambda) \qquad (1222)$$

with :

$$\left.\begin{array}{l} \lambda = r/R < \alpha^{-1} \\ q = m_3(m_1 + m_2)/M^* < 1 \\ f(\lambda) = \left[q + \lambda^2(1 - q + q\alpha\beta)\right]^{1/2} \cdot \left[\dfrac{1 - q}{\lambda} + \dfrac{q}{\sqrt{1 + \lambda^2\alpha\beta}}\right] \\ F(\lambda) = \left[q + \lambda^2(1 - q + q\alpha\beta)\right]^{1/2} \cdot \left[\dfrac{1 - q}{\lambda} + \dfrac{q\alpha}{1 + \lambda\beta} + \dfrac{q\beta}{1 - \lambda\alpha}\right] \end{array}\right\} \qquad (1223)$$

The curves $f(\lambda)$ and $F(\lambda)$ are presented in Fig. 173. They always follow a similar picture with :

$$f(\lambda) < F(\lambda) \text{ for all } \lambda\text{'s} \qquad (1224)$$

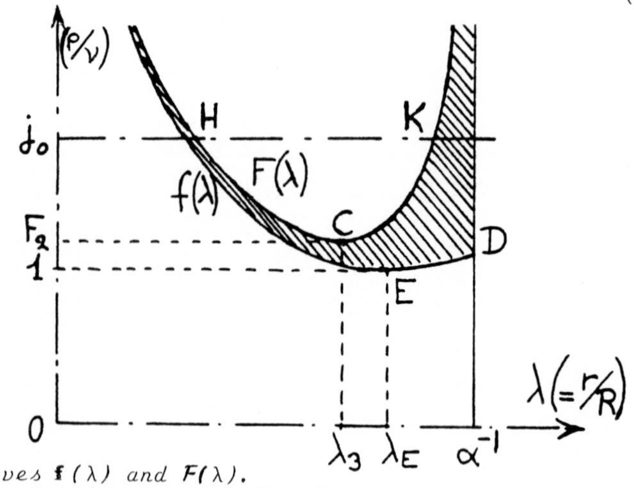

Fig. 173. The curves $f(\lambda)$ *and* $F(\lambda)$.
$1 < F_C = (\rho/\nu)_{L_2} \leq 1.1881\ldots$; $1 < (\rho/\nu)_D \leq 1.045\ldots$;
$0 < \lambda_3 < 1 < \lambda_E = (1 - \alpha\beta)^{-1/2} < \alpha^{-1} \leq 2.$

and for λ small :

$$f(\lambda) \sim F(\lambda) \sim (1 - q)\sqrt{q}/\lambda \tag{1225}$$

$\lambda \to \alpha^{-1}$ implies :

$$\left.\begin{array}{l} f(\lambda) \to (\rho/\nu)_D \in (1 \;;\; 1.045..) \\ F(\lambda) \to +\infty \end{array}\right\} \tag{1226}$$

The curves $f(\lambda)$ and $F(\lambda)$ have one an only one minimum in the $(0;\alpha^{-1})$ interval.

The minimum of $f(\lambda)$ is obtained at E (where $r/R = (1 - \alpha\beta)^{-1/2}$; $\rho/\nu = 1$), it corresponds to the equilateral central configuration of Lagrange.

The minimum of $F(\lambda)$ is obtained at C (where $r/R = \lambda_3$; $\rho/\nu = F_C = (\rho/\nu)_{L_2} \in [1 \;;\; 1.1881...]$) ; it corresponds to the central configuration with m_2 between m_1 and m_3.

Let us now consider a large value of the Sundman function j such as j_0 (Fig. 173). We know that if $r/R < \alpha^{-1}$ the point (r/R ; ρ/ν) will fall into the shaded region of Fig. 173, we also know the Sundman inequality $\rho/\nu \geq j$. Hence $j = j_0$ implies either $r/R \leq \lambda_H$ or $r/R \geq \lambda_K$. In the former case the mass m_3 is "isolated".

If initially the mass m_3 is isolated it will remain isolated as long as $j \geq F_C$. Notice that the case $\{j = F_C \;;\; r/R = \lambda_3\}$ only corresponds to $\rho/\nu = F_C = (\rho/\nu)_{L_2}$ and to a Euler motion with fixed r/R and ρ/ν, hence the point C cannot be crossed with $j \geq F_C$.

The interest of the motion of isolation comes from the expression (1217) of the Sundman function j with the gravitational constant $\mu = GM$, the two fixed lengths a and p and the variable length ρ with its derivative ρ'. Since $(dj/dt).\rho' \geq 0$ if j is initially large and $\rho' > 0$ the function j will remain non--decreasing and large at least as long as ρ is non-decreasing and, because of the expression of j itself, there will be a large interval of time with large j.

The topology of generalized Hill curves (Fig. 125, Section 11.6.3) shows that one of the three bodies is necessarily isolated when the Sundman function j is larger than or equal to $\sup(\rho/\nu)_{L_K}$ with K = 1, 2, 3.

F)- The test of escape

Let us assume that at some initial time t_0 :

$$\left.\begin{array}{l} \rho'_0 \geqq 0 \quad ; \quad R'_0 \geqq 0 \\ \left.\begin{array}{l} r_0/R_0 \leqq \lambda_3 \\ j_0 \geqq (\rho/\nu)_{L_2} \end{array}\right\} \text{The mass } m_3 \text{ is isolated} \end{array}\right\} \tag{1227}$$

With Fig. 173 these conditions imply :

$$r_0/R_0 \leqq K \quad \text{with} \left\{\begin{array}{l} F(K) = j_0 \\ K \leqq \lambda_3 \end{array}\right\} \tag{1228}$$

This coefficient K smaller than λ_3 gives a smaller ε :

$$\varepsilon = \frac{\alpha}{(1 + K\beta)^2} + \frac{\beta}{(1 - K\alpha)^2} - 1 \tag{1229}$$

and hence a smaller escape velocity :

$$V_{OE+} = V_{OE+}(\mu, R_0, \varepsilon, \theta_0) = \sqrt{\mu/R_0} \cdot f(\varepsilon, \theta_0) \tag{1230}$$

and we finally reach the following escape condition, if at t_0 :

$$V_0 \geqq V_{OE+} \tag{1231}$$

i.e. if the initial velocity is larger than or equal to the escape velocity, the ratio r/R will forever remain less than or equal to K ; the length R and the mass m_3 will escape to infinity when $t \to \infty$.

We have already seen in (1200), (1205) that if the ratio r/R remains forever less than or equal to K the condition (1231) is a sufficient condition of escape. Hence it only remains to verify that $r/R \leqq K$ will be forever true.*

* Notice that if $h \geqq 0$ the mean quadratic distance ρ will remain forever increasing and the condition $R'_0 \geqq 0$ will be useless and unecessary. The test of this Section is then almost contained inside the Hill-type stability for systems of positive or zero energy (Section 11.6.5). Its only new elements are related to the relations (1214), (1215) that allow often to discriminate between hyperbolic and hyperbolic-elliptic final evolutions.

If the mean quadratic distance ρ is forever non-decreasing the same applies to the Sundman function j and, with $\rho/\nu \geq j$, Fig. 173 shows that indeed $r/R \leq K$ will be forever true.

Let us then consider the final case, the case in which ρ reaches a local maximum ρ_M at some ulterior time t_M. This case requires a negative energy integral h and then a positive generalized semi-major axis a. During the closed interval $[t_0, t_M]$ the condition $r/R \leq k$ remains satisfied and ρ is non-decreasing, hence at $t = t_M$:

$$r_M/R_M \leq K \quad ; \quad R'_M \geq 0 \quad ; \quad (R^2)''_M \geq \frac{2\mu(1 + \varepsilon)}{R_M} > 0 \tag{1232}$$

and :

$$\rho'_M = 0 \quad ; \quad (\rho^2)''_M \leq 0 \quad ; \quad j_M = \frac{\rho_M}{2a} + \frac{p}{2\rho_M} \geq j_0 \tag{1233}$$

However the Lagrange-Jacobi identity (946) and the Sundman inequality (948) are $(\rho^2)'' = 2\mu\big((1/\nu) - (1/a)\big)$ and $\rho/\nu \geq j$ hence :

$$\nu_M \geq a \quad ; \quad \rho_M \geq aj_M = (\rho_M/2) + (ap/2\rho_M) \tag{1234}$$

which implies :

$$\rho_M \geq \sqrt{ap} \tag{1235}$$

We are looking for the existence of a limit time t_L such that :

$$\left.\begin{array}{l} t_L \geq t_M \\ \text{in the closed interval } [t_0;t_L] : r/R \leq K \\ \text{immediately after } t_L : r/R > K \end{array}\right\} \tag{1236}$$

Hence at t_L :

$$\left.\begin{array}{l} r_L/R_L = K \\ R_L \geq R_M \quad ; \quad r_L = KR_L \geq KR_M \geq r_M \end{array}\right\} \tag{1237}$$

However the mean quadratic distance ρ is given in terms of r and R by :

$$M^*\rho^2 = \frac{Mm_1m_2}{m_1 + m_2} r^2 + m_3(m_1 + m_2)R^2 \tag{1238}$$

Hence (1237) implies :

$$\rho_L \geqq \rho_M \tag{1239}$$

and, with (1235) :

$$\rho_L \geqq \rho_M \geqq \sqrt{ap} \quad ; \quad j_L \geqq \frac{\rho_L}{2a} + \frac{p}{2\rho_L} \geqq \frac{\rho_M}{2a} + \frac{p}{2\rho_M} = j_M \geqq j_0 \tag{1240}$$

Since with j_0 the limit ratio of r/R is already K we can have $r_L/R_L = K$ only if $j_L = j_M = j_0$ and then $\rho_L = \rho_M$; $R_L = R_M$, and since after t_0 the function R(t) is increasing the equality $R_L = R_M$ necessarily implies $t_L = t_M$.

However even if $t_L = t_M$ it is impossible that immediately after t_L we have r/R > K, indeed, from (1232), (1233) :

$$\left.\begin{aligned} &R'_M \geqq 0 \quad ; \quad (R^2)''_M > 0 \\ &\rho'_M = 0 \quad ; \quad (\rho^2)''_M \leqq 0 \end{aligned}\right\} \tag{1241}$$

Hence in a vicinity of t_M the ratio ρ/R is decreasing and so is r/R because of (1238) which can be written :

$$M^* \cdot (\rho^2/R^2) = \frac{Mm_1m_2}{m_1 + m_2} \cdot \frac{r^2}{R^2} + m_3(m_1 + m_2) \tag{1242}$$

Hence the test (1227)-(1231) is a true test of escape. Its efficiency will pay for its complexity (see Sections 11.7.10.4 and 11.7.10.5).

G)- Improvements of the test

The test (1227)-(1231) can be summarized by the following :

"The mass m_3 and the length R will escape to infinity when $t \to \infty$ if at some arbitrary time t_0 :

A) The mass m_3 is isolated and has an escape velocity.

B) The derivatives ρ'_0 and R'_0 are both non-negative.

However the condition $R_0' \geq 0$ is here only to ensure that $R_M' \geq 0$ at $t = t_M$ and this result can be obtained by several other means. For instance, since in $[t_0, t_M]$ the second derivative $(R^2)''$ is positive, it is sufficient that R_0 be smaller than or equal to the length R_S, the smallest possible value of R_M.

For negative energies the smallest possible value of ρ_M is given by (1233) and (1235) :

$$\rho_{M\ \text{minimum}} = aj_0 + \sqrt{a^2 j_0^2 - ap} \tag{1243}$$

The relations $r_M/R_M \leq K$ and (1238) give R_S the smallest possible value of R_M for the initial conditions :

$$R_S = \left[aj_0 + \sqrt{a^2 j_0^2 - ap}\right] \cdot \sqrt{\frac{M^*(m_1 + m_2)}{Mm_1 m_2 K^2 + m_3(m_1 + m_2)^2}} \tag{1244}$$

Conversely if at the initial time t_0 we have $R_0 \geq R_S$ and $R_0' \geq 0$ the condition $\rho_0' \geq 0$ becomes useless and the demonstration can be effected as if $t_M = t_0$.

These improvements lead to the following.

Let us assume that at some initial time t_0 the body m_3 is "isolated", i.e :

$$\left.\begin{aligned} & r_0/R_0 \leq \lambda_3 \\ & j_0 \geq (\rho/\nu)_{L_2} \end{aligned}\right\} \tag{1245}$$

This implies, with the function $F(\lambda)$ of (1223) :

$$\left. r_0/R_0 \leq K \quad \text{with} \begin{cases} F(K) = j_0 \\ K \leq \lambda_3 \end{cases} \right\} \tag{1246}$$

With the corresponding coefficient :

$$\varepsilon = \alpha(1 + K\beta)^{-2} + \beta(1 - K\alpha)^{-2} - 1$$

we define the two escape velocities as in (1203)-(1210) :

$$V_{0E+} = (\sqrt{\mu/R_0}) \cdot f(\varepsilon,\theta_0) \quad ; \quad V_{0E-} = (\sqrt{\mu/R_0}) \cdot f(\varepsilon,\pi-\theta_0) \qquad (1247)$$

There is an escape of m_3 for $t \to +\infty$ if :

$$\left.\begin{array}{l} V_0 \geq V_{0E+} \\ \text{and :} \\ \left\{\left[R_0 \leq R_S \ ; \ \rho'_0 \geq 0\right] \text{ and/or } \left[R_0 \geq R_S \ ; \ R'_0 \geq 0\right]\right\} \end{array}\right\} \qquad (1248)$$

Symmetrically there is an escape of m_3 for $t \to -\infty$ if :

$$\left.\begin{array}{l} V_0 \geq V_{0E-} \\ \text{and :} \\ \left\{\left[R_0 \leq R_S \ ; \ \rho'_0 \leq 0\right] \text{ and/or } \left[R_0 \geq R_S \ ; \ R'_0 \leq 0\right]\right\} \end{array}\right\} \qquad (1249)$$

All these escapes are with $r/R \leq K$ and $(R^2)'' \geq 2\mu(1+\varepsilon)/R$ and we can almost write that an isolated body with an escape velocity undergoes an escape in the past and/or in the future.

Several improvements have been given to this test (Ref. 235) and it seems that the following simplification is true : If the conditions of isolation (1245) are satisfied and if $V_0 \geq \inf\{V_{0+}, V_{0E-}\}$, with V_{0E+}, V_{0E-} defined in (1246)-(1247), the mass m_3 escapes for $t \to (\text{sign } R'_0)_\infty$ and on both sides if $R' = 0$.

However neither a demonstration nor a counter example has been found.

11.7.10.4 An application of the very efficient test. Analysis in the (ρ, ρ') half-plane.

Let us consider a given three-body system with its three point-masses, its axes of the center of mass, its energy h related to the generalized semi-major axis a and its angular momentum $\vec{c}$ related to the generalized semi-latus rectum p :

$$\left.\begin{aligned}
&M = m_1 + m_2 + m_3 = \text{total mass}\\
&GM = \mu = \text{gravitational constant of the system}\\
&M^* = m_1m_2 + m_1m_3 + m_2m_3\\
&h = 0.5(m_1V_1^2 + m_2V_2^2 + m_3V_3^2) -\\
&\qquad - G\left(\frac{m_1m_2}{r_{12}} + \frac{m_1m_3}{r_{13}} + \frac{m_2m_3}{r_{23}}\right) = \text{energy integral}\\
&a = - GM^*/2h = \text{generalized semi-major axis}\\
&\vec{c} = m_1\vec{r}_1 \times \vec{V}_1 + m_2\vec{r}_2 \times \vec{V}_2 + m_3\vec{r}_3 \times \vec{V}_3 = \text{angular momentum}\\
&p = Mc^2/GM^{*2} = \text{generalized semi-latus rectum}
\end{aligned}\right\} \quad (1250)$$

The mean quadratic distance ρ and its derivative ρ' are related to the semi-moment of inertia I :

$$\left.\begin{aligned}
&I = 0.5(m_1r_1^2 + m_2r_2^2 + m_3r_3^2) = 0.5\left\{\frac{m_1m_2r^2}{m_1+m_2} + \frac{m_3(m_1+m_2)R^2}{M}\right\}\\
&M^*\rho^2 = m_1m_2r_{12}^2 + m_1m_3r_{13}^2 + m_2m_3r_{23}^2 = 2MI
\end{aligned}\right\} \quad (1251)$$

It seems that the analysis in the ρ, ρ' half-plane is the most fruitful two-dimensional analysis of the three-body problem.

We must first adapt the efficient test of the previous section to this analysis and find the possible velocities $\vec{V} = d\vec{R}/dt$ for given masses, a, p, ρ, ρ', r, R .

The result is given in Fig. 174 in terms of the radial velocity $x = R' = V\cos\theta$ and the circumferential velocity $y = V\sin\theta$. It is a circle of center $\vec{V}_1$:

$$\vec{V}_1 = (x_1, y_1) \quad ; \quad x_1 = R\rho'/\rho \quad ; \quad y_1 = R\sqrt{\mu\rho}/\rho^2 \quad (1252)$$

and of the following radius W :

$$W = \left\{2\mu M m_1m_2r^2\left(F\left(\frac{r}{R}\right) - j\right)\Big/ m_3(m_1 + m_2)^2\rho^3\right\}^{1/2} \quad (1253)$$

The function $F(r/R)$ gives the maximum value of the ratio ρ/ν (Section 11.7.10.3, equation 1223) and Fig. 173.

With the Jacobi notations these geometrically simple results are easy to demonstrate*.

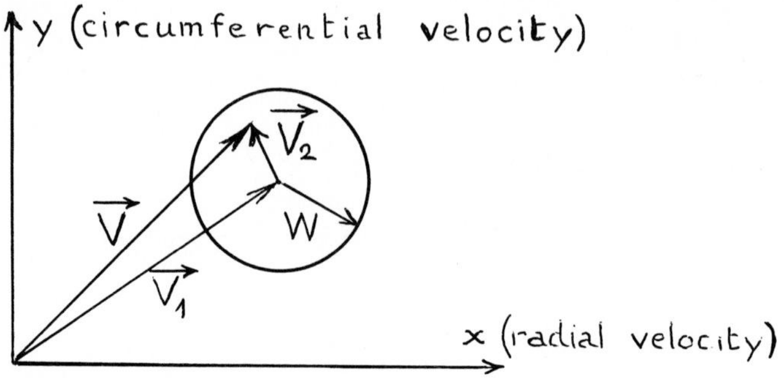

Fig. 174. Decomposition of the velocity $\vec{V} = \vec{V}_1 + \vec{V}_2$.

$\vec{V}_1 = (x_1, y_1)$; $x_1 = R\rho'/\rho$; $y_1 = R\sqrt{\mu p}/\rho^2$.

$$||\vec{V}_2|| \leq W = r\sqrt{\{(2\mu M m_1 m_2)(F(r/R) - j)\} / m_3(m_1+m_2)^2\rho^3\}}$$

---

* $\vec{r}$ and $\vec{R}$ are the usual Jacobi vectors and :

$d\vec{r}/dt = \vec{v}$; $d\vec{R}/dt = \vec{V}$; $\vec{v} = (v_r, v_c)$; $\vec{V} = (x,y)$: radial and circumferential components :

$\vec{c} = m\vec{r} \times \vec{v} + \mathcal{M}\vec{R} \times \vec{V}$

$h = 0.5(mv^2 + \mathcal{M}V^2) - (GM^*/\nu)$

$M^*\rho\rho'/M = mrv_r + \mathcal{M}Rx$

with :

$m = (m_1 m_2)/(m_1+m_2)$; $\mathcal{M} = (m_3(m_1+m_2))/M$: the two reduced masses and

$GM^*/\nu = U = \text{potential} = G((m_1 m_2/r_{12}) + (m_1 m_3/r_{13}) + (m_2 m_3/r_{23}))$

The equation of the angular momentum $\vec{c}$ gives the following triangular inequality :

$|c - \mathcal{M}Ry| \leq mrv_c \leq c + \mathcal{M}Ry$; with $v_c \geq 0$; $y \geq 0$.

The lower bound of v_c given by the first inequality and v_r given by the equation for ρ' lead to the following in the equation of h :

$$2h \geq m\left[(c-\mathcal{M}Ry)^2/m^2r^2 + (M^*\rho\rho'/M - \mathcal{M}Rx)^2/m^2r^2\right] + \mathcal{M}(x^2 + y^2) - 2GM^*/\nu$$

That is, with (1252) :

$$(x - x_1)^2 + (y - y_1)^2 \leq (2\mu m r^2/\mathcal{M}\rho^3)((\rho/\nu) - j)$$

where j is the Sundman function $\rho/2a + p/2\rho + \rho\rho'^2/2\mu$.

For given masses and r/R the largest value of ρ/ν is $F(r/R)$ given in (1223) and in Fig. 173 and we thus reach the circular domain (1252)-(1253).

If θ_1 is the argument of V_1 (Fig. 174) and if $0 \leq \theta_1 \leq \pi/2$ the equation (1210) with $\theta_0 = \theta_1$ shows that all escapes velocities V_{0E+} satisfy :

$$V_{0E+}\cos(\theta - \theta_1) \leq \sqrt{(\mu/R)(2 + 2\varepsilon g(\theta_1))} \tag{1254}$$

Hence all velocities of the circle of Fig. 174 are :

$$\left.\begin{array}{l} \text{escape velocities for the future if :} \\ 0 \leq \theta_1 \leq \pi/2 \quad ; \quad V_1 - W \geq \sqrt{(\mu/R)(2 + 2\varepsilon g(\theta_1))} \\ \text{escape velocities for the past if :} \\ \pi/2 \leq \theta_1 \leq \pi \quad ; \quad V_1 - W \geq \sqrt{(\mu/R)(2 + 2\varepsilon g(\pi - \theta_1))} \end{array}\right\} \tag{1255}$$

Notice that the constants m_1, m_2, m_3, μ, a, p and the variables ρ, ρ' give the Sundman function $j = \rho/2a + p/2\rho + \rho\rho'^2/2\mu$ and the main informations about the isolation of the "third mass" m_3. But they do not give the Jacobi lengths r and R related to ρ by only one relation :

$$M^*\rho^2 = Mm_1m_2r^2/(m_1+m_2) + m_3(m_1+m_2)R^2 \tag{1256}$$

Hence we must discuss the escape condition (1255) for all r/R ratios from 0 to the ratio K for which $F(K) = j$ and $W = 0$.

This analysis has been done for several cases and lead to the following figures, either in the (ρ,ρ') half-plane or in the equivalent $(\rho/a \; ; \; p/2\rho + \rho\rho'^2/2\mu)$ or $(\rho/a \; ; \; j)$ planes.

I)- Systems with three equal masses

Figure 175 presents a general analysis of three-equal-mass systems of negative energy.

The limits given by the conditions (1255) are the three full curves plotted for five values of the angle θ_1. Above these curves all motions are unbounded and have a hyperbolic-elliptic escape for $t \to -\infty$ and/or $t \to +\infty$.

The mixed curve represents the lower limit of the test (1245)-(1249). Below this mixed curve the escape conditions (1245)-(1249)

Figure 175. Analysis for three equal masses

Full curves {
upper curve : limit for $\theta_1 = 90°$
middle curve : limit for $\theta_1 = 60°$ or $120°$
lower curve : limit for $\theta_1 = 0°$ or $180°$
$\tan \theta_1 = \sqrt{\mu} \, \rho / \rho'$
}

Above the full curves at least one of the escape conditions (1245 – 1249) is satisfied; below the mixed curve they cannot be satisfied.

Unbounded motions

(The isolated mass escape when $t \to -\infty$ and/or $t \to +\infty$, with for ever $(R^2)'' \geq 2\mu/R$)

Possibility of bounded motions

elliptic Euler motions

elliptic Lagrange motions $(\mu = 1)$

Asymptote of the full and mixed curves
$(\rho/2\rho + \rho\rho'^2/2\mu = (2/3)^{3/2} = 0.544\,33)$

$(\mu = 5\sqrt{2/6} = 1.178\,51)$

$\dfrac{\rho}{2\rho} + \dfrac{\rho\rho'^2}{2\mu}$

1.5

1

0.5

0

Interplay

0.5   1   1.5   2   2.5

$\dfrac{e}{a}$

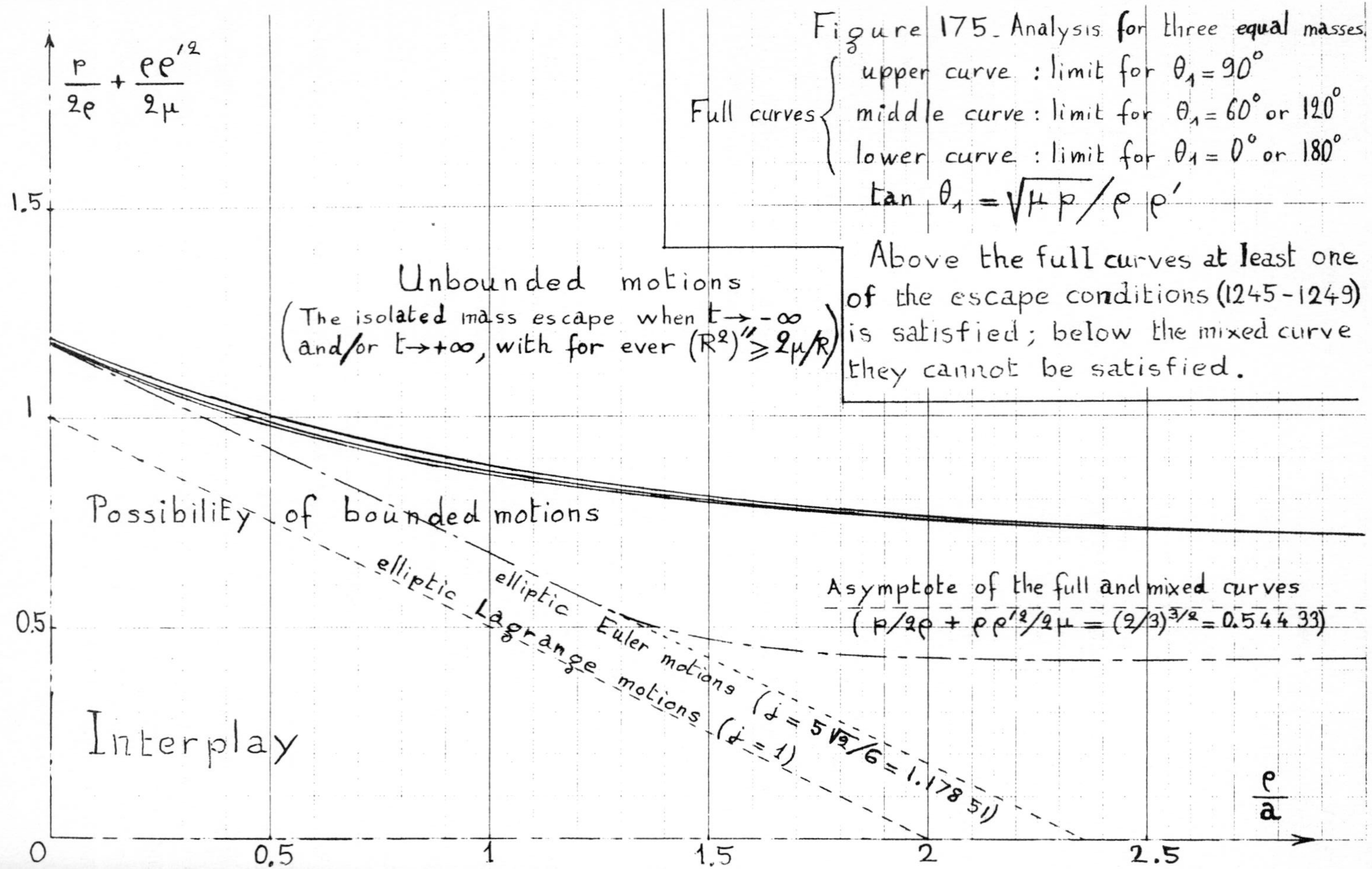

cannot be satisfied (no isolation and/or $V_1 + W < \sqrt{\mu(2 + 2\varepsilon)/R}$ for all r/R ratios).

The two inclined segments in dotted line (or dot-dash line) correspond to the elliptic and periodic Euler and Lagrange motions. The former come very close to the limit of the zone of unbounded motions as shown in the following Table.

TABLE XI

Ordinates of the full limit curves of Fig. 175 and of the segments of elliptic Euler and Lagrange motions.

$\rho/a =$	0	0.5	1	2	∞
Curve $\theta_1 = 90°$	1.187893	1.001482	0.880094	0.761944	0.544331 = $(2/3)^{3/2}$
Curve $\theta_1 = 0°$ or $\theta_1 = 180°$	1.178617	0.979468	0.860645	0.752209	0.544331 = $(2/3)^{3/2}$
Euler motion	1.178511 = $5\sqrt{2}/6$	0.928511	0.678511	0.178511	
Lagrange motion	1	0.75	0.5	0	

We can notice the following points :

I) The parameters (ρ/a ; $p/2\rho + \rho\rho'^2/2\mu$) or the equivalent parameters (ρ/a ; j) are excellent for a general presentation. Indeed the parameters p and ρ' then appear individually only through the angle θ_1 (with $\tan\theta_1 = \sqrt{\mu p}/\rho\rho'$) and this angle θ_1 has only a very little influence.

We can write that with these parameters the limit curves (either full or mixed) are almost independent of the integrals of motion.

II) Following Szebehely's classification of three-body motions (Ref. 237), we will describe as "zone of interplay" the zone without isolated body. This zone goes up to the segment of Euler motions.

III) The proximity of the full limit curves and the segment of elliptic-Euler motions shows both the excellence of the test and the smallness of the zone of bounded or oscillatory motions.

This smallness is emphasized by the relation (991) of Section 1.7.3.2, that is :

$$\left.\begin{array}{l} t_2 - t_1 = T/2 \quad \text{implies} \quad \rho(t_2) + \rho(t_1) \geq 2a \\ \text{with : } T = 2\pi\sqrt{a^3/\mu} \end{array}\right\} \qquad (1257)$$

Hence no motion can remain in Fig. 175 in the vicinity of the origin. In all intervals of time of duration T/2 the ratio ρ/a has a maximum at least equal to one.

Figure 176 and its enlarged portion in Fig. 177 present, in the (ρ/a ; j) axes, a particular case of Fig. 175, the case of three equal masses for which the ratio p/a is 8/9. For instance $G = 1$; $m_1 = m_2 = m_3 = 1$; $c^2h = -4$.

Since $j = (\rho/2a) + (p/2\rho) + (\rho\rho'^2/2\mu)$ it is easy to go from Fig. 175 to Figs 176, 177 and we found successively :

I) A straight line OA where $j = \rho/2a$.

II) A branch of hyperbola BCDEE'D'F along which $j = (\rho/2a) + (p/2\rho)$ that is here : $j = (\rho/2a) + (4a/9\rho)$.

The j-axis and the straight line OA are the asymptotes of this branch of hyperbola.

Along that branch ρ' is zero and the ($\rho/a, j$) motion occurs on or above that branch with :

$$d\rho/dt = \rho' = \pm\sqrt{(\mu/\rho)\{2j - (\rho/a) - (p/\rho)\}} \qquad (1258)$$

III) The horizontal dotted segments DD' and EE' correspond to elliptic Euler and Lagrange motions (with $j = 5\sqrt{2}/6$ and $j = 1$) for the particular value $p/a = 8/9$ of interest.

IV) The two dotted curves GG' and HH' correspond to the upper and lower full limit curves of Fig. 175 (with $\theta_1 = 90°$ and $\theta_1 = 0°$ or $180°$).

V) The full curve C"C'H' (see enlargement in Fig. 177) between the two dotted curves is the limit curve for the escape conditions (1255) and $p/a = 8/9$, with a $\sin\theta_1$ decreasing from one to zero along that curve.

VI) The mixed limit is made up from a horizontal segment C'I tangent to the previous curve C"C'H' and a mixed curve IJ corresponding to the mixed curve of Fig. 175.

Notice that C'I is above the segment of elliptic Euler motions. It corresponds to $j = 1.235897$ and not to the mixed segment of Fig. 175.

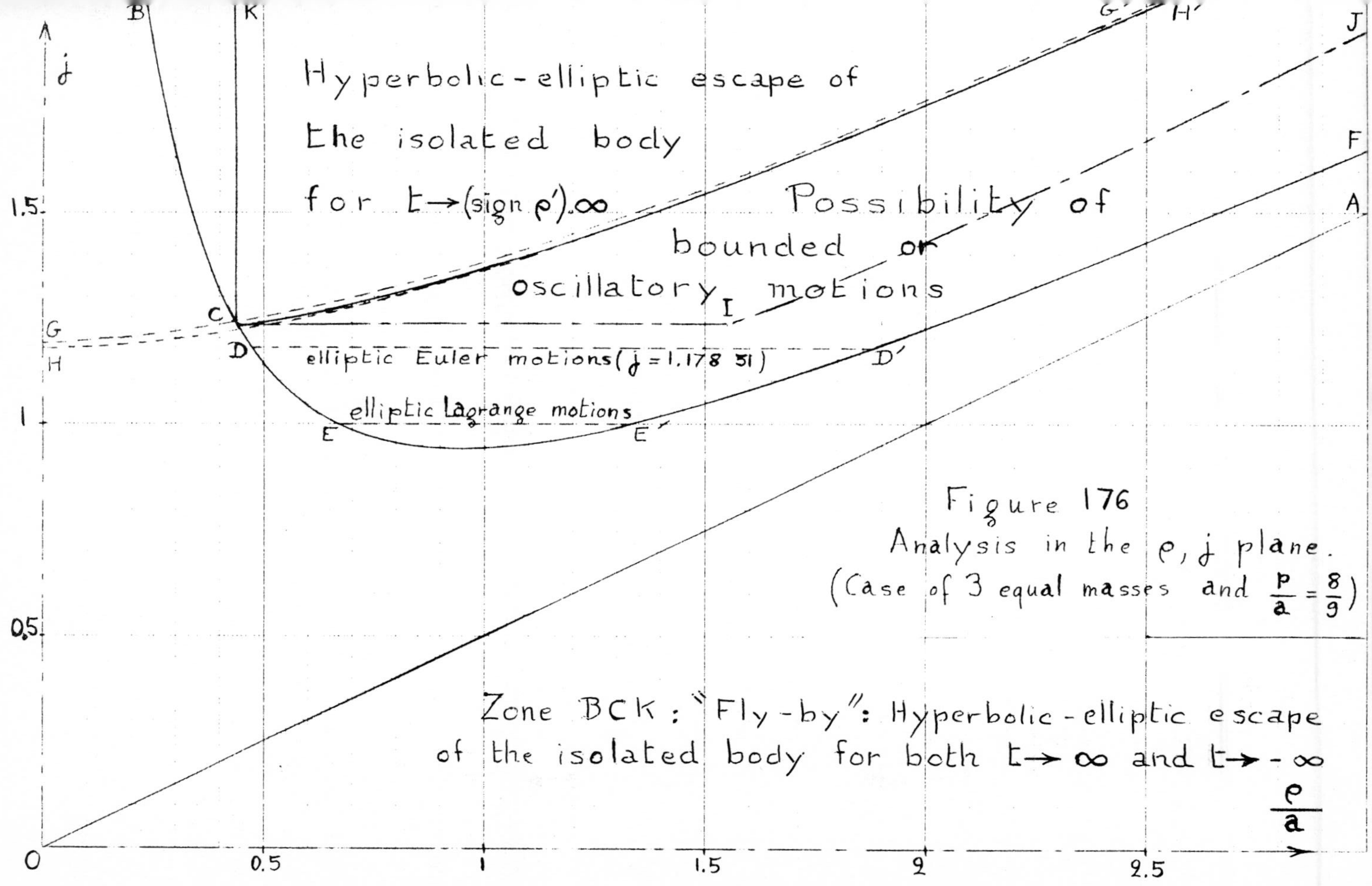

Figure 176
Analysis in the ρ, j plane.
(Case of 3 equal masses and $\frac{p}{a} = \frac{8}{9}$)

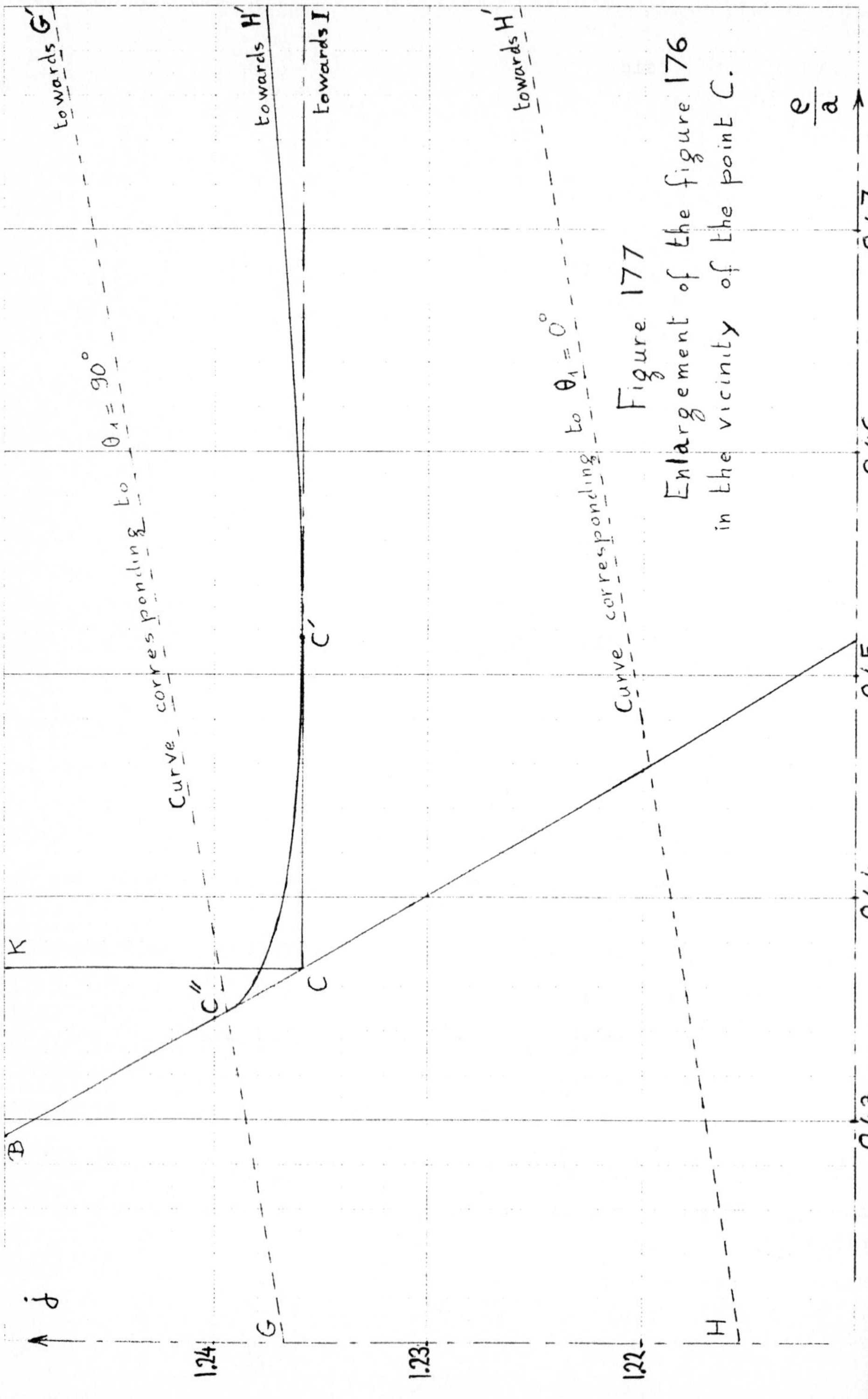

Figure 177
Enlargement of the figure 176 in the vicinity of the point C.

VII) Finally we come to the vertical full half straight line CK on which $\rho/a = 0.436802$.

Let us now consider the motion in this (ρ,j) plane.

The derivative $d\rho/dt$ is given in (1258) while dj/dt was given in (903) :

$$dj/dt = (\rho'/\rho)\big((\rho/\nu) - j\big) \qquad (1259)$$

with the lower bounds (897) and (902) :

$$\rho/\nu \geqq 1 \quad ; \quad \rho/\nu \geqq j \qquad (1260)$$

and then :

$$dj/d\rho \geqq \sup\left\{\frac{1-j}{\rho} \; ; \; 0\right\} \qquad (1261)$$

Thus the point $(\rho/a;j)$ always move according to (1258) on or above the branch of hyperbola BCDEE'D'F on curves of non negative slope with change of sign ρ', and direction of motion, at points X, Y, Z on the hyperbola (Fig. 178).

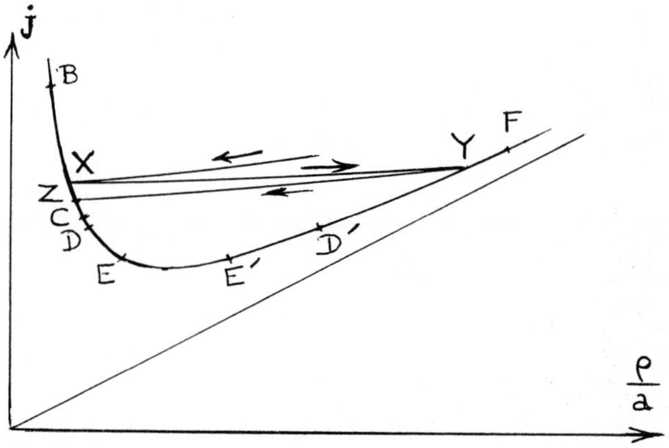

Fig. 178. Motions in the $(\rho/a;j)$ plane.

We are then led to the following results.

I) For points on or above the limit curve BC"C'H' of Figs 176-177 the escape conditions (1255) are satisfied and the motion is unbounded.

II) For points below IJ the escape conditions (1245)-(1249) are never satisfied.

III) These esape conditions (1245)-(1249) remain satisfied during the totality of the escape as soon as they are satisfied at some time, hence the orbits starting with positive ρ' above the curve BC"C'H' must remain above IJ in the future. These orbits then have forever an increasing ρ and a non-decreasing j, they have an escape in the future above the CIJ mixed line.

The same occurs symmetrically for negative ρ' and decreasing time, and then increasing ρ.

IV) Orbits starting in the small curvi-linear triangle CC'C" of Fig. 177 also have these properties : their branch with increasing ρ always cross the escape zone BC"C'H'.

V) Finally the zone BCK corresponds to hyperbolic-elliptic escapes for both $t \to +\infty$ and $t \to -\infty$ with only one minimum of ρ, along BC.

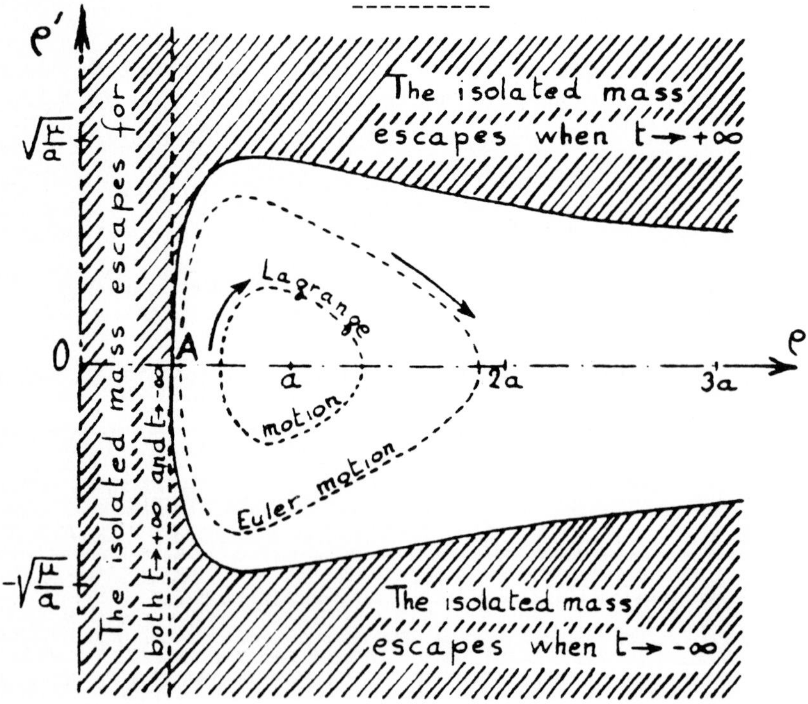

Fig. 179. Analysis of three-body systems for which $m_1 = m_2 = m_3$ and $p/a = 8/9$.
All bounded motions remain in the non-shaded region.

Figure 179 and its enlargement in Fig. 180 are exactly Figs 177-178 presented in the (ρ,ρ') half-plane with the transformation (1258).

The full lines of Fig. 180 divide that figure into six zones.

The zone BCK of Figs 177-178 with $\rho/a \leq 0.436802$ corresponds to the zone VI of Fig. 180.

The zone KCC'H' corresponds to the zone I (if $\rho' > 0$) and to the zone V (if $\rho' < 0$).

The zone between C'H' and C'IJ corresponds to the zone II (if $\rho' > 0$) and to the zone IV (if $\rho' < 0$).

Finally the zone below the mixed line CC'IJ corresponds to the zone III that contains the elliptic Euler motions and the elliptic Lagrange motions.

The results already described for Figs 177-178 then become the following :

I) Orbits without hyperbolic-elliptic escapes remain forever in the zones II, III, IV (bounded orbits, oscillatory orbits, orbits with parabolic-elliptic escapes).

II) Orbits entering the zone VI have one and only one minimum of ρ (somewhere on the segment OA).

These orbits have hyperbolic-elliptic escapes at both ends, without exchange, they arrive through the zone V and possibly the zone IV with a negative and bounded original $\rho'_{-\infty}$ and they escape through the zone I and possibly the zone II with a positive and bounded final ρ'_{∞}.

The passage in the zone VI has always a short duration an upper bound of which can be given by (1258) and (1261) :

$$(\Delta t)_{VI} \leq 0.316705\sqrt{a^3/\mu} \tag{1262}$$

III) All orbits enter either the zone VI or the zone III but none can go from either of these two zones to the other.

IV) Orbits entering the zone III have the following properties :

A) They can remain for all time in the zone III as Euler and Lagrange motions. If not between two passages in zone III they remain either in zone II (if $\rho' > 0$) or in zone IV (if $\rho' < 0$).

B) If there is a final passage in zone III the length ρ will remain forever increasing after that passage and the isolated mass will escape either parabolically (if ρ,ρ' remains in zone II) or hyperbolically (if ρ,ρ' reaches the zone I).

Symmetrically before the first passage in zone III the isolated mass arrives from infinity either parabolically (arrival through the zone IV) or hyperbolically (arrival through the zone V and possibly IV).

These properties are summarized by the following graphs :

I) Graph of orbits entering the zone VI :

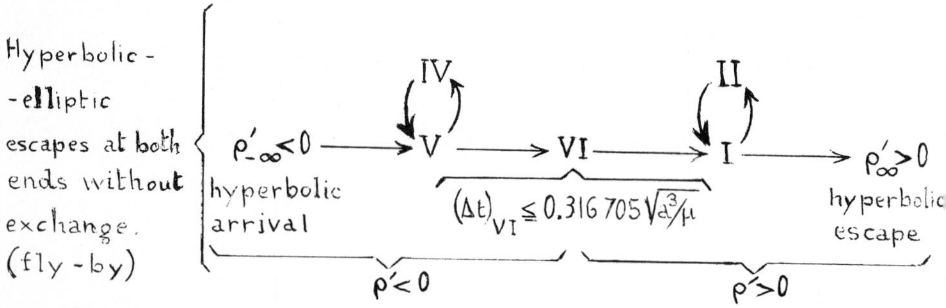

II) Graph of orbits entering the zone III :

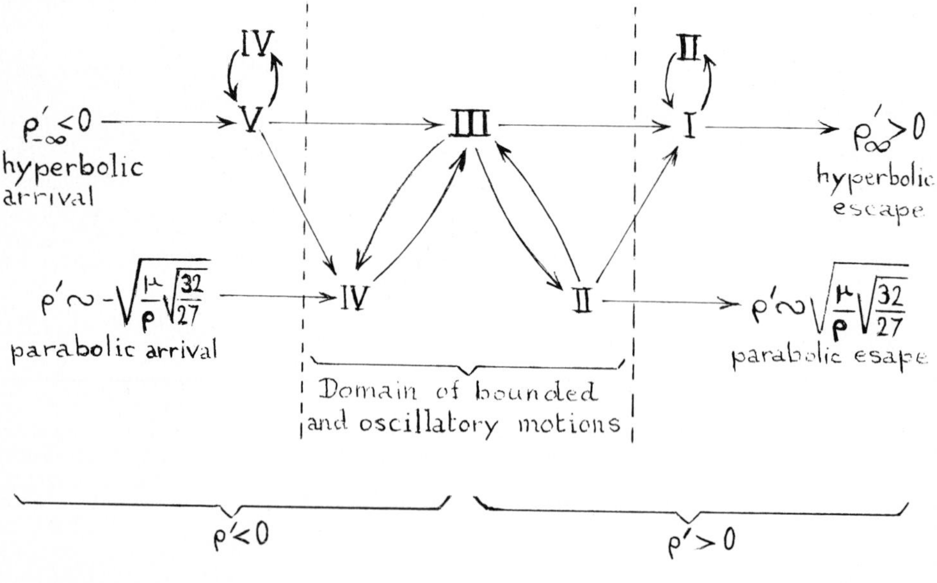

Orbits with escapes at both ends can now be either with or without exchange. The application of the test of Ref. 230 shows that these escapes have a limited velocity :

$$-2.29904\sqrt{\mu/a} \leq \rho'_{-\infty} \leq 0 \leq \rho'_{\infty} \leq 2.29904\sqrt{\mu/a} \qquad (1263)$$

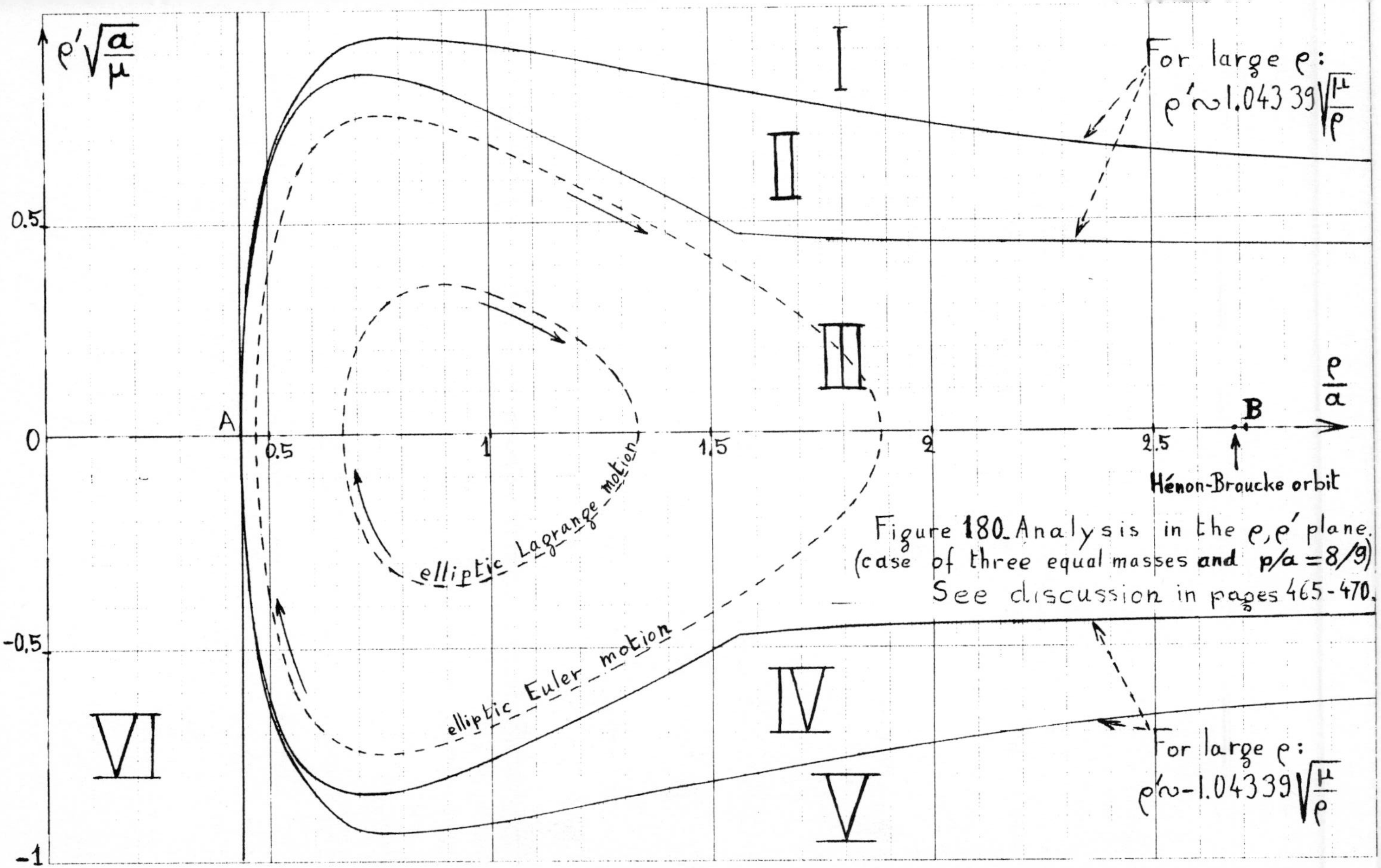

Figure **180**. Analysis in the ρ, ρ' plane. (case of three equal masses **and** $p/a = 8/9$) See discussion in pages 465-470.

This second graph is not exactly true because in the very particular case of a parabolic-elliptic escape with a binary having an angular momentum going to zero, the escape is essentially in the zone II (or IV in the past) but with an infinite number of very short passages in zone III (one per revolution of the binary).

It seems that bounded orbits can be met in almost all zones II, III, IV. The analysis of suitable periodic orbits can certainly improve this question.

Many other qualitatives results can be obtained, for instance :

For the masses and integrals of motion of Figs 179-180 the quantity d^2R/dt^2 is always negative when $\rho/a > 2.709629$.

The limit case (point B) corresponds to a plane motion with $r_{13} = r_{23}$; $R' = \rho' = r' = 0$; $r/a = 0.443621$; $R/a = 3.296291$.

Hence :

I) All orbits have at least one local minimum of ρ on the segment OB of Fig. 180.

II) All bounded and oscillatory orbits have an infinite number of local minimums of ρ on the segment AB. They always come back to small sizes.

III) We have seen in Section 11.7.3.2 that two successive local maximums and minimums of ρ satisfy :

$$\rho_m + \rho_M \geqq 2a \quad ; \quad \rho_m\rho_M \geqq ap \tag{1264}$$

hence here, since $p < a$, the greatest lower bound of local maximums is a.

On the contrary the limit given by the point B is only an upper bound of the absolute minimum of ρ for the orbits of Fig. 180. However that upper bound is very near to the least upper bound, indeed the corresponding "Hénon-Broucke orbit" or "pseudo-circular-retrograde orbit" (Section 10.8.1) has a minimum ρ_m equal to 2.684419 a, which is only a little less than at point B.

This orbit is presented in Fig. 181, it seems that it corresponds to the least upper bound.

Let us close this qualitative analysis of the three-equal-mass systems by considerations on triple close approaches that correspond to very small values of ρ.

In the case of Fig. 180, i.e. $m_1 = m_2 = m_3$, $p/a = 8/9$; either the orbit of interest enter the zone III and ρ is for all times larger than $\rho(A)$ that is 0.436802 a, or the orbit enter the zone VI

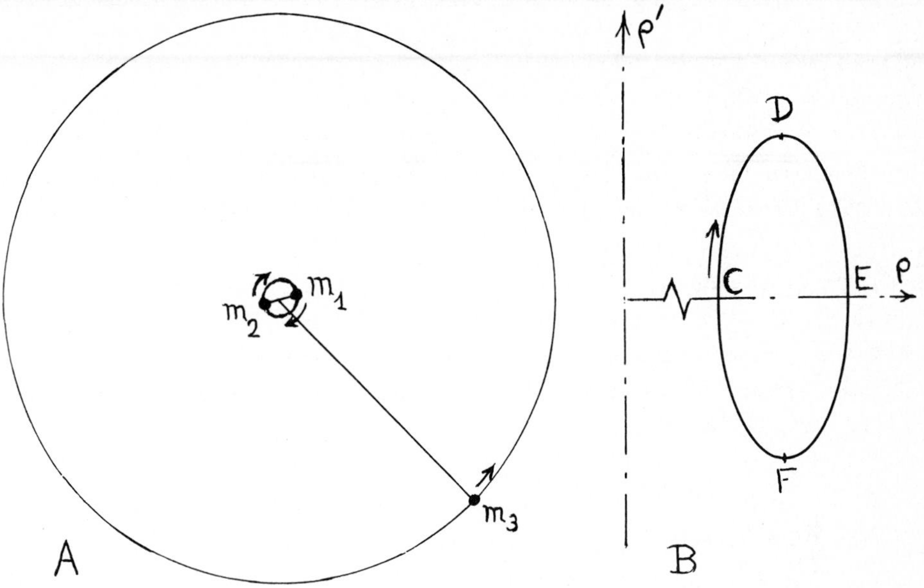

Fig. 181. The pseudo-circular retrograd orbit (or Hénon-Broucke orbit) for $m_1 = m_2 = m_3$ and $p/a = 8/9$:

A) In the non rotating set of axes of the center of mass of the small binary.

B) In the ρ, ρ' plane (a very small quasi-ellipse).

The point C of the ρ, ρ' plane corresponds to the collinear figure, **R** is maximum, r, ρ and ν are minimum. The point E corresponds to the isosceles figure ($r_{13} = r_{23}$), **R** is minimum, r, ρ and ν are maximum :

	R/a	r/a	ρ/a	ν/a
At C :	3.267690	0.418507	2.684419	0.998661
At E :	3.267648	0.419257	2.684444	1.001342

The maximum and minimum of ρ' are very small :

$$\rho'_D = -\rho'_F = 0.00007848\sqrt{\mu/a}$$

The period in the (ρ, ρ') plane is $0.9875\sqrt{a^3/\mu}$ but the three bodies come back to the same relative positions after twice that period.

The smallness of the variations of **R** and r justifies the expression "pseudo-circular retrograde orbit".

and ρ has only one minimum, ρ_m, that corresponds to the minimum j_m of j :

$$(\rho_m/2a) + (p/2\rho_m) = j_m \leqq j \qquad (1265)$$

Hence in the two cases the absolute minimum ρ_m of ρ satisfies the following in terms of the initial conditions :

$$\rho_m/a \geqq \inf\left\{0.436802 \; ; \; j - \sqrt{j^2 - (8/9)}\right\} \qquad (1266)$$

This result will be extended to the other values of the ratio p/a in Section 11.7.10.5 and to n-body systems of equal or unequal masses in section 11.8.

II)- Systems with unequal masses

The general analysis of Fig. 175, valid for all values of the angular momentum and all negative values of the energy, has been extended to the following pairs of mass ratios, with $m_A \leqq m_B \leqq m_C$:

m_A/m_B	1/3	0.1	0.001	0	1/3	0.1	1	1	1	1	0.5
m_B/m_C	1/3	0.1	arbitrary		1	1	1/3	0.1	0.001	0	0
Fig.	182	183	184	185	186	187	188	189	190	191	192

With again using the coordinates ρ/a and $(p/2\rho) + (\rho\rho'^2/2\mu)$ these eleven figures present only the segment of Lagrangian motion (at the same place on all figures), the three segments of Eulerian motion at the limit of the zone of interplay and the highest limit curves above which all motions have a hyperbolic-elliptic escape in the past and/or in the future (limit curves for $\theta_1 = 90°$)**.

If all the three masses are unequal there are three such limit curves, according to the isolated mass. Their horizontal asymptotes are drawn on the right in mixed lines ; for an isolated mass m_3 the corresponding ordinate is :

$$\{p/2\rho + \rho\rho'^2/2\mu\}_\infty = \{m_3(m_1 + m_2)/M^*\}^{3/2} \qquad (1267)$$

**With their coordinates the figures 182-192 are almost insensitive to the angle θ_1 and thus to the integrals of motion.

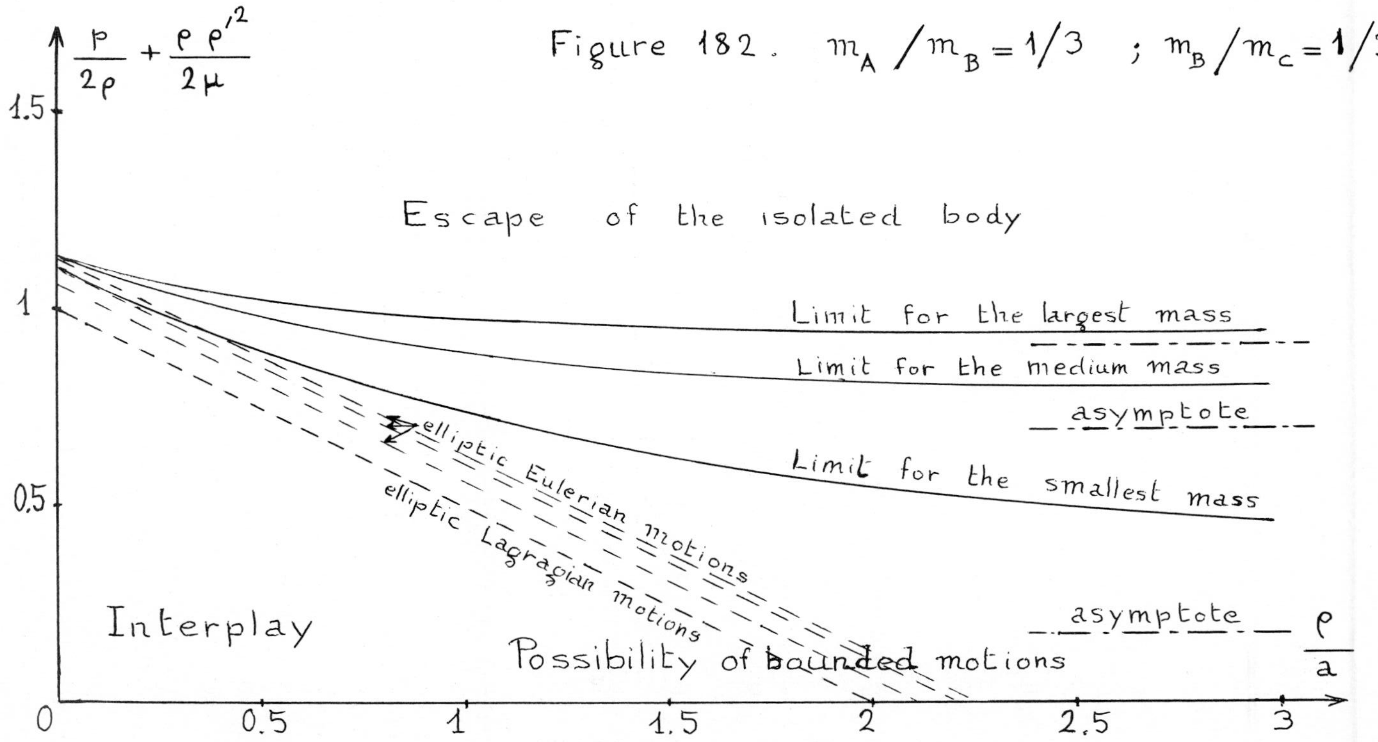
$\frac{p}{2\rho} + \frac{\rho \rho'^2}{2\mu}$
1.5
1
0.5
0
Escape of the isolated body
Limit for the largest mass
Limit for the medium mass
asymptote
Limit for the smallest mass
elliptic Eulerian motions
elliptic Lagragian motions
asymptote
Interplay
Possibility of bounded motions
$\frac{\rho}{a}$
0.5
1
1.5
2
2.5
3

Figure 182. $m_A / m_B = 1/3$; $m_B / m_C = 1/3$

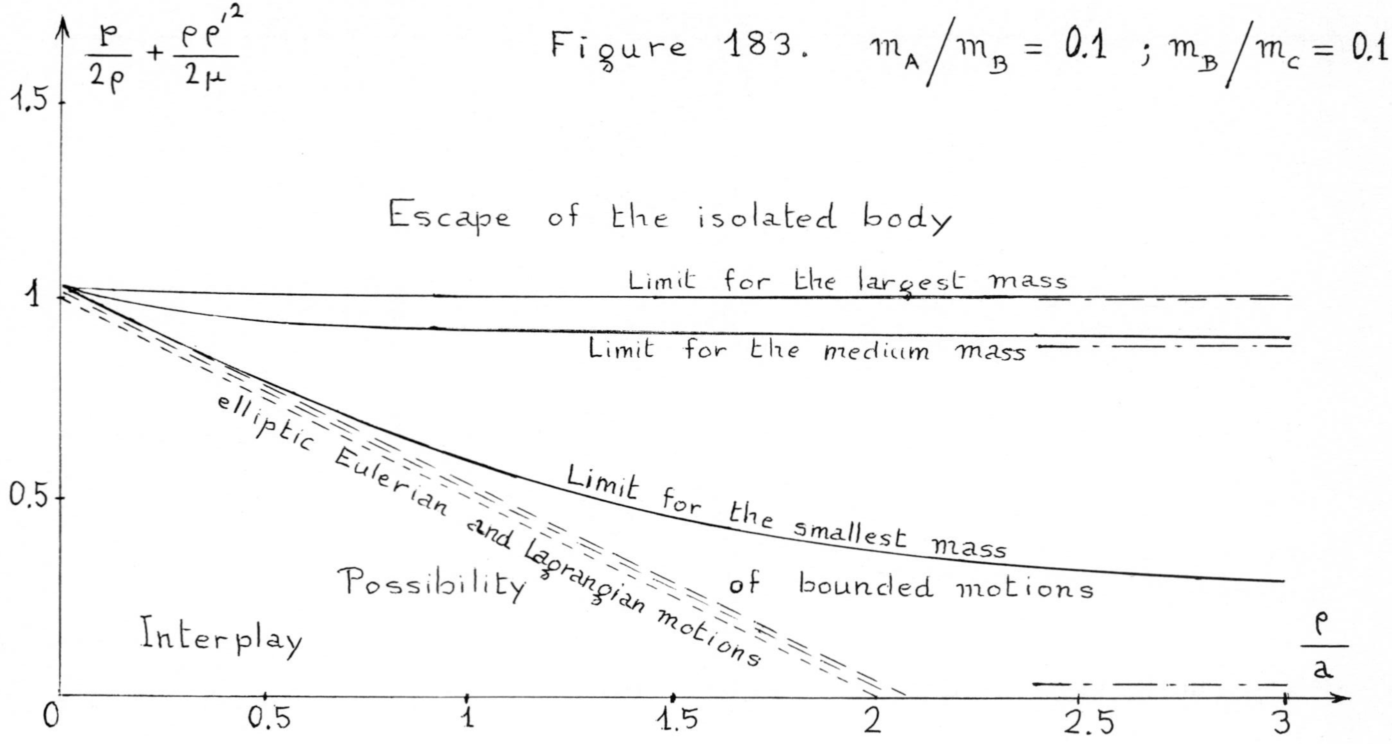
$\frac{P}{2\rho} + \frac{\rho\rho'^2}{2\mu}$
1.5
1
0.5
0
0.5
1
1.5
2
2.5
3
$\frac{\rho}{a}$
Escape of the isolated body
Limit for the largest mass
Limit for the medium mass
Limit for the smallest mass
elliptic Eulerian and Lagrangian motions
Possibility of bounded motions
Interplay

Figure 183. $m_A/m_B = 0.1$; $m_B/m_C = 0.1$

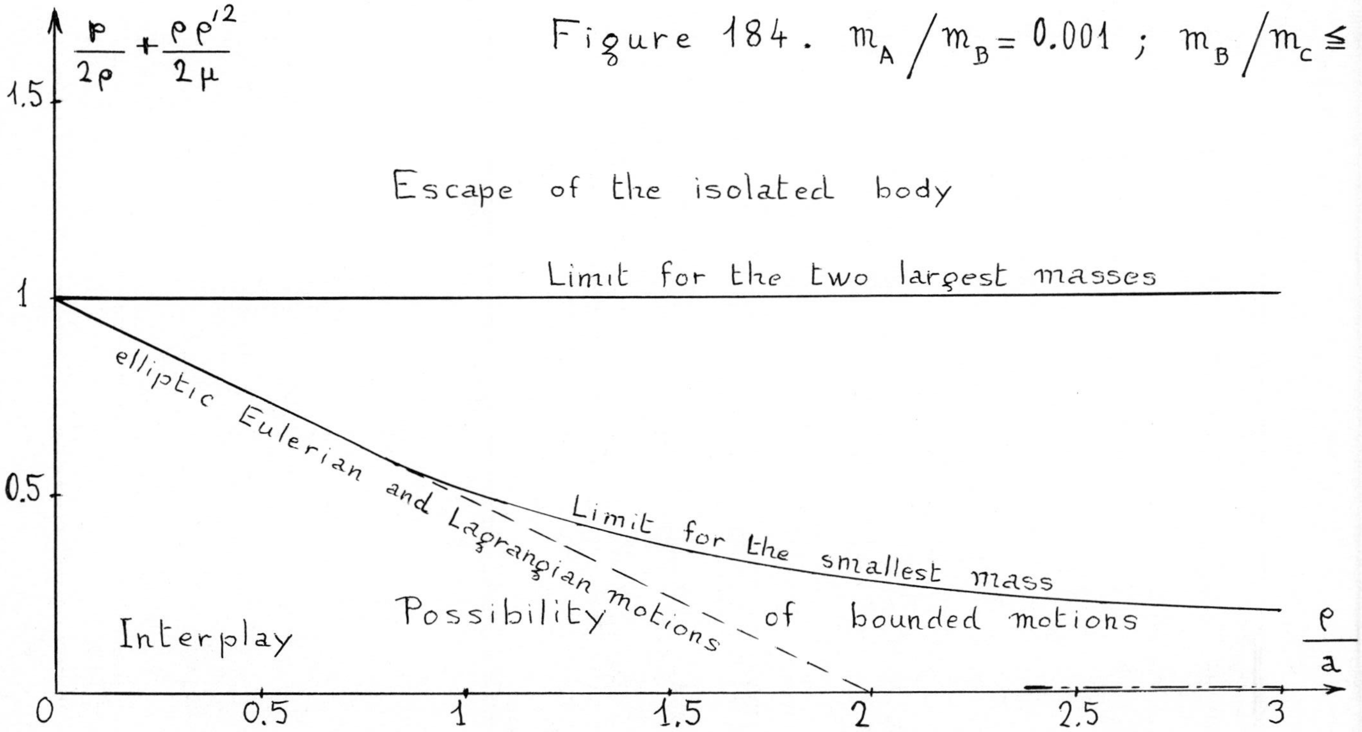

Figure 184. $m_A / m_B = 0.001$; $m_B / m_C \leq 1$

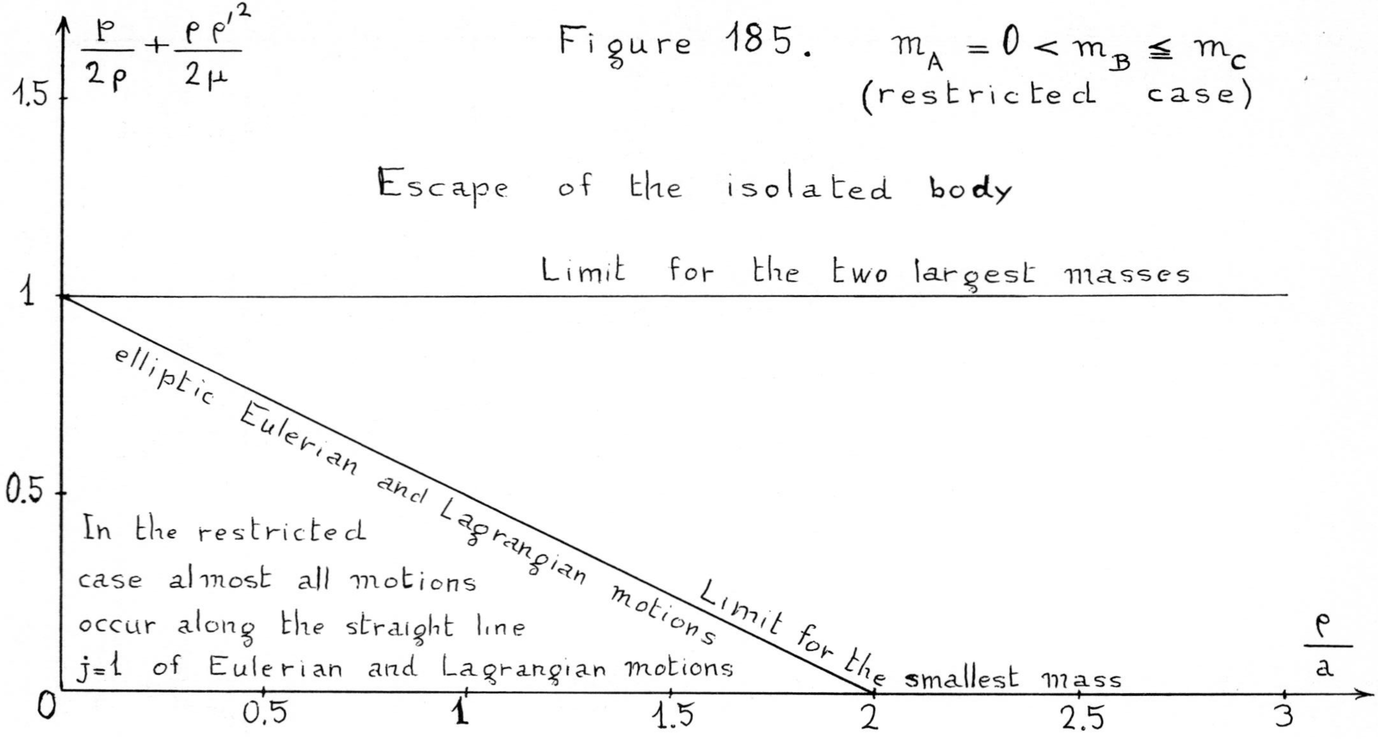

Figure 185. $m_A = 0 < m_B \leqq m_C$ (restricted case)

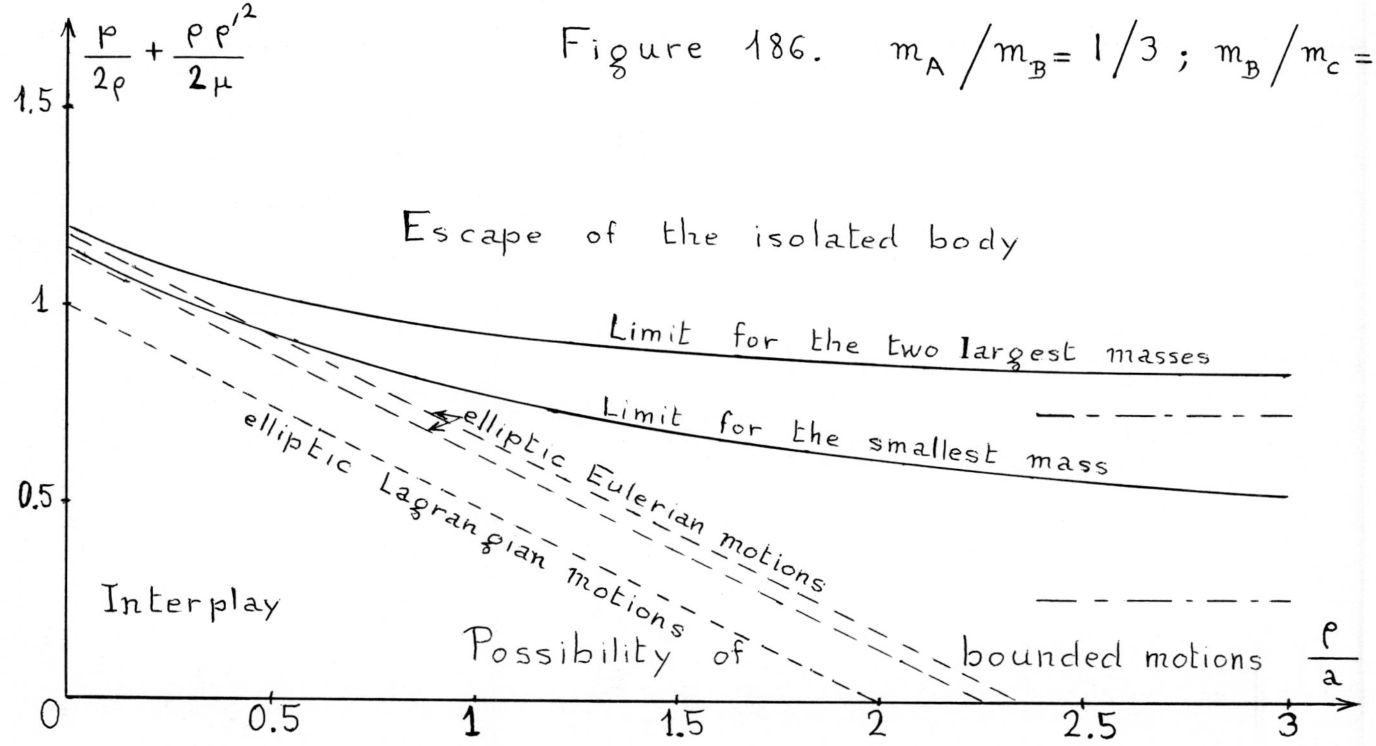

Figure 186. $m_A / m_B = 1/3$; $m_B / m_C = 1$

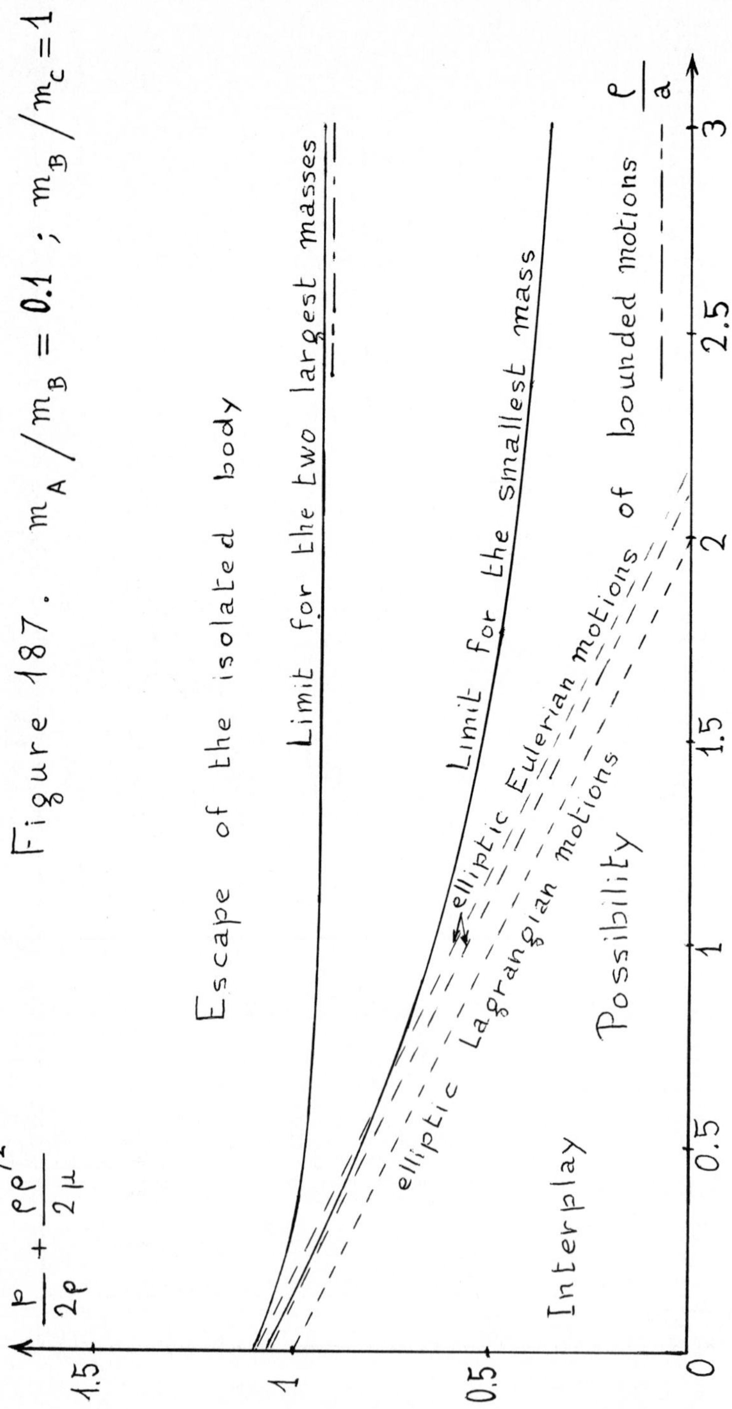

Figure 187. $m_A/m_B = 0.1$; $m_B/m_C = 1$

Figure 188. $m_A / m_B = 1$; $m_B / m_C = 1/3$

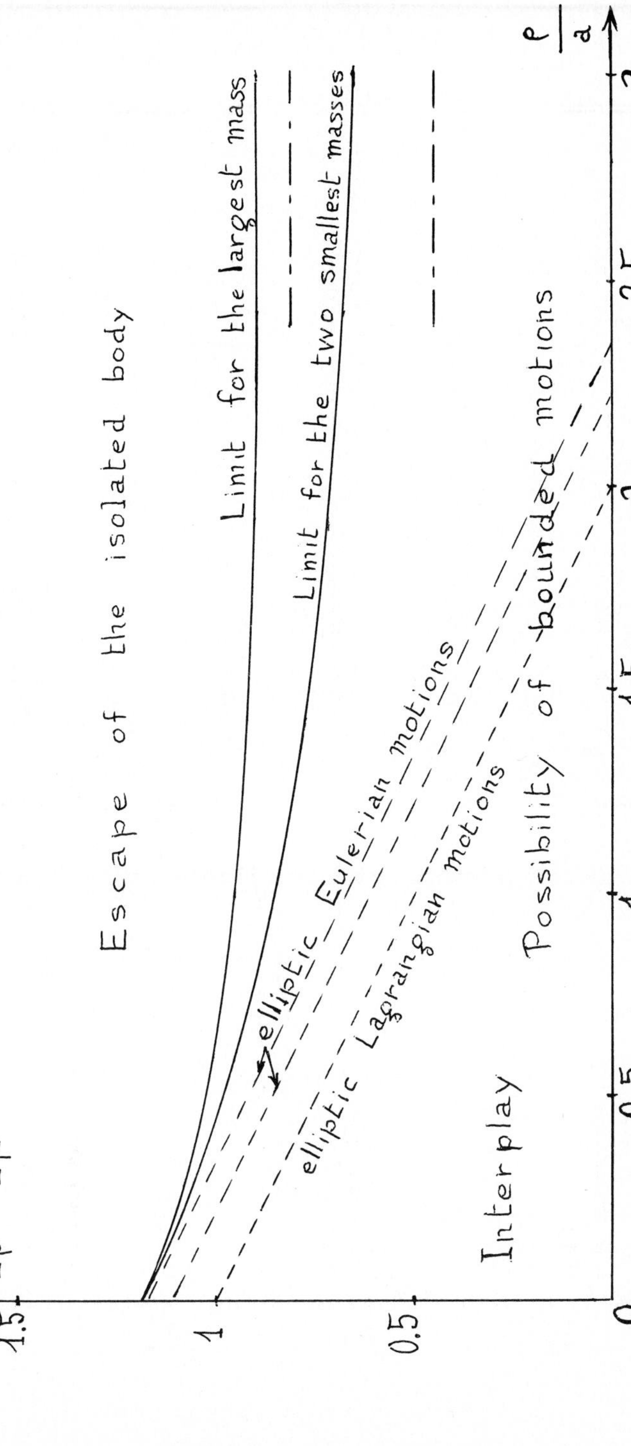

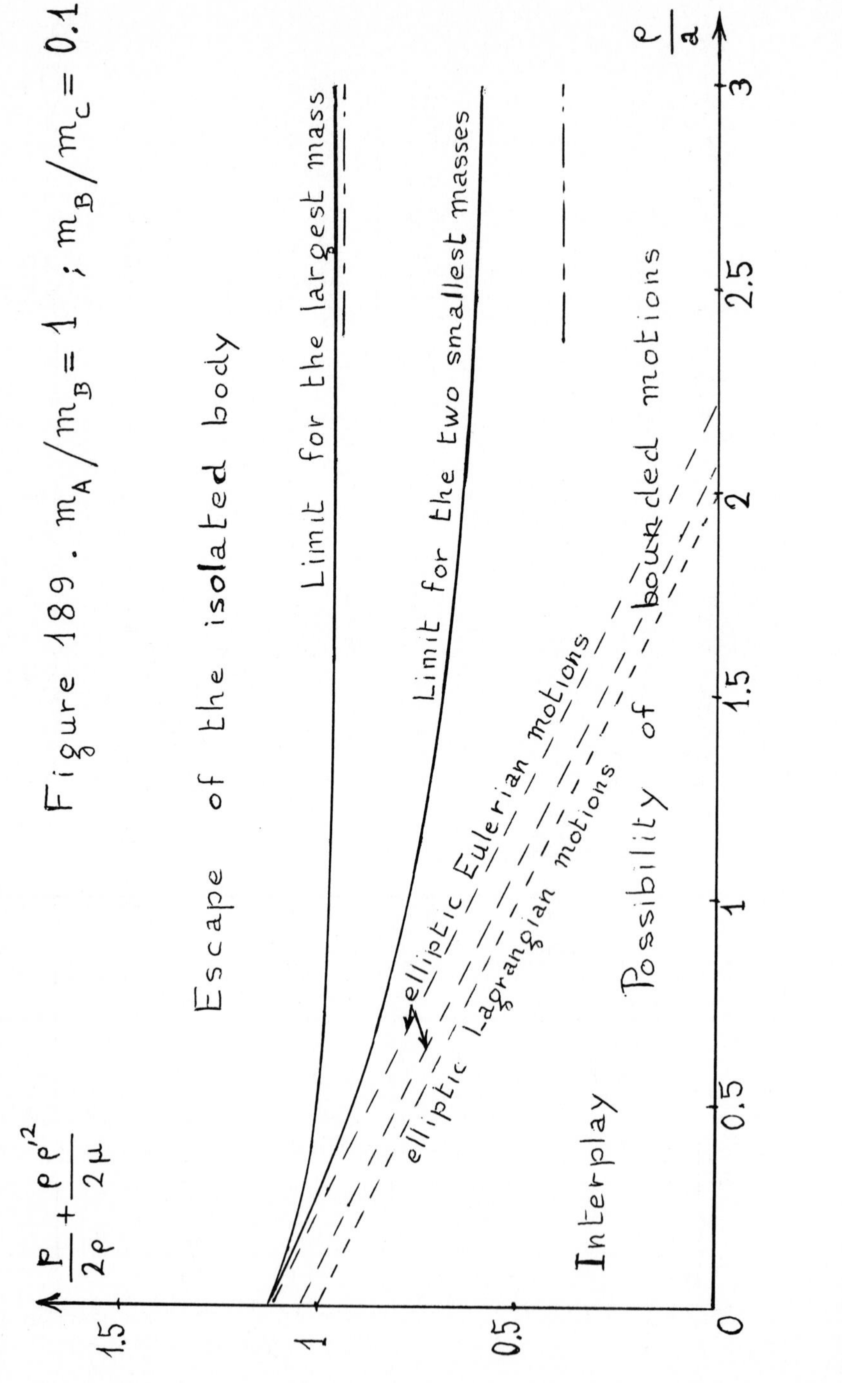

Figure 189. $m_A/m_B = 1$; $m_B/m_C = 0.1$

Figure 190. $m_A / m_B = 1$; $m_B / m_C = 0.001$

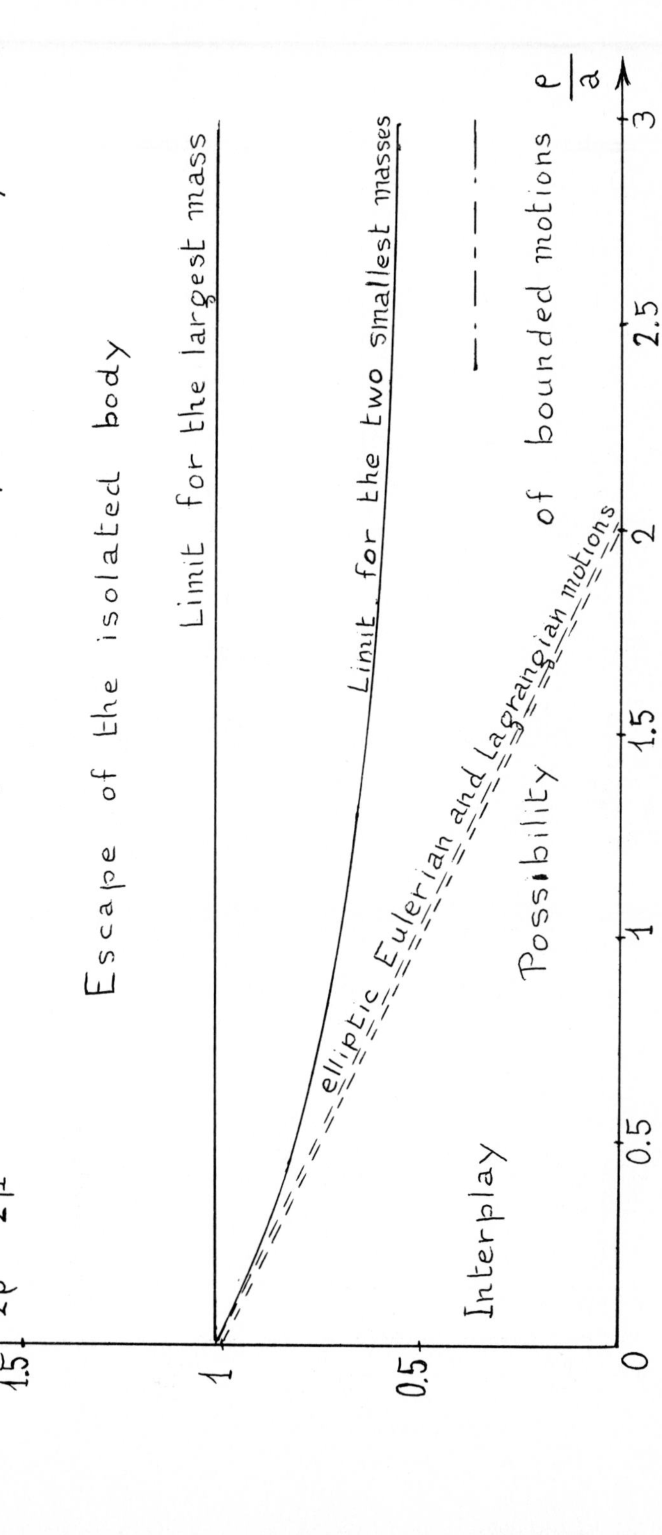

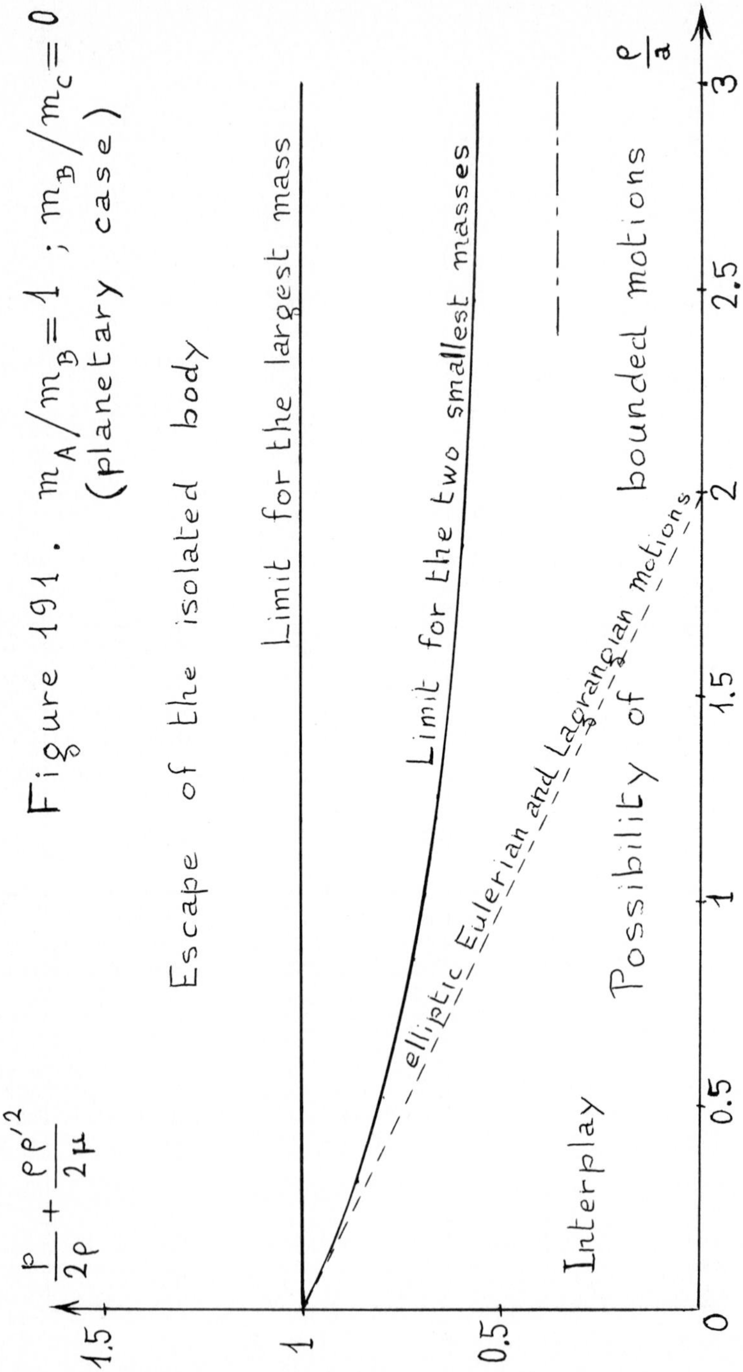

Figure 191. $m_A/m_B = 1$; $m_B/m_C = 0$ (planetary case)

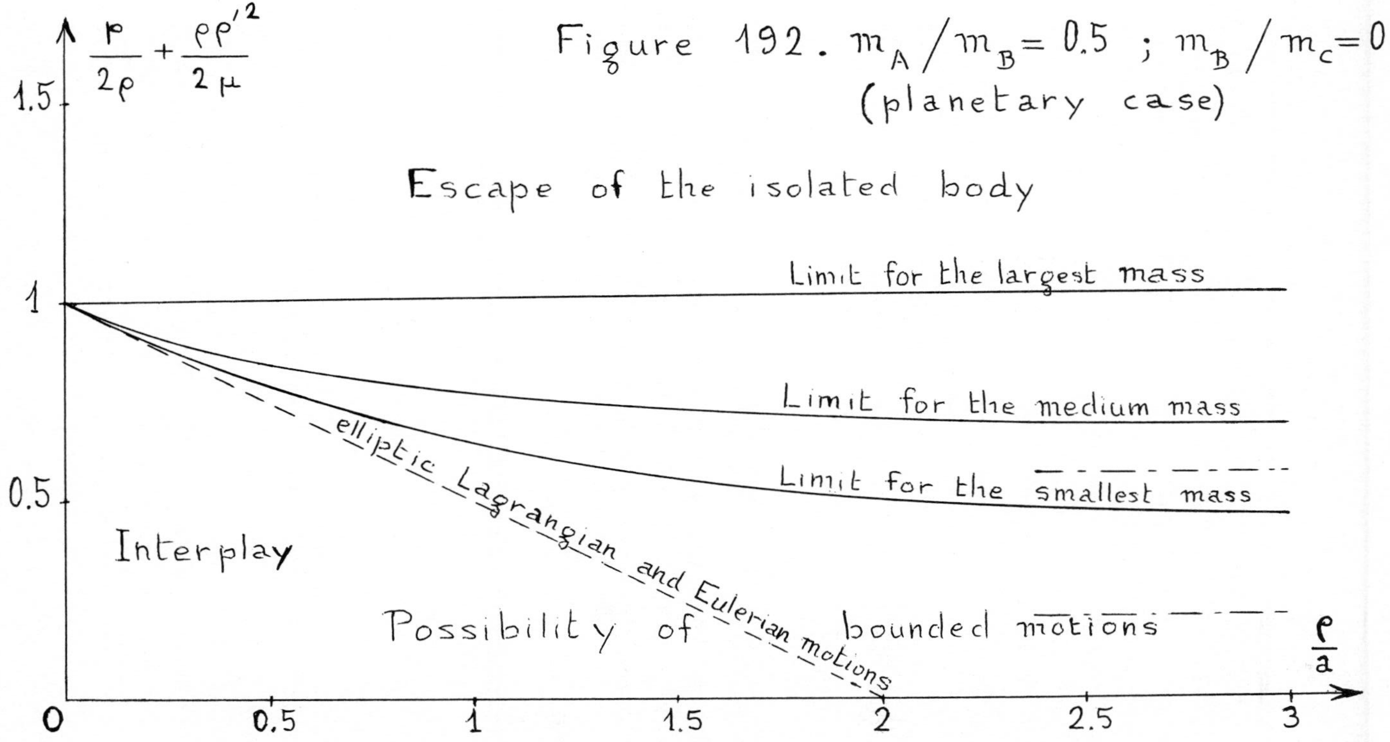

Figure 192. $m_A / m_B = 0.5$; $m_B / m_C = 0$
(planetary case)

These figures lead to two main results, the first is logical and the second unexpected.

I) The stability is a direct function of the mass : the smallest mass has a large zone of escape and a large risk of being expelled, whereas the largest mass is very stable. It is very easy to disrupt a light binary by the passage of heavy body and it is almost hopeless to try to disrupt a heavy binary by the passage of a light body... All this agree with the numerical results of ref 136,171.

II) The zones where bounded three-body motions are possible may seem large in the figures but :

A) These zones also contain all oscillatory motions and all motions with at most parabolic escapes... and many motions with faster escapes.

B) No motion can remain in the vicinity of the origin and, even for systems with unequal masses (Section 11.7.3.2) :

$$\left.\begin{array}{l} t_2 - t_1 = \pm T/2 \quad \text{implies} \quad \rho(t_1) + \rho(t_2) \geqq 2a \\ \text{with } T = 2\pi\sqrt{a^3/\mu} \end{array}\right\} \qquad (1268)$$

The true zone of three-body bounded motions turns out to be much smaller than was widely expected and, except for some very particular cases, the stability of a triple system requires a well defined hierarchy with a small binary and a distant "third mass".

Three-body systems survive large perturbations with difficulty.

We notice some final points.

A) The figures are more sensitive to the ratio m_A/m_B of the two smallest masses than to the ratio m_B/m_C of the two largest masses. That ratio m_B/m_C has almost no influence in Fig. 184 and no influence at all in Fig. 185 while Fig. 191 and 192 are very different.

B) In all the figures the limit curves approach very near to the corresponding segment of elliptic Eulerian motion (we recall that, for the two largest masses, the isolation begins at the highest Eulerian motion but for the smallest mass it begins at the second highest as clearly shown by Figs 182, 186, 187). Since all elliptic Eulerian motions are bounded this property confirm the efficiency of the test.

C) A surprising feature is the intersection of the limit curves of the two largest masses for small values of ρ/a, i.e. for triple close approaches (Ref. 236 for details). However this minor point seems to have no influence on the stability of the largest mass and triple close approaches generally lead to the formation of a very small binary (Ref. 38) and to the escape of either the smallest mass or the second smallest (Ref. 49).

It would be very interesting to improve and extend these analyses, especially those of Fig. 180 which gives so many qualitative results. The cases of unequal masses, these of systems with Hill-type stability may lead to some unexpected phenomena.

11.7.10.5 A survey of recent progress in tests of escape. Analysis of triple close approaches.

Sundman has demonstrated (Ref. 20) that three-body systems with a non-zero angular momentum cannot approach a triple collision.

Triple collision are characterized by a mean quadratic distance ρ equal to zero and the equation (1266) gives a simple lower bound of ρ in terms of the initial conditions when $m_1 = m_2 = m_3$ and $p/a = 8/9$. Figs 179 and 180 show that all bounded or oscillatory motions then satisfy $\rho \geqq 0.436802\mathbf{a}$ for all time.

The same analysis can be done for other values of the integrals of motion (that is for other values of the ratio p/a) and for other values of the mass ratios.

The results of the three-equal-mass case are presented in Fig. 193 and in the more classical Fig. 194 with the same curves.

Fig. 193 uses the notations of this book and presents some recent tests and their successive lower bounds of the ratio ρ/p in terms of p/a.

Fig. 194 describes a particular three-equal mass case, the case in which $G = 1$, $m_1 = m_2 = m_3 = 1$, and presents the same results in classical notations with the lower bounds of $\sqrt{I}/c^2$ in terms of $(-c^2h)$.

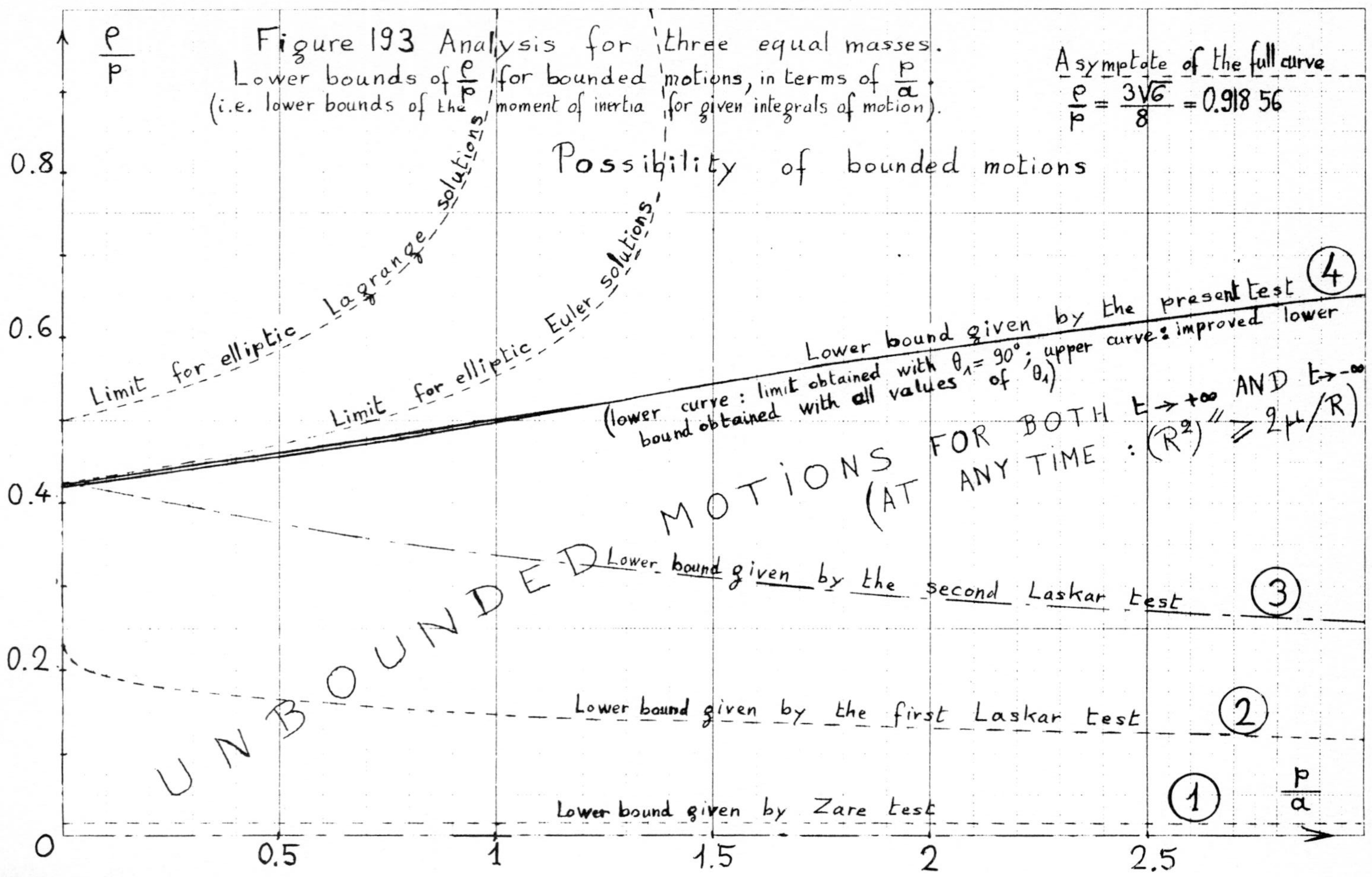

Figure 193 Analysis for three equal masses.
Lower bounds of $\frac{\rho}{p}$ for bounded motions, in terms of $\frac{p}{a}$.
(i.e. lower bounds of the moment of inertia for given integrals of motion).

We then have :

$$I = \text{semi-moment of inertia} = M^{*}\rho^{2}/2M = \rho^{2}/2$$

$$c = \text{angular momentum} = \{GM^{*}p/M\}^{1/2} = \sqrt{3p}$$

$$h = \text{energy} = -\,GM^{*}/2a = -\,3/2a$$

(I and c together give:) $$\frac{\sqrt{I}}{c^{2}} = \frac{\rho}{p}\cdot\frac{\sqrt{2}}{6}$$

(c and h together give:) $$(-\,c^{2}h) = 4.5\,\frac{p}{a}$$

The limits given by Sundman himself and by Birkhoff (Ref. 220) were very small, they only intent to give examples.

Zare (Ref. 232) gave in 1981 the first significant result (curve (1)).

The two Laskar tests (Ref. 233, 234) correspond to the curves (2) and (3).

The test of Section 11.7.10.3 gives the curve (4) and is certainly very near to the true greatest lower bound since the two mixed curves of Fig. 194 correspond to known bounded motions and are then above the greatest lower bound.

In less than ten years the progress has been very large...

Notice that :

A) The second Laskar test (curve (3)) remains the best for small values of p/a or $(-\,c^2h)$. At the ordinate-axis it reaches the limit of elliptic Euler motions with $\rho/p = 3\sqrt{2}/10$ and $\sqrt{I}/c^2 = 0.1$.

B) The two very close full curves of Fig. 193 correspond to the difference between points C" and C in Fig. 177.

This difference would have little interest if it were not so near to the greatest lower bound as shown by the enlargement presented in Fig. 195 with the limit of elliptic Euler motions above that greatest lower bound.

C) If we disregard the energy h and only consider three given masses m_1, m_2, m_3 and a given non-zero angular momentum $\vec{c}$ we can already write the following.

The bounded three-body motions corresponding to m_1, m_2, m_3, $\vec{c}$, and even the motions without hyperbolic-elliptic escape, have a positive greatest lower bound I_m of their semi-moment of inertia I. This greatest lower bound I_m is very near to the greatest lower bound I_E obtained when we restrict the analysis to elliptic Euler motions.

We always obtain $0.999\ I_E < I_m \leq I_E$ and it is conjectured that $I_m = I_E$ for all mass ratios and all angular momenta.

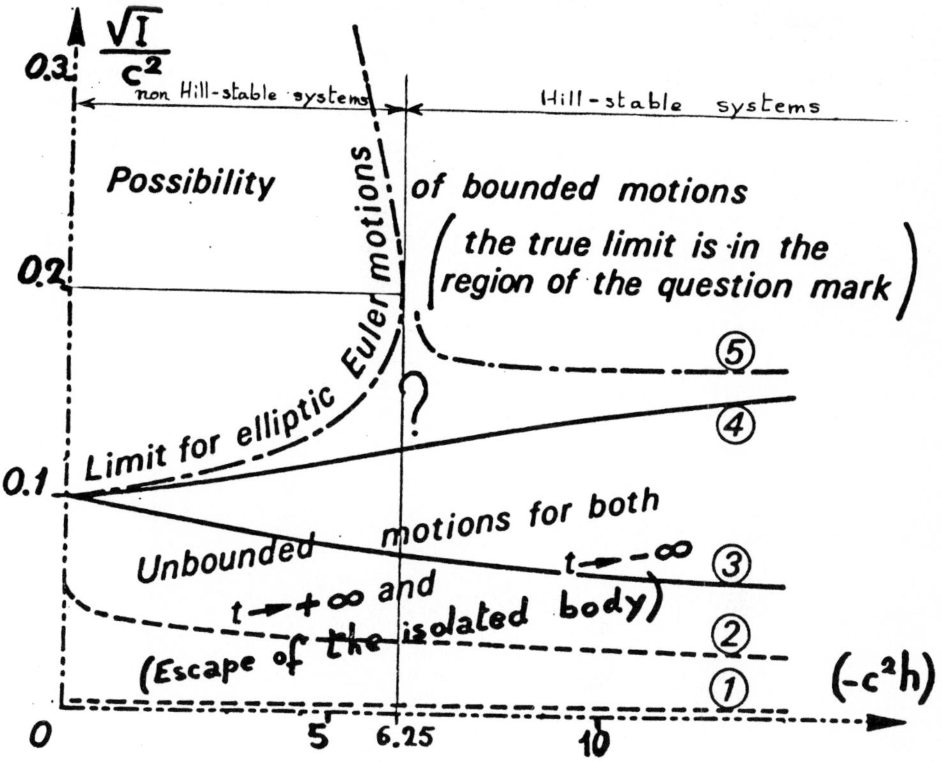

Fig. 194. Progress in tests of escape.
Case of the figure : $G = 1$; $m_1 = m_2 = m_3 = 1$.

For bounded and for oscillatory motions the energy integral h is negative and the semi-moment of inertia I has a greatest lover bound I_m function of h and of the angular momentum $\vec{c}$:

$I_m = c^4 . f(u)$, with $u = (-c^2 h)$.

The first significant lower bound (curve (1)) was given in 1981 by the Zare test (Ref. 232).

The first Laskar test ((2) and 1982) and the second Laskar test ((3) and 1984) are given in Ref. 233 and 234 .

The test of Section 11.7.10.3 give the curve (4) .

The two mixed curves correspond to bounded motions and are above the greatest lower bound.

The curves (4) and (5) have the same horizontal asymptotic on the right at the ordinate $\sqrt{3}/8$.

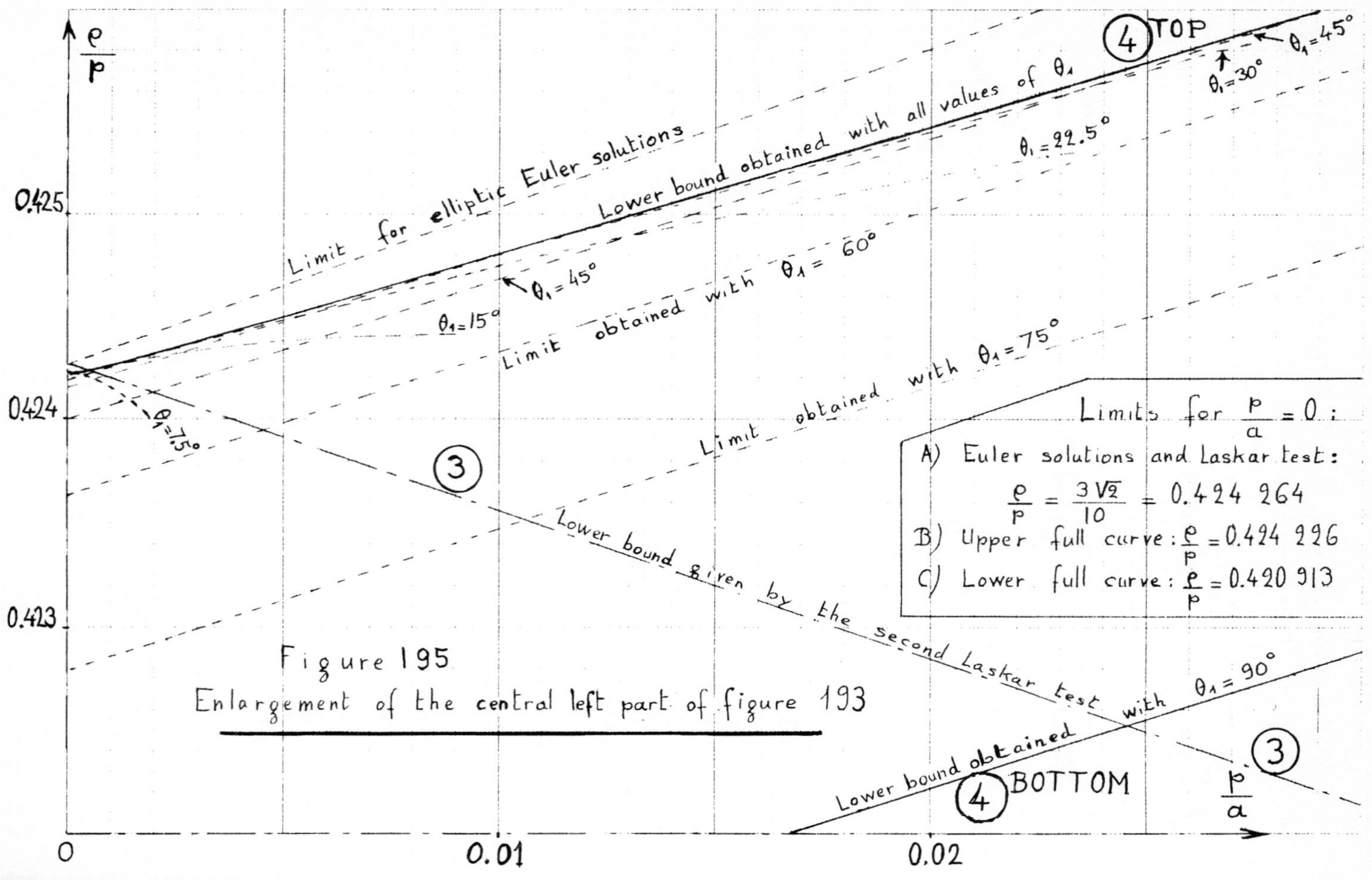

Figure 195

Enlargement of the central left part of figure 193

11.8 N-BODY MOTIONS AND COMPLETE COLLAPSES. AN EXTENSION OF THE SUNDMAN THREE-BODY RESULT.

A three-body system with a non-zero angular momentum cannot approach triple collisions (Sections 11.7.10.5 and Ref. 20). Similarly let us demonstrate that a n-body system with a non-zero angular momentum cannot approach complete collapses.

A complete collapse implies a mean quadratic distance ρ equal to zero and Section 11.7.3.2 shows that systems with a non-zero angular momentum cannot go to $\rho = 0$. We must now demonstrate that the function $\rho(t)$ is bounded away from zero.

This is easy if the energy integral h is positive or zero : $\rho(t)$ then has only one minimum that is larger than or equal to the minimum ρ_m of the orbit of the families (976) or (977) starting at the initial values ρ_0, ρ_0'.

This lower bound ρ_m is given by :

$$j_0 = (\rho_0/2a) + (p/2\rho_0) + (\rho_0\rho_0'^2/2\mu)$$

$$\left.\begin{array}{ll}
\text{In the case } h = 0 : & \left\{\begin{array}{l} \rho_m = p/2j_0 \;;\; \text{if } j_0 \geqq 1 \\ \rho_m = \frac{p}{2} + \rho_0(1 - j_0) \;;\; \text{if } j_0 \leqq 1 \end{array}\right. \\
\begin{array}{l}\text{In the case } h > 0 \\ \text{and then} \\ a = - GM^*/2h < 0 :\end{array} & \left\{\begin{array}{l} \rho_m = aj_0 + \sqrt{a^2 j_0^2 - ap} \;;\; \text{if } j_0 \geqq 1 \\ \rho_m = a + \sqrt{a^2 + 2a\rho_0(j_0 - 1) - ap} \;;\; \text{if } j_0 \leqq 1 \end{array}\right.
\end{array}\right\} \quad (1269)$$

The angular momentum $\vec{c}$ is related to the generalized semi-latus rectum p by $p = Mc^2/GM^{*2}$, hence $\vec{c} \neq 0$ implies $p > 0$ and all above $\rho_m > 0$; complete collapses cannot be approached.

Notice that $j_0 < 1$ is also a sufficient condition of $\rho_m > 0$.

It remains to analyse the cases of systems with a negative energy integral h.

Notice that it is sufficient to consider the motion of n non-infinitesimal masses : the complete collapse of a system includes of course that of its non-infinitesimal masses.

We will use the length λ defined in (1036)-(1040) for the problem of ten equal masses and in (1060)-(1062) for the general n-body problem.

The lengths λ can be defined in any direction of space and we will denote by Λ (capital λ) its supremum for all space directions*.

The lengths λ always satisfy :

$$d^2\lambda/dt^2 \geqq - K/\lambda^2 \qquad (1270)$$

with K given in (1065), (1066). Hence the supremum Λ always satisfies :

$$d^2\Lambda/dt^2 \geqq - K/\Lambda^2 \qquad (1271)$$

Let us recall that the functions $\lambda(t)$ and $\Lambda(t)$ are continuous but their derivatives can have positive discontinuities that do not contradict (1270) and (1271).

The length Λ thus has an "escape velocity" equal to $\sqrt{2K/\Lambda}$ and the bounded motions never reach that escape velocity ; for them the function $\Lambda^{3/2} - t\sqrt{4.5K}$ is forever decreasing.

Finally we will need an upper and a lower bound of the ratio (ρ/Λ) in terms of the n non-infinitesimal masses m_1, m_2, m_3, ..., m_n. We will assume that m_1 is the smallest mass.

For a given distance Λ the minimum of ρ requires that all n masses be gathered at only two points A and B separated by the distance Λ (Fig. 196).

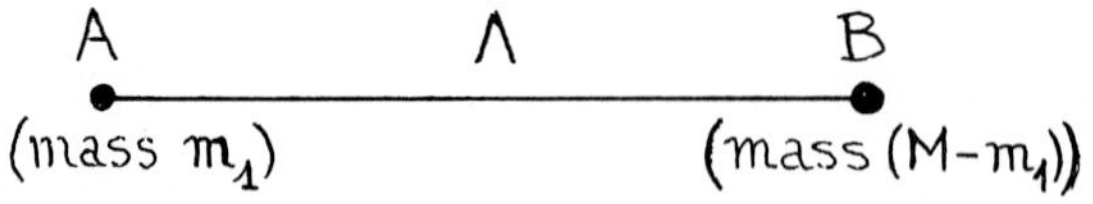

Fig. 196. For a given Λ the minimum of ρ is obtained when all masses but m_1 are gathered at the same point.

It is then easy to see that ρ is minimum for the most unequal sharing : the mass m_1 on one side and the remainder on the other side.

Hence in all cases ρ and Λ satisfy :

$$M^*\rho^2 \geqq (M - m_1)m_1\Lambda^2 \qquad (1272)$$

* Λ is the largest possible distance between the center of mass of q of the n bodies and the center of mass of the (n-q) remaining bodies.

On the other hand the length Λ being the largest possible distance between the center of mass of q of the n bodies and the center of mass of the (n-q) remaining bodies we have with q = 1 :

$$\left(M/(M - m_j)\right) r_j \leq \Lambda \quad ; \quad j = \{1, 2, \ldots, n\} \tag{1273}$$

Hence :

$$M^* \rho^2/M = \sum_{j=1}^{n} m_j r_j^2 \leq \frac{\Lambda^2}{M^2} \sum_{j=1}^{n} m_j (M - m_j)^2 \tag{1274}$$

that is :

$$\rho^2 \leq \Lambda^2 \left[2 - \frac{1}{MM^*} \sum_{j=1}^{n} m_j^2 (M - m_j)\right] \tag{1275}$$

(1272) and (1275) give the two bounds of the ratio ρ/Λ :

$$K_1 \leq \rho/\Lambda \leq K_2 \tag{1276}$$

with :

$$\left.\begin{array}{l} \sqrt{2m_1/M} \leq \sqrt{\left(m_1(M-m_1)/M^*\right)} = K_1 \quad ; \\ K_2 = \sqrt{2 - \left(\sum_{j=1}^{n} m_j^2 (M-m_j)\right)/MM^*} \leq \sqrt{2} \end{array}\right\} \tag{1277}$$

The demonstration also need the following elements of Section 11.7.3.2.

A) The inequation (969) that is :

$$d^2(\rho^2)/dt^2 \geq \sup\left\{\left(\frac{2\mu}{\rho} - \frac{2\mu}{a}\right) \quad ; \quad \left(\frac{\mu p}{\rho^2} - \frac{\mu}{a} + \rho'^2\right)\right\} \tag{1278}$$

Hence when $\rho < \sup(a, \sqrt{ap})$ the second derivative $(\rho^2)''$ is positive, the function $\rho^2(t)$ is concave up and has locally one and only one minimum before $\rho(t)$ goes back up to $\sup(a, \sqrt{ap})$.

B) The relation (922) that is :

$$\left.\begin{array}{l} t - t_0 = \pm T\left((\rho_0/4a) + (p/4\rho_0)\right) \text{ implies } \rho_0 \cdot \rho(t) \geq ap \\ \text{with} \left\{\begin{array}{l} a = \text{generalized semi-major axis} = -GM^*/2h > 0 \\ T = 2\pi\sqrt{a^3/\mu} \end{array}\right. \end{array}\right\} \tag{1279}$$

Let us now consider a very small local minimum ρ_m of the curve $\rho(t)$, smaller than $\sup(a,\sqrt{ap})$, and let us see if this minimum is compatible with bounded motions.

At t_0 we have :

$$\rho_0 = \rho_m \quad ; \quad \rho_0' = 0 \quad ; \quad \rho_m/K_2 \leqq \Lambda_0 \leqq \rho_m/K_1 \tag{1280}$$

At $t_1 = t_0 + T\big((\rho_m/4a) + (p/4\rho_m)\big)$ we have, with (1279) :

$$\rho_1 = \rho(t_1) \geqq ap/\rho_m \quad ; \quad \Lambda_1 \geqq \rho_1/K_2 \geqq ap/\rho_m K_2 \tag{1280.1}$$

If the motion is bounded the function $\Lambda^{3/2} - t\sqrt{4.5K}$ is decreasing and hence :

$$\Lambda_0^{3/2} > \Lambda_1^{3/2} - T\big((\rho_m/4a) + (p/4\rho_m)\big)\sqrt{4.5K} \tag{1280.2}$$

which requires :

$$(\rho_m/K_1)^{3/2} > (ap/K_2\rho_m)^{3/2} - T\big((\rho_m/4a) + (p/4\rho_m)\big)\sqrt{4.5K} \tag{1281}$$

This condition is impossible for too small positive ρ_m and there is one and only one positive ρ_L limit such that :

$$(\rho_L/K_1)^{3/2} = (ap/K_2\rho_L)^{3/2} - T\big((\rho_L/4a) + (p/4\rho_L)\big)\sqrt{4.5K} \tag{1282}$$

For bounded and also for oscillatory motions the function $\rho(t)$ cannot go below ρ_L and complete collapses cannot be approached.

The limit ρ_L is below $\sup(a,\sqrt{ap})$ and even below $\sqrt{(apK_1/K_2)}$ and if $\rho(t)$ goes below ρ_L it has there a minimum ρ_m at some time t_m.

The Sundman function $j(t)$ is locally minimum at $t = t_m$ and is larger than or equal to j_m throughout the interval $t_m \pm$ $\pm\ T((\rho_m/4a) + (p/4\rho_m))$. Outside this interval the length Λ has a velocity above its escape velocity, it is monotonic, goes to infinity for both $t \to +\infty$ and $t \to -\infty$ and, with (1276), imply $\rho > ap/\rho_m$ and $j > j_m$.

Hence a function $\rho(t)$ can go below ρ_L at most one time and for a very short interval. Since then the absolute minimum of the Sundman function j is obtained at the minimum we can give the following lower bound of the mean harmonic distance ρ in terms of the initial conditions ρ_0, ρ'_0 :

$$\left.\begin{array}{l} \text{At any time } t : \rho(t) \geq \inf\left(\rho_L \; ; \; aj_0 - \sqrt{a^2 j_0^2 - ap}\right) \\ \text{with :} \\ j_0 = (\rho_0/2a) + (p/2\rho_0) + (\rho_0\rho_0'^2/2\mu) \end{array}\right\} \quad (1283)$$

The limit ρ_L is given in (1282) and if the angular momentum $\vec{c}$ is not zero, that is if $p > 0$, the conditions (1282), (1283) give a positive lower bound of $\rho(t)$. Complete collapses cannot be approached.

ρ_L is generally very small and it is possible to improve that lower bound very much either by the use of functions as L in (1056) or (1067) instead of Λ (they improve K_1, K_2 and K), or by consideration of intervals of time different from $\{t_m \; ; \; t_m + T\left((\rho_m/4a) + (p/4\rho_m)\right)\}$, or by several other means.

The final purpose, like that in the three-body problem, is to demonstrate that for bounded or oscillatory motions of given masses and angular momentum the greatest lower bounds of the semi-moment of inertia and of the mean quadratic distance are given by the families of elliptic Euler motions.

However we are still very far from that result, much further than in the three-body case.

11.9 ORIGINAL AND FINAL EVOLUTIONS

The past-future symmetry of the three-body problem leads to tables of original evolutions almost identical to Tables VI, VII, VIII of final evolutions (Section 11.7.9). The only differences are the signs of the time t.

The main question are of course such as the following : "A three-body motion had in the past an original evolution of, say, hyperbolic type, what are its possible final evolutions ?"

The main elements of the answer are given by :

The integrals of motion

The ergodic theorem (Section 11.7.10.2)

The Hill-type stability (Section 11.6.3 and 11.6.5)

Sometimes the evolution of the mean quadratic distance ρ (Section 11.7.3.2).

For instance a motion of hyperbolic type in the past (H_-) cannot give a motion of bounded type in the future (B_+) because the first requires a positive energy integral and the second a negative energy integral.

Notice that if a three-body motion has hyperbolic-elliptic escapes at both ends its nature is very different if the same body is escaping in the past and in the future (case called fly-by) or if these two bodies are different (case called exchange). The latter case is impossible for systems having Hill-type stability.

We are thus led to the following subdivisions.

11.9.1 General three-body systems of positive energy and non-zero angular momentum

The final evolutions are of hyperbolic, hyperbolic-parabolic or hyperbolic-elliptic types and, according to the escaping body being either A or B or C, there are seven different types :

$$\left.\begin{array}{lll} H_+ & & \\ HP_{A+} & HP_{B+} & HP_{C+} \\ HE_{A+} & HE_{B+} & HE_{C+} \end{array}\right\} \qquad (1284)$$

All these types correspond to connected sets of phase space and the corresponding topology is given in Fig. 197 with the hyperbolic-parabolic types at the boundaries of the other types.

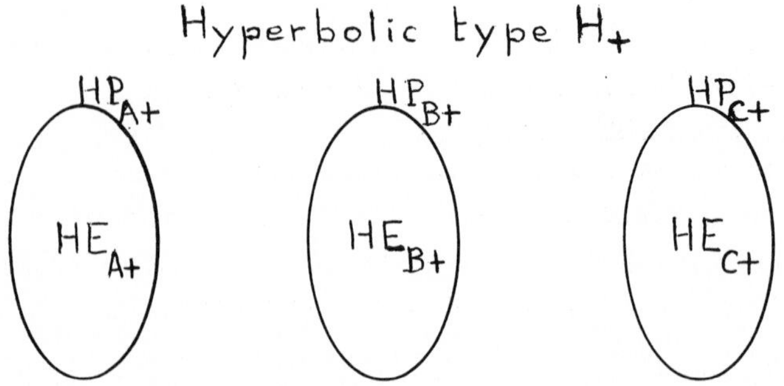

Fig. 197. Topology of the seven types of final evolutions for three positive masses and positive c^2h.

The seven original evolution H_- to HE_{C-} have the same topology and there are theoretically 49 possible original-final evolutions.

The main types are the following :

Type	Name
$H_- \longrightarrow H_+$	
$H_- \longrightarrow HE_{J+}$; $J = \{A,B,C\}$	partial capture
$HE_{J-} \longrightarrow H_+$; $J = \{A,B,C\}$	ionization
$HE_{J-} \longrightarrow HE_{J+}$; $J = \{A,B,C\}$	fly-by
$HE_{J-} \longrightarrow HE_{K+}$; $J \neq K$	exchange

The name "ionization" is of course an analogy with quantum mechanics. The first type has no name and can perhaps be called "super fly-by".

Are the 49 theoretically possible types really possible ? This is the first Chazy conjecture :

"If $c^2h > 0$ the 49 possible types of original-final evolutions exist for any positive values of the masses and any values of the integral of motion".*

This conjecture is always true even if the product c^2h is very large. However the motions of exchange type may then require a large number of very close binary encounters (Section 10.4.3 and Fig. 46) and are thus exceptional.

1.9.2 General three-body system of positive energy and zero angular momentum

Because of the zero angular momentum the motion is then a plane motion (Section 5.1.2 and Ref. 8). The only difference from the previous Section is the existence of "triple collisions" of four possible types according to the limit central configuration.

---

* Initially Chazy had conjectured that original and final evolutions were always of the same type and that partial capture, ionization and exchange were impossible. He has kept this conjecture for systems of negative energy (see the second Chazy conjecture in Section 11.9.5). The possibility of partial capture was demonstrated by Sitnikov in 1953 (Ref. 220).

Notice that in a given case only one of these four types is possible : the triple collision of Lagrange type (TC_L) if the motion is not collinear and the suitable triple collision of Euler type (TC_A, TC_B, TC_C according to the central mass) if the motion is collinear : for collinear motions the order of succession of the three bodies is constant.

The corresponding original evolutions are sometimes called "triple collisions" (for decreasing times) but I prefer to call them "triple explosions" : TE_L, TE_A, TE_B, TE_C.

When the energy integral is positive or zero it is impossible to go from a triple explosion to a triple collision because the mean quadratic distance ρ then has no maximum. Hence in the non collinear case there are at most $8 \times 8 - 1 = 63$ possible original-final evolutions and indeed these 63 possible evolutions always exist for any positive values of the three masses and the energy integral.

In the collinear case, with for instance B between A and C only six original and also six final evolutions are possible : HE_{C-}, HP_{C-}, H_-, HP_{A-}, HE_{A-}, TE_B and HE_{C+}, HP_{C+}, H_+, HP_{A+}, HE_{A+}, TC_B. Are the corresponding $6 \times 6 - 1 = 35$ original-final evolutions possible ? The answer is yes for most types, it is yes for all types begining by H_- or TE_B and/or ending by H_+ or TC_B ; but it may be no for the other types, according to the mass ratios. This is a beautiful study that has no yet been done.

11.9.3 General three-body systems of zero energy and non-zero angular momentum

This case has seven possible final evolutions :

3 Hyperbolic-elliptic evolutions : HE_{A+}, HE_{B+}, HE_{C+}.

4 Tri-parabolic evolutions : TP_{L+}, TP_{A+}, TP_{B+}, TP_{C+}, according to the asymptotic central configuration either Lagrangian (L) or Eulerian with body A, B or C between the two other bodies.

The corresponding original evolutions are of course HE_{A-}, HE_{B-}, HE_{C-}, TP_{L-}, TP_{A-}, TP_{B-}, TP_{C-}.

The topology of final evolution is presented in Fig. 198 and the analysis of the vicinity of parabolic Lagrangian solutions shows that all 49 original-final evolutions are always possible for any positive mass ratios and any non-zero angular momentum (Ref. 47 , 54).

The special test for motions with h = 0 developped at the end of Section 11.7.10.1 is very useful for the analysis of motions of this Section and the next one.

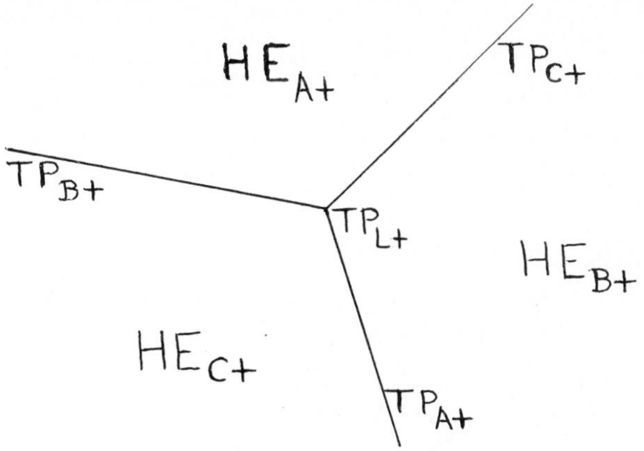

Fig. 198. Topology of the seven types of final evolutions when $h = 0\ ;\ \vec{c} \neq 0$.

11.9.4 General three-body systems of zero energy and zero angular momentum

The considerations of Section 11.9.2 on triple collisions and "triple explosions" are also valid here.

The solutions from a triple explosion to a tri-parabolic escape of the same limit configuration are necessarily radial parabolic Lagrangian or Eulerian solutions.

There are solutions from a triple explosion of Lagrange type to a tri-parabolic escape of Euler type such as the numerical solution given in Ref. 48.

As in Section 11.9.2 plane and non-collinear motions are those with possible triple explosion and triple collision of Lagrangian type (TE_L and TC_L) and among the 64 original-final evolutions theoretically possible from (TE_L, $HE_{A,B,C-}$, $TP_{L,A,B,C-}$) to (TC_L, HE_{ABC+}, TP_{LABC+}) the type $TE_L \rightarrow TC_L$ never exists. However it seems that there are also problems in a second case, at least for some particular mass ratios, and the type $TP_{L-} \rightarrow TP_{L+}$ may be impossible.

Collinear three-body motions of zero energy have been the subject of many studies (as ref 40,45-47,50,54). If B is the central body the only possible original evolutions are TE_B, HE_{A-}, TP_{B-},

HE_{C-}, and the only possible final evolutions are TC_B, HE_{A+}, TP_{B+}, HE_{C+}.

Again the type $TE_B \to TC_B$ is impossible while the six types $TE_B \to escape_+$ and $escape_- \to TC_B$ are possible for all mass ratios.

The discussion of the nine types $escape_- \to escape_+$ depends very much on the mass ratios.

The mass ratios for which the type $TP_{B-} \to TP_{B+}$ is possible are given in Fig. 199 and have been computed by Carlès Simò in Ref. 50 .

These mass ratios are either on one of the symmetrical full curves numbered 4, 6, 8, 10, etc... or on the symmetrical pairs of dotted curves numbered 3, 5, 7, 9 etc... (according to the number of binary collisions in the motion $TP_{B-} \to TP_{B+}$ of interest).

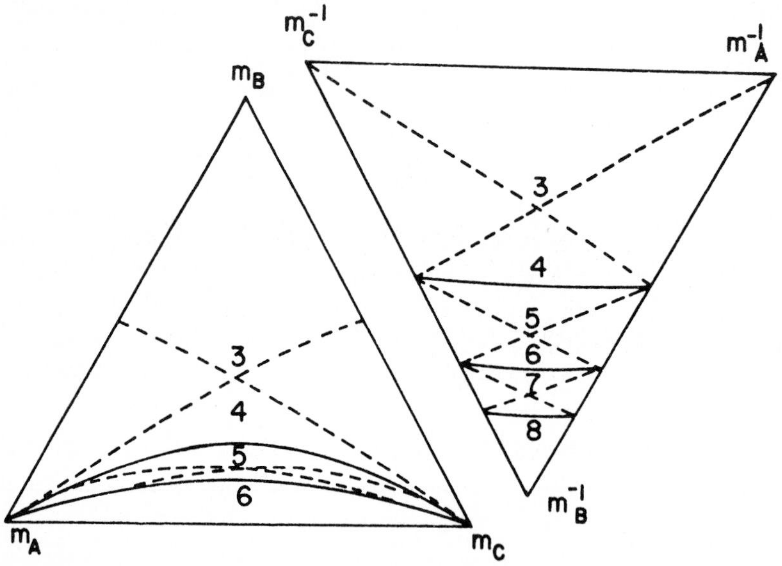

Fig. 199. Mass diagram and equivalent (but more readable) inverse mass diagram of the mass ratios for which the collinear solutions of zero energy $TP_{B-} \to TP_{B+}$ are possible. Source : Simo, C. [50].

The numbers 3, 4, 5, 6, 7, 8, 9, etc... are the numbers of binary collisions of the motions $TP_{B-} \to TP_{B+}$ of interest.

At the left end of the full curves the mass ratios are the following :

$$m_B/m_A = 0 \quad ; \quad m_C/m_B = \begin{cases} 1 & \longrightarrow \text{curve } 4 \\ 2.3014 & \longrightarrow \text{curve } 6 \\ 3.7501 & \longrightarrow \text{curve } 8 \\ \dots\dots\dots\dots\dots\dots & \\ \phi(n) & \longrightarrow \text{curve } n \; ; \; \phi(n) \sim n^2/4\pi^2 \end{cases}$$

The possible types of original-final evolutions depend on the position in the mass diagram or the inverse mass diagram and we find the following simple rules.

A) The exchange types $HE_{A-} \to HE_{C+}$ and $HE_{C-} \to HE_{A+}$ are impossible on the full curves and possible anywhere else.

B) The fly-by $HE_{A-} \to HE_{A+}$ is impossible in the left triangles of the inverse mass diagram and along their dotted boundaries, but it is possible anywhere else and then it is always possible if $m_A < m_C$.

The symmetrical rule is true for the fly-by $HE_{C-} \to HE_{C+}$.

C) The four remaining types as $HE_{A-} \to TP_{B+}$ are possible if and only if their two major adjacent types are possible (the two major adjacent types of $HE_{A-} \to TP_{B+}$ are $HE_{A-} \to HE_{A+}$ and $HE_{A-} \to HE_{C+}$).

Hence at the intersection of two dotted curves we obtain $m_A = m_C$ and there are only three possible $escape_- \to escape_+$ types : the type $TP_{B-} \to TP_{B+}$ and the two exchange types.

11.9.5 General three-body systems of negative energy and non-zero angular momentum

This case is much more complex than the previous ones, but it is also much more interesting.

The nine possible final evolutions are :

- HE_{A+}, HE_{B+}, HE_{C+} : open sets in phase-space.
- PE_{A+}, PE_{B+}, PE_{C+} : on the boundaries of the corresponding HE_{ABC+}
- B_+ : the bounded type.
- $Os1_+$, $Os2_+$: the two oscillatory types.

The ergodic theorem (Section 11.7.10.2) gives many informations :

The escape solutions $HE_{A,B,C+}$ and PE_{ABC+} are of open type for the future, they must then "almost always" be of open type for the past and hence the closure in phase-space of the open sets HE_{A+}, HE_{B+}, HE_{C+} is identical to the closure in phase-space of the open sets HE_{A-}, HE_{B-}, HE_{C-}.

The orbits of B_+ are either "abnormal" or "bounded recurrent" and apart from a set of measure zero they have the Poisson stability and are of type B_- for the past.

The orbits of $Os1_+$ and $Os2_+$ are either "abnormal" or "oscillatory and recurrent", they also almost always have the Poisson stability.

The ergodic theorem and the limit character of parabolic-elliptic escapes drastically reduce the number of original-final evolutions that may correspond to a set of positive measure in phase-space.

These original-final evolutions are :

- The fly-by as $HE_{A-} \to HE_{A+}$
- The exchanges as $HE_{A-} \to HE_{B+}$
- The bounded recurrent orbits $B_- \to B_+$
- The oscillatory recurrent orbits $Os1_- \to Os1_+$ and $Os2_- \to Os2_+$.

Hence the type $HE_{A-} \to B_+$ called "complete capture" is infinitely rare even if some examples are given with the Sitnikov motions (Section 11.7.8), as well as examples $B_- \to Os1_+$; $B_- \to PE_{A+}$, etc...

The structure of the set of exchange motions is necessarily very complex. Consider for instance Fig. 200 and let us study the motion in terms of the modulus of the velocity $\vec{V}_c$ of a given direction with initial conditions that are not very far from the Sitnikov conditions of Section 11.7.8.

For large V_c the motion is of the fly-by type $HE_{C-} \to HE_{C+}$ but the limit cases of types PE_{C-} and PE_{C+} do not necessarily correspond to the same velocity V_c.

What happens when V_c is between these two limits on the segment QT of Fig. 200 ?

The body C arrives hyperbolically from the left, passes between the two bodies A and B and loses in that passage a part of its energy ; it is ejected towards the right but has not enough energy to escape. It falls back toward the binary AB.

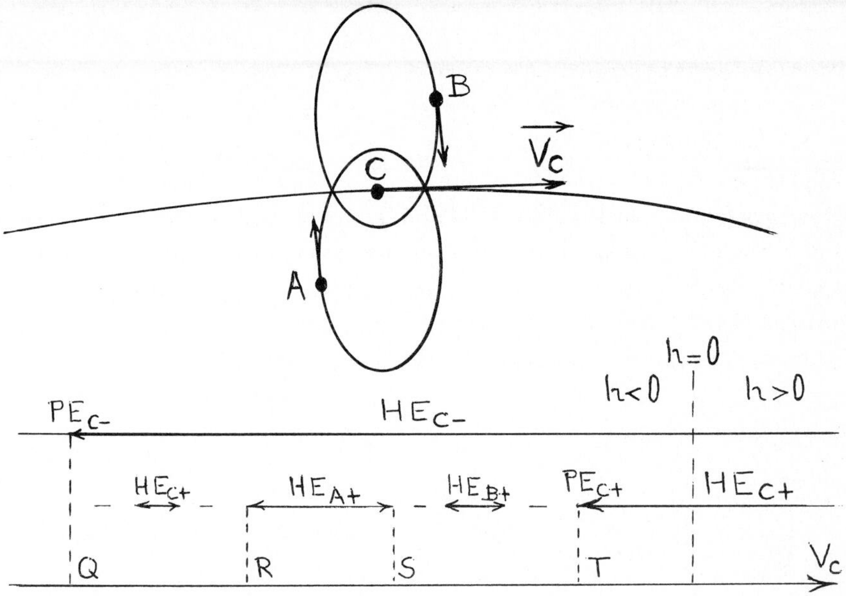

Fig. 200. The structure of the sets of motions of exchange type. Discussion in terms of the magnitude of $\vec{V_c}$.

This return gives a complex motion but the ergodic theorem tells us that in almost all cases one of the three bodies will escape since the motion is already "open" in the past. For instance when the velocity V_c corresponds to a point of the open segment RS of Fig. 200 the body A escapes and the motion is of the exchange type $HE_{C-} \to HE_{A+}$ (the hyperbolic type is here impossible since $h<0$).

We saw in Section 11.7.8 how complex the structure of the set of symmetrical Sitnikov solutions can be, the complexity is of course larger for the more general solutions of Fig. 200. The QT segment is filled with an infinite number of open segments of the three types HE_{A+}, HE_{B+}, HE_{C+} as **RS** and the structure is a typical fractal structure as a Cantor set (Ref. 238).

Such a complexity was considered unlikely by Chazy in his study of final evolutions (Ref. 217,239) and he proposed what is called the second Chazy conjecture : "When the energy integral h is negative, motions of exchange type are impossible".

Notice that this conjecture is at least partially true : when the three-body system of interest has the Hill-type stability, i.e. when its product c^2h is below a negative critical value which is a function of the masses, there is forever a small binary that the third body can neither approach nor disrupt and exchange motions are impossible (Section 11.6.3). The ergodic theorem then implies that almost all motions have a final evolution of the type of their original evolution.

After more than twenty years of general consensus Schmidt, Alexeev and Khilmi (Ref.19,218,220,240-256) where the first to challenge the second Chazy conjecture in planetary problems (one large mass and two small masses, Fig. 201) but the numerical examples presented by L. Becker and later by Alexeev remained inconclusive because of the poor accuracy of the computers of that time.

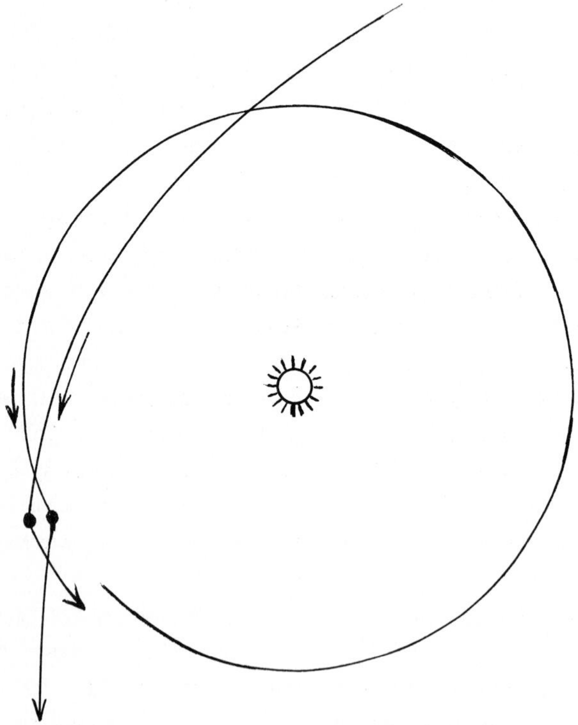

Fig. 201. The objection of Schmidt, Alexeev and Khilmi.

A planet revolves around the Sun on an almost circular orbit. A fast outer body with a mass identical to that of the planet arrives from infinity and undergoes a very close approach to the planet. These two bodies almost exchange their orbits and the motion is of the "exchange" type even if the total energy is negative.

Finally, in 1975, Szebehely (Ref. 153) has computed the beautiful symmetrical exchange motion presented in Section 10.4.1 and in Figs 39-40. The second Chazy conjecture must be forsaked.

The initial conditions of this symmetrical exchange motion were chosen as follows.

A) The initial conditions satisfy the space-time symmetry conditions III and IV of Section 11.2.1 and Fig. 109 with symmetry of the two equal masses m_2 and m_3. Hence only one half of the solution need be computed and if the final evolution is of type HE_{3+} the original evolution is of type HE_{2-} and the motion is an exchange motion.

B) The energy integral h is zero, instead of $h < 0$, because analysis of the motion is then simpler and because the validity of the result is the same : If we get an exchange motion of the type $HE_{2-} \to HE_{3+}$ for $h = 0$ there will be in the vicinity exchange motions of the same type for $h < 0$ since this type of motion is an open set in phase space.

C) The special test for the case $h = 0$ developed at the end of Section 11.7.10.1 is especially useful here (Fig. 202).

The initial conditions correspond to $\rho' = 0$ and then to the absolute minimum of the mean quadratic distance ρ and the Sundman function j, with $j = j_m = p/2\rho$.

Since always $\rho/\nu \geqq j \geqq j_m$ if the minimum j_m is sufficiently high the informations given by the ρ/ν = constant curves (generalized Hill curves of Fig. 202) restrict very much the possibilities of final evolutions.

The choice of the initial position of m_1 in the vicinity of L_1 with $j_m > (\rho/\nu)_{L_2}$; (which requires $(\rho/\nu)_{L_1} \geqq \rho/\nu > (\rho/\nu)_{L_2}$ and then $m_1 < m_2$) forces the final evolution to be either of type TP_{1+}, type HE_{2+} or type HE_{3+}. With the space-time symmetry the two last types are convenient for the demonstration of the existence of exchange motions and the limit character of the first type reduces its probability.

The symmetrical exchange motion of Section 10.4.1 and Figs 39-40 is planar with m_1 initially at L_1, it has the following initial conditions :

$$\left.\begin{aligned} &G = m_1 = 1 \;;\; m_2 = m_3 = 2 \\ &x_1 = y_1 = y_2 = y_3 = 0 \;;\; x_2 = -x_3 = -3 \\ &x_1' = -\frac{28}{123} \;;\; y_1' = 0 \;;\; x_2' = x_3' = \frac{7}{123} \;;\; y_2' = -y_3' = -\frac{122}{123} \end{aligned}\right\} \quad (1285)$$

These conditions give indeed :

$$\left.\begin{aligned} &h = 0 \;;\; \rho = \sqrt{22.5} \;;\; p = 18605/1681 = 11.0678\ldots \\ &j_m = p/2_p = 1.16665\ldots > (\rho/\nu)_{L_2} = 1.15165\ldots \end{aligned}\right\} \quad (1286)$$

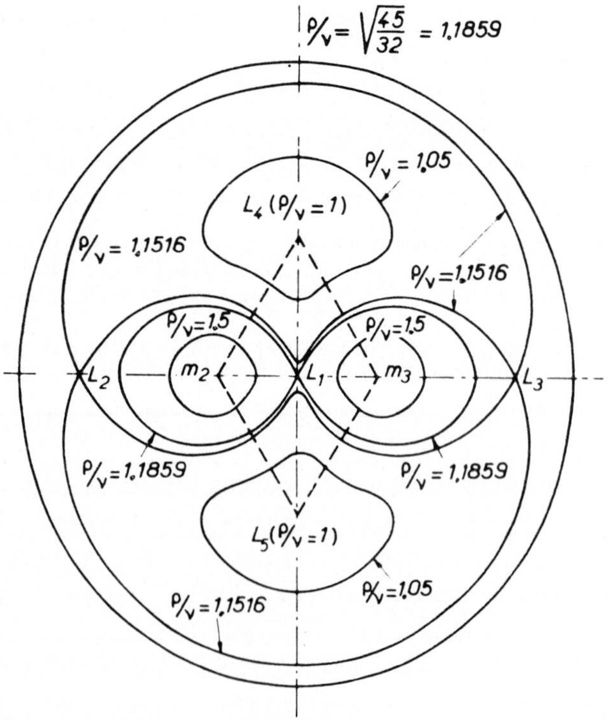

Fig. 202. Generalized Hill curves for $2m_1 = m_2 = m_3$.
(Curves of constant ratio ρ/ν in terms of the position of m_1 with respect to m_2 and m_3).

This procedure give many possible initial conditions of exchange motion : a three non-trivial-parameters family for planar motions and two families with four non-trivial parameters for three-dimensional motions (one family with space-time symmetry of type III, Fig. 109, and the other with type IV)*.

The outstanding numerical study of Hut and Bahcall presented in Section 10.4.3 and Fig. 47 shows the importance of exchange motions, especially for systems of negative energy.

As we have already notice, systems with Hill-type stability cannot have an exchange motion but Fig. 47 shows that, in the conditions of the Hut and Bahcall studies, about 15% of non Hill-stable systems of slow incoming velocity undergo an exchange.

Thus we know that exchange motions exist for all mass ratios when the product c^2h is above a critical negative value, function of the masses, and we also know that for negative h the structure of the set of exchange motions is extremely complex.

This complexity also exist for the structure of the set of bounded recurrent orbits (Sections 11.10 and 11.11) and a fortiori for that of the set of oscillatory orbits of either of the two types. But at least we known that the set of bounded recurrent orbits has a positive measure (Section 11.10 and Ref. 182, 257).

Let us only mention the main questions that remain to be solved.

What is the critical value of c^2h for given masses ? We only know that it is between zero and the limit given by the Hill-type stability.

What is the importance of motions of the types $Os1_- \to Os1_+$ and $Os2_- \to Os2_+$? It seems that the set of motions of type $Os1_- \to Os1_+$ is of measure zero while the other set seems to be of positive measure.

Are the escaping motions of type $HE_{A,B,C+}$ dense everywhere ?

Are the periodic orbits dense in the set of bounded and oscillatory orbits ? (Poincaré conjecture).

---

* However among these families, sub-families exists with one non--trivial parameter fewer and with original and final tri-parabolic evolution.

11.9.6 Remaining cases.
Restricted cases.

We will not consider the general three-body systems with zero angular momentum and negative energy. Because of the possibilities of triple collisions they are even more complex than the systems of the above Section 11.9.5.

The restricted three-body problems reflect the simplicity or the complexity of the corresponding general three-body problems.

If $h > 0$ and $\vec{c} \neq 0$ (hyperbolic restricted three-body problem) there are five possible final evolutions : HE_{A+}, HP_{A+}, H_+, HP_{B+}, HE_{B+} (the two primaries being A and B) and their topology is similar to that of Fig. 197. The corresponding 25 possible original-final evolutions always exist for all values of the masses of primaries, of the energy and of the angular momentum.

If $h = 0$ and $\vec{c} \neq 0$ (parabolic restricted three-body problem) there are seven possible final evolutions : PE_{A+}, PE_{B+}, HP_{C+} and the four tri-parabolic evolutions TP_L, TP_{A+}, TP_{B+}, TP_{C+} with a topology like that of Fig. 198. The corresponding 49 possible original-final evolutions always exist in the vicinity of the parabolic Lagrangian solution for all values of the angular momentum and of the masses of the primaries A and B.

If $h < 0$ and $\vec{c} \neq 0$ (elliptic or circular restricted three-body problem) we again find the complexity of the general three-body problem when the energy is negative. Notice that exchange motions are impossible but there is, as a compensation, the theoretical possibility of sub-parabolic escapes.

The ergodic theorem then implies that almost all motions have a final evolution of the type of their original evolution while the Hill-stability only exists in the circular restricted case and is related to the Jacobi integral.

If $\vec{c} = 0$ (collinear motion of the primaries) we must take account of the supplementary integral of motion, the component of angular momentum of infinitesimal mass along the axis of motion of the primaries.

If that component is not zero the motion is three-dimensional and the distance between the infinitesimal mass and the axis of motion of the primaries has a positive lower bound (a simple function of the initial conditions).

These cases of three-dimensional motions are simple, they are only an extension of the above restricted cases of the same energy with the same discussion.

Finally if $\vec{c} = 0$ and if the motion is planar the picture of the solutions becomes very complex because of the existence of triple collisions and especially of special triple collisions as described in Section 11.7.6.4 (Fig. 151) and Section 11.7.9 (Fig. 163). However notice that if $h < 0$ the ergodic theorem again leads to the conclusion that the final evolution is almost always of the type of the original evolution.

11.10 ON THE KOLMOGOROV-ARNOLD-MOSER THEOREM

The Kolomogorov-Arnold-Moser theorem or K.A.M. theorem (Ref. 182 257) represented a major advance in the knowledge of the topology of Hamiltonian systems of differential equations such as those of the general n-body problem.

This theorem states that in the vicinity of most periodic and "linearly stable" solutions (those that are sufficiently far from resonance conditions) there exist an infinite number of "Arnold-tori" of quasi-periodic solutions. Furthermore these Arnold-tori fill a set of positive measure in phase-space.

In the three-body problem the Kolmogorov-Arnold-Moser theorem can be applied to the autonomous Hamiltonian system (90)-(91) with four degrees of freedom obtained after the elimination of the nodes (before this elimination the conditions on resonances cannot be satisfied).

In these conditions the K.A.M. theorem leads to the construction of four-dimensional tori of quasi-periodic solutions in the vicinity of stable periodic solutions and in the eight-dimensional space of Delaunay's variables $\mathcal{L}_i$, $\mathcal{G}_i$, $\mathcal{L}_e$, $\mathcal{G}_e$, ℓ_i, ℓ_e, g_i, g_e given in (31). The "ignorable parameters" $\mathcal{H}_i$, $\mathcal{H}_e$, h_i, h_e are given by (92) that is by :

$$\left.\begin{aligned} \mathcal{H}_i &= (c^2 + \mathcal{G}_i^2 - \mathcal{G}_e^2)/2c \\ \mathcal{H}_e &= (c^2 + \mathcal{G}_e^2 - \mathcal{G}_i^2)/2c = c - \mathcal{H}_i \\ h_e &= h_i + \pi \quad ; \quad dh_i/dt = dh_e/dt = \partial H/\partial c \end{aligned}\right\} \qquad (1287)$$

with :

$$\left.\begin{array}{l} c = \text{angular momentum} \\ H = H(\mathcal{L}_i,\ \mathcal{G}_i,\ \mathcal{L}_e,\ \mathcal{G}_e,\ \ell_i,\ g_i,\ \ell_e,\ g_e,\ c) = \\ \quad = \text{Hamiltonian function} = \text{energy integral } h \end{array}\right\} \qquad (1288)$$

Notice that in the planar three-body problem the "elimination of the nodes" (or more precisely the elimination of the absolute longitudes) leads to the autonomous Hamiltonian system (100)-(101) with only three degrees of freedom and we have only two degrees of freedom in the collinear three-body problem. Hence the Arnold--tori have three dimensions in the planar case and two dimensions in the collinear case.

Another result of the K.A.M. theorem is interesting.

Consider the question of oscillating orbits of the second kind, i.e. orbits with $\liminf_{t\to\pm\infty} r = 0$ and with an infinite number of very close approaches.

The equations of motion at collision can be regularized (see for instance the equations (88.1) to (88.3) of Chapter 6) and instead of the eight conjugate parameters of (1288) we can use eight equivalent conjugate parameters without singularity at collision.

For these latter eight parameters the manifold of collisions is a five-dimensional manifold that intersect very naturally the four-dimensional Arnold tori. On these tori some quasi-periodic orbits will go to a strict collision but most of them will avoid strict collision of point masses and will be of the oscillatory type of the second kind with $\liminf_{t\to\pm\infty} r = 0$.

It is certainly possible to demonstrate in this way that orbits of the oscillatory type of the second kind fill a set of positive measure in phase space.

11.11 THE ARNOLD DIFFUSION CONJECTURE, THE TEMPORARY CHAOTIC MOTIONS, THE TEMPORARY CAPTURES.

Arnold (Ref. 182) gave the first known example of the strange phenomenon called "Arnold diffusion" and we met it in Section 10.7.7.5.

The Arnold diffusion is an extremely slow instability that appears in the vicinity of periodic solutions without a strict stability but with "all order stability".

It seems that Arnold diffusion is a very common phenomenon among Hamiltonian systems with a large number of degrees of freedom.

The Arnold diffusion conjecture is the following :

Let us consider an analytic autonomous Hamiltonian system* and let us simplify it as much as possible with its integrals of motions and its eventual decomposability.

In a subset of phase-space defined by given values of the integrals of motion we find the following.

A) The periodic orbits and the tori of quasi-periodic orbits that will be the backbone of the set of solutions.

B) Between these first solutions the holes of infinite measure are almost everywhere filled with open solutions coming from infinity and going back to infinity, while the holes of finite measure are almost everywhere filled with "chaotic solutions" dense in the hole (Fig. 203).

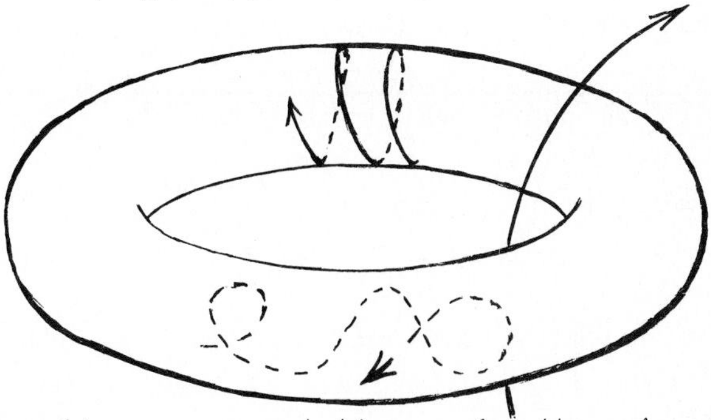

Fig. 203. Schematic representation of the Arnold diffusion conjecture with a torus of quasi-periodic solutions, an open solution and an enclosed chaotic solution.

---

* The conjecture can certainly be extended to analytic Hamiltonian systems with an Hamiltonian function periodic in terms of the time t.

If the problem of interest is not integrable the tori of quasi--periodic orbits are nowhere dense, because of the abundance of resonances that destroy many of them, and the union of open and chaotic solutions is everywhere dense even if its measure can be very small in the parts of phase-space where the perturbations are small and the problem "almost integrable".

Notice that if the utmost reduction of the Hamiltonian problem of interest leads to an autonomous Hamiltonian system with Q degrees of freedom :

If Q = 1 the problem is integrable.

If Q = 2 the Arnold tori are 2 dimensional manifolds that enclose bounded parts of phase-space in the three-dimensional invariant manifolds determined by a given value of the Hamiltonian.

If $Q \geq 3$ the Arnold tori are Q-dimensional manifolds and their dimensionality is too small to enclose parts of the (2Q - 1) dimensional invariant manifolds of phase-space determined by a given value of the Hamiltonian.

In the last case only the integrals of motion can be used to enclose bounded parts of phase-space. Section 9.4 and Figs 26 to 33 present these phenomena in the "Copenhagen problem" (i.e. the planar circular restricted three-body problem with equal primaries). Figures 27 and 28 present simple periodic and quasi--periodic orbits, Figs 29 and 30 present more complex quasi-periodic orbits and give a taste of the complexity of the phenomenon, while Figs 31 and 32 present the erratic behavior of a chaotic orbit that goes everywhere in the accessible domain and finally Fig. 33 shows that as soon as the accessible domain is open the former chaotic trajectories escape to infinity while bounded quasi-periodic solutions continue to exist.

This escape to infinity can be extremely slow and can be hidden by other phenomena.

Consider for instance, the "chaotic" motions of Figs 204 and 205. These motions in full lines are limited by the nearest tori of quasi-periodic motions in dotted lines.

The small passage between the two large zones leads to the usual phenomenon of "intermittencies" : the motion remains for a very long duration in one of these large zones and finally passes to the next large zone where it will also remain for a very long period.

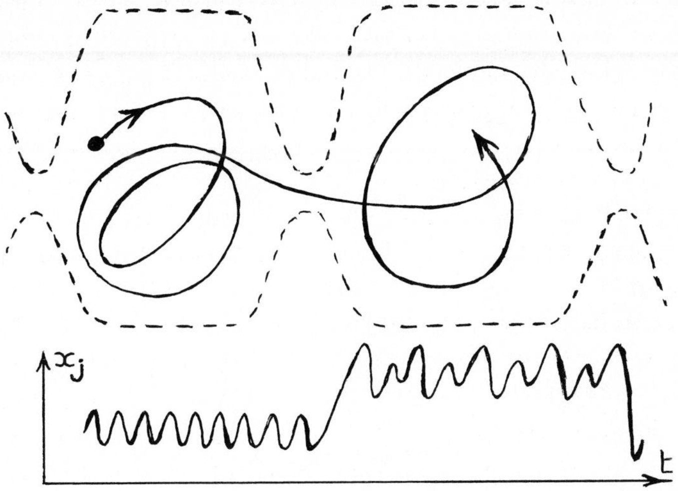

Fig. 204. Chaotic motions and intermittencies.

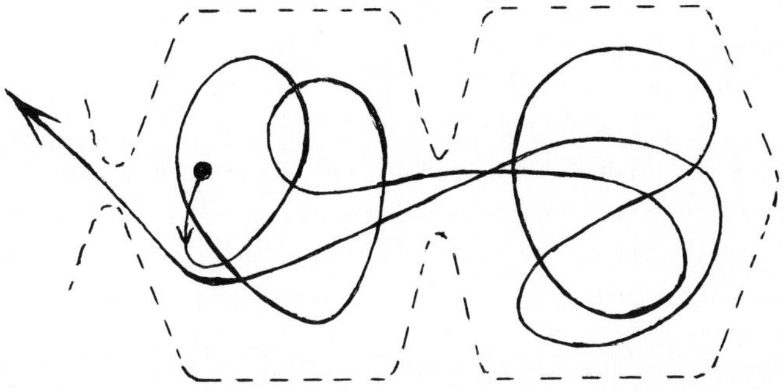

Fig. 205. A temporary chaotic motion that finally escape to infinity.

The average characteristics of the motion, and especially its "Liapunov characteristic exponents" related to the motions of neighbouring trajectories, are very difficult to compute : they differ very much depending on whether only one long intermittency or many of them are considered.

Furthermore it very often happens that after a large number of long inttermitencies the motion finds a way of escape to infinity (Fig. 205), it is then an "open motion" even if it was considered for a very long duration as an apparently chaotic motion. Such a motion is called "temporary chaotic motion" and the usual statistical properties of chaotic motion can be applied only for a long but limited period of time.

The first known example of temporary chaotic motion is the "temporary capture" of the three-body problem : A comet arrives from infinity, passes ahead of Jupiter and is sent on an elliptic orbit around the Sun (Fig. 206).

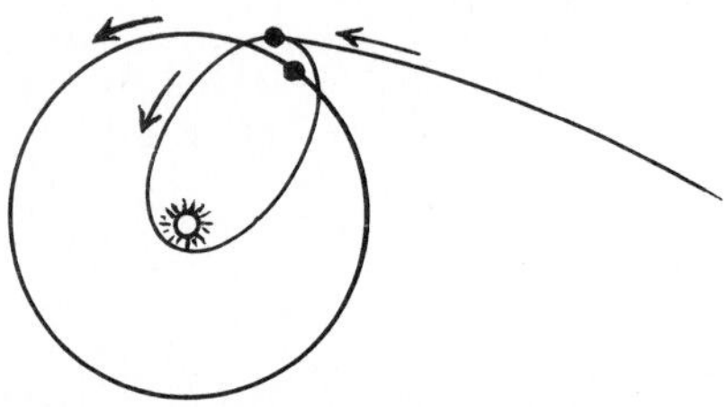

Fig. 206. A temporary capture of a comet by Jupiter: the prototype of temporary chaotic motions.

Further close approaches to Jupiter may modify the cometary orbit but sooner or later one of these modifications will send the comet back to infinity (if collisions are avoided).

The Sitnikov motions of Section 11.7.8 can provide examples of temporary chaotic motions of any duration.

When applied to the general three-body problem the Arnold diffusion conjecture leads to a surprising result very different from that of the circular restricted three-body problem : In the planar and in the three-dimensional problems the integrals of motion and the tori of quasi-periodic motions are insufficient to enclose a bounded part of phase-space and then almost all apparently chaotic motions are open, they are only temporary chaotic motions and the open solutions are everywhere dense while the bounded and oscillatory solutions are "nowhere dense" and have a Cantor set structure of positive measure (see Fig 207 page 526).

This is also true for the general n-body problem. Fortunately this does not mean that we will see the escape of a planet for the following reasons.

A) An open set can be everywhere dense even with an extremely small measure ; the corresponding events thus have a very small probability.

B) The Sun and its nine major planets present an excellent hierarchy with a main mass (99.85% of the total) and with well separated, almost circular and almost coplanar planetary orbits. The planetary perturbations are small, the chance of a quasi-periodic motion are high and if the true motion was temporarily chaotic, the instability would require billions of years (and even much more) to develop.

As already stated in Section 10.7.7.5 the differences between the Arnold diffusion and the true stability are sometimes purely theoretical.

11.12 AN APPLICATION OF QUALITATIVE METHODS. THE CONTROVERSY BETWEEN Mrs KAZIMIRCHAK-POLONSKAYA AND Mr R. DVORAK.

In a series of beautiful numerical studies Mrs Kazimirchak--Polonskaya and her colleagues have shown how comets can undergo drastic perturbations in the vicinity of planets and can, for instance, jump in a few years from the "Saturn family" to the "Jupiter family" or to any other family((for instance ref 258,259).

However a very surprising and much more intriguing example was found, the example of the comet "N-4" that was captured by Neptune in a few years and reached a satellite orbit of period 11.23 days, very near to Neptune (Ref. 260).

Mr Dvorak (Ref. 261) has attempted to verify this unexpected capture with the same hypotheses (solar and planetary attractions, no other attractions, no drag or other non-conservative influence, no oblateness of Sun and planets). However his computations were unsuccessful and, as usual among numericists, the controversy began by considerations on suitable initial conditions, method and accuracy of computation, planetary masses and model of planetary motion, etc ...

Fortunately a very simple qualitative test is able to clarify the question (Ref. 262).

Let us define the function h_N by :

$$h_N = (V^2/2) - (\mu_N/r) \qquad (1289)$$

with :

$$\left.\begin{aligned}
&V = ||\vec{V}|| = ||\vec{V}_{comet} - \vec{V}_{Neptune}|| = \\
&\qquad = \text{neptunocentric velocity of the comet.} \\
&\mu_N = \text{gravitational constant of Neptune} \\
&r = \text{distance Neptune-comet} \\
&r = ||\vec{r}|| \;;\; \vec{r} = \vec{r}_{comet} - \vec{r}_{Neptune} \;;\; \vec{V} = d\vec{r}/dt
\end{aligned}\right\} \qquad (1290)$$

The philosophy of the test is similar to that of usual tests of escape : we look for the possible values of dh_N/dt for a given value of h_N.

The equations of motion are :

$$\left.\begin{aligned}
&\vec{V} = d\vec{r}/dt \\
&d^2\vec{r}/dt^2 = -\frac{\mu_N\vec{r}}{r^3} + \sum_{j=1}^{K} Gm_j\left[\frac{\vec{r}_j - \vec{r}_c}{||\vec{r}_j - \vec{r}_c||^3} - \frac{\vec{r}_j - \vec{r}_N}{||\vec{r}_j - \vec{r}_N||^3}\right]
\end{aligned}\right\} \qquad (1291)$$

where :

$$\left.\begin{aligned}
&m_1 = \text{mass of the Sun} \\
&m_2,\, m_3,\, \ldots\, m_K = \text{mass of the remaining planets} \\
&\vec{r}_c = \vec{r}_{comet} \;;\; \vec{r}_N = \vec{r}_{Neptune} \\
&\vec{r}_j = \text{radius-vectors of the Sun and the remaining planets}
\end{aligned}\right\} \qquad (1292)$$

Hence :

$$\left.\begin{aligned} dh_N/dt &= \vec{V}(d^2\vec{r}/dt^2) + \mu_N(\vec{r}\vec{V}/r^3) = \\ &= \vec{V}\sum_{j=1}^{K} Gm_j\left[\frac{\vec{r}_{Nj} - \vec{r}}{||\vec{r}_{Nj} - \vec{r}||^3} - \frac{\vec{r}_{Nj}}{r_{Nj}^3}\right] \end{aligned}\right\} \quad (1293)$$

with :

$$\vec{r}_{Nj} = \vec{r}_j - \vec{r}_N \quad ; \quad r_{Nj} = ||\vec{r}_{Nj}|| \quad (1294)$$

Let us define the following critical value h_c, with A.U. = = Astronomical Unit :

$$h_c = -\mu_N/(1.A.U.) \quad (1295)$$

Hence :

$$h_N \leqq h_c \quad \text{implies} \quad r \leqq 1.A.U. \quad (1296)$$

and then :

$$\left|\left| \frac{\vec{r}_{Nj} - \vec{r}}{||r_{Nj} - r||^3} - \frac{\vec{r}_{Nj}}{r_{Nj}^3} \right|\right| \leqq \frac{2r}{r_{Nj}(r_{Nj} - 1.A.U)^2} \quad (1297)$$

and thus :

$$\left.\begin{aligned} &h_N \leqq h_c \quad \text{implies} \quad |dh_N/dt| \leqq QrV \\ &\text{with :} \\ &Q = 2G\sum_{j=1}^{K} \frac{m_j}{R_{Nj}\left[R_{Nj} - 1.A.U\right]^2} \\ &R_{Nj} = \text{a lower bound of } r_{Nj} \text{ for the period of interest} \end{aligned}\right\} \quad (1298)$$

For a given negative h_N the maximum of rV is $\mu_N/\sqrt{-2h_N}$ obtained for $r = -\mu_N/2h_N$ and $V = \sqrt{-2h_N}$.

Hence we arrive to :

$$h_N \leq h_c \quad \text{implies} \quad |dh_N/dt| \leq Q\mu_N/\sqrt{-2h_N} \tag{1299}$$

that can be integrated to :

$$\left.\begin{aligned} &\{h_N(t_1) \leq h_c \;;\; h_N(t_2) \leq h_c\} \Longrightarrow |t_2 - t_1| \geq \\ &\quad \geq \left((2\sqrt{2})/3Q\mu_N\right) \cdot \left| \left(-h_N(t_2)\right)^{3/2} - \left(-h_N(t_1)\right)^{3/2} \right| \end{aligned}\right\} \tag{1300}$$

We can simplify (1300) with the osculating neptunocentric mean angular motion n_N defined by the following usual Keplerian relation

$$n_N = \sqrt{(-8h_N^3/\mu_N^2)} \tag{1301}$$

Then :

$$\{h_N(t_1) \leq h_c \;;\; h_N(t_2) \leq h_c\} \Longrightarrow |t_2 - t_1| \geq \frac{|n_N(t_1) - n_N(t_2)|}{3Q} \tag{1302}$$

We need an evaluation of Q. Its main term is the solar term $2Gm_1/R_{N1}(R_{N1} - 1\ \text{A.U.})^2$. On the other hand the third Keplerian law gives $Gm_1 \simeq N^2A^3$ where N is the mean angular motion of Neptune and A its semi-major axis.

Since $A \simeq R_{N1}$ we will have $Q \simeq 2N^2$ and realistic evaluations of masses and distances in the Solar system give :

$$Q = 2.24\ N^2$$

Let us apply these results to the problem of interest.

$$\left.\begin{aligned} &h_N(t_1) \geq h_c \;;\; n_c = \sqrt{(-8h_c^3/\mu_N^2)} = 2\pi/49.14 \text{ years} \\ &h_N(t_2) < h_c \;;\; n_N(t_2) = \frac{2\pi}{11.23 \text{ days}} \;;\; \text{period } 11.23 \text{ days} \\ &N = 2\pi/164.76 \text{ years} \;;\; 164.76 \text{ years} = \text{orbital period of Neptune} \\ &|t_2 - t_1| \geq \frac{(2\pi/11.23 \text{ days}) - (2\pi/49.14 \text{ years})}{3 \times 2.24 \times (2\pi/164.76 \text{ years})^2} = \\ &\qquad = 20\ 987 \text{ years.} \end{aligned}\right\} \tag{1303}$$

Thus the solar and planetary perturbations will require at least 20 000 years to lead a comet from an initial neptunocentric distance larger than one A.U. to a satellite orbit about Neptune with a period of 11.23 days.

A lower bound ten times larger can easily be obtained with a better test using a kind of Sun-Neptune Jacobi integral instead of the faster function h_N defined in (1289).

11.13 THE LAGRANGIAN AND THE QUALITATIVE METHODS.

The Lagrangian L of a n-body problem is given by the classical expression :

$$L = 0.5 \sum_{K=1}^{n} m_K V_K^2 + G \sum_{1 \leq j < K \leq n} (m_j m_K)/(r_{jK}) \tag{1304}$$

that is, with the expressions of section 11.1 :

$$\left.\begin{aligned} &L = h + 2U \\ &h = \text{energy integral} \quad ; \quad U = \text{potential} \end{aligned}\right\} \tag{1305}$$

The n-body problem can be considered as the following problem of "optimization" :

$$\left.\begin{aligned} &\text{Parameter of description : the time } t \\ &\text{State parameters : the radius-vectors } \vec{r}_1, \vec{r}_2, \ldots, \vec{r}_n \\ &\text{Control parameters : the velocities } \vec{V}_1, \vec{V}_2, \ldots, \vec{V}_n \\ &\text{Given : } G, \text{ the n masses } m_1, m_2, \ldots, m_n, \\ &\qquad \text{the initial conditions } t_0, \vec{r}_{10}, \vec{r}_{20}, \ldots, \vec{r}_{n0} \text{ and} \\ &\qquad \text{the final conditions } t_f, \vec{r}_{1f}, \vec{r}_{2f}, \ldots, \vec{r}_{nf}. \end{aligned}\right\} \tag{1306}$$

Hamilton has shown that if all masses are positive the trajectories that <u>minimize</u> the integral :

$$\int_{t_0}^{t_f} L(t)dt \tag{1307}$$

are n-body trajectories.

We will not demonstrate this proposition which require elements of optimality theory, but let us show that it has many applications.

For instance if the final positions $\vec{r}_{1f}, \ldots, \vec{r}_{nf}$ are arbitrary the minimization of the integral (1307) obviously leads to $\vec{V}_{1f} = \vec{V}_{2f} = \ldots = \vec{V}_{nf} = 0$ and thus for any given masses and initial positions exist initial velocities leading to velocities equal to zero at a given final time t_f.

Let us now assume that $\vec{r}_{4f}, \vec{r}_{5f}, \ldots, \vec{r}_{nf}$ are given as well as t_0, t_f and the initial positions, and that we look for the minimization of :

$$\vec{\lambda}_1\vec{r}_{1f} + \vec{\lambda}_2\vec{r}_{2f} + \vec{\lambda}_3\vec{r}_{3f} + \int_{t_0}^{t_f} L(t)dt \tag{1308}$$

with some given vectors $\vec{\lambda}_1$, $\vec{\lambda}_2$, $\vec{\lambda}_3$.

This minimization leads to a n-body trajectory with the final velocities :

$$\vec{V}_1 = -\vec{\lambda}_1/m_1 \quad ; \quad \vec{V}_2 = -\vec{\lambda}_2/m_2 \quad ; \quad \vec{V}_3 = -\vec{\lambda}_3/m_3 \tag{1309}$$

Hence for any given masses and initial positions at least one set of initial velocities exists leading at a given final time t_f to given final positions or velocities of the n masses. This is a beautiful proposition applicable even to problems with 5000 bodies !

There are many other applications of this method of the Lagrangian. However we notice that along a given n-body trajectory the integral (1307) of the Lagrangian is minimal only if $t_f - t_0$ is sufficiently small. For large $t_f - t_0$ the integral is stationary but no longer minimal.

In the planar two-body problem the limit of local minimality is one revolution and the limit of global minimality is half a revolution (i.e. $\vec{r}_0$ and $\vec{r}_f$ of opposite directions). This is also the limit of both local and global minimality in the three--dimensional two-body problem.

Chapter 12

MAIN CONJECTURES AND FURTHER INVESTIGATIONS

We have met many conjectures of which a few only have rencently been solved.

A) The two Chazy conjectures on original and final evolutions (Sections 11.9.1 and 11.9.5).

The first Chazy onjecture on general three-body systems of positive energy is always true. The second Chazy conjecture on the impossibility of exchange motions for three-body systems of negative energy is only partially true, it requires a product c^2h below a negative critical value (function of the masses).

The determination of this critical value somewhere between zero and the limit of Hill-type stability is one of the most interesting further investigations.

B) The conjectures on regularization of triple or multiple collisions are almost completely solved.

The triple or multiple collisions of the planar or three-dimensional n-body problem are never Easton-regularizable (regularization by continuity) and almost never Siegel-regularizable (analytical regularization).

The same is true for collinear multiple collision of positive masses unless B1) The number of colliding masses is only three, B2) These three masses have Simò mass ratios (full lines and intersection of dotted lines of Fig. 199), B3) The analysis is restricted to collinear motions of zero-energy.

There are several classes of regularization of triple or multiple collisions of rectilinear systems with infinitesimal masses (Ref. 50, 54).

C) The three Brown conjectures on the planar circular restricted three-body problem (Sections 9.1 and 11.5, Ref. 138-143).

C.1) "The long-period family of periodic orbits about L_4 ends in an homoclinic orbit doubly asymptotic to the point L_3 of Fig. 120 (Section 11.5)".

This conjecture was proven false by Deprit, Henrard and Garfinkel (Sections 9.1 and 11.5, ref. 138, 139, 141, 142). The limit is not L_3 but at any rate it is a very small periodic orbit about L_3.

C.2) "Horseshoe-shaped periodic orbits similar to the large Hill's curve exist" (see Hill's curves Fig. 23, Section 9.1).

This conjecture was proven true by the numerical computations of Taylor (Ref. 140) and the analysis of Garfinkel (Ref. 141, part 1).

C.3) "The family of horseshoe-shaped periodic orbits ends in homoclinic orbits doubly asymptotic to the collinear Lagrangian point beyond the minor primary".

This conjecture is true only for particular values of the mass ratio m_2/M (Ref. 141, part 6.).

Among the conjectures non-yet solved we find :

A) The Poincaré conjecture on periodic orbits : the periodic orbits are dense in the set of bounded orbits.

This conjecture is certainly true and can have perhaps the two following improvements in the three or n-body problem :

A.1) The periodic orbits are dense in the sets of orbits without hyperbolic or superhyperbolic expansion (bounded orbits, oscillatory orbits of the two kinds, orbits with parabolic escapes).

A.2) In the general (i.e. non restricted) three or n-body problem the periodic orbits <u>stable in the linear approximation</u> are dense in a set that contains almost all bounded and oscillatory solutions.

B) The Arnold diffusion conjecture on the set of solutions.

This conjecture, presented in Section 11.11, divide the solutions into three main types : the periodic or quasi-periodic solutions, the open solutions coming from infinity and going back to infinity and the chaotic solutions. There are also some exceptional solutions filling a set of measure zero as the asymptotic solutions to an unstable periodic orbit.

This conjecture can certainly be applied to all analytic autonomous Hamiltonian systems (which is not the case for the Poincaré conjecture).

Notice that when, after reduction and simplification, the remaining autonomous Hamiltonian system has more than two degrees of freedom its Arnold tori of quasi-periodic orbits cannot enclose a part of phase-space (the same happens to periodic and non-

-autonomous Hamiltonian systems with more than one degree of freedom). In these cases only the integrals of motion can limit the expansion of chaotic motions.

This latter point explains the qualitative differences between the plane circular restricted three-body problem (Hill stability, two degrees of freedom) the three-dimensional circular restricted three-body problem (Hill-stability, three degrees of freedom), the rectilinear general three-body problem (no Hill-type stability, two degrees of freedom) and most of the other three or n-body problems (no boundeness given by the integral of motions, even in the case of Hill-type stability, more than two degrees of freedom). In this last case almost all chaotic motions disappear and the Arnold diffusion conjectue leads to the following surprising conjecture on the structure of solutions : the open solutions are <u>everywhere dense</u> in phase-space even if they are rare and "temporarily chaotic" in the regions of phase-space where dominate the periodic and quasi-periodic solutions.

C) A second conjecture related to the Arnold diffusion conjecture is the following : the linearly unstable periodic and quasi-periodic orbits seems to be infinitely rare. Hence almost all bounded orbits with positive Liapunov characteristic exponents would be chaotic orbits (Ref. 263).

As the Arnold diffusion conjecture these two adjacent conjectures are certainly true for all analytic autonomous Hamiltonian systems.

D) The minor conjectures are the following.

D.1) Conjectures on the measure in phase-space of the two kinds of oscillatory motions :

Oscillatory motions of the first kind ($\lim\sup R = \infty$, $\lim\inf R < \infty$) are certainly infinitely rare.

Oscillatory motions of the second kind (R bounded ; $\lim\inf r = 0$; $\lim\sup r > 0$) correspond certainly to a set of positive measure.

D.2) The total dispersion conjecture (Chapter 6, Fig. 6).

This conjecture is exactly opposite to the idea of regularization. It seems to be true in the planar and the three-dimensional general three-body problems.

D.3) Conjectures on the limits of bounded motions.

Figures (193)-(195) present lower bounds of the moment of inertia for three-body bounded motions.

In the vicinity of the ordinate axis these bounds are so near to the limit of elliptic Euler motions that it is conjectured that, for given masses and angular momentum, the elliptic Euler motions give the greatest lower bound of the moment of inertia for bounded motions (and also for oscillatory motions).

This conjecture can certainly be extended to the n-body problem even if the present lower bounds are much lower.

A very different conjecture depending on the integral energy h and independent of the angular momentum is the following :

It seems that bounded three-body motions bear uneasily triple close approaches and it is suspected that for these motions, apart from a set of measure zero, the product Ih^2 has a positive lower bound function only of the masses.

According to the Kolmogorov-Arnold-Moser theorem positive sets of bounded motion can be found in the vicinity of linearly stable periodic orbits. Hence this conjecture must be tested with such periodic orbits.

For three equal masses the best of these periodic orbits is at present a pseudo-circular orbit of the family of Table I, page 283, between the orbits 31 and 32, with Ih^2 minimum = $= 1.33261\ G^2m_1^5$, that is $\rho/a = 1.08837$ in the notation of Section 11.6.1.

It is thus suspected that for almost all bounded orbits with three equal masses the mean quadratic distance ρ remain for all time larger than the generalized semi-major axis a and similar conjectures for unequal masses further reduce the already small domain of bounded orbits.

All these conjectures lead to many further investigations, there are also many other possible investigations :

Improvements of tests of escape, exact limits of bounded motions.

Analytico-numerical studies as the outstanding studies of Heggie, Hut and Bahcall (Section 10.4.3), e.g. extension to unequal masses (mass effects in formation and dispersion of binaries).

Computation of main families as the most symmetrical family (Section 10.8.2).

Original-final evolutions in the still unknown cases (e.g. collinear motions of non-zero energy), etc...

Remember the surprise given by the family of retrograde pseudo-circular orbits (Section 10.9.1)...

CONCLUSIONS

Recent years have seen considerable progress in the three and n-body problems and in the general theory of systems of differential equations.

These progress appears in all domains.

A) In numerical computating with the construction of ever better and faster computers.

Anyone is now able to do in a single day more computations than all mankind before 1950. These new possibilities have of course allowed to solve many old questions, even if they have raised new ones, and their most surprising success is the rediscovery in the 1960s of Poincaré chaotic motions.

These researches have underlined the very general character of chaotic motions that were once considered as mathematical curiosities. These motions now appear not only in the restricted three-body problem where Michel Hénon rediscovered them but also in almost all conservative systems of differential equations and in all domains of science and technology. At the same time their counterparts, the strange attractors and the fractal structures, have appeared in dissipative systems and are even more general.

B) In theoretical analysis the greatest recent progress are the Kolmogorov-Arnold-Moser theorem with all the subsequent developments, the analysis of the different transitions between quasi-periodic and chaotic motions, the near-resonance theorem and the theory of fractals by B. Mandelbrot.

All these progress are strongly connected with the three or n-body problem, including even the fractals theory since the structure of the set of bounded solutions appears to be a fractal.

C) For the three-body problem it is perhaps in topological and qualitative analysis that the recent progress have been the greatest with the extensions of the ergodic theorem, the Hill--type stability, the new tests of escape that almost reach the extreme bounds and the construction of a new and accurate picture of the set of solutions.

This new picture has upset many old ideas on Celestial Mechanics.

The bounded motions are much rarer than was widely expected and furthermore in the general three or n-body problems these bounded motions are certainly <u>nowhere dense</u> in phase-space ; their structure in succesive Arnold tori seems to be a fractal or a Cantor set of positive measure , see Fig 207 page 526.

The motions with hyperbolic escapes are certainly everywhere dense even if some of these escapes are extremely slow and "temporarily chaotic".

In these conditions the old definition of Liapunov stability becomes useless : all general three or n-body orbits are Liapunov--unstable (and even Poincaré-unstable) and we must now use a "long--term stability" related to the process of Arnold diffusion : the Solar-System has the long-term stability, it will undergo very few change in billions of years and its major sources of evolution are the small non-conservative forces such as tidal forces.

The small size of the zone of bounded motions in phase-space has strong consequences for the evolution and the stability of many-body systems.

I) With the exception of very particular cases, such as the vicinity of the Schubart orbit (see Section 10.9.1), the long--term stability of a many-body system require a small level of perturbations.

A first possibility is given by the Solar-System with its well defined hierarchy of Sun-planets-satellites and its well sparated, nearly circular and nearly coplanar orbits : this situation certainly reflects the origin of the system but different origins have led to collapse or dispersion.

A second possibility is that of multi-stellar systems : the larger mass-ratios are compensated by much wider separations of successive orbits.

II) A second consequence is the brittle character of the stability : the stability of the Solar-System would probably not very long survive to the passage of a star at two or three times the distance of Pluto.

Another surprise is the existence of a set of "oscillatory motions of the second kind". This set certainly has a positive measure in phase-space and these motions lead either to collisions, and thus perhaps to a significant part of the nova phenomenon, or more probably to the formation of very close binary stars by tidal energy losses inside triple or multiple stellar systems. The large number of very close binary stars is certainly related to this type of motion.

These oscillatory motions are necessarily transitory, but all other main types of motions can be found among usual astronomical motions :

A) The motions of most triple stellar systems are of quasi-periodic type.

B) The motion of the nine major planets is essentially of quasi-periodic type but seems to show a chaoticity of small amplitude (Ref. 264).

This phenomenon looks like the motion of sea level : at the decimetric scale the tides have the regularity of astronomical phenomena and can be predicted well in advance, but the passage of barometric depressions modifies the sea level by a few centimeters and introduce the chaoticity of meteorology.

C) The motion of comets is either open or temporarily chaotic: after some centuries or millenia a dangerous close approach to a major planet throw them away out of the Solar System.

D) Asteroids and minor planets have either a quasi-periodic motions or a temporary chaotic motion but here "temporary" means "for some billions of years" or even much more.

E) The motion of galaxies and star clusters is of open type and they are continuously losing some stars.

F) A true typical chaotic motion can be found in the rotation of Hyperion. This small satellite of Saturn has strong and erratic swings that cannot be of temporary chaotic type since the domain of possible orientations is bounded and no "escape to infinity" is possible.

These different types of motion have very different statistical properties especially for long-terms correlations. They are now discovered in almost all problems of conservative motions.

Thus once more the Celestial Mechanics and the study of the three-body problem have been the source of major progress : the discoveries of the importance of chaotic motions and of oscillatory motions of the second kind.

The chaotic motions do not exist in integrable problems and become very general far from integrability. They are certainly the reason for the difference between the mathematical and the physical meanings of the word reversibility : the kinetic theory of gases is mathematically reversible and physically irreversible. This is because it is impossible to reverse or even to reproduce accurately a given chaotic motion of millions or billions of molecules.

The deep and fascinating insights provided by study of the three-body problem ensure that it will certainly continue to accompany the progress of Mathematics, Mechanics, Physics, Astronomy.

Fig 207. Example of the construction of a nowhere dense Cantor set of positive measure on the segment [0;1].

A) Exclusion of the central third

B) Exclusion of the central ninth of the two remaining segments

C) Exclusion of the central twenty-seventh of the four remaining segments

D) Etc...,with always the proportion 3^{-n} *of the* $2^{(n-1)}$ *remaining segments*

The resulting nowhere dense Cantor set has a measure equal to:

$$\prod_{n=1}^{\infty}\left(1-3^{-n}\right) = 0.560\ 126$$

APPENDIX I

SOME ASYMPTOTIC MOTIONS IN A VERY PARTICULAR CASE OF INTEGRABILITY OF THE THREE-BODY PROBLEM

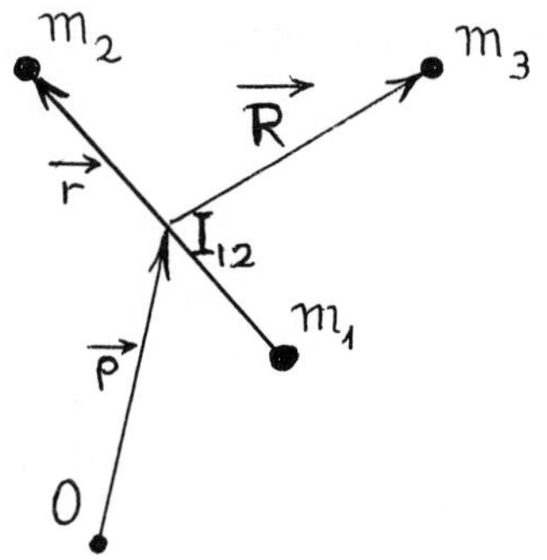

Fig. A.1 - The Jacobi decomposition with I_{12} at the center of masses m_1 and m_2. When the total mass is zero the general center of mass disappears and the absolute motion is related to the evolution of the vector $\overrightarrow{O.I_{12}} = \vec{\rho}$

Let us consider the very unrealistic case of negative masses and the three-body problem when the total mass M is zero.

The Jacobi equations of motions (18)-(23) presented in Section 4.3 remain unchanged and the equation (22) gives then :

$$d^2\vec{R}/dt^2 = -GM(\alpha\vec{r}_{13}\, r_{13}^{-3} + \beta\vec{r}_{23}\, r_{23}^{-3}) = 0 \qquad \text{(A1)}$$

Hence when the total mass M is zero, the $\vec{R}$ vector of the Jacobi decomposition (Fig A.1) has a uniform and rectilinear motion:

$$\vec{R} = \vec{A} + \vec{B}t \quad ; \quad \vec{A} \text{ and } \vec{B} : \text{constant vectors} \qquad \text{(A2)}$$

This result is also a consequence of the usual "integral of the center of mass" :

$$\left.\begin{array}{l} d^2(m_1\vec{r}_1 + m_2\vec{r}_2 + m_3\vec{r}_3)/dt^2 = 0 \\ M = \text{total mass} = 0 \text{ implies : } \quad m_1\vec{r}_1 + m_2\vec{r}_2 + m_3\vec{r}_3 = m_3\vec{R} \end{array}\right\} \qquad \text{(A3)}$$

The problem of the relative motion of the three bodies is then reduced to that of the evolution of the Jacobi vector $\vec{r}$:

$$\left.\begin{aligned} &d^2\vec{r}/dt^2 = Gm_3\left[\frac{\vec{r}}{r^3} + \frac{\vec{r}_{23}}{r_{23}^3} - \frac{\vec{r}_{13}}{r_{13}^3}\right] \\ &\text{with : } \vec{r}_{13} = \vec{R} + \beta\vec{r} \quad ; \vec{r}_{23} = \vec{R} - \alpha\vec{r} \quad ; \quad \vec{R} = \vec{A} + \vec{B}t \\ &\alpha = m_1/(m_1 + m_2) \qquad \beta = m_2/(m_1 + m_2) \end{aligned}\right\} \quad (A4)$$

Finally the problem of the absolute motion cannot be reduced to that of the center of mass, since when M = 0 the center of mass disappears, and we can use the vector $\vec{\rho} = \overrightarrow{OI_{1,2}}$ of Fig. A.1 :

$$\left.\begin{aligned} &d^2\vec{\rho}/dt^2 = -G(m_1\vec{r}_{13}\, r_{13}^{-3} + m_2\vec{r}_{23}r_{23}^{-3}) \\ &\vec{r}_1 = \vec{\rho} - \beta\vec{r} \quad ; \quad \vec{r}_2 = \vec{\rho} + \alpha\vec{r} \quad ; \quad \vec{r}_3 = \vec{\rho} + \vec{R} \end{aligned}\right\} \quad (A5)$$

This final problem requires only three quadratures as soon as the problem (A4) of relative motion is solved. These quadratures are sometimes simplified by the integrals of motion :

$$\left.\begin{aligned} \vec{c} &= \text{angular momentum} = \sum_{k=1}^{3} m_k\vec{r}_k \times \vec{V}_k \\ &= m\vec{r} \times \vec{v} + m_3\left[\vec{A}\times\vec{B} + \vec{\rho}\times\vec{B} + (\vec{A} + \vec{B}t) \times \frac{d\vec{\rho}}{dt}\right] \\ &\text{with : } \vec{V}_k = d\vec{r}_k/dt \quad ; \quad m = m_1 m_2/(m_1 + m_2) \quad ; \quad \vec{v} = d\vec{r}/dt \end{aligned}\right\} \quad (A6)$$

$$h = \text{energy} = \frac{1}{2}(mv^2 + m_3 B^2) + m_3\vec{B}.\frac{d\vec{\rho}}{dt} - G\left[\frac{m_1 m_2}{r} + \frac{m_1 m_3}{r_{13}} + \frac{m_2 m_3}{r_{23}}\right] \quad (A7)$$

Notice that, if $\vec{A} \times \vec{B} = 0$, the component of $\vec{r} \times \vec{v}$ in the direction of $\vec{A}$ and $\vec{B}$ is an integral of relative motion. Furthermore, if $\vec{B} = 0$, the energy integral becomes another integral of relative motion.

There are three main cases of integrability :

I) The case when $\vec{A} = \vec{B} = 0$

The configuration of the three bodies remains collinear and constant. It is a central configuration and the motion is a Eulerian motion

II) The case of collinear motions with $\vec{B} = 0$

The energy integral (A 7) is then sufficient for the integration of the motion.

III) The case of Sitnikov motions with $\vec{B} = 0$

In Sitnikov motions (Fig. A2 and Section 11.7.8) two bodies with the same mass have symmetrical motions about an axis and the third body moves along this axis of symmetry.

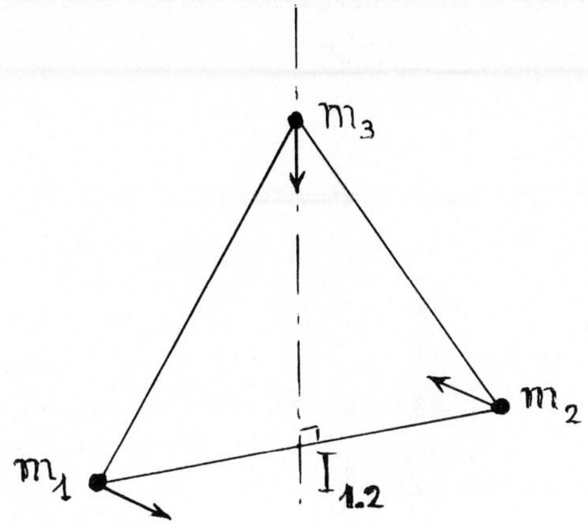

Fig. A.2 - The symmetry of Sitnikov motions. $m_1 = m_2$; $r_{13} \equiv r_{23}$

Since the total mass is here zero, the three masses m_1, m_2, m_3 are proportionnal to 1 ; 1 ; -2 and the relative motion is given by the following.

$$\left.\begin{array}{l} \overrightarrow{I_{12}, m_3} = \vec{R} = \vec{A} = \text{constant vector} \\ \vec{r}.\vec{R} \equiv 0 \quad ; \quad r_{13}(t) \equiv r_{23}(t) \equiv \left[A^2 + \dfrac{r^2(t)}{4}\right]^{1/2} \\ d^2\vec{r}/dt^2 = Gm_3\,\vec{r}\left(r^{-3} - r_{13}^{-3}\right) \end{array}\right\} \quad (A8)$$

This system has two integrals of motion related to (A6) and (A7) :

$$\left.\begin{array}{l} \vec{r} \times \vec{v} = \vec{D} \ ; \ \text{with } \vec{v} = d\vec{r}/dt \\ K = v^2 + 4Gm_1\left[\dfrac{4}{r_{13}} - \dfrac{1}{r}\right] = \text{constant} = 4h/m_1 \end{array}\right\} \quad (A9)$$

These integrals lead to the following quadrature with the five constants K, D, G, m_1 , A :

$$\left(\frac{dr}{dt}\right)^2 = K + 4Gm_1\left[\frac{1}{r} - \frac{8}{\sqrt{r^2+4A^2}}\right] - \frac{D^2}{r^2} \qquad (A10)$$

The study of quadratures as (A.10) is classical, it leads usually either to escape motions (only if $K \geq 0$) or to periodic motions between a pericenter distance r_{min} and an apocenter distance r_{max}

If we call f(r) the right member of (A.10) we can have asymptotic motions with $r \to r_\infty$ when $t \to \infty$ if f(r) verifies the following:

$$f(r_\infty) = 0 \quad ; \quad f(r) \text{ has a local minimum for } r = r_\infty \qquad (A11)$$

This happens in the following conditions :

$$\left.\begin{aligned} &0.684219\ldots < r_\infty/A \leqq 2/\sqrt{3} = 1.1547005.. \\ &Gm_1 > 0 \\ &K = 2Gm_1\left[\frac{24r_\infty^2 + 64A^2}{(r_\infty^2 + 4A^2)^{3/2}} - \frac{3}{r_\infty}\right] \\ &D^2 = 2Gm_1\left[\frac{8r_\infty^4}{(r_\infty^2 + 4A^2)^{3/2}} - r_\infty\right] \end{aligned}\right\} \quad (A12)$$

These examples of asymptotic motions in the three-body problem are the only known examples of integrable asymptotic three-body motions but they are not realistic since they require a negative mass.

APPENDIX II

THE HALO ORBITS
COMPLEMENTS
THE MICHALODIMITRAKIS FAMILY OF HALO ORBITS

The twelve parameters F_0 to G_5 of equations (841) - (847), Section 10.8.3, are used for the description of Halo orbits. They are complicate rational fractions of the four parameters K_2, K_3, K_4, k.

K_2, K_3, K_4 are functions of the mass ratio and of the Lagrangian point of interest (equations (805) - (807) and Figs 81, 82) while k is the pulsation of the Halo orbit of interest (equations (812),(813)). The period of the Halo orbit is equal to that of the primaries divided by k.

The twelve parameters F_0 to G_5 are given by the following expressions with the auxilliary parameters D_1, D_2, E, F, G, H, A_0 to A_3, B_0 to B_3.

$$D_1 = K_3^2\left\{\frac{2}{1+2K_2} + \frac{1-K_2+4k^2}{(1-4k^2)^2+K_2(1+4k^2)-2K_2^2}\right\} - K_4 \qquad (A13)$$

D_1 is the denominator of the expression of c_1^2 in (840)

$$D_2 = \frac{3K_3}{4\{2K_2^2 - K_2(1+4k^2)-(1-4k^2)^2\}} \qquad (A14)$$

D_2 appears in the expressions (836) and (837) of a_2 and b_2.

$$E = 8(k^2 - K_2)/9D_1 \qquad (A15)$$

$$F = 2 + \frac{1}{D_1}\left\{\frac{4K_3^2}{K_2} - \frac{2K_3^2}{K_2-4k^2} - 2K_4\right\} \qquad (A16)$$

$$G = 8kK_3^2/D_1\{(1-4k^2)^2 + K_2(1+4k^2) - 2K_2^2\} \qquad (A17)$$

$$H = \frac{1}{D_1}\left\{\frac{(1-K_2+4k^2)K_3^2}{(1-4k^2)^2+K_2(1+4k^2)-2K_2^2} - \frac{2K_3^2}{1+2K_2} + \frac{K_4}{3}\right\} \qquad (A18)$$

The expression of c_1^2 given in (840) is then :

$$c_1^2 = E + Fa_1^2 + Ga_1 b_1 + Hb_1^2 \tag{A19}$$

$$\left.\begin{aligned} A_0 &= (1 - K_2 + 4k^2)\, ED_2 \\ A_1 &= (1 - K_2 + 4k^2)\,(F - 2)D_2 \\ A_2 &= [(1 - K_2 + 4k^2)G - 8k]\, D_2 \\ A_3 &= (1 - K_2 + 4k^2)(H - 1)D_2 \end{aligned}\right\} \tag{A20}$$

$$\left.\begin{aligned} B_0 &= -4kED_2 \\ B_1 &= 4kD_2\,(2 - F) \\ B_2 &= 2D_2\,(1 + 2K_2 - 2kG + 4k^2) \\ B_3 &= 4kD_2\,(1 - H) \end{aligned}\right\} \tag{A21}$$

The expression of a_2 and b_2 given in (836), (837) are then:

$$a_2 = A_0 + A_1 a_1^2 + A_2 a_1 b_1 + A_3 b_1^2 \quad ; \quad b_2 = B_0 + B_1 a_1^2 + B_2 a_1 b_1 + B_3 b_1^2$$

and the expressions of the twelve parameters F_0 to G_5 arise from the comparison of (818) and (841) ; of (821) and (842) :

$$\left.\begin{aligned} F_0 &= 1+2K_2+k^2-3K_3A_0+4.5K_3^2E\left\{K_2^{-1}+(1+2K_2)^{-1}+(2K_2-8k^2)^{-1}\right\}-4.5K_4E \\ F_1 &= 2k+1.5K_3B_0 \\ F_2 &= -3K_3A_1+4.5K_3^2\left\{\frac{F}{K_2}+\frac{F-2}{1+2K_2}+\frac{F}{2K_2-8k^2}\right\}+1.5K_4(2-3F) \\ F_3 &= 1.5K_3(B_1-2A_2)+4.5K_3^2G\left\{K_2^{-1}+(1+2K_2)^{-1}+(2K_2-8k^2)^{-1}\right\}-4.5K_4G \\ F_4 &= 1.5K_3(B_2-2A_3)+4.5K_3^2\left\{\frac{H}{K_2}+\frac{H+1}{1+2K_2}+\frac{H}{2K_2-8k^2}\right\}-1.5K_4(1+3H) \\ F_5 &= 1.5K_3B_3 \end{aligned}\right\} \tag{A22}$$

$$\left.\begin{aligned} G_0 &= F_1 = 2k+1.5K_3B_0 \\ G_1 &= 1-K_2+k^2-1.5K_3A_0-9K_3^2E(4+8K_2)^{-1}+0.375K_4E \\ G_2 &= 1.5K_3B_1 \\ G_3 &= 1.5K_3(B_2-A_1)+9K_3^2(2-F)(4+8K_2)^{-1}+0.375K_4(F-4) \\ G_4 &= 1.5K_3(B_3-A_2)-9K_3^2G(4+8K_2)^{-1}+0.375K_4G \\ G_5 &= -1.5K_3A_3-9K_3^2(H+1)(4+8K_2)^{-1}+0.375K_4(H+3) \end{aligned}\right\} \tag{A23}$$

Thus the third-order Lindstedt-Poincaré method exposed in this appendix gives a long but also straightforward solution for the Fourier development (812) of Halo orbits.

The accuracy of this method is good for small Halo orbits but poor for large ones and numerical computations are then necessary.*

---

* The results of this Lindstedt-Poincaré method are even better for the families of orbits of Fig.84, p.268.

Let us consider the simplest of these numerical computations: the Halo orbits of the Hill problem.

THE MICHALODIMITRAKIS FAMILY OF HALO ORBITS

In the natural two-body systems most mass ratios are very small :

$$\left.\begin{array}{l}\text{Mass of the Moon / Mass of the Earth} = 0.012\ 300\ 02\\ \text{Mass of Jupiter / Mass of the Sun} = 0.000\ 954\ 786\\ \text{Mass of the Earth / Mass of the Sun} = 0.000\ 002\ 999\ 1\end{array}\right\} \quad (A24)$$

Only binary stars present large mass-ratios.

In these conditions it is interesting to study the limit case when the mass-ratio is going to zero. This limit is of course useless far from the small primary (the motions are there usual Keplerian motions) but it becomes very useful, with suitable units, in the vicinity of the small primary.

This problem is the Hill-problem presented in Section 9.2 with the small primary at the origin and with the following equations of motion :

$$\left.\begin{array}{l}\vec{r} = \begin{pmatrix} x \\ y \\ z \end{pmatrix} \quad \text{: in the rotating set of axes} \\[2ex] \dfrac{d^2\vec{r}}{dt^2} = \vec{r}\,'' = \begin{pmatrix} x'' \\ y'' \\ z'' \end{pmatrix} = -\dfrac{\vec{r}}{r^3} + \begin{pmatrix} 3x + 2y' \\ -2x' \\ -z \end{pmatrix}\end{array}\right\} \quad (A25)$$

There is an integral of motion, the Jacobi integral Γ :

$$\Gamma = \frac{2}{r} + 3x^2 - z^2 - x'^2 - y'^2 - z'^2 \qquad (A26)$$

There are two equilibrium points : the Lagrangian points L_1 and L_2 along the x-axis at the symmetrical abscissas $\pm 3^{-1/3}$ that is $\pm 0.693\ 361...$

Notice the following .

A) The Hill problem (A25) has the (z/-z) and the (y,t/-y,-t)symmetries of all circular restricted three-body problems, but it has also the (x,y/-x,-y)symmetry. Hence it is sufficient to consider Halo orbits in the vicinity of L_1 only.

B) The values of the coefficients K_2, K_3, K_4 are presented in Section 10.8.3, between the equations (807) and (808) for several usual two-body systems, and in Figs. 81, 82.

With $K_2 = 4$; $K_3 = K_4 = 3$ the Hill problem appears to be intermediate between the problems of motions about L_1 (larger K_n) and motions about L_2 (smaller K_n).

C) The set of axes of Section 10.8.3 (Fig. 80) is not that of (A25) used by Michalodimitrakis for his computations (Ref 71).

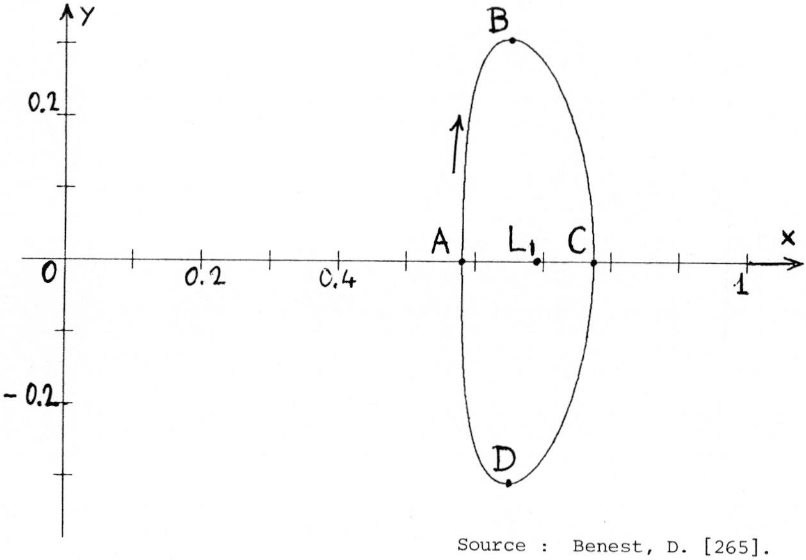

Source : Benest, D. [265].

Fig. A.3 - The first orbit of the Michalodimitrakis family of Halo orbits.

This orbit is in the Oxy plane and has a y/-y symmetry.

	t	x	x'	y	y'
A	0	0.581 26	0	0	0.670 12
B	T/4	0.653 79	0.208 59	0.308 19	-0.023 75
C	T/2	0.774 66	0	0	-0.613 80

Period : $T = 3.081\ 44$

Jacobi constant : $\Gamma = 4.005\ 32$

The first orbit of the Michalodimitrakis family is a plane periodic orbit (Fig A.3 and Ref 265)that belongs also to the Strömgren family "a" that is the Matukuma family "F" (the planar symmetric family of periodic orbits beginning at L , réf 104).

This first orbit is thus at the bifurcation of two families.

The "Michalodimitrakis family" of Halo orbits in the Hill problem is the family a1v of ref 71 . Fig.A.4. presents the three main projections of an average orbit of this family, while Fig.A.5 presents the main points ABC for the 14 orbits computed by Michalodimitrakis.

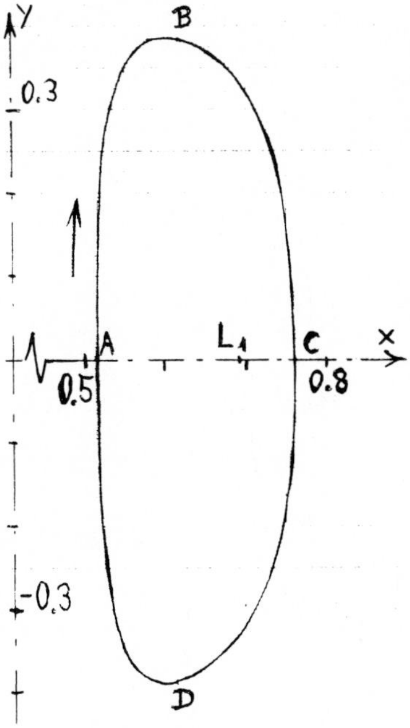

Fig.A.4 - The three main projections of an Halo orbit of the Michalodimitrakis family (i.e. the Halo orbits of the Hill problem).

This orbit has the y/-y symmetry and the following points and velocities :

	t	x	y	z
A	0	0.51322	0	0.2
B	T/4	0.61702	0.38574	-0.06632
C	T/2	0.76148	0	-0.26858
		x'	y'	z'
	A	0	0.89008	0
	B	0.26803	-0.06179	-0.45856
	C	0	-0.74525	0

Period : T = 3.06730

Jacobi constant :   = 3.58894

Source : Michalodimitrakis, M. [71].

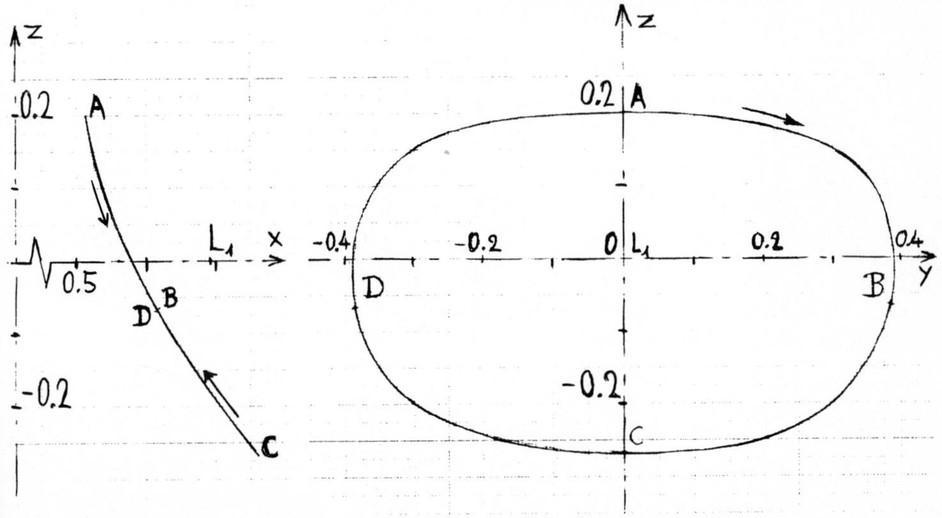

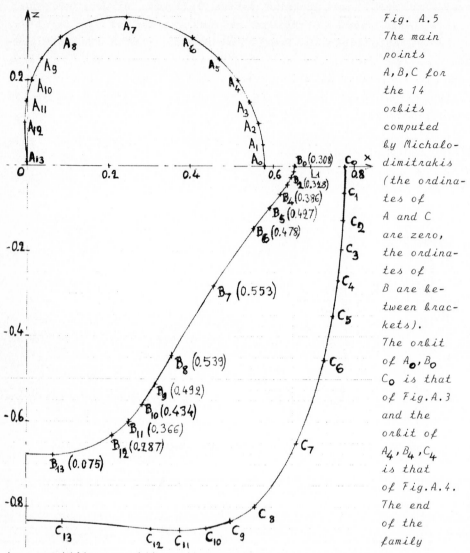

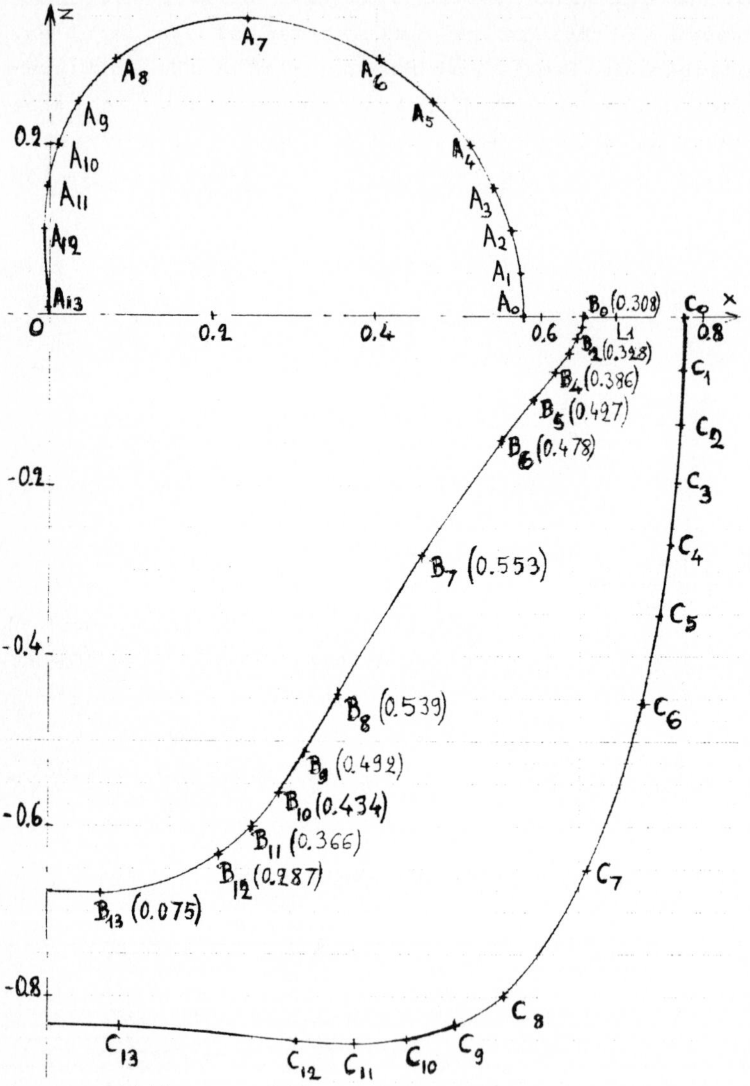

Fig. A.5 The main points A, B, C for the 14 orbits computed by Michalodimitrakis (the ordinates of A and C are zero, the ordinates of B are between brackets). The orbit of A_0, B_0, C_0 is that of Fig. A.3 and the orbit of A_4, B_4, C_4 is that of Fig. A.4. The end of the family is a rectilinear orbit on the z-axis.

The period T decreases continously along the family from $T_0 = 3.08144$ for the orbit A_0, B_0, C_0 to $T_{12} = 2.06573$ for the orbit A_{12}, B_{12}, C_{12} and $T_{13} = 1.47186$ for the orbit A_{13}, B_{13}, C_{13}. (Let us recall that the period of the primaries is 2π).

The figure presents only one-fourth of the whole family. It must be completed by the z/-z symmetry and by the x,y/-x,-y symmetry.

These two figures show the wide choice of Halo orbits for missions as those of Fig. 79 in Section 10.8.3 : radio-communication with the lunar far side, continuous observation of the Sun etc...

Let us add the two following final comments

A) The differences between the Hill problem and the usual restricted three-body problem, either circular or elliptic, lead to small modifications of Halo orbits (Ref. 73).

B) Most of the Halo orbits are unstable and require an ordinary "station keeping procedure" as those already used for geostationnary satellites.

There is a small zone of stable orbits in the vicinity of orbits A_{10} , B_{10} , C_{10} and A_{11} , B_{11} , C_{11} of Fig. A.5. This zone extends from $x_A = 0.011\ 62$; $z_A = 0.207\ 61$ to $x_A = -0.000\ 58$; $z_A = 0.166\ 29$

APPENDIX III

FULL DEMONSTRATION OF THE NEAR-RESONANCE THEOREM

The demonstrations of Sections 10.7.7.1 and 10.7.7.2 were valid only in the general case, i.e. when all coefficients ε_k of (454) or (516) are zero. This case is called "semi-simple" and includes all cases in which the n eigenvalues $\lambda_1,\dots,\lambda_n$ are different.

Hence for a full demonstration of the near resonance theorem we also need to consider the cases when some ε_k are equal to one, which corresponds to multiple eigenvalues.

The dynamical system of interest will be written :

$$\left.\begin{array}{l} \vec{X} = (x_1, x_2, \dots, x_n) = \text{state vector} \\ t = \text{time} = \text{description parameter} \\ d\vec{X}/dt = F(\vec{X},t) : \text{equation of motion} \end{array}\right\} \quad (A27)$$

The function F will be either autonomous ($F = F(\vec{X})$) or at least periodic in terms of t with the period P :

$$F(\vec{X},t) \equiv F(\vec{X},t+P) \quad (A28)$$

The near resonance theorem helps to analyse the vicinity and the stability of a periodic solution $\vec{X}(t)$ with the period T :

$$\left.\begin{array}{l} d\,\vec{X}(t)/dt \equiv F\left(\vec{X}(t),t\right) \\ \vec{X}(t) \equiv \vec{X}(t+T) \end{array}\right. \quad (A29)$$

In the case (A28) we will need : T multiple of P, which is usual.

The first classical simplifications (reduction of the solution of interest to the origin, suitable linear and periodic transformation) were presented in Section 10.7.7.2, they lead to the following equations of motion already presented in (516) :

$$\left.\begin{array}{l} dx_k/dt = \lambda_k x_k + \varepsilon_k x_{k+1} + g_k(x_1, x_2, \dots, x_n, t) \\ \lambda_k = k^{th} \text{ eigenvalue} \quad ; \quad \exp(\lambda_k T) = \mu_k = k^{th} \text{ Floquet multiplier} \\ \varepsilon_k = k^{th} \text{ Jordan coefficient} \\ \varepsilon_n = 0 \quad ; \quad \lambda_k \neq \lambda_{k+1} \text{ implies } \varepsilon_k = 0 \\ \lambda_k = \lambda_{k+1} \text{ implies } \varepsilon_k = \text{either 0 or 1} \end{array}\right\} k = \{1,2,\dots,n\} \quad (A30)$$

$$Re(\lambda_k) = k^{th} \text{ Liapunov characteristic exponent} \qquad \text{(A 30)}$$

The n functions $g_k(x_1, x_2, ..x_n, t)$ have the period T and have neither zero-order nor first-order term. The near-resonance theorem is applicable if these n functions are analytic or if, at least, they have in a small neighbourhood of the origin a converging development in "power-Fourier series" :

$$g_k(x_1, x_2, ..., x_n, t) = \sum_{\vec{\alpha},m} G(\vec{\alpha}, m) x_1^{\alpha_1} x_2^{\alpha_2} \cdots x_n^{\alpha_n} \exp\{m\, \omega_0 i t\} \qquad \text{(A31)}$$

with :

$$\left.\begin{array}{l} k = \{1, 2, ...n\} \quad ; \quad \vec{\alpha} = (\alpha_1, \alpha_2, ...\alpha_n) \quad ; \\ \alpha_1, \alpha_2, ..., \alpha_n \in \mathbb{N} \; ; \quad \alpha_1 + \alpha_2 + ... + \alpha_n \geq 2 \\ m \in \mathbb{Z} \; ; \quad G_k(\vec{\alpha}, m) \in \mathbb{C} \; ; \quad \omega_0 = 2\pi/T \end{array}\right\} \qquad \text{(A32)}$$

The near-resonance theorem is the natural prolongation of the transformations and simplifications that have led from the equation of motion (A27) to the equations of motion (A30) without zero-order term and with simple and constant first-order terms. The near-resonance theorem allows to build the converging power-Fourier series of transformations $\vec{x} \rightarrow \vec{u}$ that leads to equations of motion with far less terms : all non near-resonant terms disappear even those of very large orders.

The near-resonance theorem specifies that for any positive scalar ε, even very small, exists a transformation $\vec{x} \rightarrow \vec{u}$ with the following properties.

I) If the functions g_k of (A30) and (A31) are analytic the transformation $\vec{x} \rightarrow \vec{u}$ is also analytic, if not, this transformation is at least developable into power-Fourier series of x and t (or of u and t). These series are converging in a sufficiently small neighbourhood of the origin.

II) $\vec{x}$ and $\vec{u}$ have the same main part :

$$\vec{u} - \vec{x} = O(x^2) = O(u^2) \qquad \text{(A33)}$$

III) The transformation $\vec{x} \rightarrow \vec{u}$ is periodic in terms of t :

$$u(\vec{x}, t) \equiv u(\vec{x}, t+T) \qquad \text{(A34)}$$

IV) The equations of motions of u have far less terms :

$$du_k/dt = \lambda_k u_k + \varepsilon_k u_{k+1} + j_k(u_1, u_2, \dots, u_n, t) \quad ; \quad k = \{1, 2, \dots, n\} \tag{A35}$$

The $\lambda_k's$ and $\varepsilon_k's$ are those of (A30) while the power-Fourier series j_k have neither zero-order nor first-order term, are periodic in terms of t, are converging in a small neighbourhood of the origin and verify the following :

The power-Fourier series $j_k(\vec{u}, t)$ have only "sufficiently resonant terms", i.e. terms $J_k(\vec{\alpha}, m)\, u_1^{\alpha_1} \dots u_n^{\alpha_n} \exp\{m\omega_0 it\}$ for which:

$$\left| \lambda_k - m\omega_0 i - \sum_{q=1}^{n} \alpha_q \lambda_q \right| \leq \varepsilon(\alpha_1 + \alpha_2 + \dots + \alpha_n) \tag{A36}$$

The demonstration has two parts : the formal construction of the transformation $\vec{x} \to \vec{u}$ and the analysis of its convergence.

<u>Construction of the transformation $\vec{x} \to \vec{u}$</u>

Let us define the transformation $\vec{x} \to \vec{u}$ by the following expressions

$$x_k = u_k + \varphi_k(u_1, u_2, \dots, u_n, t) \quad ; \quad k = \{1, 2, \dots, n\} \tag{A37}$$

The n functions φ_k will be developped in power-Fourier series of period T :

$$\varphi_k(u_1, u_2, \dots, u_n, t) = \sum_{\vec{\alpha}, m} F(\vec{\alpha}, m) . u_1^{\alpha_1}, u_2^{\alpha_2} \dots u_n^{\alpha_n} \exp(m\omega_0 it) \tag{A38}$$

with, as in (A32) :

$$\left.\begin{array}{l} k = \{1, 2, \dots, n\} \quad ; \quad \vec{\alpha} = (\alpha_1, \alpha_2, \dots, \alpha_n) \quad ; \\ \alpha_1, \alpha_2, \dots, \alpha_n \in \mathbb{N} \quad ; \quad \alpha_1 + \alpha_2 + \dots + \alpha_n \geq 2 \quad ; \\ m \in \mathbb{Z} \quad ; \quad F(\vec{\alpha}, m) \in \mathbb{C} \quad ; \quad \omega_0 = 2\pi/T \end{array}\right\} \tag{A39}$$

The equations (A35) of motion of $\vec{u}$ should be deduced from (A30) and (A37) :

$$dx_k/dt = du_k/dt + d\varphi_k/dt \tag{A40}$$

that is :

$$\left.\begin{array}{l}\lambda_k x_k+\varepsilon_k x_{k+1}+g_k(x_1,\ldots,x_n,t) = \lambda_k u_k+\varepsilon_k u_{k+1}+j_k(u_1,\ldots,u_n,t)+\\ +\partial\varphi_k/\partial t+\sum_{q=1}^{n}\frac{\partial\varphi_k}{\partial u_q}\left[\lambda_q u_q+\varepsilon_q u_{q+1}+j_q(u_1,\ldots,u_n,t)\right]\end{array}\right\} \quad (A41)$$

that is, with (A37) for all $k = \{1,2,..n\}$:

$$\left.\begin{array}{l}\lambda_k\varphi_k(\vec{u},t)+\varepsilon_k\varphi_{k+1}(\vec{u},t)+g_k(\vec{x},t) = j_k(\vec{u},t)+\partial\varphi_k/\partial t+\\ +\sum_{q=1}^{n}\frac{\partial\varphi_k}{\partial u_q}\left(\lambda_q u_q+\varepsilon_q u_{q+1}+j_q(\vec{u},t)\right)\end{array}\right\} \quad (A42)$$

We can write (A42) in the following way :

$$\left.\begin{array}{l}g_k(\vec{x},t)-\sum_{q=1}^{n}\frac{\partial\varphi_k}{\partial u_q}.j_q(\vec{u},t) = j_k(\vec{u},t)+\frac{\partial\varphi_k}{\partial t}+\sum_{q=1}^{n}\frac{\partial\varphi_k}{\partial u_q}(\lambda_q u_q+\varepsilon_q u_{q+1})-\\ -\lambda_k\varphi_k(\vec{u},t)-\varepsilon_k\varphi_{k+1}(\vec{u},t) \qquad ; \qquad k = 1,..,n\end{array}\right\} \quad (A43)$$

Let us build the functions $\varphi_k(u,t)$ and $j_k(u,t)$ steps by steps in the following way :

Let us assume that the functions $\varphi_k(u,t)$ and $j_k(u,t)$ are known up to the order (p-1) and let us examine the terms of order p.

The left member of (A43) is known up to the order p in terms of u and t, indeed :

I) $g_k(x,t)$ is given in terms of $\vec{x}$ and t, it begins with second-order terms and, with (A37), $\vec{x}$ is known up to the order (p-1) in terms of u and t.

II) $\partial\varphi_k/\partial u_q$ begins at the order one and is known up to the order (p-2) while $j_q(u,t)$ begins at the order two and is known up to the order (p-1).

We can decompose this left member into power-Fourier series beginning at the second-order :

$$g_k(\vec{x},t)-\sum_{q=1}^{n}\frac{\partial\varphi_k}{\partial u_q}j_q(\vec{u},t) = \sum_{\vec{\alpha},m} B_k(\vec{\alpha},m)u_1^{\alpha_1}u_2^{\alpha_2}..\,u_n^{\alpha_n}\exp(m\omega_0 it) \qquad (A44)$$

and for the term in $u_1^{\alpha_1}u_2^{\alpha_2}\ldots u_n^{\alpha_n}\exp(m\omega_0 it)$ the equation (A43) gives :

$$\left.\begin{aligned} B_k(\vec{\alpha},m) &= J_k(\vec{\alpha},m)+F_k(\vec{\alpha},m).\left(m\omega_0 i+\sum_{q=1}\alpha_q\lambda_q-\lambda\right)+ \\ &+\sum_{q=1}^{n-1}(\alpha_q+1).\varepsilon_q F_k\left(\alpha_1,\alpha_2,\ldots,(\alpha_q+1)(\alpha_{q+1}-1),\alpha_{q+2},\ldots,\alpha_n,m\right)- \\ &-\varepsilon_k F_{k+1}(\vec{\alpha},m) \end{aligned}\right\} \quad (A45)$$

We will use the rule already used in Section 10.7.7.2 for the definition of the coefficients $J_k(\vec{\alpha},m)$ and $F_k(\vec{\alpha},m)$, that is :

If the near-resonance condition (A36) is satisfied the coefficient of $F_k(\vec{\alpha},m)$ in (A45) is considered as too small and the choice will be :

$$\left.\begin{aligned} F_k(\vec{\alpha},m) &= 0 \\ J_k(\vec{\alpha},m) &= B_k(\vec{\alpha},m)+\varepsilon_k F_{k+1}(\vec{\alpha},m)- \\ &-\sum_{q=1}^{n-1}(\alpha_q+1)\varepsilon_q F_k\left(\alpha_1,\ldots,(\alpha_q+1),(\alpha_{q+1}-1),\alpha_{q+2}\ldots\alpha_n,m\right) \end{aligned}\right\} \quad (A46)$$

On the contrary, if the near-resonance condition (A36) is not satisfied the near-resonance theorem specifies $J_k(\vec{\alpha},m)=0$ and the choice will be :

$$\left.\begin{aligned} J_k(\vec{\alpha},m) &= 0 \\ F_k(\vec{\alpha},m) &= \left(m\omega_0 i+\sum_{q=1}^{n}\alpha_q\lambda_q-\lambda_k\right)^{-1}\cdot\left[B_k(\vec{\alpha},m)+\varepsilon_k F_{k+1}(\vec{\alpha},m)-\right. \\ &\left.-\sum_{q=1}^{n-1}(\alpha_q+1)\varepsilon_q F_k\left(\alpha_1\ldots,(\alpha_q+1),(\alpha_{q+1}-1),\ldots\alpha_n,m\right)\right] \end{aligned}\right\} \quad (A47)$$

These conditions (A46), (A47) allow a unique and well-defined formal development of the functions $j_k(\vec{u},t)$ and $\varphi_k(\vec{u},t)$, indeed:

I) For the order p all $B_k(\vec{\alpha},m)$ are known.

II) Since $\varepsilon_n = 0$, the conditions (A46), (A47) give first the coefficients $J_n(p,0,0,\ldots,0,m)$ and $F_n(p,0,0,\ldots 0,m)$, then $J_n(p-1,1,0\ldots 0,m)$ and $F_n(p-1,1,0,\ldots 0,m)$ etc... up to $J_n(0,0,\ldots 0,p,m)$ and $F_n(0,0,\ldots 0,p,m)$.

III) All J_n and F_n of order p being known so are the terms $\varepsilon_{n-1}F_n$ and we can now obtain the coefficients J_{n-1} and F_{n-1} of order p.

We can proceed up to the coefficients J_1 and F_1 of order p, and since all coefficients of order p are now known we can look for the coefficients of order (p+1).

And so on...

Hence conditions (A46), (A47) associated with the near resonance condition (A36) allow a unique and well-defined formal development of the functions $j_k(\vec{u},t)$ and $\varphi_k(\vec{u},t)$ in power-Fournier series and the only remaining question is that of the convergence of these series.

DEMONSTRATION OF THE CONVERGENCE OF THE SERIES OF THE FUNCTIONS $j_k(\vec{u},t)$ AND $\varphi_k(\vec{u},t)$

We want to demonstrate the convergence of the 2n power-Fourier series $j_k(\vec{u},t)$ and $\varphi_k(\vec{u},t)$, at least in a small neighbourhood of the origin, and we will use the technique of the majorizing series.

The power-Fourier series $s_1(\vec{u},t)$ will majorize the power-Fourier series $s_2(\vec{u},t)$ if and only if :

$$\left.\begin{aligned} s_1(\vec{u},t) &= \sum_{\vec{\alpha},m} S_1(\vec{\alpha},m)\,u_1^{\alpha_1}\ldots u_n^{\alpha_n}\exp\{m\omega_0 it\} \\ s_2(\vec{u},t) &= \sum_{\vec{\alpha},m} S_2(\vec{\alpha},m)\,u_1^{\alpha_1}\ldots u_n^{\alpha_n}\exp\{m\omega_0 it\} \\ &\text{For all } \{\vec{\alpha},m\} : \ |S_2(\vec{\alpha},m)| \leq S_1(\vec{\alpha},m) \end{aligned}\right\} \quad (A48)$$

This property is written :

$$s_1(\vec{u},t) \succ s_2(\vec{u},t) \quad , \text{ or } \quad s_2(\vec{u},t) \prec s_1(\vec{u},t) \qquad (A49)$$

That is : "the series s_1 majorizes the series s_2" or "the series s_2 is majorized by the series s_1".

This implies of course :

I) All coefficients $S_1(\vec{\alpha},m)$ are real and non-negative.

II) The convergence of the series $s_1(\vec{u},t)$ implies that of the series $s_2(\vec{u},t)$.

These definitions are valid for the series of $\vec{x}$ and t, and the convergence of the power-Fourier series $g_k(x,t)$ is equivalent to the following :

There exists a positive scalar b and a converging series of positive numbers a_0, a_1, ..., a_q,... such that for all $k \in \{1,\ldots,n\}$:

$$g_k(\vec{x},t) \prec \varepsilon^k.g(t).\underbrace{(X^2 + bX^3 + b^2X^4 + b^3X^5 + \ldots)}_{\text{series of } \frac{X^2}{1-b\,X}} \qquad (A50)$$

with :

$$\left.\begin{aligned} g(t) &= a_o + \sum_{q=1}^{\infty} a_q \left(\exp\{q\omega_o it\} + \exp\{-q\omega_o it\}\right) \\ X &= x_1 + x_2 + \ldots + x_n \end{aligned}\right\} \quad (A50)$$

The domain of convergence of the functions $g_k(\vec{x},t)$ includes all points for which :

$$\sum_{q=1}^{n} |x_q| < b^{-1} \quad (A51)$$

Let us now define the n functions $h_k(\vec{u},t)$ by :

$$j_k(\vec{u},t) = \sum_{q=1}^{n} u_q \frac{\partial h_k}{\partial u_q} \quad (A52)$$

The corresponding coefficients $J_k(\vec{\alpha},m)$ and $H_k(\vec{\alpha},m)$ are related by :

$$J_k(\vec{\alpha},m) = (\alpha_1 + \alpha_2 + \ldots + \alpha_n).H_k(\vec{\alpha},m) \quad (A53)$$

and for any k the functions $j_k(\vec{u},t)$ and $h_k(\vec{u},t)$ have the same radius of convergence.

The equations (A43) and the rules related to (A46), (A47) give step by step the 2n functions $j_k(\vec{u},t)$ and $\varphi_k(\vec{u},t)$ and we can get majorizing series of these functions $j_k(\vec{u},t)$ and $\varphi_k(\vec{u},t)$ if we rewrite (A43) in the following way :

$$\left.\begin{aligned} & g_k(\vec{x},t) - \sum_{q=1}^{n} \frac{\partial \varphi_k}{\partial u_q}.\, j_q(\vec{u},t) - \sum_{q=1}^{n-1} \varepsilon_q \frac{\partial \varphi_k}{\partial u_q} u_{q+1} + \varepsilon_k \varphi_{k+1}(\vec{u},t) = \\ & = j_k(\vec{u},t) + \frac{\partial \varphi_k}{\partial t} - \lambda_k \varphi_k + \sum_{q=1}^{n} \lambda_q \frac{\partial \varphi_k}{\partial u_q} u_q \end{aligned}\right\} \quad (A54)$$

and if we majorize the left member of (A54) and minorize its right member.

Hence for all k the functions $\varepsilon \Psi_k(\vec{u},t)$ majorize $h_k(\vec{u},t)$; $\varepsilon \varphi_k(\vec{u},t)$ and the sums $h_k + \varepsilon \varphi_k$ if these functions $\varepsilon \Psi_k(\vec{u},t)$ are defined by (A50) and by :

$$\left.\begin{aligned} & x_k = u_k + \Psi_k(\vec{u},t) \quad ; \quad k = \{1,2,\ldots n\} \\ & \frac{\varepsilon^k.g(t).X^2}{1 - b\,X} + \varepsilon \sum_{q=1}^{n} \frac{\partial \Psi_k}{\partial u_q} \left\{ \sum_{p=1}^{n} \frac{\partial \Psi_q}{\partial u_p} u_p \right\} + \sum_{q=1}^{n-1} \varepsilon_q \frac{\partial \Psi_k}{\partial u_q} u_{q+1} + \\ & + \varepsilon_k \Psi_{k+1}(\vec{u},t) = \varepsilon \sum_{q=1}^{n} \frac{\partial \Psi_k}{\partial u_q} u_q \end{aligned}\right\} \quad (A55)$$

The step by step construction of the functions $\Psi_k(\vec{u},t)$ leads indeed to always positive and larger coefficients, because of the rules (A46), (A47) and because of the inequality (A36).

Since $\varepsilon_n = 0$ and since all $\varepsilon_k =$ either 0 or 1 the functions $f_k(\vec{u},t)$ are majorized by $\varepsilon^k f(\vec{u},t)$ defined by :

$$x_k = u_k + \varepsilon^k f(\vec{u},t) \quad ; \quad k = \{1,2,\dots n\} \tag{A56}$$

$$\left.\begin{aligned} &\frac{\varepsilon^k g(t).X^2}{1 - b X} + \varepsilon^{k+1} \sum_{q=1}^{n} \frac{\partial f}{\partial u_q}\left\{\sum_{p=1}^{n} \varepsilon^q \frac{\partial f}{\partial u_p} u_p\right\} + \varepsilon^k . \sum_{q=1}^{(n-1)} \frac{\partial f}{\partial u_q} u_{q+1} = \\ &= \varepsilon^{(k+1)}\left[\sum_{q=1}^{n} \frac{\partial f}{\partial u_q} u_q - f(\vec{u},t)\right] \end{aligned}\right\} \tag{A57}$$

The equation (A57) is independent of k and the unique majorizing series $f(\vec{u},t)$, beginning at the second-order, is then defined by :

$$\left.\begin{aligned} &X = \sum_{k=1}^{n}\left\{u_k + \varepsilon^k f(\vec{u},t)\right\} \\ &\frac{g(t)X^2}{1 - b X} + \left[\sum_{q=1}^{n} \varepsilon^{(q+1)} \frac{\partial f}{\partial u_q}\right] . \left[\sum_{p=1}^{n} \frac{\partial f}{\partial u_p} u_p\right] + \sum_{q=1}^{n-1} \frac{\partial f}{\partial u_q} u_{q+1} = \\ &\qquad = \varepsilon\left[\sum_{q=1}^{n} \frac{\partial f}{\partial u_q} u_q - f(\vec{u},t)\right] \end{aligned}\right\} \tag{A58}$$

The series $f(\vec{u},t)$ has only real and positive coefficients, hence it is sufficient to verify its convergence for the small values of a new parameter u with :

$$u_k = (\varepsilon/4)^k u \quad ; \quad k = \{1,2,\dots,n\} \tag{A59}$$

Hence :

$$\left.\begin{aligned} &X = u \sum_{k=1}^{n} (\varepsilon/4)^k + f(u,t) \sum_{k=1}^{n} \varepsilon^k \\ &\frac{g(t)X^2}{1 - bX} + \left[\sum_{q=1}^{n} \varepsilon^{(q+1)} \frac{\partial f}{\partial u_q}\right] . u\frac{\partial f}{\partial u} + \sum_{q=1}^{n-1} \frac{\partial f}{\partial u_q} u_{q+1} = \varepsilon\left[\frac{\partial f}{\partial u} u - f\right] \end{aligned}\right\} \tag{A60}$$

The sum $\sum_{q=1}^{n} \varepsilon^{(q+1)} \frac{\partial f}{\partial u_q}$ is majorized by $\sum_{q=1}^{n} 4^{(n-q)} . \varepsilon^{(q+1)} \frac{\partial f}{\partial u_q}$ that is $4^n \varepsilon\, \partial f/\partial u$, and the sum $\sum_{q=1}^{n-1} \frac{\partial f}{\partial u_q} u_{q+1}$ is $\sum_{q=1}^{n-1} \frac{\varepsilon}{4} \frac{\partial f}{\partial u_q} u_q$ that is majorized by $\sum_{q=1}^{n} \frac{\varepsilon}{4} \frac{\partial f}{\partial u_q} u_q$ that is $\frac{\varepsilon}{4} \frac{\partial f}{\partial u}$.

Hence the two-parameter series f(u,t) beginning with u^2 terms is majorized by the two-parameter series e(u,t) also beginning with u^2 terms and defined by :

$$\left.\begin{aligned} X &= u \sum_{k=1}^{n} \left(\frac{\varepsilon}{4}\right)^k + e(u,t) \sum_{k=1}^{n} \varepsilon^k \\ \frac{g(t)X^2}{1 - bX} &+ 4^n \varepsilon . u\left(\frac{\partial e}{\partial u}\right)^2 = \varepsilon\left(\frac{3}{4}u\frac{\partial e}{\partial u} - e\right) \end{aligned}\right\} \qquad (A61)$$

Since the power-Fourier series e(u,t) begins with u^2 terms, it is majorized by $\frac{u}{2}\frac{\partial e}{\partial u}$ and these two power-Fourier series are both majorized by the power-Fourier series S(u,t) defined by:

$$X = Au + BS \quad ; \quad A = \sum_{k=1}^{n} \left(\frac{\varepsilon}{4}\right)^k \quad ; \quad B = \sum_{k=1}^{n} \varepsilon^k \qquad (A62)$$

$$g(t) \frac{(Au + BS)^2}{1 - b(Au + BS)} + 4^{(n+1)} . \varepsilon . \frac{S^2}{u} = \frac{\varepsilon}{2} S(u,t) \qquad (A63)$$

This algebraic equation of S(u,t) is very similar to the equation (535) of Section 10.7.7.2 and can be analysed in the same way. The power-Fourier development of S(u,t) is always converging when $|u| < \varepsilon/(2\varepsilon bA + (8AB + 4^{(n+3)} A^2).g(0))$

Thus from majorizing series to majorizing series, this long demonstration leads to the desired result : the near-resonance theorem is true even when some eigenvalues λ_k are multiple eigenvalues and when the corresponding ε_k are equal to one.

Notice that the demonstration is valid for analytic dynamical systems (either autonomous or periodic) and even for dynamical systems developable into converging power-Fourier series.

REFERENCES

1 SZEBEHELY V. - "Theory of orbits. The restricted problem of three bodies". Academic Press New-York (1967)

2 HAGIHARA Y. - "Celestial Mechanics". Vol.1 and 2, MIT Press (1970-1972). Vol.3-5. Japan Society for the promotion of Science (1974-1976)

3 САРЫЧЕВ В.А., -"ИТОГИ НАУКИ И ТЕХНИКИ" СЕРИЯ АСТРОНОМИЯ - ТОМ 20 - ВИНИТИ - МОСКВА (1982).

4 LORENZ E.N. - "Deterministic Non Periodic Flow". J. Atmos. Sci. Vol.20, p.130-141 (1963)

5 HENON M. - "Exploration numérique du problème restreint" III.IV. Bulletin Astronomique (Série 3). Vol.1, Fasc.1, p.57-79. Fasc.2, p.49-66 (1966)

6 SZEBEHELY V., PETERS C.F. - "Complete Solution of a General Problem of Three Bodies". Astronomical Journal. Vol.72, p.876-883 (1967)

7 SAARI D.G. - "Restriction on the motion of the 3-body problem" SIAM J. Appl. Math. 26, p.806-815 (1974)

8 DZIOBECK O. - "Die Mathematischen Theorien der Planeten Bewegungen". Leipzig. Johann Ambrosius Barth, p.68 (1888)

9 BRUNS H. - Berichte der Königlischen Sächs. Gesellschaft der Wiss. p.1-55 (1887)

10 BRUNS H. - Acta Mathematica 11, p.25-96 (1887-88)

11 PAINLEVE P. - Bulletin Astronomique 15, p.81 (1898)

12 POINCARE H. - Acta Mathematica 13, p.259 (1890-91)

13 POINCARE H. - "Les méthodes nouvelles de la mécanique céleste" Dover Publications New-York. Vol.1, p.233-268 (1957)

14 PAINLEVE P. - Compte-rendu à l'Académie des Sciences, 80, p. 1699 (1900)

15 WHITTAKER E.T. - A treatise on the Analytical Dynamics of particles and rigid bodies with an introduction to the problem of 3-bodies". Cambridge at the University Press (1927)

16 SIEGEL C.L. - "Über die algebraischen Integrale des restringierten Dreikörperproblems". Transactions of American Mathematical Society 39,p.225 (1936)

17 CONTOPOULOS G. - "Resonance phenomena and the non applicability of the "third" integral". Bulletin Astronomique, Série 3, Vol.2, p.223-241 (1967)

18 FROESCHLE C. - "On the number of isolating integrals in systems with 3 degrees of freedom". Gravitational N-body Problem. Report on the IAU Colloquim - Cambridge - England 12-15 August 1970. D. Reidel Publishing Company. Dordrecht Holland (1972)

19 SAGNIER J.L. - "Aspects qualitatifs du problème des trois corps". Seminarios do Observatorio Astronomico, N.T. 07/69, 10-17 Juin (1969)

20 SUNDMAN K.F. - "Mémoire sur le problème des trois corps". Acta Mathematica 36, p.105-179 (1912)

21 KHILMI G.F. - "Qualitative methods in the many-body problem". Russian Tracts on Advanced Mathematics and Physics, Vol. 3. Gordon and Breach Science Publishers, New-York (1961)

22 POINCARE H. - "Les méthodes nouvelles de la Mécanique Céleste" Dover Publications. New-York. Vol.3, p.1-139 (1957)

23 POINCARE H. - "Les méthodes nouvelles de la Mécanique Céleste" Dover Publications. New-York. Vol.3, p.140-174 (1957)

24 BIRKHOFF G.D. - "Proof of a recurrence theorem for strongly transitive systems ; proof of the ergodic theorem". Proc. Nat. Acad. Sciences (1931)

25 HALMOS P.R. - "Lectures on ergodic theory". Chelsea publishing company, New-York (1958)

26 NEMYTSKII V.V., STEPANOV V.V. - "Qualitative theory of differential equations". Princeton Mathematical series, p.307-486 (1960)

27 OXTOBY J.C. - "Measure and Category". Graduate texts in Mathematics. Springer Verlag, New-York, Heidelberg, Berlin (1970)

28 MARCHAL C. - "Etude topologique des équations différentielles conservatives (indépendantes du temps ou périodiques). Application à la Mécanique Céleste". Celestial Mechanics, Vol. 4, N°3-4, p.406-422 (1971)

29 MARCHAL C. - "Time-asymptotic behavior of solutions of systems of conservative differential equations and of measure preserving transformations". Journal of differential equations, Vol.23, N°3, p.387-401 (1977)

30 PAINLEVE P. - "Leçons sur la théorie analytique des équations différentielles". Paris. A. Hermann (1897)

31 Von ZEIPEL - "Sur les singularités du problème des n-corps". Ark. Mat. Astr. Fys. 4, N°32 (1908)

32 POLLARD H. - "Gravitational Systems". Journal of Mathematics and Mechanics, 17, p.601-612 (1967)

33 POLLARD H., SAARI D.G. - "Singularities of the n-body problem I". Arch Rational Mech. Anal. 30, p.263-269 (1968)

34 POLLARD H., SAARI D.G. - "Singularities of the n-body problem II" in Inequabilities II. p.255-259, Academic Press, New-York (1970)

35 SPERLING H. - "On the real singularities of the n-body problem". J. Reine Angew. Math. 245, p.15-40 (1970)

36 SAARI D.G. - "Singularities and collisions of Newtonian gravitational systems". Archive for rational Mechanics and Aanlysis. Vol.49, N°4, p.311-320 (1973)

37 SIEGEL C.L. - "Der Dreierstoss". Ann. Math. 2-42, p.127-168 (1941)

38 SZEBEHELY V. - "Numerical integration of a one-parameter family of triple close approaches occuring in stellar systems". Astronomical Journal 79. p.981-982 and p.1449-1454 (1974)

39 MAC GEHEE R. - "Triple Collisions in Newtonian Gravitational Systems". Dynamical Systems, Theory and Applications, J. Moser, ed., p.550-572 (1975)

40 WALDVOGEL J. - "The rectilinear Restricted Problem of Three Bodies". Celestial Mechanics, 8, p.189-198 (1973)

41 WALDVOGEL J. - Celestial Mechanics, 11, p.429 (1975)

42 WALDVOGEL J. - Celestial Mechanics, 14, p.287 (1976)

43 WALDVOGEL J. - Bulletin de l'Académie Royale de Belgique 5, 63, p.34 (1977)

44 LOSCO L. - Celestial Mechanics, 15, p.477 (1977)

45 NAHON F. - "Trajectoires rectilignes du problème des 3 corps lorsque la constante des forces vives est nulle". Celestial Mechanics, 18, p.169-175 (1973)

46 IRIGOYEN M., NAHON F. - "An Integrable Case of the Rectilinear Problem of Three Bodies". Astronomy and Astrophysics, 17, p.286-295 (1972)

47 RICHA C. - "Cas de régularisation des singularités du problème des n-corps". Thèse de 3° cycle à l'Université Pierre et Marie Curie (Paris 6), (15 Octobre 1980)

48 MARCHAL C., LOSCO L. - "Analysis of the Neighbourhood of Triple Collisions and of Tri-Parabolic Escapes in the 3 Body-Problem". Astronomy and Astrophysics 84, p.1-6 (1980)

49 MARCHAL C. - "n-Body problem : selective effects of triple close approaches". Acta Astronautica 7, p.123-126 (1980)

50 SIMO C. - "Masses for which Triple Collision is Regularizable" Celestial Mechanics, 21, p.25-36 (1980)

51 WALDVOGEL J. - "Coordonnées symétriques sur la variété de Collision Triple du Problème Plan des Trois Corps". in Application of Modern Dynamics to Celestial Mechanics and Astrodynamics. Szebehely V. Ed. D. Reidel. Publ. Co, p.249-266 (1982)

52 IRIGOYEN M. - "Sur la régularisation des singularités dans le problème plan des trois corps". Thèse de doctorat d'état - Paris 6 - (7 Février 1985)

53 EASTON R. - "Regularization of a vector field by surgery". Journal of Differential Equations 10, p.92-99 (1971)

54 MARCHAL C. - "Regularization of the singularities of the n-body problem". Nato Advanced Study Institute - Cortina d'Ampezzo (August 1981), also : Application of Modern Dynamics in Celestial Mechanics and Astrodynamics. Szebehely V. Ed.D. Reidel. Publ. Co, p.201-236 (1982)

55 LEVI-CIVITA T. - "Sur la régularisation du problème des trois corps". C.R.A.S. 162, p.1 (1916)

56 KUSTAANHEIMO P., STIEFEL E. - "Perturbation theory of Keplerian motion based on Spinor regularization". Journal of Math. 218, p.204 (1965)

57 BURDET C.A. - Z.A.M.P. 18, p.434 (1967)

58 HEGGIE D.C. - In Recent Advances in Dynamical Astronomy. Tapley B.D. and Szebehely V. Eds. D. Reidel Publishing Company (1973)

59 TISSERAND F. - "Traité de Mécanique Céleste", Tome 1, Gauthier Villars (1960)

60 MARCHAL C. - "Etude de la stabilité des solutions de Lagrange du problème des 3 corps. Cas où l'excentricité et les 3 masses sont quelconques". Séminaire du Bureau des Longitudes (7 Mars 1968)

61 BENNETT A. - "Characteristic Exponents of the Five Equilibrium

Solutions in the Elliptically Restricted Problem". Icarus, 4, p.177 (1965)

62 DANBY J.M.A. - " Stability of Triangular Points in the Elliptic Restricted Problem of Three Bodies". Astronomical Journal, 69, p.165 (1964)

63 TSCHAUNER J. - "Die Bewegung in der Nähe der Dreieckspunkte des elliptischen eingeschränkten Dreikörperproblems". Celestial Mechanics 3, p.189-196 (1971)

64 YOSHIDA HARUO - Private Correspondance (December 8, 1981)

65 DEPRIT A, DEPRIT-BARTOLOME A. - "Stability of the triangular Lagrangian points". Astronomical Journal, 72, N°2, p.173-179 (1967)

66 LEONTOVITCH A.M. - "On the stability of the Lagrange periodic solution of the restricted three-body problem". Dokl. Akad. Nauk. SSSR. Vol.143, N°3, p.525-528 (1962)

67 ALOTHMAN-ALRAGHEB A. - "Influence des perturbations d'ordre élevé sur la stabilité des systèmes hamiltoniens (cas à deux degrés de liberté)". Thèse de doctorat d'état. Observatoire de Paris (1er Juillet 1986)

68 FARQUHAR R.W., KAMEL A.A. - "Quasi periodic orbits about the translunar libration point". Celestial Mechanics 7, p.458-473 (1973)

69 BREAKWELL J.V., BROWN J.V. - "The Halo family of three-dimensional periodic orbits in the Earth-Moon restricted three-body problem". Celestial Mechanics, Vol.20, p.389-404 (1979)

70 RICHARDSON D.L. - "Analytic construction of periodic orbits about the collinear points". Celestial Mechanics 22, p.241-253 (1980)

71 MICHALODIMITRAKIS M. - "Hill's problem : Families of three-dimensional periodic orbits (Part 1)" Astroph. and Space Sc. 68, p.253-268 (1980)

72 ZAGOURAS C., MARKELLOS V.V. - "Three-dimensional periodic solutions around equilibrium points in Hill's problem". Celestial Mechanics 35, p.257-267 (1985)

73 MARCHAL C. - "Study on analytic representation of Halo orbits" ESOC Contract N°5647/83/D/JS (SC) ONERA (July 22,1985)

74 GOMEZ G., LLIBRE J., MARTINEZ R., SIMO C. - "Station keeping of libration point orbits". ESOC Contract N°5648/83/D/JS (SC). (November 1985)

75 MOULTON F.R. - "Periodic orbits". Carnegie Foundation. Washington (1920)

76 CHARLIER C.L. - "Die Mechanik des Himmels", 2. Auflage. Vol.II. Walter de Gruyter. Berlin (1927)

77 MATUKUMA T. - Proceedings of the Imperial Academy of Japan. Vol.6, p.6 (1930)

78 MATUKUMA T. - Proceedings of the Imperial Academy of Japan. Vol.8, p.147 (1932)

79 MATUKUMA T. - Proceedings of the Imperial Academy of Japan. Vol.9, p.364 (1933)

80 STROMGREN E. - Bulletin Astronomique, Vol.2, N°9, p.87-130 (1933)

81 STROMGREN E. - Publication of the Copenhagen Observatory, N°100 (1935)

82 UNO T. - Japan J. Astronomy and Astrophysics, Vol. 15, p.149-191 (1937)

83 MOSER J.D. - Mathematical Analysis, Vol.126, p.325 (1953)

84 BROUCKE R. - "Recherches d'orbites périodiques dans le problème restreint plan, Système Terre-Lune". Doctoral Dissertation (1962)

85 ARENSTORF R.F. - "Periodic solutions of the restricted three-body problem representing analytic continuations of Keplerian elliptic motions". American Journal of Mathematics, 85, p.27-35 (1963)

86 ARENSTORF R.F. - "Existence of periodic solutions passing near both masses of the restricted three-body problem". AIAA J. 1, p.238-240 (1963)

87 ARENSTORF R.F. - "Periodic trajectories passing near both masses of the restricted three-body problem". Proceedings of the 14th International Astronautical Congress Springer-Verlag, Berlin (1963)

88 HENON M. - "Exploration numérique du problème restreint I". Annales d'Astrophysique, Vol.28, p.499 (1965)

89 HENON M. - "Exploration numérique du problème restreint II". Annales d'Astrophysique, Vol.28, p.992 (1965)

90 BARTLETT J.H., WAGNER C.A. - Copenhagen Observatory Publication N°183 (1965)

91 BARRAR R.B. - Astronomical Journal, Vol.70, p.3-6 (1965)

92 DEPRIT A., HENRARD J. - Astronomical Journal, Vol.70, p.271 (1965)

93 SZEBEHELY V., PETERS C.F. -Astronomical Journal, Vol.72,

p.1187-1190 (1967)

4 GIACAGLIA G.E.O. - Astronomical Journal, Vol.72, p.386-391 (1967)

5 DEPRIT A., ROM A. - Boeing Science Research Laboratory, D1-820651 (1967)

6 DEPRIT A. HENRARD J. - "A manifold of periodic orbits". Boeing Science Research Laboratory, D1-820622, p.205 (1967)

7 DEPRIT A., PALMORE J. - Boeing Science Research Laboratory, Document D1-820754 (1968)

8 DEPRIT A., HENRARD J. - Astronomical Journal, Vol.74, N°2, p.308 (1968)

9 DEPRIT A., RABE E. - Astronomical Journal, Vol.74, N°2, p.317 (1968)

00 BROUCKE R. - JPL Technical Report 32-1168 (1968)

01 HENON M. - Bulletin Astronomique, Vol.3, p.377 (1968)

02 GUILLAUME P. - Astronomy and Astrophysics, Vol.3, p.57-76 (1969)

03 ROELS J. - "Nouvelle famille d'orbites périodiques autour des points équilatéraux pour la résonance 1/4. III. Astronomy and Astrophysics, Vol.2, p.52-62 (1969)

04 HENON M. - "Numerical exploration of the restricted problem V. Hill's case, periodic orbits and their stability". Astronomy and Astrophysics, Vol.1, p.223-238 (1969)

105 BROUCKE R. - "Periodic orbits in the elliptic restricted 3-body problem". NASA. JPL. Technical Report 32-1360 (July 15th 1969)

106 BROUCKE R. - "Stability of periodic orbits in the elliptic restricted 3-body problem". AIAA Journal, Vol.7, p.1003 -1009 (1969)

107 SCHMIDT D.S. - "Families of periodic orbits in the restricted problem of 3 bodies". Ph.D. Dissertation (1970)

108 HENON M., GUYOT M. - Proceedings of the Symposium on periodic orbits, stability and resonances. Sao Paulo, 4-12 September 1969 (1970)

109 HENON M., GUYOT M. - "Stability of periodic orbits in the restricted 3-body problem" in Periodic orbits, stability and resonances. Ciagaglia GEO editor; Reidel publishing company. Dordrecht Holland, p.349-374 (1970)

110 CIACAGLIA GEO - "Periodic orbits, stability and resonances" Reidel publishing company. Dordrecht Holland (1970)

111 SINCLAIR A.T. - "Periodic solutions in the commensurable

3-body problem". Celestial Mechanics, Vol.2, p.350 (1970)

112 BOZIS G. - "Proceedings of the Sao Paulo Symposium on periodic orbits, stability and resonances (Sao Paulo, 4-12 Sept.1969), p.176 (1970)

113 BOZIS G. - "On the problem of consecutive collisions periodic orbits". Astronomy and Astrophysics, Vol.II, p.320 (1971)

114 BROUCKE R. - "Periodic collisions orbits in the elliptic restricted 3-body problem". Celestial Mechanics, Vol.3, p.461-477 (1971)

115 SIEGEL C.L., MOSER J.K. - Lectures on Celestial Mechanics. Springer-Verlag, Berlin (1971)

116 SHELUS P.J. - "A two parameter survey of periodic orbits in the restricted problem of 3 bodies". Celestial Mechanics, Vol.5, p.483-489 (1972)

117 GOUDAS C.L. - Ph. D. Thesis, University of Manchester (1972)

118 KOZAI Y., HIROSHI K. - "Periodic solutions of the third sort for the restricted Problem of 3-bodies and their stability". Celestial Mechanics, Vol.7, p.156-176 (1973)

119 HENON M. - "Vertical stability of periodic orbits in the restricted problem". Astronomy and Astrophysics, Vol.28 p.415 (1973)

120 HENON M. - "Vertical stability of periodic orbits in the restricted problem". Astronomy and Astrophysicsn Vol.30 p.317-321 (1974)

121 HENON M. - "Families of periodic orbits in the 3-body problem" Celestial Mechanics, Vol.10, p.375 (1974)

122 GUILLAUME P. - "Families of symmetric periodic orbits of the restricted 3-body problem when the perturbing mass is small". Astronomy and Astrophysics, Vol.37, N°2, p.209-218 (1974)

123 GUILLAUME P. - "New periodic solutions of the 3-dimensionnal restricted problem". Celestial Mechanics, Vol.10, N°4, p.475-496 (1974)

124 MARKELLOS V.V., BLACK W., MORAN P.E. - "A grid search for families of periodic orbits in the restricted problem of 3-bodies". Celestial Mechanics, Vol.9, N°4, p.507-512 (1974)

125 - MARKELLOS V.V. - "Numerical investigation of the planar restricted three-body problem". Celestial Mechanics,

Vol.9, p.365-380 (1974)

126 PERKO L.M. - "Periodic orbits in the restricted 3-body problem : existence and asymptotic approximation". SIAM Journal on Applied Mathematics, Vol.27, N°1, p.200 (1974)

127 BROUCKE R., BOGGS D. - "Periodic orbits in the planar general 3-body problem". Celestial Mechanics, Vol.II, N°1, p.13-38 (1975)

128 SERGYSELS-LAMY A. - "Existence of periodic orbit of the second kind in the elliptic restricted problem of 3 bodies". Celestial Mechanics, Vol.11, N°1, p.43-52 (1975)

129 BROUWER D., CLEMENCE G.M. - "Methods of Celestial Mechanics". Academic Press. New-York and London (1961)

130 VINH N.X. - "Sur la stabilité des points d'équilibre triangulaires dans le problème restreint elliptique". Celestial Mechanics, Vol.6, p.305-321 (1972)

131 HADJIDEMETRIOU J.D. - "The present status of periodic orbits". Celestial Mechanics 23, p.277-286 (1980)

132 HADJIDEMETRIOU J.D. - "Periodic orbits and stability". The Few Body problem. Valtonen M.J. Ed. Kluwer Academic Publishers, p.31-48 (1988)

133 HENON M. - "A family of periodic solutions of the planar three-body problem, and their stability". Celestial Mechanics 13, p.267-285 (1976)

134 BROUCKE R. - Celestial Mechanics, 12, p.439 (1975)

135 SCHUBART J. - Astron. Nachr. 283, p.17 (1956)

136 АНОСОВА Ж.П., ОРЛОВ В.В. - "Динамическая эволюция тройных систем" ТРУДЫ АСТРОНОМИЧЕСКОЙ ОБСЕРВАТОРИИ, ТОМ 40, ИЗДАТЕЛЬСТВО ЛЕНИНГРАДСКОВО УНИВЕРСИТЕТА, СЕРИЯ МАТЕМ. НАУК, ВЫПУСК 62, p.66-144 (1985).

137 DAVOUST E., BROUCKE R. - "A manifold of periodic orbits in the planar general three-body problem with equal masses". Astronomy and Astrophysics. 112, p.305-320 (1982)

138 DEPRIT A., HENRARD J. - "The Trojan manifold. Survey and conjectures" in : Periodic Orbits, Stability and Resonances. Ciacaglia G.E.O. Ed.D. Reidel Publ. Co. p.1-18 (1970)

139 HENRARD J. - "On Brown's conjecture". Celestial Mechanics 31, p.115-122 (1983)

140 TAYLOR D. - "Horseshoe Periodic Orbits in the Restricted

Problem of Three Bodies for a Sun-Jupiter Mass Ratio". Astronomy and Astrophysics 103, p.288 (1981)

141 GARFINKEL B. - "Theory of the Trojan asteroids".
Part 1 - Astronomical Journal 82, 5, p.386-379 (1977)
Part 2 - Celestial Mechanics 18, p.259-275 (1978)
Part 3 - Celestial Mechanics 22, p;267-287 (1980)
Part 4 - Celestial Mechanics 30, p.373-383 (1983)
Part 5 - Celestial Mechanics 36, p.19-45 (1985)
Part 6 - Celestial Mechanics, to appear

142 GARFINKEL B. - "On the Brown conjecture". in Stability of the Solar System and its minor natural and artificial bodies. Szebehely V. Ed. D. Reidel Publ. Co, p.33-37 (1985)

143 BROWN E.W. - "On the oscillating orbits about the triangular equilibrium points in the problem of the three bodies". Monthly Notices, Royal Astronomical Society 71, p.438 (1911)

144 HENON M., PETIT J.M. - "Series expansions for encounter-type solutions of Hill's problem". Celestial Mechanics 38, p.67-100 (1986)

145 BROWN E.W. - "An Introductory Treatise on the Lunar Theory". Cambridge University Press. London and New-York (1896)

146 MURAKAMI H. - "Theory of the Perturbations of the Moon". Murakami T. Ed. Hiroshima University 1940-1943 and 1945-1946. Nakamotohonten Printing Co, Hiroshima (1982)

147 CHAPRONT-TOUZE M., CHAPRONT J. - "ELP 2000-85 : a semi-analytical lunar ephemeris adequate for historical times". Astronomy and Astrophysics 190, p.342-352 (1988)

148 POINCARE H. - "Les méthodes nouvelles de la Mécanique Céleste" Dover Publications. New-York. Vol.3, p.389 (1957)

149 VON ZEIPEL H. - "Recherches sur le mouvement des petites planètes". Ark. Matematik Astronomi och Fysik 11 (1); 12 (9) ; 13 (3) ; (1916-1917)

150 HORI G. - Publications of the Astronomical Society of Japan 18, 4, p.287-296 (1966)

151 BROWN E.W. - "The Stellar Problem of Three Bodies I, II and III" Monthly Notices, Royal Astronomical Society 97, p.56-61, p.62-66, p.116-127 (1937)

152 KOVALEVSKY J. - "Sur la théorie du mouvement d'un satellite à fortes inclinaison et excentricité". in The Theory of Orbits in the Solar System and in Stellar Dynamics

Contopoulos G. Ed., Int. Astron. Union Symposium N°25, Academic Press London, p.326-344 (1966)

153 SZEBEHELY V. - "Numerical integration of a three-body symmetrical solution of exchange type". Private correspondance (February 1975)

154 HADJIDEMETRIOU J. - "Numerical integration of an oscillatory motion of the second kind". Private correspondance (September 27, 1977)

156 HEGGIE D.C. - "Binary evolution in Stellar Dynamics". Monthly Notices Royal Astr. Soc. 173, p.729-787 (1975)

157 HEGGIE D.C. - In Dynamics of Stellar Systems, IAU Symposium, N°69, Hayli A. Ed. Reidel, p.73 (1975)

158 HEGGIE D.C. - Revista Mex. astron. y astrofisika. 3, p.169-173 (1977)

159 HEGGIE D.C. - In Globular Clusters, Hanes D. and Madore D. Ed., Cambridge University Press, p.281 (1980)

160 HUT P. - "Binary-Single Star Scattering II. Analytic Approximations for High Velocity". Astrophys. J. 283, p.319-341 (1983)

161 MONAGHAN J.J. - Monthly Notices, R. Ast. Soc. 176, p.63-72; 177, p.583 (1976) and 179, p.31 (1977)

162 NASH P.E., MONAGHAN J.J. - Monthly Notices, R. Ast. Soc. 184, p.119 (1978)

163 SASLAW W.C., VALTONEN M.J., AARSETH S.J. - Astrophysical J. 190, p.253-270 (1974)

164 VALTONEN M.J., HEGGIE D.C. - "Three-body gravitational scattering : Comparison between theory and experiment" Celestial Mechanics 19, p.53-58 (1979)

165 HILLS J.G. - Astronomical J. 10, p.809-825 (1975)

166 HILLS J.G., FULLERTON L.W. - Astronomical J. 85, p.1281 (1980)

167 FULLERTON L.W., HILLS J.G. - Astronomical J. 87, p.175 (1982)

168 HUT P., BAHCALL J.N. - "Binary-Single Star Scattering. I. Numerical Experiments for Equal Masses". Astrophysical J.268, p.319-341 (1983)

169 HEGGIE D.C., HUT P. - "Binary-Single Star Scattering. IV. Analytic Approximations and Fitting Formulae for Cross Sections and Reaction Rates". Astrophysical J. to appear (1989)

170 HUT P. - "A Laboratory for Gravitational Scattering Experiments". Astrophysics Preprint Series. IASSNS-AST 88/50 (1988)

171 АНОСОВА Ж.П. - "ЧИСЛЕННО-ЭКСПЕРИМЕНТАЛЬНЫЕ МЕТОДЫ В ЗВЗДНОЙ ДИНАМИКЕ" ИТОГИ НАУКИ И ТЕХНИКИ, АСТРОНОМИЯ, ТОМ 26, МОСКВА, p57-112 (1985).

172 MARKELLOS V.V. - "Bifurcation and trifurcation of asymmetric periodic orbits". Astronomy and Astrophysics. Vol.61, N°2, p.195-198 (1977)

173 FLOQUET G. - "Sur les équations différentielles linéaires à coefficients périodiques". Annales de l'Ecole Normale Supérieure, Paris 12, p.47-89 (1883)

174 NEMYTSKII V.V., STEPANOV V.V. - "Qualitative theory of Differential Equations". Princeton Mathematical Series, p.307-486 (1960)

175 HADJIDEMETRIOU J.D. - "The continuation of periodic orbits from the restricted to the general three-body problem". Celestial Mechanics 12, p.155-174 (1975)

176 HENON M. - "Stability of interplay motions". Celestial Mechanics 15, p.243-261 (1977)

177 PRADEEP S., SHRIVASTAVA S.K. - "Stability of dynamical systems : an overview". International Workshop on Space Dynamics and Celestial Mechanics, New-Delhi (November 14-16, 1985)

178 SZEBEHELY V. - "Review of Concepts of Stability". Celestial Mechanics, 34, p.49-64 (1984)

179 SIEGEL C.L. - "Über die Normalform analytischer Differentialgleichungen in der Nähe einer Gleichgewichtslösung". Nachr. Akad. Wiss. Göttingen, math-phys. Kl, p.21-30 (1952)

180 MARCHAL C. - "Upper bounds on the Arnold diffusion", Communication at the Düsseldorf Dynamic Days, June 15-18 (1988), to appear.

181 BIRKHOFF G.D. "Dynamical systems". Am. Math. Soc. Colloq. Publ. IX, New-York, Am. Math. Soc. VIII (1927)

182 ARNOLD V.I., AVEZ A. - "Problèmes ergodiques de la Mécanique classique". Gauthier-Villars, Paris (1967)

183 SIMO C. - "Stability of Degenerate Fixed Points of Analytic Area Preserving Mappings". Revue Mathématique Astérisque. Société Mathématique de France 99, p.184-194 (1982)

184 HADJIDEMETRIOU J.D. - "The stability of periodic orbits in the three-body problem". Celestial Mechanics 12, p.255-276 (1975)

185 BURRAU C. - Astron. Nachr. 195, p.113 (1913)
186 MERMAN G.A. - Bull. Inst. Theoret. Astron. Leningrad, 6, p.687 (1958)
187 ALEXEEV V.M. - Astron. J. USSR. 38, 6, p.1099 (1961) and Soviet Astronomy, A.J. 5, 6, p.841 (1962)
188 STANDISH E.M. - Private communication to Dr Szebehely (1966)
189 SPINELLI R. - Private communication to Dr Szebehely (1966)
190 STANEK L. - Private communication to Dr Szebehely (1966)
191 HARRINGTON R.S. - "Stability criteria for triple stars" Celestial Mechanics 6, p.322-327 (1972)
192 PETIT J.M., HENON M. - "Satellite encounters", Icarus 66, p.536-555 (1986)
193 DERMOTT S.F., MURRAY C.D. - "The dynamics of tadpole and horseshoe orbits. I. Theory" Icarus 48, p.1-11 (1981)
194 YODER C.F., COLOMBO G., SYNNOTT S.P., YODER K.A. - "Theory of motion of Saturn's coorbiting satellites" Icarus 53, p.431-443 (1983)
195 SPIRIG F., WALDVOGEL J. - "The three-body problem with two small masses : a singular perturbation approach to the problem of Saturn's coorbiting satllites". In Stability of the Solar System and its Minor Natural and Artificial Bodies. Nato Advanced Study Institute Series C. Szebehely V. Ed. Reidel (1985)
196 SITNIKOV K.A. - "The existence of oscillatory motions in the 3-body problem" Dokladij Akad. Nauk SSSR, Vol.133, 2, p.303-306 (1960). Soviet Physics Dokl.5, p.647-650 (1961)
197 PALMORE J. - "Classifying relative equilibria" I and II Bull. Amer. Math. Soc. 79, p.904-908 (1973). 81, p.489-491 (1975). III Lett. Math. Phys. 1, p.71-73 (1975)
198 SIMO C. - "Relative equilibrium solutions in the four-body problem" Celestial Mechanics 18, p.165-184 (1978)
199 POINCARE H. - "Les méthodes nouvelles de la Mécanique Céleste" Dover Publications, New-York, Vol.1 (1957)
200 GOLUBEV V.G. - Soviet Phys. Dokl. 13, p.373 (1968)
201 SMALE S. - "Topology and Mechanics - II" Inventiones Math. 11, p.45-64 (1970)
202 EASTON R. - "Some topology of the 3-body problem" Journal of Differential Equations 10, p.371-377 (1971)
203 MARCHAL C. - "Qualitative study of a n-body system ; a new condition of complete scattering" Astronomy and Astrophysics, 10, 2, p.278-289 (1971)
204 EL MABSOUT B. - "Espace des phases dans le problème plan des 3 corps". Comptes-rendus à l'Académie des Sciences. A 276, p.495 (1973). A 278, p.459 (1974)

205 TUNG CHIN-CHU - Scienta Sinica 17, N°3 (1974)

206 MARCHAL C., SAARI D.G. - "Hills regions for the general three-body problem" Celestial Mechanics 12, p.115 (1975)

207 BOZIS G. - "Zero velocity surfaces for the general planar three-body problem". Astrophysics and Space Sciences 43, p.355-368 (1976)

208 ZARE K. - Celestial Mechanics 14, p.73-83 (1976). 16, p.35 (1977)

209 CHEN XIANG-YAN, SUN YI-SUI, LUO DIN-JUN - Acta Astronomica Sinica 19, p.119 (1978)

210 MARCHAL C., SUN YI-SUI - "Evolution of the moment of inertia in the many-body problem" Scienta Sinica , Series A, Vol.28, N°6, p.638-647 (1985)

211 SZEBEHELY V., ZARE K. - Astron. and Astroph. 58, p.145 (1977)

212 MARCHAL C., BOZIS G. - "Hill stability and distance curves for the general three-body problem" Celestial Mechanics 26, p.311-333 (1982)

213 CHAZY J. - "Sur l'allure finale du mouvement dans le problème des 3 corps quand le temps croît indéfiniment" Annales de l'Ecole Normale Supérieure, 3ème Série, 39, p.29-130 (1922)

214 MATHER J.N., MAC GEHEE R. - "Solutions of the collinear four-body problem which become unbounded in finite time" National Science Foundation, Grant N°GP4313X and GP38955 (1974)

215 SIMO C. - "Masses for which triple collision is regularizable" Celestial Mechanics 21, p.25-36 (1980)

216 MARCHAL C., SAARI D.G. - "On the final evolution of the n-body problem" Journal of Differential Equations 20, 1, p.150-186 (1976)

217 CHAZY J. - Journal de Mathématique pure et appliquée. Vol.8, p.353 (1929)

218 KHILMI G.F. - Doklady Akad. Nauk SSSR, 78, p.653 (1951)

219 MERMAN G.A. - Byull. Inst. Teor. Astron. 5, p.594 (1954)

220 KHILMI G.F. - "Qualitative methods in the many-body problem". Russian Tracts on Advanced Mathematics and Physics, Vol.3, Gordon and Breach Science Publishers, New-York (1961)

221 SITNIKOV K.A. - "The existence of oscillatory motions in the three-body problem" Sov. Phys. Dokl. 5, p.647-650 (1961)

222 MARCHAL C. - "Collisions of stars by oscillating orbits of the second kind" Acta Astronautica, Vol.5, p.745-764 (1978)

223 CRAIG WHEELER J. - Private correspondance (February 23 1984)

224 MARCHAL C. - "Survey paper. Qualitative methods and results in Celestial Mechanics" ONERA T.P., N°1975-77 (1977)

225 YOSHIDA J. - Publ. of Astronomical Society of Japan, 24, 3, p.391-408 (1972)

226 TEVZADZE G.A. - Izv. Akad. Nauk Armyan. SSR 15, N°5, p.67 (1962)

227 STANDISH E.M. Jr - "Sufficient conditions for escape in the 3-body problem"

Celestial Mechanics 4, 1, p.44-48 (1971)
228 STANDISH E.M. Jr - "Sufficient conditions for return in the 3-body problem" Celestial Mechanics 6, 3, p.352-355 (1972)
229 YOSHIDA J. - "Improved criteria for hyperbolic-elliptic motion in the general three-body problem. II" Publication of the Astronomical Society of Japan, 26, p.367-377 (1974)
230 MARCHAL C. - "Sufficient conditions for hyperbolic-elliptic escape and for ejection without escape in the 3-body problem" Celestial Mechanics, Vol.9, p.381-393 (1974)
231 POLLARD H. - "Disintegration and escape" in Periodic Orbits, Stability and Resonances. Giacaglia G.E.O. Ed. Reidel Publ. Co. p.53-55 (1970)
232 ZARE K. - "Properties of the moment of inertia in the problem of the three bodies" Celestial Mechanics 24, p.345-354 (1981)
233 LASKAR J. - "Approche triple dans le domaine des trois corps. Une limite pour orbites bornées" Rapport de stage de D.E.A., p.1-30. Observatoire de Paris (1982)
234 LASKAR J., MARCHAL C. - "Triple close approach in the three-body problem, a limit for the bounded orbits" Celestial Mechanics 32, p.15-28 (1984)
235 MARCHAL C., YOSHIDA J., SUN YI-SUI - "Three-body problem. A test of escape valid even for very small mutual distances. Part I. The acceleration and the escape velocities of the third-body" Celestial Mechanics 33 p.193-207 (1984). Part II Celestial Mechanics 34, p.65-93 (1984). (Short presentation in Acta Astronautica 11, 7-8, p.415-422, 1984)
236 MARCHAL C. - "Three-body problem. Some applications of the tests of escape" in Stability of the Solar System and its Minor Natural and Artificial Bodies. Szebehely V.G. Ed. D. Reidel Publ. Co. p.115-138 (1985)
237 SZEBEHELY V. - "Classification of the motions of three bodies in a plane" Celestial Mechanics 4, p.116-118 (1971)
238 MANDELBROT B. - "The fractal geometry of nature" W.H. Freeman and Co. Ed. San Francisco (1982). "Les objets fractals : forme, hasard et dimension" Paris, Flammarion (1984)
239 CHAZY J. - Bulletin Astronomique 8, p.403 (1932)
240 SCHMIDT O.J. - Dokl. Akademii Nauk SSSR 58, 2, p.213 (1947)
241 SCHMIDT O.J. - "Chetyre lekcii o teorii proiskhozhdenija zemli" Izd. vo Akad. Nauk SSSR (1951)
242 SCHMIDT O.J. - "Quatre leçons sur la théorie de l'origine de la Terre" Editions en langue étrangère, Moscou (1959)
243 KHILMI G.F. - "Kachestvennye metody v problem n tel" Izd. vo Akad. Nauk SSSR (1958)
244 KHILMI G.F. - Doklady Akad. Nauk SSSR, 62, 1, p.39 (1948)

245 SCHMIDT O.J., KHILMI G.F. - Uspekhi mat. Nauk 3, 4 (26), p.157 (1948)

246 KHILMI G.F. - "Problema n tel v nebesnoy mekhanike i kosmogonii" Izd. vo Akad. Nauk SSSR (1951)

247 ALEXEEV V.M. - Doklady Akad. Nauk SSSR, 108, 4, p.599 (1956)

248 ALEXEEV V.M. - Astron. Zh., 38, 2, p.325 (1961)

249 ALEXEEV V.M. - Soviet Astron. AJ. 5, 2, p.242 (1961)

250 ALEXEEV V.M. - Astron. Zh., 38, 4, p.726 (1961)

251 ALEXEEV V.M. - Astron. Zh., 38, 6, p.1099 (1961)

252 ALEXEEV V.M. - Vestn MGU. ser. fiz. i. astr. 1, p.67 (1961)

253 ALEXEEV V.M. - Soviet Astr. AJ. 5, 4, p.550 (1962)

254 ALEXEEV V.M. - Soviet Astr. AJ. 5, 6, p.841 (1962)

255 ALEXEEV V.M. - Vestn. MGU. ser. matem. 4, p.17 (1962)

256 ALEXEEV V.M. - Problemy dvizhenija isk. neb. tel. Izd. vo Akad. Nauk SSSR, p.50 (1963)

257 ARNOLD V.I. - "Dynamical Systems. III" Arnold V.I. Ed. Springer Verlag. 291 pages (1988)

258 KAZIMIRCHAK-POLONSKAYA E.I., SHAPOREV S.D. - Soviet Astron. Zh. 53, p.1306-1314 (1976)

259 KAZIMIRCHAK-POLONSKAYA E.I. - Trudy Inst. Teor. Astron. 18, p.3-77, p.91-106 (1982)

260 KAZIMIRCHAK-POLONSKAYA E.I. - "Asteroids, Comets, Meteoritic Matter" Critescu C., Klepczynski W.J., Milet B., Ed. p.205-221 (1974)

261 DVORAK R. - Astron. Astroph. 49, p.293-298 (1976)

262 DVORAK R., MARCHAL C. - "Duration of escape or capture of a satellite : A simple lower bound" Astron. Astroph. 69, p.373-374 (1978)

263 MARCHAL C. - "Round table discussion on chaotic motions" in The Few Body Problem, Valtonen M.J. Ed. Kluwer Acad. Publ. p.91-97 (1988)

264 LASKAR J. - "Numerical experiment on the chaotic behaviour of the solar system" Letters to Nature. To appear (1989)

265 BENEST D. - "Orbite initiale de la famille d'orbites de halo autour des points de Lagrange" Private correspondance (April 24 1987)

BIBLIOGRAPHY.

I must apologize for my poor knowledge of foreign languages that has considerably biaised the list of references, most of them being from western Europe and United States.

Fortunately, the prime importance of the Russian School of Mathematics and Celestial mechanics is world renown (Liapunov, Kolmogorov, Arnold, Khilmi, Sitnikov, Brumberg, Sarychev and so many younger people ...). Also note the numerous and careful works of the Japanese School (Matukuma, Murakami, Hori, Hagihara, Kinoshita, Kozai, the Yoshida's, Koda, Fujitsu...), and the great interest of so many works from all around the world and especially central Europe, China India.....

It is of course impossible to give even a small idea of the variety and the depth of all these works even in the only domain of the three-body problem and even only in U.S and western Europe.

I would only notice the following :

A)- Dynamical systems. Ref. 115, 181-183, 257, 263 ; Section 10.7.7 and :

KOLMOGOROV A.N., "General theory of dynamical systems and classical mechanics". Proc.Int.Congr.Math, 1954, Amsterdam 1 p.315-333 (1957).

ARNOLD V.I., "Proof of A.N.KOLMOGOROV's theorem on the preservation of quasi-periodic motions under small perturbations of the Hamiltonian" Usp.Mat.Nauk 18, No 5, 13-40 (1963) ; English transl. : Russ.Math.Surv. Zbl. 129, 166 Vol. 18, No 5 p.9-36 (1963).

ARNOLD V.I., "Small denominators and problems of stability of motion in classical and celestial mechanics". Usp.Mat.Nauk 18, No 6, 91-192 (1963) ; English transl. : Russ.Math.Surv. Zbl. 135, 427 Vol. 18, No 6, 85-192 (1963).

ARNOLD V.I., "Mathematical methods of classical mechanics".Moskva Nauka. 431 p.(1974) ; English transl. : New-York - Heidelberg - Berlin: Springer-Verlag. Zbl. 386.70001. Vol.X. 462 p. (1978).

MOSER J., "Lectures on Hamiltonian systems". Mem.Am.Math.Soc. Vol.81, 60 p. (1968).

MOSER J., "Stable and random motions in dynamical systems". Ann.Math.Stud. 77, VIII 199 p.(1973).

MOSER J., "Various aspects of integrable Hamiltonian systems . Dynamical systems". C.I.M.E. Lect., Bressanone 1978. Prog.Math. 8, 233-290 (1980).

RÜSSMANN H., "Über das Verhalten analytischer Hamiltonscher Differentialgleichungen in der Nähe einer Gleichgewichtslösung. Math.Ann. 154, p. 285-300 (1964).

B)- Some recent general studies of the Russian School of Celestial Mechanics.

Ref. 3, 136, 171 and :

ГОЛУБЕВ В.Г., ГРЕБЕНИКОВ Е.А., "ПРОБЛЕМА ТРЕХ ТЕЛ В НЕБЕСНОЙ МЕХАНИКЕ (Three-body problem in Celestial Mechanics)". ИЗДАТЕЛЬСТВО МОСКОВСКОГО УНИВЕРСИТЕТА (1985).

ХОЛШЕВНИКОВ К.В., "АСИМПТОТИЧЕСКИЕ МЕТОДЫ НЕБЕСНОЙ МЕХАНИКИ (Asymptotical methods in Celestial Mechanics)" ИЗДАТЕЛЬСТВО ЛЕНИНГРАДСКОВО УНИВЕРСИТЕТА (1985).

АГЕКЯН Т.А. ИТОГИ НАУКИ И ТЕХНИКИ, АСТРОНОМИЯ, ТОМ 26, "ЗВЕЗДНАЯ АСТРОНОМИЯ" ВИНИТИ, МОСКВА (1985).

C)- Resonances and Kirkwood gaps. Ref. 138-143 and :

WISDOM J., Astron.J. 87 p. 577 (1982), Icarus 56 p. 51 (1983), Icarus 63 p. 272 (1985), Celestial Mechanics 38 p. 175 (1986).

MURRAY C.D, FOX K., Icarus 59 p. 221 (1984).

HENRARD J. LEMAITRE A., MILANI A., MURRAY C.D, Celestial Mechanics 38 p.235 (1986)

LEMAITRE A., Celestial Mechanics 34 p. 329 (1984). "The 2/1 Jovian resonance in the elliptic problem". In The Few Body Problem Valtonen MJ.Ed, Kluwer Acad. Publ. p. 129 (1988).

SIDLICHOVSKY M., MELENDO B., Bull Astron., Inst. Czechosl. 37, p. 37 (1986).

SIDLICHOVSKY M., "On the origin of the 5/2 Kirkwood gap", in The Few Body Problem, Valtoven M.J. Ed, Kluwer Acad. Publ. p. 117-121 (1988).

SCHOLL H., FROESCHLE Ch., Astron. and Astroph. 42 p. 157 (1975), Astron. and Astroph. 170, p. 138 (1986).

LIU L., LIAO X.H., "The problem of evolution on orbital resonance" in The Few Body Problem, Valtonen M.J. Ed, Kluwer Acad. Publ. p. 135-139 (1988).

ELST E.W., "Trojan search at ESO", El Mensajero N° 53, p. 53-54 (september 1988).

D)- Recent surveys. Stability : ref. 177, 178, Periodic orbits : ref. 131, 132, Asteroids and resonances :

FROESCHLE Cl. and Ch., FARINELLA P., CARPINO M., GONCZI R., PAOLICCHI P., ZAPPALA V., KNEZEVIC Z., "Asteroid families", in The Few Body Problem, Valtonen M.J. Ed, Kluwer Acad. Publ., p. 101-116 (1988).

Qualitative methods, ref. 220, 224 and :

MARCHAL C., "Qualitative analysis in the few body problem" in the The Few Body Problem, Valtonen M.J. Ed, Kluwer Acad. Publ. p. 5-25 (1988)

E)- Future questions.

BRUMBERG V.A., KOVALEVSKY J., "Unsolved problems of Celestial Mechanics", Celestial Mechanics 39, p. 133-140 (1986).

SEIDELMANN P.K., "Unsolved problems of Celestial Mechanics-The Solar System", Celestial Mechanics 39, p. 141-146 (1986).

SUBJECT INDEX

ELEMENTS OF THE N-BODY PROBLEM

AUTHOR INDEX